# A COURSE IN PROBABILITY THEORY

THIRD EDITION

# A COURSE IN PROBABILITY THEORY

THIRD EDITION

**Kai Lai Chung**
*Stanford University*

ACADEMIC PRESS

A Harcourt Science and Technology Company

San Diego San Francisco New York
Boston London Sydney Tokyo

This book is printed on acid-free paper. ♾



ACADEMIC PRESS
*A Harcourt Science and Technology Company*
525 B Street, Suite 1900, San Diego, CA 92101-4495, USA
http://www.academicpress.com

ACADEMIC PRESS
Harcourt Place, 32 Jamestown Road, London, NW1 7BY, UK
http://www.academicpress.com

Library of Congress Cataloging in Publication Data: 00-106712

International Standard Book Number: 0-12-174151-6

PRINTED IN THE UNITED STATES OF AMERICA
**00 01 02 03 IP 9 8 7 6 5 4 3 2 1**

# Contents

## 9 | Conditioning. Markov property. Martingale



## | Supplement: Measure and Integral

# Preface to the third edition

In this new edition, I have added a Supplement on Measure and Integral. The subject matter is first treated in a general setting pertinent to an abstract measure space, and then specified in the classic Borel-Lebesgue case for the real line. The latter material, an essential part of real analysis, is presupposed in the original edition published in 1968 and revised in the second edition of 1974. When I taught the course under the title "Advanced Probability" at Stanford University beginning in 1962, students from the departments of statistics, operations research (formerly industrial engineering), electrical engineering, etc. often had to take a prerequisite course given by other instructors before they enlisted in my course. In later years I prepared a set of notes, lithographed and distributed in the class, to meet the need. This forms the basis of the present Supplement. It is hoped that the result may as well serve in an introductory mode, perhaps also independently for a short course in the stated topics.

The presentation is largely self-contained with only a few particular references to the main text. For instance, after (the old) §2.1 where the basic notions of set theory are explained, the reader can proceed to the first two sections of the Supplement for a full treatment of the construction and completion of a general measure; the next two sections contain a full treatment of the mathematical expectation as an integral, of which the properties are recapitulated in §3.2. In the final section, application of the new integral to the older Riemann integral in calculus is described and illustrated with some famous examples. Throughout the exposition, a few side remarks, pedagogic, historical, even judgmental, of the kind I used to drop in the classroom, are approximately reproduced.

In drafting the Supplement, I consulted Patrick Fitzsimmons on several occasions for support. Giorgio Letta and Bernard Bru gave me encouragement for the uncommon approach to Borel's lemma in §3, for which the usual proof always left me disconsolate as being too devious for the novice's appreciation.

A small number of additional remarks and exercises have been added to the main text.

Warm thanks are due: to Vanessa Gerhard of Academic Press who deciphered my handwritten manuscript with great ease and care; to Isolde Field of the Mathematics Department for unfailing assistence; to Jim Luce for a mission accomplished. Last and evidently not least, my wife and my daughter Corinna performed numerous tasks indispensable to the undertaking of this publication.

# Preface to the second edition

This edition contains a good number of additions scattered throughout the book as well as numerous voluntary and involuntary changes. The reader who is familiar with the first edition will have the joy (or chagrin) of spotting new entries. Several sections in Chapters 4 and 9 have been rewritten to make the material more adaptable to application in stochastic processes. Let me reiterate that this book was designed as a basic study course prior to various possible specializations. There is enough material in it to cover an academic year in class instruction, if the contents are taken seriously, including the exercises. On the other hand, the ordering of the topics may be varied considerably to suit individual tastes. For instance, Chapters 6 and 7 dealing with limiting distributions can be easily made to precede Chapter 5 which treats almost sure convergence. A specific recommendation is to take up Chapter 9, where conditioning makes a belated appearance, before much of Chapter 5 or even Chapter 4. This would be more in the modern spirit of an early weaning from the independence concept, and could be followed by an excursion into the Markovian territory.

Thanks are due to many readers who have told me about errors, obscurities, and inanities in the first edition. An incomplete record includes the names below (with apology for forgotten ones): Geoff Eagleson, Z. Govindarajulu, David Heath, Bruce Henry, Donald Iglehart, Anatole Joffe, Joseph Marker, P. Masani, Warwick Millar, Richard Olshen, S. M. Samuels, David Siegmund, T. Thedéen, A. González Villa lobos, Michel Weil, and Ward Whitt. The revised manuscript was checked in large measure by Ditlev Monrad. The galley proofs were read by David Kreps and myself independently, and it was fun to compare scores and see who missed what. But since not all parts of the old text have undergone the same scrutiny, readers of the new edition are cordially invited to continue the fault-finding. Martha Kirtley and Joan Shepard typed portions of the new material. Gail Lemmond took charge of the final page-by-page revamping and it was through her loving care that the revision was completed on schedule.

In the third printing a number of misprints and mistakes, mostly minor, are corrected. I am indebted to the following persons for some of these corrections: Roger Alexander, Steven Carchedi, Timothy Green, Joseph Horowitz, Edward Korn, Pierre van Moerbeke, David Siegmund.

In the fourth printing, an oversight in the proof of Theorem 6.3.1 is corrected, a hint is added to Exercise 2 in Section 6.4, and a simplification made in (VII) of Section 9.5. A number of minor misprints are also corrected. I am indebted to several readers, including Asmussen, Robert, Schatte, Whitley and Yannaros, who wrote me about the text.

# Preface to the first edition

A mathematics course is not a stockpile of raw material nor a random selection of vignettes. It should offer a sustained tour of the field being surveyed and a preferred approach to it. Such a course is bound to be somewhat subjective and tentative, neither stationary in time nor homogeneous in space. But it should represent a considered effort on the part of the author to combine his philosophy, conviction, and experience as to how the subject may be learned and taught. The field of probability is already so large and diversified that even at the level of this introductory book there can be many different views on orientation and development that affect the choice and arrangement of its content. The necessary decisions being hard and uncertain, one too often takes refuge by pleading a matter of "taste." But there is good taste and bad taste in mathematics just as in music, literature, or cuisine, and one who dabbles in it must stand judged thereby.

It might seem superfluous to emphasize the word "probability" in a book dealing with the subject. Yet on the one hand, one used to hear such specious utterance as "probability is just a chapter of measure theory"; on the other hand, many still use probability as a front for certain types of analysis such as combinatorial, Fourier, functional, and whatnot. Now a properly constructed course in probability should indeed make substantial use of these and other allied disciplines, and a strict line of demarcation need never be drawn. But PROBABILITY is still distinct from its tools and its applications not only in the final results achieved but also in the manner of proceeding. This is perhaps best seen in the advanced study of stochastic processes, but will already be abundantly clear from the contents of a general introduction such as this book.

Although many notions of probability theory arise from concrete models in applied sciences, recalling such familiar objects as coins and dice, genes and particles, a basic mathematical text (as this pretends to be) can no longer indulge in diverse applications, just as nowadays a course in real variables cannot delve into the vibrations of strings or the conduction of heat. Incidentally, merely borrowing the jargon from another branch of science without treating its genuine problems does not aid in the understanding of concepts or the mastery of techniques.

A final disclaimer: this book is not the prelude to something else and does not lead down a strait and righteous path to any unique fundamental goal. Fortunately nothing in the theory deserves such single-minded devotion, as apparently happens in certain other fields of mathematics. Quite the contrary, a basic course in probability should offer a broad perspective of the open field and prepare the student for various further possibilities of study and research. To this aim he must acquire knowledge of ideas and practice in methods, and dwell with them long and deeply enough to reap the benefits.

A brief description will now be given of the nine chapters, with some suggestions for reading and instruction. Chapters 1 and 2 are preparatory. A synopsis of the requisite "measure and integration" is given in Chapter 2, together with certain supplements essential to probability theory. Chapter 1 is really a review of elementary real variables; although it is somewhat expendable, a reader with adequate background should be able to cover it swiftly and confidently — with something gained from the effort. For class instruction it may be advisable to begin the course with Chapter 2 and fill in from Chapter 1 as the occasions arise. Chapter 3 is the true introduction to the language and framework of probability theory, but I have restricted its content to what is crucial and feasible at this stage, relegating certain important extensions, such as shifting and conditioning, to Chapters 8 and 9. This is done to avoid overloading the chapter with definitions and generalities that would be meaningless without frequent application. Chapter 4 may be regarded as an assembly of notions and techniques of real function theory adapted to the usage of probability. Thus, Chapter 5 is the first place where the reader encounters *bona fide* theorems in the field. The famous landmarks shown there serve also to introduce the ways and means peculiar to the subject. Chapter 6 develops some of the chief analytical weapons, namely Fourier and Laplace transforms, needed for challenges old and new. Quick testing grounds are provided, but for major battlefields one must await Chapters 7 and 8. Chapter 7 initiates what has been called the "central problem" of classical probability theory. Time has marched on and the center of the stage has shifted, but this topic remains without doubt a crowning achievement. In Chapters 8 and 9 two different aspects of (discrete parameter) stochastic processes are presented in some depth. The random walks in Chapter 8 illustrate the way probability theory transforms other parts of mathematics. It does so by introducing the trajectories of a process, thereby turning what was static into a dynamic structure. The same revolution is now going on in potential theory by the injection of the theory of Markov processes. In Chapter 9 we return to fundamentals and strike out in major new directions. While Markov processes can be barely introduced in the limited space, martingales have become an indispensable tool for any serious study of contemporary work and are discussed here at length. The fact that these topics are placed at the end rather than the beginning of the book, where they might very well be, testifies to my belief that the student of mathematics is better advised to learn something old before plunging into the new.

A short course may be built around Chapters 2, 3, 4, selections from Chapters 5, 6, and the first one or two sections of Chapter 9. For a richer fare, substantial portions of the last three chapters should be given without skipping any one of them. In a class with solid background, Chapters 1, 2, and 4 need not be covered in detail. At the opposite end, Chapter 2 may be filled in with proofs that are readily available in standard texts. It is my hope that this book may also be useful to mature mathematicians as a gentle but not so meager introduction to genuine probability theory. (Often they stop just before things become interesting!) Such a reader may begin with Chapter 3, go at once to Chapter 5 with a few glances at Chapter 4, skim through Chapter 6, and take up the remaining chapters seriously to get a real feeling for the subject.

Several cases of exclusion and inclusion merit special comment. I chose to construct only a sequence of independent random variables (in Section 3.3), rather than a more general one, in the belief that the latter is better absorbed in a course on stochastic processes. I chose to postpone a discussion of conditioning until quite late, in order to follow it up at once with varied and worthwhile applications. With a little reshuffling Section 9.1 may be placed right after Chapter 3 if so desired. I chose not to include a fuller treatment of infinitely divisible laws, for two reasons: the material is well covered in two or three treatises, and the best way to develop it would be in the context of the underlying additive process, as originally conceived by its creator Paul Lévy. I took pains to spell out a peripheral discussion of the logarithm of characteristic function to combat the errors committed on this score by numerous existing books. Finally, and this is mentioned here only in response to a query by Doob, I chose to present the brutal Theorem 5.3.2 in the original form given by Kolmogorov because I want to expose the student to hardships in mathematics.

There are perhaps some new things in this book, but in general I have not striven to appear original or merely different, having at heart the interests of the novice rather than the connoisseur. In the same vein, I favor as a rule of writing (euphemistically called "style") clarity over elegance. In my opinion the slightly decadent fashion of conciseness has been overwrought, particularly in the writing of textbooks. The only valid argument I have heard for an excessively terse style is that it may encourage the reader to think for himself. Such an effect can be achieved equally well, for anyone who wishes it, by simply omitting every other sentence in the unabridged version.

This book contains about 500 exercises consisting mostly of special cases and examples, second thoughts and alternative arguments, natural extensions, and some novel departures. With a few obvious exceptions they are neither profound nor trivial, and hints and comments are appended to many of them. If they tend to be somewhat inbred, at least they are relevant to the text and should help in its digestion. As a bold venture I have marked a few of them with $*$ to indicate a "must," although no rigid standard of selection has been used. Some of these are needed in the book, but in any case the reader's study of the text will be more complete after he has tried at least those problems.

Over a span of nearly twenty years I have taught a course at approximately the level of this book a number of times. The penultimate draft of the manuscript was tried out in a class given in 1966 at Stanford University. Because of an anachronism that allowed only two quarters to the course (as if probability could also blossom faster in the California climate!), I had to omit the second halves of Chapters 8 and 9 but otherwise kept fairly closely to the text as presented here. (The second half of Chapter 9 was covered in a subsequent course called "stochastic processes.") A good fraction of the exercises were assigned as homework, and in addition a great majority of them were worked out by volunteers. Among those in the class who cooperated in this manner and who corrected mistakes and suggested improvements are: Jack E. Clark, B. Curtis Eaves, Susan D. Horn, Alan T. Huckleberry, Thomas M. Liggett, and Roy E. Welsch, to whom I owe sincere thanks. The manuscript was also read by J. L. Doob and Benton Jamison, both of whom contributed a great deal to the final revision. They have also used part of the manuscript in their classes. Aside from these personal acknowledgments, the book owes of course to a large number of authors of original papers, treatises, and textbooks. I have restricted bibliographical references to the major sources while adding many more names among the exercises. Some oversight is perhaps inevitable; however, inconsequential or irrelevant "name-dropping" is deliberately avoided, with two or three exceptions which should prove the rule.

It is a pleasure to thank Rosemarie Stampfel and Gail Lemmond for their superb job in typing the manuscript.

# A
# COURSE IN
# PROBABILITY
# THEORY

THIRD EDITION

# 1 Distribution function

## 1.1 Monotone functions

We begin with a discussion of distribution functions as a traditional way of introducing probability measures. It serves as a convenient bridge from elementary analysis to probability theory, upon which the beginner may pause to review his mathematical background and test his mental agility. Some of the methods as well as results in this chapter are also useful in the theory of stochastic processes.

*In this book we shall follow the fashionable usage of the words "positive", "negative", "increasing", "decreasing" in their loose interpretation. For example, "$x$ is positive" means "$x \geq 0$"; the qualifier "strictly" will be added when "$x > 0$" is meant.* By a "function" we mean in this chapter a real finite-valued one unless otherwise specified.

Let then $f$ be an increasing function defined on the real line $(-\infty, +\infty)$. Thus for any two real numbers $x_1$ and $x_2$,

$$x_1 < x_2 \Rightarrow f(x_1) \leq f(x_2). \tag{1}$$

We begin by reviewing some properties of such a function. The notation "$t \uparrow x$" means "$t < x, t \to x$"; "$t \downarrow x$" means "$t > x, t \to x$".

(**i**) For each $x$, both *unilateral limits*

$$\lim_{t \uparrow x} f(t) = f(x-) \text{ and } \lim_{t \downarrow x} f(t) = f(x+) \tag{2}$$

exist and are finite. Furthermore the *limits at infinity*

$$\lim_{t \downarrow -\infty} f(t) = f(-\infty) \text{ and } \lim_{t \uparrow +\infty} f(t) = f(+\infty)$$

exist; the former may be $-\infty$, the latter may be $+\infty$.

This follows from monotonicity; indeed

$$f(x-) = \sup_{-\infty < t < x} f(t), \; f(x+) = \inf_{x < t < +\infty} f(t).$$

(**ii**) For each $x$, $f$ is continuous at $x$ if and only if

$$f(x-) = f(x) = f(x+).$$

To see this, observe that the continuity of a monotone function $f$ at $x$ is equivalent to the assertion that

$$\lim_{t \uparrow x} f(t) = f(x) = \lim_{t \downarrow x} f(t).$$

By (i), the limits above exist as $f(x-)$ and $f(x+)$ and

$$f(x-) \le f(x) \le f(x+), \tag{3}$$

from which (ii) follows.

In general, we say that the function $f$ has a *jump* at $x$ iff the two limits in (2) both exist but are unequal. The value of $f$ at $x$ itself, viz. $f(x)$, may be arbitrary, but for an increasing $f$ the relation (3) must hold. As a consequence of (i) and (ii), we have the next result.

(**iii**) The only possible kind of discontinuity of an increasing function is a jump. [The reader should ask himself what other kinds of discontinuity there are for a function in general.]

If there is a jump at $x$, we call $x$ a *point of jump* of $f$ and the number $f(x+) - f(x-)$ the *size of the jump* or simply "the jump" at $x$.

It is worthwhile to observe that points of jump may have a finite point of accumulation and that such a point of accumulation need not be a point of jump itself. Thus, the set of points of jump is not necessarily a closed set.

**Example 1.** Let $x_0$ be an arbitrary real number, and define a function $f$ as follows:

$$
\begin{aligned}
f(x) &= 0 && \text{for } x \le x_0 - 1; \\
&= 1 - \frac{1}{n} && \text{for } x_0 - \frac{1}{n} \le x < x_0 - \frac{1}{n+1},\ n = 1, 2, \dots; \\
&= 1 && \text{for } x \ge x_0.
\end{aligned}
$$

The point $x_0$ is a point of accumulation of the points of jump $\{x_0 - 1/n, n \ge 1\}$, but $f$ is continuous at $x_0$.

Before we discuss the next example, let us introduce a notation that will be used throughout the book. For any real number $t$, we set

$$
(4) \qquad \delta_t(x) = \begin{cases} 0 & \text{for } x < t, \\ 1 & \text{for } x \ge t. \end{cases}
$$

We shall call the function $\delta_t$ the *point mass at* $t$.

**Example 2.** Let $\{a_n, n \ge 1\}$ be any given enumeration of the set of all rational numbers, and let $\{b_n, n \ge 1\}$ be a set of positive ($\ge 0$) numbers such that $\sum_{n=1}^{\infty} b_n < \infty$. For instance, we may take $b_n = 2^{-n}$. Consider now

$$
(5) \qquad f(x) = \sum_{n=1}^{\infty} b_n \delta_{a_n}(x).
$$

Since $0 \le \delta_{a_n}(x) \le 1$ for every $n$ and $x$, the series in (5) is absolutely and *uniformly* convergent. Since each $\delta_{a_n}$ is increasing, it follows that if $x_1 < x_2$,

$$
f(x_2) - f(x_1) = \sum_{n=1}^{\infty} b_n [\delta_{a_n}(x_2) - \delta_{a_n}(x_1)] \ge 0.
$$

Hence $f$ is increasing. Thanks to the uniform convergence (why?) we may deduce that for each $x$,

$$
(6) \qquad f(x+) - f(x-) = \sum_{n=1}^{\infty} b_n [\delta_{a_n}(x+) - \delta_{a_n}(x-)].
$$

But for each $n$, the number in the square brackets above is 0 or 1 according as $x \ne a_n$ or $x = a_n$. Hence if $x$ is different from all the $a_n$'s, each term on the right side of (6) vanishes; on the other hand if $x = a_k$, say, then exactly one term, that corresponding to $n = k$, does not vanish and yields the value $b_k$ for the whole series. This proves that the function $f$ has jumps at all the rational points and nowhere else.

This example shows that the set of points of jump of an increasing function may be everywhere dense; in fact the set of rational numbers in the example may be replaced by an arbitrary countable set without any change of the argument. We now show that the condition of countability is indispensable. *By "countable" we mean always "finite (possibly empty) or countably infinite".*

(**iv**) The set of discontinuities of $f$ is countable.

We shall prove this by a topological argument of some general applicability. In Exercise 3 after this section another proof based on an equally useful counting argument will be indicated. For each point of jump $x$ consider the open interval $I_x = (f(x-), f(x+))$. If $x'$ is another point of jump and $x < x'$, say, then there is a point $\tilde{x}$ such that $x < \tilde{x} < x'$. Hence by monotonicity we have

$$f(x+) \leq f(\tilde{x}) \leq f(x'-).$$

It follows that the two intervals $I_x$ and $I_{x'}$ are disjoint, though they may abut on each other if $f(x+) = f(x'-)$. Thus we may associate with the set of points of jump in the domain of $f$ a certain collection of pairwise disjoint open intervals in the range of $f$. Now any such collection is necessarily a countable one, since each interval contains a rational number, so that the collection of intervals is in one-to-one correspondence with a certain subset of the rational numbers and the latter is countable. Therefore the set of discontinuities is also countable, since it is in one-to-one correspondence with the set of intervals associated with it.

(**v**) Let $f_1$ and $f_2$ be two increasing functions and $D$ a set that is (everywhere) dense in $(-\infty, +\infty)$. Suppose that

$$\forall x \in D: f_1(x) = f_2(x).$$

Then $f_1$ and $f_2$ have the same points of jump of the same size, and they coincide except possibly at some of these points of jump.

To see this, let $x$ be an arbitrary point and let $t_n \in D$, $t'_n \in D$, $t_n \uparrow x$, $t'_n \downarrow x$. Such sequences exist since $D$ is dense. It follows from (i) that

$$\begin{aligned} f_1(x-) &= \lim_n f_1(t_n) = \lim_n f_2(t_n) = f_2(x-), \\ f_1(x+) &= \lim_n f_1(t'_n) = \lim_n f_2(t'_n) = f_2(x+). \end{aligned} \tag{6}$$

In particular

$$\forall x: f_1(x+) - f_1(x-) = f_2(x+) - f_2(x-).$$

The first assertion in (v) follows from this equation and (ii). Furthermore if $f_1$ is continuous at $x$, then so is $f_2$ by what has just been proved, and we have

$$f_1(x) = f_1(x-) = f_2(x-) = f_2(x),$$

proving the second assertion.

How can $f_1$ and $f_2$ differ at all? This can happen only when $f_1(x)$ and $f_2(x)$ assume different values in the interval $(f_1(x-), f_1(x+)) = (f_2(x-), f_2(x+))$. It will turn out in Chapter 2 (see in particular Exercise 21 of Sec. 2.2) that the precise value of $f$ at a point of jump is quite unessential for our purposes and may be modified, subject to (3), to suit our convenience. More precisely, given the function $f$, we can define a new function $\tilde{f}$ in several different ways, such as

$$\tilde{f}(x) = f(x-),\ \tilde{f}(x) = f(x+),\ \tilde{f}(x) = \frac{f(x-) + f(x+)}{2},$$

and use one of these instead of the original one. The third modification is found to be convenient in Fourier analysis, but either one of the first two is more suitable for probability theory. We have a free choice between them and we shall choose the second, namely, right continuity.

**(vi)** If we put

$$\forall x: \tilde{f}(x) = f(x+),$$

then $\tilde{f}$ is increasing and right continuous everywhere.

Let us recall that an arbitrary function $g$ is said to be *right continuous* at $x$ iff $\lim_{t \downarrow x} g(t)$ exists and the limit, to be denoted by $g(x+)$, is equal to $g(x)$. To prove the assertion (vi) we must show that

$$\forall x: \lim_{t \downarrow x} f(t+) = f(x+).$$

This is indeed true for any $f$ such that $f(t+)$ exists for every $t$. For then: given any $\epsilon > 0$, there exists $\delta(\epsilon) > 0$ such that

$$\forall s \in (x, x + \delta): |f(s) - f(x+)| \leq \epsilon.$$

Let $t \in (x, x + \delta)$ and let $s \downarrow t$ in the above, then we obtain

$$|f(t+) - f(x+)| \leq \epsilon,$$

which proves that $\tilde{f}$ is right continuous. It is easy to see that it is increasing if $f$ is so.

Let $D$ be dense in $(-\infty, +\infty)$, and suppose that $f$ is a function with the domain $D$. We may speak of the monotonicity, continuity, uniform continuity, and so on of $f$ on its domain of definition if in the usual definitions we restrict ourselves to the points of $D$. Even if $f$ is defined in a larger domain, we may still speak of these properties "on $D$" by considering the "restriction of $f$ to $D$".

**(vii)** Let $f$ be increasing on $D$, and define $\tilde{f}$ on $(-\infty, +\infty)$ as follows:

$$\forall x: \tilde{f}(x) = \inf_{x<t\in D} f(t).$$

Then $\tilde{f}$ is increasing and right continuous everywhere.

This is a generalization of (vi). $\tilde{f}$ is clearly increasing. To prove right continuity let an arbitrary $x_0$ and $\epsilon > 0$ be given. There exists $t_0 \in D$, $t_0 > x_0$, such that

$$f(t_0) - \epsilon \le \tilde{f}(x_0) \le f(t_0).$$

Hence if $t \in D$, $x_0 < t < t_0$, we have

$$0 \le f(t) - \tilde{f}(x_0) \le f(t_0) - \tilde{f}(x_0) \le \epsilon.$$

This implies by the definition of $\tilde{f}$ that for $x_0 < x < t_0$ we have

$$0 \le \tilde{f}(x) - \tilde{f}(x_0) \le \epsilon.$$

Since $\epsilon$ is arbitrary, it follows that $\tilde{f}$ is right continuous at $x_0$, as was to be shown.

## EXERCISES

**1.** Prove that for the $f$ in Example 2 we have

$$f(-\infty) = 0, \quad f(+\infty) = \sum_{n=1}^{\infty} b_n.$$

**2.** Construct an increasing function on $(-\infty, +\infty)$ with a jump of size one at, each integer, and constant between jumps. Such a function cannot be represented as $\sum_{n=1}^{\infty} b_n \delta_n(x)$ with $b_n = 1$ for each $n$, but a slight modification will do. Spell this out.

****3.** Suppose that $f$ is increasing and that there exist real numbers $A$ and $B$ such that $\forall x: A \le f(x) \le B$. Show that for each $\epsilon > 0$, the number of jumps of size exceeding $\epsilon$ is at most $(B - A)/\epsilon$. Hence prove (iv), first for bounded $f$ and then in general.

** indicates specially selected exercises (as mentioned in the Preface).

**4.** Let $f$ be an arbitrary function on $(-\infty, +\infty)$ and $L$ be the set of $x$ where $f$ is right continuous but not left continuous. Prove that $L$ is a countable set. [HINT: Consider $L \cap M_n$, where $M_n = \{x \mid O(f;x) > 1/n\}$ and $O(f;x)$ is the oscillation of $f$ at $x$.]

**$^\star$5.** Let $f$ and $\tilde{f}$ be as in (vii). Show that the continuity of $f$ on $D$ does not imply that of $\tilde{f}$ on $(-\infty, +\infty)$, but uniform continuity does imply uniform continuity.

**6.** Given any extended-valued $f$ on $(-\infty, +\infty)$, there exists a countable set $D$ with the following property. For each $t$, there exist $t_n \in D$, $t_n \to t$ such that $f(t) = \lim_{n\to\infty} f(t_n)$. This assertion remains true if "$t_n \to t$" is replaced by "$t_n \downarrow t$" or "$t_n \uparrow t$". [This is the crux of "separability" for stochastic processes. Consider the graph $(t, f(t))$ and introduce a metric.]

## 1.2 Distribution functions

Suppose now that $f$ is bounded as well as increasing and not constant. We have then

$$\forall x: -\infty < f(-\infty) \le f(x) \le f(+\infty) < +\infty.$$

Consider the "normalized" function:

$$\tilde{f}(x) = \frac{f(x) - f(-\infty)}{f(+\infty) - f(-\infty)} \tag{1}$$

which is bounded and increasing with

$$\tilde{f}(-\infty) = 0, \quad \tilde{f}(+\infty) = 1. \tag{2}$$

Owing to the simple nature of the linear transformation in (1) from $f$ to $\tilde{f}$ and vice versa, we may without loss of generality assume the normalizing conditions (2) in dealing with a bounded increasing function. To avoid needless complications, we shall also assume that $\tilde{f}$ is right continuous as discussed in Sec. 1.1. These conditions will now be formalized.

DEFINITION OF A DISTRIBUTION FUNCTION. A real-valued function $F$ with domain $(-\infty, +\infty)$ that is increasing and right continuous with $F(-\infty) = 0$, $F(+\infty) = 1$ is called a distribution function, to be abbreviated hereafter as "d.f." A d.f. that is a point mass as defined in (4) of Sec. 1.1 is said to be "degenerate", otherwise "nondegenerate".

Of course all the properties given in Sec. 1.1 hold for a d.f. Indeed the added assumption of boundedness does not appreciably simplify the proofs there. In particular, let $\{a_j\}$ be the countable set of points of jump of $F$ and

$b_j$ the size at jump at $a_j$, then

$$F(a_j) - F(a_j-) = b_j$$

since $F(a_j+) = F(a_j)$. Consider the function

$$F_d(x) = \sum_j b_j \delta_{a_j}(x)$$

which represents the sum of all the jumps of $F$ in the half-line $(-\infty, x]$. It is clearly increasing, right continuous, with

$$F_d(-\infty) = 0, \quad F_d(+\infty) = \sum_j b_j \leq 1. \tag{3}$$

Hence $F_d$ is a bounded increasing function. It should constitute the "jumping part" of $F$, and if it is subtracted out from $F$, the remainder should be positive, contain no more jumps, and so be continuous. These plausible statements will now be proved—they are easy enough but not really trivial.

**Theorem 1.2.1.** Let

$$F_c(x) = F(x) - F_d(x);$$

then $F_c$ is positive, increasing, and continuous.

PROOF. Let $x < x'$, then we have

$$F_d(x') - F_d(x) = \sum_{x < a_j \leq x'} b_j = \sum_{x < a_j \leq x'} [F(a_j) - F(a_j-)] \tag{4}$$

$$\leq F(x') - F(x).$$

It follows that both $F_d$ and $F_c$ are increasing, and if we put $x = -\infty$ in the above, we see that $F_d \leq F$ and so $F_c$ is indeed positive. Next, $F_d$ is right continuous since each $\delta_{a_j}$ is and the series defining $F_d$ converges uniformly in $x$; the same argument yields (cf. Example 2 of Sec. 1.1)

$$F_d(x) - F_d(x-) = \begin{cases} b_j & \text{if } x = a_j, \\ 0 & \text{otherwise.} \end{cases}$$

Now this evaluation holds also if $F_d$ is replaced by $F$ according to the definition of $a_j$ and $b_j$, hence we obtain for each $x$:

$$F_c(x) - F_c(x-) = F(x) - F(x-) - [F_d(x) - F_d(x-)] = 0.$$

This shows that $F_c$ is left continuous; since it is also right continuous, being the difference of two such functions, it is continuous.

**Theorem 1.2.2.** Let $F$ be a d.f. Suppose that there exist a continuous function $G_c$ and a function $G_d$ of the form

$$G_d(x) = \sum_j b'_j \delta_{a'_j}(x)$$

[where $\{a'_j\}$ is a countable set of real numbers and $\sum_j |b'_j| < \infty$], such that

$$F = G_c + G_d,$$

then

$$G_c = F_c, \quad G_d = F_d,$$

where $F_c$ and $F_d$ are defined as before.

PROOF. If $F_d \neq G_d$, then either the sets $\{a_j\}$ and $\{a'_j\}$ are not identical, or we may relabel the $a'_j$ so that $a'_j = a_j$ for all $j$ but $b'_j \neq b_j$ for some $j$. In either case we have for at least one $j$, and $\tilde{a} = a_j$ or $a'_j$:

$$[F_d(\tilde{a}) - F_d(\tilde{a}-)] - [G_d(\tilde{a}) - G_d(\tilde{a}-)] \neq 0.$$

Since $F_c - G_c = G_d - F_d$, this implies that

$$F_c(\tilde{a}) - G_c(\tilde{a}) - [F_c(\tilde{a}-) - G_c(\tilde{a}-)] \neq 0,$$

contradicting the fact that $F_c - G_c$ is a continuous function. Hence $F_d = G_d$ and consequently $F_c = G_c$.

DEFINITION. A d.f. $F$ that can be represented in the form

$$F = \sum_j b_j \delta_{a_j}$$

where $\{a_j\}$ is a countable set of real numbers, $b_j > 0$ for every $j$ and $\sum_j b_j = 1$, is called a *discrete* d.f. A d.f. that is continuous everywhere is called a *continuous* d.f.

Suppose $F_c \not\equiv 0$, $F_d \not\equiv 0$ in Theorem 1.2.1, then we may set $\alpha = F_d(\infty)$ so that $0 < \alpha < 1$,

$$F_1 = \frac{1}{\alpha} F_d, \quad F_2 = \frac{1}{1-\alpha} F_c,$$

and write

(5) $$F = \alpha F_1 + (1-\alpha) F_2.$$

Now $F_1$ is a discrete d.f., $F_2$ is a continuous d.f., and $F$ is exhibited as a *convex combination* of them. If $F_c \equiv 0$, then $F$ is discrete and we set $\alpha = 1$, $F_1 \equiv F$, $F_2 \equiv 0$; if $F_d \equiv 0$, then $F$ is continuous and we set $\alpha = 0$, $F_1 \equiv 0$,

$F_2 \equiv F$; in either extreme case (5) remains valid. We may now summarize the two theorems above as follows.

**Theorem 1.2.3.** Every d.f. can be written as the convex combination of a discrete and a continuous one. Such a decomposition is unique.

EXERCISES

**1.** Let $F$ be a d.f. Then for each $x$,

$$\lim_{\epsilon \downarrow 0}[F(x+\epsilon) - F(x-\epsilon)] = 0$$

unless $x$ is a point of jump of $F$, in which case the limit is equal to the size of the jump.

***2.** Let $F$ be a d.f. with points of jump $\{a_j\}$. Prove that the sum

$$\sum_{x-\epsilon<a_j<x} [F(a_j) - F(a_j-)]$$

converges to zero as $\epsilon \downarrow 0$, for every $x$. What if the summation above is extended to $x - \epsilon < a_j \leq x$ instead? Give another proof of the continuity of $F_c$ in Theorem 1.2.1 by using this problem.

**3.** A plausible verbal definition of a discrete d.f. may be given thus: "It is a d.f. that has jumps and is constant between jumps." [Such a function is sometimes called a "step function", though the meaning of this term does not seem to be well established.] What is wrong with this? But suppose that the set of points of jump is "discrete" in the Euclidean topology, then the definition is valid (apart from our convention of right continuity).

**4.** For a general increasing function $F$ there is a similar decomposition $F = F_c + F_d$, where both $F_c$ and $F_d$ are increasing, $F_c$ is continuous, and $F_d$ is "purely jumping". [HINT: Let $a$ be a point of continuity, put $F_d(a) = F(a)$, add jumps in $(a, \infty)$ and subtract jumps in $(-\infty, a)$ to define $F_d$. Cf. Exercise 2 in Sec. 1.1.]

**5.** Theorem 1.2.2 can be generalized to any bounded increasing function. More generally, let $f$ be the difference of two bounded increasing functions on $(-\infty, +\infty)$; such a function is said to be *of bounded variation* there. Define its purely discontinuous and continuous parts and prove the corresponding decomposition theorem.

***6.** A point $x$ is said to belong to the support of the d.f. $F$ iff for every $\epsilon > 0$ we have $F(x+\epsilon) - F(x-\epsilon) > 0$. The set of all such $x$ is called *the support* of $F$. Show that each point of jump belongs to the support, and that each isolated point of the support is a point of jump. Give an example of a discrete d.f. whose support is the whole line.

**7.** Prove that the support of any d.f. is a closed set, and the support of any continuous d.f. is a perfect set.

## 1.3 Absolutely continuous and singular distributions

Further analysis of d.f.'s requires the theory of Lebesgue measure. Throughout the book this measure will be denoted by $m$; "almost everywhere" on the real line without qualification will refer to it and be abbreviated to "a.e."; an integral written in the form $\int \ldots dt$ is a Lebesgue integral; a function $f$ is said to be "integrable" in $(a, b)$ iff

$$\int_a^b f(t)\,dt$$

is defined and finite [this entails, of course, that $f$ be Lebesgue measurable]. The class of such functions will be denoted by $L^1(a, b)$, and $L^1(-\infty, \infty)$ is abbreviated to $L^1$. The complement of a subset $S$ of an understood "space" such as $(-\infty, +\infty)$ will be denoted by $S^c$.

DEFINITION. A function $F$ is called *absolutely continuous* [in $(-\infty, \infty)$ and with respect to the Lebesgue measure] iff there exists a function $f$ in $L^1$ such that we have for every $x < x'$:

$$F(x') - F(x) = \int_x^{x'} f(t)\,dt. \tag{1}$$

It follows from a well-known proposition (see, e.g., Natanson [3]*) that such a function $F$ has a derivative equal to $f$ a.e. In particular, if $F$ is a d.f., then

$$f \geq 0 \text{ a.e.} \quad \text{and} \quad \int_{-\infty}^{\infty} f(t)\,dt = 1. \tag{2}$$

Conversely, given any $f$ in $L^1$ satisfying the conditions in (2), the function $F$ defined by

$$\forall x\colon F(x) = \int_{-\infty}^{x} f(t)\,dt \tag{3}$$

is easily seen to be a d.f. that is absolutely continuous.

DEFINITION. A function $F$ is called *singular* iff it is not identically zero and $F'$ (exists and) equals zero a.e.

* Numbers in brackets refer to the General Bibliography.

The next theorem summarizes some basic facts of real function theory; see, e.g., Natanson [3].

**Theorem 1.3.1.** Let $F$ be bounded increasing with $F(-\infty) = 0$, and let $F'$ denote its derivative wherever existing. Then the following assertions are true.

(a) If $S$ denotes the set of all $x$ for which $F'(x)$ exists with $0 \leq F'(x) < \infty$, then $m(S^c) = 0$.

(b) This $F'$ belongs to $L^1$, and we have for every $x < x'$:

$$\int_x^{x'} F'(t)\,dt \leq F(x') - F(x). \tag{4}$$

(c) If we put

$$\forall x\colon F_{ac}(x) = \int_{-\infty}^{x} F'(t)\,dt, \qquad F_s(x) = F(x) - F_{ac}(x), \tag{5}$$

then $F'_{ac} = F'$ a.e. so that $F'_s = F' - F'_{ac} = 0$ a.e. and consequently $F_s$ is singular if it is not identically zero.

DEFINITION. Any positive function $f$ that is equal to $F'$ a.e. is called a *density* of $F$. $F_{ac}$ is called *the absolutely continuous part*, $F_s$ *the singular part of* $F$. Note that the previous $F_d$ is part of $F_s$ as defined here.

It is clear that $F_{ac}$ is increasing and $F_{ac} \leq F$. From (4) it follows that if $x < x'$

$$F_s(x') - F_s(x) = F(x') - F(x) - \int_x^{x'} f(t)\,dt \geq 0.$$

Hence $F_s$ is also increasing and $F_s \leq F$. We are now in a position to announce the following result, which is a refinement of Theorem 1.2.3.

**Theorem 1.3.2.** Every d.f. $F$ can be written as the convex combination of a discrete, a singular continuous, and an absolutely continuous d.f. Such a decomposition is unique.

EXERCISES

**1.** A d.f. $F$ is singular if and only if $F = F_s$; it is absolutely continuous if and only if $F \equiv F_{ac}$.

**2.** Prove Theorem 1.3.2.

***3.** If the support of a d.f. (see Exercise 6 of Sec. 1.2) is of measure zero, then $F$ is singular. The converse is false.

***4.** Suppose that $F$ is a d.f. and (3) holds with a continuous $f$. Then $F' = f \geq 0$ everywhere.

**5.** Under the conditions in the preceding exercise, the support of $F$ is the closure of the set $\{t \mid f(t) > 0\}$; the complement of the support is the *interior* of the set $\{t \mid f(t) = 0\}$.

**6.** Prove that a discrete distribution is singular. [Cf. Exercise 13 of Sec. 2.2.]

**7.** Prove that a singular function as defined here is (Lebesgue) measurable but need not be of bounded variation even locally. [HINT: Such a function is continuous except on a set of Lebesgue measure zero; use the completeness of the Lebesgue measure.]

The remainder of this section is devoted to the construction of a singular continuous distribution. For this purpose let us recall the construction of the *Cantor (ternary) set* (see, e.g., Natanson [3]). From the closed interval [0,1], the "middle third" open interval $(\frac{1}{3}, \frac{2}{3})$ is removed; from each of the two remaining disjoint closed intervals the middle third, $(\frac{1}{9}, \frac{2}{9})$ and $(\frac{7}{9}, \frac{8}{9})$, respectively, are removed and so on. After $n$ steps, we have removed

$$1 + 2 + \cdots + 2^{n-1} = 2^n - 1$$

disjoint open intervals and are left with $2^n$ disjoint closed intervals each of length $1/3^n$. Let these removed ones, in order of position from left to right, be denoted by $J_{n,k}$, $1 \leq k \leq 2^n - 1$, and their union by $U_n$. We have

$$m(U_n) = \frac{1}{3} + \frac{2}{3^2} + \frac{4}{3^3} + \cdots + \frac{2^{n-1}}{3^n} = 1 - \left(\frac{2}{3}\right)^n.$$

As $n \uparrow \infty$, $U_n$ increases to an open set $U$; the complement $C$ of $U$ with respect to [0,1] is a perfect set, called the Cantor set. It is of measure zero since

$$m(C) = 1 - m(U) = 1 - 1 = 0.$$

Now for each $n$ and $k$, $n \geq 1$, $1 \leq k \leq 2^n - 1$, we put

$$c_{n,k} = \frac{k}{2^n};$$

and define a function $F$ on $U$ as follows:

$$F(x) = c_{n,k} \quad \text{for } x \in J_{n,k}. \tag{7}$$

This definition is consistent since two intervals, $J_{n,k}$ and $J_{n',k'}$, are either disjoint or identical, and in the latter case so are $c_{n,k} = c_{n',k'}$. The last assertion becomes obvious if we proceed step by step and observe that

$$J_{n+1,2k} = J_{n,k}, \quad c_{n+1,2k} = c_{n,k} \quad \text{for } 1 \le k \le 2^n - 1.$$

The value of $F$ is constant on each $J_{n,k}$ and is strictly greater on any other $J_{n',k'}$ situated to the right of $J_{n,k}$. Thus $F$ is increasing and clearly we have

$$\lim_{x \downarrow 0} F(x) = 0, \quad \lim_{x \uparrow 1} F(x) = 1.$$

Let us complete the definition of $F$ by setting

$$F(x) = 0 \quad \text{for } x \le 0, \quad F(x) = 1 \quad \text{for } x \ge 1.$$

$F$ is now defined on the domain $D = (-\infty, 0) \cup U \cup (1, \infty)$ and increasing there. Since each $J_{n,k}$ is at a distance $\ge 1/3^n$ from any distinct $J_{n,k'}$ and the total variation of $F$ over each of the $2^n$ disjoint intervals that remain after removing $J_{n,k}$, $1 \le k \le 2^n - 1$, is $1/2^n$, it follows that

$$0 \le x' - x \le \frac{1}{3^n} \Rightarrow 0 \le F(x') - F(x) \le \frac{1}{2^n}.$$

Hence $F$ is uniformly continuous on $D$. By Exercise 5 of Sec. 1.1, there exists a continuous increasing $\tilde{F}$ on $(-\infty, +\infty)$ that coincides with $F$ on $D$. This $\tilde{F}$ is a continuous d.f. that is constant on each $J_{n,k}$. It follows that $\tilde{F}' = 0$ on $U$ and so also on $(-\infty, +\infty) - C$. Thus $\tilde{F}$ is singular. Alternatively, it is clear that none of the points in $D$ is in the support of $F$, hence the latter is contained in $C$ and of measure 0, so that $\tilde{F}$ is singular by Exercise 3 above. [In Exercise 13 of Sec. 2.2, it will become obvious that the *measure* corresponding to $\tilde{F}$ is singular because there is no mass in $U$.]

## EXERCISES

The $F$ in these exercises is the $\tilde{F}$ defined above.

**8.** Prove that the support of $F$ is exactly $C$.

***9.** It is well known that any point $x$ in $C$ has a ternary expansion without the digit 1:

$$x = \sum_{n=1}^{\infty} \frac{a_n}{3^n}, \quad a_n = 0 \text{ or } 2.$$

Prove that for this $x$ we have

$$F(x) = \sum_{n=1}^{\infty} \frac{a_n}{2^{n+1}}.$$

**10.** For each $x \in [0, 1]$, we have

$$2F\left(\frac{x}{3}\right) = F(x), \quad 2F\left(\frac{2}{3} + \frac{x}{3}\right) - 1 = F(x).$$

**11.** Calculate

$$\int_0^1 x\, dF(x), \quad \int_0^1 x^2\, dF(x), \quad \int_0^1 e^{itx}\, dF(x).$$

[HINT: This can be done directly or by using Exercise 10; for a third method see Exercise 9 of Sec. 5.3.]

**12.** Extend the function $F$ on [0,1] trivially to $(-\infty, \infty)$. Let $\{r_n\}$ be an enumeration of the rationals and

$$G(x) = \sum_{n=1}^{\infty} \frac{1}{2^n} F(r_n + x).$$

Show that $G$ is a d.f. that is strictly increasing for all $x$ and singular. Thus we have a singular d.f. with support $(-\infty, \infty)$.

***13.** Consider $F$ on [0,1]. Modify its inverse $F^{-1}$ suitably to make it single-valued in [0,1]. Show that $F^{-1}$ so modified is a discrete d.f. and find its points of jump and their sizes.

**14.** Given any closed set $C$ in $(-\infty, +\infty)$, there exists a d.f. whose support is exactly $C$. [HINT: Such a problem becomes easier when the corresponding measure is considered; see Sec. 2.2 below.]

***15.** The Cantor d.f. $F$ is a good building block of "pathological" examples. For example, let $H$ be the inverse of the homeomorphic map of [0,1] onto itself: $x \to \frac{1}{2}[F(x) + x]$; and $E$ a subset of [0,1] which is not Lebesgue measurable. Show that

$$1_{H(E)} \circ H = 1_E$$

where $H(E)$ is the image of $E$, $1_B$ is the indicator function of $B$, and $\circ$ denotes the composition of functions. Hence deduce: (1) a Lebesgue measurable function of a strictly increasing and continuous function need not be Lebesgue measurable; (2) there exists a Lebesgue measurable function that is not Borel measurable.

# 2 Measure theory

## 2.1 Classes of sets

Let $\Omega$ be an "abstract space", namely a nonempty set of elements to be called "points" and denoted generically by $\omega$. Some of the usual operations and relations between sets, together with the usual notation, are given below.

| | | |
|---|---|---|
| Union | : | $E \cup F, \quad \bigcup_n E_n$ |
| Intersection | : | $E \cap F, \quad \bigcap_n E_n$ |
| Complement | : | $E^c = \Omega \backslash E$ |
| Difference | : | $E \backslash F = E \cap F^c$ |
| Symmetric difference | : | $E \,\Delta\, F = (E \backslash F) \cup (F \backslash E)$ |
| Singleton | : | $\{\omega\}$ |

Containing (for subsets of $\Omega$ as well as for collections thereof):

$$E \subset F, \qquad F \supset E \qquad \text{(not excluding } E = F\text{)}$$

$$\mathscr{A} \subset \mathscr{B}, \qquad \mathscr{B} \supset \mathscr{A} \qquad \text{(not excluding } \mathscr{A} = \mathscr{B}\text{)}$$

Belonging (for elements as well as for sets):

$$\omega \in E, \qquad E \in \mathscr{A}$$

Empty set: $\varnothing$

The reader is supposed to be familiar with the elementary properties of these operations.

A nonempty collection $\mathscr{A}$ of subsets of $\Omega$ may have certain "closure properties". Let us list some of those used below; note that $j$ *is always an index for a countable set* and *that commas as well as semicolons are used to denote "conjunctions" of premises.*

**(i)** $E \in \mathscr{A} \Rightarrow E^c \in \mathscr{A}$.

**(ii)** $E_1 \in \mathscr{A}, E_2 \in \mathscr{A} \Rightarrow E_1 \cup E_2 \in \mathscr{A}$.

**(iii)** $E_1 \in \mathscr{A}, E_2 \in \mathscr{A} \Rightarrow E_1 \cap E_2 \in \mathscr{A}$.

**(iv)** $\forall n \geq 2 : E_j \in \mathscr{A}, 1 \leq j \leq n \Rightarrow \bigcup_{j=1}^{n} E_j \in \mathscr{A}$.

**(v)** $\forall n \geq 2 : E_j \in \mathscr{A}, 1 \leq j \leq n \Rightarrow \bigcap_{j=1}^{n} E_j \in \mathscr{A}$.

**(vi)** $E_j \in \mathscr{A}; E_j \subset E_{j+1}, 1 \leq j < \infty \Rightarrow \bigcup_{j=1}^{\infty} E_j \in \mathscr{A}$.

**(vii)** $E_j \in \mathscr{A}; E_j \supset E_{j+1}, 1 \leq j < \infty \Rightarrow \bigcap_{j=1}^{\infty} E_j \in \mathscr{A}$.

**(viii)** $E_j \in \mathscr{A}, 1 \leq j < \infty \Rightarrow \bigcup_{j=1}^{\infty} E_j \in \mathscr{A}$.

**(ix)** $E_j \in \mathscr{A}, 1 \leq j < \infty \Rightarrow \bigcap_{j=1}^{\infty} E_j \in \mathscr{A}$.

**(x)** $E_1 \in \mathscr{A}, E_2 \in \mathscr{A}, E_1 \subset E_2 \Rightarrow E_2 \backslash E_1 \in \mathscr{A}$.

It follows from simple set algebra that under (i): (ii) and (iii) are equivalent; (vi) and (vii) are equivalent; (viii) and (ix) are equivalent. Also, (ii) implies (iv) and (iii) implies (v) by induction. It is trivial that (viii) implies (ii) and (vi); (ix) implies (iii) and (vii).

DEFINITION. A nonempty collection $\mathscr{F}$ of subsets of $\Omega$ is called a field iff (i) and (ii) hold. It is called a monotone class (M.C.) iff (vi) and (vii) hold. It is called a Borel field (B.F.) iff (i) and (viii) hold.

**Theorem 2.1.1.** A field is a B.F. if and only if it is also an M.C.

PROOF. The "only if" part is trivial; to prove the "if" part we show that (iv) and (vi) imply (viii). Let $E_j \in \mathscr{A}$ for $1 \leq j < \infty$, then

$$F_n = \bigcup_{j=1}^{n} E_j \in \mathscr{A}$$

by (iv), which holds in a field, $F_n \subset F_{n+1}$ and

$$\bigcup_{j=1}^{\infty} E_j = \bigcup_{j=1}^{\infty} F_j;$$

hence $\bigcup_{j=1}^{\infty} E_j \in \mathscr{A}$ by (vi).

The collection $\mathscr{S}$ of all subsets of $\Omega$ is a B.F. called the *total B.F.*; the collection of the two sets $\{\emptyset, \Omega\}$ is a B.F. called the *trivial B.F.* If $A$ is any index set and if for every $\alpha \in A$, $\mathscr{F}_\alpha$ is a B.F. (or M.C.) then the intersection $\bigcap_{\alpha\in A} \mathscr{F}_\alpha$ of all these B.F.'s (or M.C.'s), namely the collection of sets each of which belongs to all $\mathscr{F}_\alpha$, is also a B.F. (or M.C.). Given any nonempty collection $\mathscr{C}$ of sets, there is a *minimal* B.F. (or field, or M.C.) containing it; this is just the intersection of all B.F.'s (or fields, or M.C.'s) containing $\mathscr{C}$, of which there is at least one, namely the $\mathscr{S}$ mentioned above. This minimal B.F. (or field, or M.C.) is also said to be *generated by* $\mathscr{C}$. In particular if $\mathscr{F}_0$ is a field there is a minimal B.F. (or M.C.) containing $\mathscr{F}_0$.

**Theorem 2.1.2.** Let $\mathscr{F}_0$ be a field, $\mathscr{G}$ the minimal M.C. containing $\mathscr{F}_0$, $\mathscr{F}$ the minimal B.F. containing $\mathscr{F}_0$, then $\mathscr{F} = \mathscr{G}$.

PROOF. Since a B.F. is an M.C., we have $\mathscr{F} \supset \mathscr{G}$. To prove $\mathscr{F} \subset \mathscr{G}$ it is sufficient to show that $\mathscr{G}$ is a B.F. Hence by Theorem 2.1.1 it is sufficient to show that $\mathscr{G}$ is a field. We shall show that it is closed under intersection and complementation. Define two classes of subsets of $\mathscr{G}$ as follows:

$$\mathscr{C}_1 = \{E \in \mathscr{G} : E \cap F \in \mathscr{G} \text{ for all } F \in \mathscr{F}_0\},$$

$$\mathscr{C}_2 = \{E \in \mathscr{G} : E \cap F \in \mathscr{G} \text{ for all } F \in \mathscr{G}\}.$$

The identities

$$F \cap \left( \bigcup_{j=1}^{\infty} E_j \right) = \bigcup_{j=1}^{\infty} (F \cap E_j)$$

$$F \cap \left( \bigcap_{j=1}^{\infty} E_j \right) = \bigcap_{j=1}^{\infty} (F \cap E_j)$$

show that both $\mathscr{C}_1$ and $\mathscr{C}_2$ are M.C.'s. Since $\mathscr{F}_0$ is closed under intersection and contained in $\mathscr{G}$, it is clear that $\mathscr{F}_0 \subset \mathscr{C}_1$. Hence $\mathscr{G} \subset \mathscr{C}_1$ by the minimality of $\mathscr{G}$

and so $\mathscr{G} = \mathscr{C}_1$. This means for any $F \in \mathscr{F}_0$ and $E \in \mathscr{G}$ we have $F \cap E \in \mathscr{G}$, which in turn means $\mathscr{F}_0 \subset \mathscr{C}_2$. Hence $\mathscr{G} = \mathscr{C}_2$ and this means $\mathscr{G}$ is closed under intersection.

Next, define another class of subsets of $\mathscr{G}$ as follows:

$$\mathscr{C}_3 = \{E \in \mathscr{G} : E^c \in \mathscr{G}\}$$

The (DeMorgan) identities

$$\left(\bigcup_{j=1}^{\infty} E_j\right)^c = \bigcap_{j=1}^{\infty} E_j^c$$

$$\left(\bigcap_{j=1}^{\infty} E_j\right)^c = \bigcup_{j=1}^{\infty} E_j^c$$

show that $\mathscr{C}_3$ is a M.C. Since $\mathscr{F}_0 \subset \mathscr{C}_3$, it follows as before that $\mathscr{G} = \mathscr{C}_3$, which means $\mathscr{G}$ is closed under complementation. The proof is complete.

**Corollary.** Let $\mathscr{F}_0$ be a field, $\mathscr{F}$ the minimal B.F. containing $\mathscr{F}_0$; $\mathscr{C}$ a class of sets containing $\mathscr{F}_0$ and having the closure properties (vi) and (vii), then $\mathscr{C}$ contains $\mathscr{F}$.

The theorem above is one of a type called *monotone class theorems*. They are among the most useful tools of measure theory, and serve to extend certain relations which are easily verified for a special class of sets or functions to a larger class. Many versions of such theorems are known; see Exercise 10, 11, and 12 below.

EXERCISES

*1. $(\bigcup_j A_j)\backslash(\bigcup_j B_j) \subset \bigcup_j (A_j\backslash B_j)$. $(\bigcap_j A_j)\backslash(\bigcap_j B_j) \subset \bigcup_j (A_j\backslash B_j)$. When is there equality?

*2. The best way to define the symmetric difference is through indicators of sets as follows:

$$1_{A \triangle B} = 1_A + 1_B \qquad (\text{mod } 2)$$

where we have arithmetical addition modulo 2 on the right side. All properties of $\triangle$ follow easily from this definition, some of which are rather tedious to verify otherwise. As examples:

$$(A \triangle B) \triangle C = A \triangle (B \triangle C),$$

$$(A \triangle B) \triangle (B \triangle C) = A \triangle C,$$

$$(A \triangle B) \triangle (C \triangle D) = (A \triangle C) \triangle (B \triangle D),$$

$$A \triangle B = C \Leftrightarrow A = B \triangle C,$$

$$A \triangle B = C \triangle D \Leftrightarrow A \triangle C = B \triangle D.$$

**3.** If $\Omega$ has exactly $n$ points, then $\mathscr{J}$ has $2^n$ members. The B.F. generated by $n$ given sets "without relations among them" has $2^{2n}$ members.

**4.** If $\Omega$ is countable, then $\mathscr{J}$ is generated by the singletons, and conversely. [HINT: All countable subsets of $\Omega$ and their complements form a B.F.]

**5.** The intersection of any collection of B.F.'s $\{\mathscr{F}_\alpha, \alpha \in A\}$ is the maximal B.F. contained in all of them; it is indifferently denoted by $\bigcap_{\alpha \in A} \mathscr{F}_\alpha$ or $\wedge_{\alpha \in A} \mathscr{F}_\alpha$.

***6.** The union of a countable collection of B.F.'s $\{\mathscr{F}_j\}$ such that $\mathscr{F}_j \subset \mathscr{F}_{j+1}$ need not be a B.F., but there is a minimal B.F. containing all of them, denoted by $\vee_j \mathscr{F}_j$. In general $\vee_{\alpha \in A} \mathscr{F}_\alpha$ denotes the minimal B.F. containing all $\mathscr{F}_\alpha$, $\alpha \in A$. [HINT: $\Omega =$ the set of positive integers; $\mathscr{F}_j =$ the B.F. generated by those up to $j$.]

**7.** A B.F. is said to be countably generated iff it is generated by a countable collection of sets. Prove that if each $\mathscr{F}_j$ is countably generated, then so is $\vee_{j=1}^{\infty} \mathscr{F}_j$.

***8.** Let $\mathscr{F}$ be a B.F. generated by an arbitrary collection of sets $\{E_\alpha, \alpha \in A\}$. Prove that for each $E \in \mathscr{F}$, there exists a countable subcollection $\{E_{\alpha_j}, j \geq 1\}$ (depending on $E$) such that $E$ belongs already to the B.F. generated by this subcollection. [HINT: Consider the class of all sets with the asserted property and show that it is a B.F. containing each $E_\alpha$.]

**9.** If $\mathscr{F}$ is a B.F. generated by a countable collection of disjoint sets $\{\Lambda_n\}$, such that $\bigcup_n \Lambda_n = \Omega$, then each member of $\mathscr{F}$ is just the union of a countable subcollection of these $\Lambda_n$'s.

**10.** Let $\mathscr{D}$ be a class of subsets of $\Omega$ having the closure property (iii); let $\mathscr{A}$ be a class of sets containing $\Omega$ as well as $\mathscr{D}$, and having the closure properties (vi) and (x). Then $\mathscr{A}$ contains the B.F. generated by $\mathscr{D}$. (This is Dynkin's form of a monotone class theorem which is expedient for certain applications. The proof proceeds as in Theorem 2.1.2 by replacing $\mathscr{F}_0$ and $\mathscr{G}$ with $\mathscr{D}$ and $\mathscr{A}$ respectively.)

**11.** Take $\Omega = \mathscr{R}^n$ or a separable metric space in Exercise 10 and let $\mathscr{D}$ be the class of all open sets. Let $\mathscr{H}$ be a class of real-valued functions on $\Omega$ satisfying the following conditions.

(a) $1 \in \mathscr{H}$ and $1_D \in \mathscr{H}$ for each $D \in \mathscr{D}$;

(b) $\mathscr{H}$ is a vector space, namely: if $f_1 \in \mathscr{H}$, $f_2 \in \mathscr{H}$ and $c_1, c_2$ are any two real constants, then $c_1 f_1 + c_2 f_2 \in \mathscr{H}$;

(c) $\mathscr{H}$ is closed with respect to increasing limits of positive functions, namely: if $f_n \in \mathscr{H}, 0 \leq f_n \leq f_{n+1}$ for all $n$, and $f = \lim_n \uparrow f_n < \infty$, then $f \in \mathscr{H}$.

Then $\mathscr{H}$ contains all Borel measurable functions on $\Omega$, namely all finite-valued functions measurable with respect to the topological Borel field (= the minimal B.F. containing all open sets of $\Omega$). [HINT: let $\mathscr{C} = \{E \subset \Omega: 1_E \in \mathscr{H}\}$; apply Exercise 10 to show that $\mathscr{C}$ contains the B.F. just defined. Each positive Borel measurable function is the limit of an increasing sequence of simple (finitely-valued) functions.]

**12.** Let $\mathscr{C}$ be a M.C. of subsets of $\mathscr{R}^n$ (or a separable metric space) containing all the open sets and closed sets. Prove that $\mathscr{C} \supset \mathscr{B}^n$ (the topological Borel field defined in Exercise 11). [HINT: Show that the minimal such class is a field.]

## 2.2 Probability measures and their distribution functions

Let $\Omega$ be a space, $\mathscr{F}$ a B.F. of subsets of $\Omega$. A probability measure $\mathscr{P}(\cdot)$ on $\mathscr{F}$ is a numerically valued set function with domain $\mathscr{F}$, satisfying the following axioms:

**(i)** $\forall E \in \mathscr{F} : \mathscr{P}(E) \geq 0$.

**(ii)** If $\{E_j\}$ is a countable collection of (pairwise) disjoint sets in $\mathscr{F}$, then

$$\mathscr{P}\left(\bigcup_j E_j\right) = \sum_j \mathscr{P}(E_j).$$

**(iii)** $\mathscr{P}(\Omega) = 1$.

The abbreviation "p.m." will be used for "probability measure".

These axioms imply the following consequences, where all sets are members of $\mathscr{F}$.

**(iv)** $\mathscr{P}(E) \leq 1$.

**(v)** $\mathscr{P}(\varnothing) = 0$.

**(vi)** $\mathscr{P}(E^c) = 1 - \mathscr{P}(E)$.

**(vii)** $\mathscr{P}(E \cup F) + \mathscr{P}(E \cap F) = \mathscr{P}(E) + \mathscr{P}(F)$.

**(viii)** $E \subset F \Rightarrow \mathscr{P}(E) = \mathscr{P}(F) - \mathscr{P}(F \backslash E) \leq \mathscr{P}(F)$.

**(ix)** *Monotone property.* $E_n \uparrow E$ or $E_n \downarrow E \Rightarrow \mathscr{P}(E_n) \to \mathscr{P}(E)$.

**(x)** *Boole's inequality.* $\mathscr{P}(\bigcup_j E_j) \leq \sum_j \mathscr{P}(E_j)$.

Axiom (ii) is called "countable additivity"; the corresponding axiom restricted to a finite collection $\{E_j\}$ is called "finite additivity".

The following proposition

$$E_n \downarrow \varnothing \Rightarrow \mathscr{P}(E_n) \to 0 \tag{1}$$

is called the "axiom of continuity". It is a particular case of the monotone property (x) above, which may be deduced from it or proved in the same way as indicated below.

**Theorem 2.2.1.** The axioms of finite additivity and of continuity together are equivalent to the axiom of countable additivity.

PROOF. Let $E_n \downarrow$. We have the obvious identity:

$$E_n = \bigcup_{k=n}^{\infty} (E_k \backslash E_{k+1}) \cup \bigcap_{k=1}^{\infty} E_k.$$

If $E_n \downarrow \varnothing$, the last term is the empty set. Hence if (ii) is assumed, we have

$$\forall n \geq 1 : \mathscr{P}(E_n) = \sum_{k=n}^{\infty} \mathscr{P}(E_k \backslash E_{k+1});$$

the series being convergent, we have $\lim_{n\to\infty} \mathscr{P}(E_n) = 0$. Hence (1) is true. Conversely, let $\{E_k, k \geq 1\}$ be pairwise disjoint, then

$$\bigcup_{k=n+1}^{\infty} E_k \downarrow \varnothing$$

(why?) and consequently, if (1) is true, then

$$\lim_{n\to\infty} \mathscr{P}\left(\bigcup_{k=n+1}^{\infty} E_k\right) = 0.$$

Now if finite additivity is assumed, we have

$$\mathscr{P}\left(\bigcup_{k=1}^{\infty} E_k\right) = \mathscr{P}\left(\bigcup_{k=1}^{n} E_k\right) + \mathscr{P}\left(\bigcup_{k=n+1}^{\infty} E_k\right)$$
$$= \sum_{k=1}^{n} \mathscr{P}(E_k) + \mathscr{P}\left(\bigcup_{k=n+1}^{\infty} E_k\right).$$

This shows that the infinite series $\sum_{k=1}^{\infty} \mathscr{P}(E_k)$ converges as it is bounded by the first member above. Letting $n \to \infty$, we obtain

$$\mathscr{P}\left(\bigcup_{k=1}^{\infty} E_k\right) = \lim_{n\to\infty} \sum_{k=1}^{n} \mathscr{P}(E_k) + \lim_{n\to\infty} \mathscr{P}\left(\bigcup_{k=n+1}^{\infty} E_k\right)$$
$$= \sum_{k=1}^{\infty} \mathscr{P}(E_k).$$

Hence (ii) is true.

*Remark.* For a later application (Theorem 3.3.4) we note the following extension. Let $\mathscr{P}$ be defined on a field $\mathscr{F}$ which is finitely additive and satisfies axioms (i), (iii), and (1). Then (ii) holds whenever $\bigcup_k E_k \in \mathscr{F}$. For then $\bigcup_{k=n+1}^{\infty} E_k$ also belongs to $\mathscr{F}$, and the second part of the proof above remains valid.

The triple $(\Omega, \mathscr{F}, \mathscr{P})$ is called a *probability space* (*triple*); $\Omega$ alone is called the *sample space*, and $\omega$ is then a *sample point*.

Let $\Delta \subset \Omega$, then the *trace* of the B.F. $\mathscr{F}$ on $\Delta$ is the collection of all sets of the form $\Delta \cap F$, where $F \in \mathscr{F}$. It is easy to see that this is a B.F. of subsets of $\Delta$, and we shall denote it by $\Delta \cap \mathscr{F}$. Suppose $\Delta \in \mathscr{F}$ and $\mathscr{P}(\Delta) > 0$; then we may define the set function $\mathscr{P}_\Delta$ on $\Delta \cap \mathscr{F}$ as follows:

$$\forall E \in \Delta \cap \mathscr{F}: \qquad \mathscr{P}_\Delta(E) = \frac{\mathscr{P}(E)}{\mathscr{P}(\Delta)}.$$

It is easy to see that $\mathscr{P}_\Delta$ is a p.m. on $\Delta \cap \mathscr{F}$. The triple $(\Delta, \Delta \cap \mathscr{F}, \mathscr{P}_\Delta)$ will be called the *trace* of $(\Omega, \mathscr{F}, \mathscr{P})$ on $\Delta$.

**Example 1.** Let $\Omega$ be a countable set: $\Omega = \{\omega_j, j \in J\}$, where $J$ is a countable index set, and let $\mathscr{F}$ be the total B.F. of $\Omega$. Choose any sequence of numbers $\{p_j, j \in J\}$ satisfying

(2)
$$\forall j \in J: p_j \geq 0; \qquad \sum_{j\in J} p_j = 1;$$

and define a set function $\mathscr{P}$ on $\mathscr{F}$ as follows:

(3)
$$\forall E \in \mathscr{F}: \mathscr{P}(E) = \sum_{\omega j \in E} p_j.$$

In words, we assign $p_j$ as the value of the "probability" of the singleton $\{\omega_j\}$, and for an arbitrary set of $\omega_j$'s we assign as its probability the sum of all the probabilities assigned to its elements. Clearly axioms (i), (ii), and (iii) are satisfied. Hence $\mathscr{P}$ so defined is a p.m.

Conversely, let any such $\mathscr{P}$ be given on $\mathscr{F}$. Since $\{\omega_j\} \in \mathscr{F}$ for every $j$, $\mathscr{P}(\{\omega_j\})$ is defined, let its value be $p_j$. Then (2) is satisfied. We have thus exhibited all the possible p.m.'s on $\Omega$, or rather on the pair $(\Omega, \mathscr{S})$; this will be called a *discrete sample space*. The entire first volume of Feller's well-known book [13] with all its rich content is based on just such spaces.

**Example 2.** Let $\mathscr{U} = (0, 1]$, $\mathscr{C}$ the collection of intervals:

$$\mathscr{C} = \{(a, b]: 0 < a < b \leq 1\};$$

$\mathscr{B}$ the minimal B.F. containing $\mathscr{C}$, $m$ the Borel–Lebesgue measure on $\mathscr{B}$. Then $(\mathscr{U}, \mathscr{B}, m)$ is a probability space.

Let $\mathscr{B}_0$ be the collection of subsets of $\mathscr{U}$ each of which is the union of a finite number of members of $\mathscr{C}$. Thus a typical set $B$ in $\mathscr{B}_0$ is of the form

$$B = \bigcup_{j=1}^{n} (a_j, b_j] \qquad \text{where } a_1 < b_1 < a_2 < b_2 < \cdots < a_n < b_n.$$

It is easily seen that $\mathscr{B}_0$ is a field that is generated by $\mathscr{C}$ and in turn generates $\mathscr{B}$.

If we take $\mathscr{U} = [0, 1]$ instead, then $\mathscr{B}_0$ is no longer a field since $\mathscr{U} \notin \mathscr{B}_0$, but $\mathscr{B}$ and $m$ may be defined as before. The new $\mathscr{B}$ is generated by the old $\mathscr{B}$ and the singleton $\{0\}$.

**Example 3.** Let $\mathscr{R}^1 = (-\infty, +\infty)$, $\mathscr{C}$ the collection of intervals of the form $(a, b]$. $-\infty < a < b < +\infty$. The field $\mathscr{B}_0$ generated by $\mathscr{C}$ consists of finite unions of disjoint sets of the form $(a, b]$, $(-\infty, a]$ or $(b, \infty)$. The Euclidean B.F. $\mathscr{B}^1$ on $\mathscr{R}^1$ is the B.F. generated by $\mathscr{C}$ or $\mathscr{B}_0$. A set in $\mathscr{B}^1$ will be called a (*linear*) *Borel set* when there is no danger of ambiguity. However, the Borel–Lebesgue measure $m$ on $\mathscr{R}^1$ is not a p.m.; indeed $m(\mathscr{R}^1) = +\infty$ so that $m$ is not a finite measure but it is *$\sigma$-finite* on $\mathscr{B}_0$, namely: there exists a sequence of sets $E_n \in \mathscr{B}_0$, $E_n \uparrow \mathscr{R}^1$ with $m(E_n) < \infty$ for each $n$.

## EXERCISES

**1.** For any countably infinite set $\Omega$, the collection of its finite subsets and their complements forms a field $\mathscr{F}$. If we define $\mathscr{P}(E)$ on $\mathscr{F}$ to be 0 or 1 according as $E$ is finite or not, then $\mathscr{P}$ is finitely additive but not countably so.

$^\star$**2.** Let $\Omega$ be the space of natural numbers. For each $E \subset \Omega$ let $N_n(E)$ be the cardinality of the set $E \cap [0, n]$ and let $\mathscr{C}$ be the collection of $E$'s for which the following limit exists:

$$\mathscr{P}(E) = \lim_{n \to \infty} \frac{N_n(E)}{n}.$$

$\mathscr{P}$ is finitely additive on $\mathscr{C}$ and is called the "asymptotic density" of $E$. Let $E =$ {all odd integers}, $F =$ {all odd integers in $[2^{2n}, 2^{2n+1}]$ and all even integers in $[2^{2n+1}, 2^{2n+2}]$ for $n \geq 0$}. Show that $E \in \mathscr{C}$, $F \in \mathscr{C}$, but $E \cap F \notin \mathscr{C}$. Hence $\mathscr{C}$ is not a field.

**3.** In the preceding example show that for each real number $\alpha$ in [0, 1] there is an $E$ in $\mathscr{C}$ such that $\mathscr{P}(E) = \alpha$. Is the set of all primes in $\mathscr{C}$ ? Give an example of $E$ that is not in $\mathscr{C}$.

**4.** Prove the nonexistence of a p.m. on $(\Omega, \mathscr{S})$, where $(\Omega, \mathscr{S})$ is as in Example 1, such that the probability of each singleton has the same value. Hence criticize a sentence such as: "Choose an integer at random".

**5.** Prove that the trace of a B.F. $\mathscr{F}$ on any subset $\Delta$ of $\Omega$ is a B.F. Prove that the trace of $(\Omega, \mathscr{F}, \mathscr{P})$ on any $\Delta$ in $\mathscr{F}$ is a probability space, if $\mathscr{P}(\Delta) > 0$.

***6.** Now let $\Delta \notin \mathscr{F}$ be such that

$$\Delta \subset F \in \mathscr{F} \Rightarrow \mathscr{P}(F) = 1.$$

Such a set is called *thick* in $(\Omega, \mathscr{F}, \mathscr{P})$. If $E = \Delta \cap F$, $F \in \mathscr{F}$, define $\mathscr{P}^*(E) = \mathscr{P}(F)$. Then $\mathscr{P}^*$ is a well-defined (what does it mean?) p.m. on $(\Delta, \Delta \cap \mathscr{F})$. This procedure is called the *adjunction* of $\Delta$ to $(\Omega, \mathscr{F}, \mathscr{P})$.

**7.** The B.F. $\mathscr{B}^1$ on $\mathscr{R}^1$ is also generated by the class of all open intervals or all closed intervals, or all half-lines of the form $(-\infty, a]$ or $(a, \infty)$, or these intervals with rational endpoints. But it is not generated by all the singletons of $\mathscr{R}^1$ nor by any finite collection of subsets of $\mathscr{R}^1$.

**8.** $\mathscr{B}^1$ contains every singleton, countable set, open set, closed set, $G_\delta$ set, $F_\sigma$ set. (For the last two kinds of sets see, e.g., Natanson [3].)

***9.** Let $\mathscr{C}$ be a countable collection of pairwise disjoint subsets $\{E_j, j \geq 1\}$ of $\mathscr{R}^1$, and let $\mathscr{F}$ be the B.F. generated by $\mathscr{C}$. Determine the most general p.m. on $\mathscr{F}$ and show that the resulting probability space is "isomorphic" to that discussed in Example 1.

**10.** Instead of requiring that the $E_j$'s be pairwise disjoint, we may make the broader assumption that each of them intersects only a finite number in the collection. Carry through the rest of the problem.

The question of probability measures on $\mathscr{B}^1$ is closely related to the theory of distribution functions studied in Chapter 1. There is in fact a one-to-one correspondence between the set functions on the one hand, and the point functions on the other. Both points of view are useful in probability theory. We establish first the easier half of this correspondence.

**Lemma.** Each p.m. $\mu$ on $\mathscr{B}^1$ determines a d.f. $F$ through the correspondence

$$\forall x \in \mathscr{R}^1: \mu((-\infty, x]) = F(x). \tag{4}$$

As a consequence, we have for $-\infty < a < b < +\infty$:

$$
(5) \qquad \begin{aligned}
\mu((a, b]) &= F(b) - F(a), \\
\mu((a, b)) &= F(b-) - F(a), \\
\mu([a, b)) &= F(b-) - F(a-), \\
\mu([a, b]) &= F(b) - F(a-).
\end{aligned}
$$

Furthermore, let $D$ be any dense subset of $\mathscr{R}^1$, then the correspondence is already determined by that in (4) restricted to $x \in D$, or by any of the four relations in (5) when $a$ and $b$ are both restricted to $D$.

PROOF. Let us write

$$\forall x \in \mathscr{R}^1 : I_x = (-\infty, x].$$

Then $I_x \in \mathscr{B}^1$ so that $\mu(I_x)$ is defined; call it $F(x)$ and so define the function $F$ on $\mathscr{R}^1$. We shall show that $F$ is a d.f. as defined in Chapter 1. First of all, $F$ is increasing by property (viii) of the measure. Next, if $x_n \downarrow x$, then $I_{x_n} \downarrow I_x$, hence we have by (ix)

$$(6) \qquad F(x_n) = \mu(I_{x_n}) \downarrow \mu(I_x) = F(x).$$

Hence $F$ is right continuous. [The reader should ascertain what changes should be made if we had defined $F$ to be left continuous.] Similarly as $x \downarrow -\infty$, $I_x \downarrow \varnothing$; as $x \uparrow +\infty$, $I_x \uparrow \mathscr{R}^1$. Hence it follows from (ix) again that

$$\lim_{x \downarrow -\infty} F(x) = \lim_{x \downarrow -\infty} \mu(I_x) = \mu(\varnothing) = 0;$$

$$\lim_{x \uparrow +\infty} F(x) = \lim_{x \uparrow +\infty} \mu(I_x) = \mu(\Omega) = 1.$$

This ends the verification that $F$ is a d.f. The relations in (5) follow easily from the following complement to (4):

$$\mu((-\infty, x)) = F(x-).$$

To see this let $x_n < x$ and $x_n \uparrow x$. Since $I_{x_n} \uparrow (-\infty, x)$, we have by (ix):

$$F(x-) = \lim_{n \to \infty} F(x_n) = \mu((-\infty, x_n)) \uparrow \mu((-\infty, x)).$$

To prove the last sentence in the theorem we show first that (4) restricted to $x \in D$ implies (4) unrestricted. For this purpose we note that $\mu((-\infty, x])$, as well as $F(x)$, is right continuous as a function of $x$, as shown in (6). Hence the two members of the equation in (4), being both right continuous functions of $x$ and coinciding on a dense set, must coincide everywhere. Now suppose, for example, the second relation in (5) holds for rational $a$ and $b$. For each real $x$ let $a_n, b_n$ be rational such that $a_n \downarrow -\infty$ and $b_n > x$, $b_n \downarrow x$. Then $\mu((a_n, b_n)) \to \mu((-\infty, x])$ and $F(b_n-) - F(a_n) \to F(x)$. Hence (4) follows.

Incidentally, the correspondence (4) "justifies" our previous assumption that $F$ be right continuous, but what if we have assumed it to be left continuous?

Now we proceed to the second-half of the correspondence.

**Theorem 2.2.2.** Each d.f. $F$ determines a p.m. $\mu$ on $\mathscr{B}^1$ through any one of the relations given in (5), or alternatively through (4).

This is the classical theory of Lebesgue-Stieltjes measure; see, e.g., Halmos [4] or Royden [5]. However, we shall sketch the basic ideas as an important review. The d.f. $F$ being given, we may define a set function for intervals of the form $(a, b]$ by means of the first relation in (5). Such a function is seen to be countably additive on its domain of definition. (What does this mean?) Now we proceed to extend its domain of definition while preserving this additivity. If $S$ is a countable union of such intervals which are disjoint:

$$S = \bigcup_i (a_i, b_i]$$

we are forced to define $\mu(S)$, if at all, by

$$\mu(S) = \sum_i \mu((a_i, b_i]) = \sum_i \{F(b_i) - F(a_i)\}.$$

But a set $S$ may be representable in the form above in different ways, so we must check that this definition leads to no contradiction: namely that it depends really only on the set $S$ and not on the representation. Next, we notice that any open interval $(a, b)$ is in the extended domain (why?) and indeed the extended definition agrees with the second relation in (5). Now it is well known that any open set $U$ in $\mathscr{R}^1$ is the union of a countable collection of disjoint open intervals [there is no exact analogue of this in $\mathscr{R}^n$ for $n > 1$], say $U = \bigcup_i (c_i, d_i)$; and this representation is unique. Hence again we are forced to define $\mu(U)$, if at all, by

$$\mu(U) = \sum_i \mu((c_i, d_i)) = \sum_i \{F(d_i-) - F(c_i)\}.$$

Having thus defined the measure for all open sets, we find that its values for all closed sets are thereby also determined by property (vi) of a probability measure. In particular, its value for each singleton $\{a\}$ is determined to be $F(a) - F(a-)$, which is nothing but the jump of $F$ at $a$. Now we also know its value on all countable sets, and so on — all this provided that no contradiction is ever forced on us so far as we have gone. But even with the class of open and closed sets we are still far from the B.F. $\mathscr{B}^1$. The next step will be the $G_\delta$ sets and the $F_\sigma$ sets, and there already the picture is not so clear. Although it has been shown to be possible to proceed this way by transfinite induction, this is a rather difficult task. There is a more efficient way to reach the goal via the notions of outer and inner measures as follows. For any subset $S$ of $\mathscr{R}^1$ consider the two numbers:

$$\mu^*(S) = \inf_{U \text{ open},\, U \supset S} \mu(U),$$

$$\mu_*(S) = \sup_{C \text{ closed},\, C \subset S} \mu(C).$$

$\mu^*$ is the *outer measure*, $\mu_*$ the *inner measure* (both with respect to the given $F$). It is clear that $\mu^*(S) \geq \mu_*(S)$. Equality does not in general hold, but when it does, we call $S$ "measurable" (with respect to $F$). In this case the common value will be denoted by $\mu(S)$. This new definition requires us at once to check that it agrees with the old one for all the sets for which $\mu$ has already been defined. The next task is to prove that: (a) the class of all measurable sets forms a B.F., say $\mathscr{L}$; (b) on this $\mathscr{L}$, the function $\mu$ is a p.m. Details of these proofs are to be found in the references given above. To finish: since $\mathscr{L}$ is a B.F., and it contains all intervals of the form $(a, b]$, it contains the minimal B.F. $\mathscr{B}^1$ with this property. It may be larger than $\mathscr{B}^1$, indeed it is (see below), but this causes no harm, for the restriction of $\mu$ to $\mathscr{B}^1$ is a p.m. whose existence is asserted in Theorem 2.2.2.

Let us mention that the introduction of both the outer and inner measures is useful for approximations. It follows, for example, that for each measurable set $S$ and $\epsilon > 0$, there exists an open set $U$ and a closed set $C$ such that $U \supset S \supset C$ and

$$\mu(U) - \epsilon \leq \mu(S) \leq \mu(C) + \epsilon. \tag{7}$$

There is an alternative way of defining measurability through the use of the outer measure alone and based on Carathéodory's criterion.

It should also be remarked that the construction described above for $(\mathscr{R}^1, \mathscr{B}^1, \mu)$ is that of a "topological measure space", where the B.F. is generated by the open sets of a given topology on $\mathscr{R}^1$, here the usual Euclidean one. In the general case of an "algebraic measure space", in which there is no topological structure, the role of the open sets is taken by an arbitrary field $\mathscr{F}_0$, and a measure given on $\mathscr{F}_0$ may be extended to the minimal B.F. $\mathscr{F}$ containing $\mathscr{F}_0$ in a similar way. In the case of $\mathscr{R}^1$, such an $\mathscr{F}_0$ is given by the field $\mathscr{B}_0$ of sets, each of which is the union of a finite number of intervals of the form $(a, b]$, $(-\infty, b]$, or $(a, \infty)$, where $a \in \mathscr{R}^1$, $b \in \mathscr{R}^1$. Indeed the definition of the outer measure given above may be replaced by the equivalent one:

$$\mu^*(E) = \inf \sum_n \mu(U_n). \tag{8}$$

where the infimum is taken over all countable unions $\bigcup_n U_n$ such that each $U_n \in \mathscr{B}_0$ and $\bigcup_n U_n \supset E$. For another case where such a construction is required see Sec. 3.3 below.

There is one more question: besides the $\mu$ discussed above is there any other p.m. $\nu$ that corresponds to the given $F$ in the same way? It is important to realize that this question is not answered by the preceding theorem. It is also worthwhile to remark that any p.m. $\nu$ that is defined on a domain strictly containing $\mathscr{B}^1$ and that coincides with $\mu$ on $\mathscr{B}^1$ (such as the $\mu$ on $\mathscr{L}$ as mentioned above) will certainly correspond to $F$ in the same way, and strictly speaking such a $\nu$ is to be considered as distinct from $\mu$. Hence we should phrase the question more precisely by considering only p.m.'s on $\mathscr{B}^1$. This will be answered in full generality by the next theorem.

**Theorem 2.2.3.** Let $\mu$ and $\nu$ be two measures defined on the same B.F. $\mathscr{F}$, which is generated by the field $\mathscr{F}_0$. If either $\mu$ or $\nu$ is $\sigma$-finite on $\mathscr{F}_0$, and $\mu(E) = \nu(E)$ for every $E \in \mathscr{F}_0$, then the same is true for every $E \in \mathscr{F}$, and thus $\mu = \nu$.

PROOF. We give the proof only in the case where $\mu$ and $\nu$ are both finite, leaving the rest as an exercise. Let

$$\mathscr{C} = \{E \in \mathscr{F}: \mu(E) = \nu(E)\},$$

then $\mathscr{C} \supset \mathscr{F}_0$ by hypothesis. But $\mathscr{C}$ is also a monotone class, for if $E_n \in \mathscr{C}$ for every $n$ and $E_n \uparrow E$ or $E_n \downarrow E$, then by the monotone property of $\mu$ and $\nu$, respectively,

$$\mu(E) = \lim_n \mu(E_n) = \lim_n \nu(E_n) = \nu(E).$$

It follows from Theorem 2.1.2 that $\mathscr{C} \supset \mathscr{F}$, which proves the theorem.

*Remark.* In order that $\mu$ and $\nu$ coincide on $\mathscr{F}_0$, it is sufficient that they coincide on a collection $\mathscr{G}$ such that finite disjoint unions of members of $\mathscr{G}$ constitute $\mathscr{F}_0$.

**Corollary.** Let $\mu$ and $\nu$ be $\sigma$-finite measures on $\mathscr{B}^1$ that agree on all intervals of one of the eight kinds: $(a, b]$, $(a, b)$, $[a, b)$, $[a, b]$, $(-\infty, b]$, $(-\infty, b)$, $[a, \infty)$, $(a, \infty)$ or merely on those with the endpoints in a given dense set $D$, then they agree on $\mathscr{B}^1$.

PROOF. In order to apply the theorem, we must verify that any of the hypotheses implies that $\mu$ and $\nu$ agree on a field that generates $\mathscr{B}$. Let us take intervals of the first kind and consider the field $\mathscr{B}_0$ defined above. If $\mu$ and $\nu$ agree on such intervals, they must agree on $\mathscr{B}_0$ by countable additivity. This finishes the proof.

Returning to Theorems 2.2.1 and 2.2.2, we can now add the following complement, of which the first part is trivial.

**Theorem 2.2.4.** Given the p.m. $\mu$ on $\mathscr{B}^1$, there is a unique d.f. $F$ satisfying (4). Conversely, given the d.f. $F$, there is a unique p.m. satisfying (4) or any of the relations in (5).

We shall simply call *$\mu$ the p.m. of F, and F the d.f. of $\mu$*.

Instead of $(\mathscr{R}^1, \mathscr{B}^1)$ we may consider its restriction to a fixed interval $[a, b]$. Without loss of generality we may suppose this to be $\mathscr{U} = [0, 1]$ so that we are in the situation of Example 2. We can either proceed analogously or reduce it to the case just discussed, as follows. Let $F$ be a d.f. such that $F = 0$ for $x \leq 0$ and $F = 1$ for $x \geq 1$. The probability measure $\mu$ of $F$ will then have support in [0, 1], since $\mu((-\infty, 0)) = 0 = \mu((1, \infty))$ as a consequence of (4). Thus the trace of $(\mathscr{R}^1, \mathscr{B}^1, \mu)$ on $\mathscr{U}$ may be denoted simply by $(\mathscr{U}, \mathscr{B}, \mu)$, where $\mathscr{B}$ is the trace of $\mathscr{B}^1$ on $\mathscr{U}$. Conversely, any p.m. on $\mathscr{B}$ may be regarded as such a trace. The most interesting case is when $F$ is the "uniform distribution" on $\mathscr{U}$:

$$F(x) = \begin{cases} 0 & \text{for } x < 0, \\ x & \text{for } 0 \leq x \leq 1, \\ 1 & \text{for } x > 1. \end{cases}$$

The corresponding measure $m$ on $\mathscr{B}$ is the usual *Borel measure* on [0, 1], while its extension on $\mathscr{L}$ as described in Theorem 2.2.2 is the usual *Lebesgue measure* there. It is well known that $\mathscr{L}$ is actually larger than $\mathscr{B}$; indeed $(\mathscr{L}, m)$ is the completion of $(\mathscr{B}, m)$ to be discussed below.

DEFINITION. The probability space $(\Omega, \mathscr{F}, \mathscr{P})$ is said to be *complete* iff any subset of a set in $\mathscr{F}$ with $\mathscr{P}(F) = 0$ also belongs to $\mathscr{F}$.

Any probability space $(\Omega, \mathscr{F}, \mathscr{P})$ can be *completed* according to the next theorem. Let us call a set in $\mathscr{F}$ with probability zero a *null set*. A property that holds except on a null set is said to hold *almost everywhere (a.e.), almost surely (a.s.)*, or *for almost every $\omega$*.

**Theorem 2.2.5.** Given the probability space $(\Omega, \mathscr{F}, \mathscr{P})$, there exists a complete space $(\Omega, \overline{\mathscr{F}}, \overline{\mathscr{P}})$ such that $\mathscr{F} \subset \overline{\mathscr{F}}$ and $\mathscr{P} = \overline{\mathscr{P}}$ on $\mathscr{F}$.

PROOF. Let $\mathcal{N}$ be the collection of sets that are subsets of null sets, and let $\overline{\mathcal{F}}$ be the collection of subsets of $\Omega$ each of which differs from a set in $\mathcal{F}$ by a subset of a null set. Precisely:

$$\overline{\mathcal{F}} = \{E \subset \Omega: E \,\Delta\, F \in \mathcal{N} \quad \text{for some } F \in \mathcal{F}\}. \tag{9}$$

It is easy to verify, using Exercise 1 of Sec. 2.1, that $\overline{\mathcal{F}}$ is a B.F. Clearly it contains $\mathcal{F}$. For each $E \in \overline{\mathcal{F}}$, we put

$$\overline{\mathcal{P}}(E) = \mathcal{P}(F),$$

where $F$ is any set that satisfies the condition indicated in (7). To show that this definition does not depend on the choice of such an $F$, suppose that

$$E \,\Delta\, F_1 \in \mathcal{N}, \qquad E \,\Delta\, F_2 \in \mathcal{N}.$$

Then by Exercise 2 of Sec. 2.1,

$$(E \,\Delta\, F_1) \,\Delta\, (E \,\Delta\, F_2) = (F_1 \,\Delta\, F_2) \,\Delta\, (E \,\Delta\, E) = F_1 \,\Delta\, F_2.$$

Hence $F_1 \,\Delta\, F_2 \in \mathcal{N}$ and so $\mathcal{P}(F_1 \,\Delta\, F_2) = 0$. This implies $\mathcal{P}(F_1) = \mathcal{P}(F_2)$, as was to be shown. We leave it as an exercise to show that $\overline{\mathcal{P}}$ is a measure on $\overline{\mathcal{F}}$. If $E \in \mathcal{F}$, then $E \,\Delta\, E = \varnothing \in \mathcal{N}$, hence $\overline{\mathcal{P}}(E) = \mathcal{P}(E)$.

Finally, it is easy to verify that if $E \in \overline{\mathcal{F}}$ and $\overline{\mathcal{P}}(E) = 0$, then $E \in \mathcal{N}$. Hence any subset of $E$ also belongs to $\mathcal{N}$ and so to $\overline{\mathcal{F}}$. This proves that $(\Omega, \overline{\mathcal{F}}, \overline{\mathcal{P}})$ is complete.

What is the advantage of completion? Suppose that a certain property, such as the existence of a certain limit, is known to hold *outside* a certain set $N$ with $\mathcal{P}(N) = 0$. Then the *exact* set on which it fails to hold is a subset of $N$, not necessarily in $\mathcal{F}$, but will be in $\overline{\mathcal{F}}$ with $\overline{\mathcal{P}}(N) = 0$. We need the measurability of the exact exceptional set to facilitate certain dispositions, such as defining or redefining a function on it; see Exercise 25 below.

## EXERCISES

In the following, $\mu$ is a p.m. on $\mathcal{B}^1$ and $F$ is its d.f.

***11.** An *atom* of any measure $\mu$ on $\mathcal{B}^1$ is a singleton $\{x\}$ such that $\mu(\{x\}) > 0$. The number of atoms of any $\sigma$-finite measure is countable. For each $x$ we have $\mu(\{x\}) = F(x) - F(x-)$.

**12.** $\mu$ is called *atomic* iff its value is zero on any set not containing any atom. This is the case if and only if $F$ is discrete. $\mu$ is without any atom or *atomless* if and only if $F$ is continuous.

**13.** $\mu$ is called *singular* iff there exists a set $Z$ with $m(Z) = 0$ such that $\mu(Z^c) = 0$. This is the case if and only if $F$ is singular. [HINT: One half is proved by using Theorems 1.3.1 and 2.1.2 to get $\int_B F'(x)\,dx \le \mu(B)$ for $B \in \mathcal{B}^1$; the other requires Vitali's covering theorem.]

**14.** Translate Theorem 1.3.2 in terms of measures.

***15.** Translate the construction of a singular continuous d.f. in Sec. 1.3 in terms of measures. [It becomes clearer and easier to describe!] Generalize the construction by replacing the Cantor set with any perfect set of Borel measure zero. What if the latter has positive measure? Describe a probability scheme to realize this.

**16.** Show by a trivial example that Theorem 2.2.3 becomes false if the field $\mathscr{F}_0$ is replaced by an arbitrary collection that generates $\mathscr{F}$.

**17.** Show that Theorem 2.2.3 may be false for measures that are $\sigma$-finite on $\mathscr{F}$. [HINT: Take $\Omega$ to be $\{1, 2, \dots, \infty\}$ and $\mathscr{F}_0$ to be the finite sets excluding $\infty$ and their complements, $\mu(E) =$ number of points in $E$, $\mu(\infty) \neq \nu(\infty)$.]

**18.** Show that the $\overline{\mathscr{F}}$ in (9) | is also the collection of sets of the form $F \cup N$ [or $F \backslash N$] where $F \in \mathscr{F}$ and $N \in \mathscr{N}$.

**19.** Let $\mathscr{N}$ be as in the proof of Theorem 2.2.5 and $\mathscr{N}_0$ be the set of all null sets in $(\Omega, \mathscr{F}, \mathscr{P})$. Then both these collections are monotone classes, and closed with respect to the operation "$\backslash$".

***20.** Let $(\Omega, \mathscr{F}, \mathscr{P})$ be a probability space and $\mathscr{F}_1$ a Borel subfield of $\mathscr{F}$. Prove that there exists a minimal B.F. $\mathscr{F}_2$ satisfying $\mathscr{F}_1 \subset \mathscr{F}_2 \subset \mathscr{F}$ and $\mathscr{N}_0 \subset \mathscr{F}_2$, where $\mathscr{N}_0$ is as in Exercise 19. A set $E$ belongs to $\mathscr{F}_2$ if and only if there exists a set $F$ in $\mathscr{F}_1$ such that $E \Delta F \in \mathscr{N}_0$. This $\mathscr{F}_2$ is called the *augmentation* of $\mathscr{F}_1$ with respect to $(\Omega, \mathscr{F}, \mathscr{P})$.

**21.** Suppose that $\tilde{F}$ has all the defining properties of a d.f. except that it is not assumed to be right continuous. Show that Theorem 2.2.2 and Lemma remain valid with $F$ replaced by $\tilde{F}$, provided that we replace $F(x)$, $F(b)$, $F(a)$ in (4) and (5) by $\tilde{F}(x+)$, $\tilde{F}(b+)$, $\tilde{F}(a+)$, respectively. What modification is necessary in Theorem 2.2.4?

**22.** For an arbitrary measure $\mathscr{P}$ on a B.F. $\mathscr{F}$, a set $E$ in $\mathscr{F}$ is called an atom of $\mathscr{P}$ iff $\mathscr{P}(E) > 0$ and $F \subset E$, $F \in \mathscr{F}$ imply $\mathscr{P}(F) = \mathscr{P}(E)$ or $\mathscr{P}(F) = 0$. $\mathscr{P}$ is called *atomic* iff its value is zero over any set in $\mathscr{F}$ that is disjoint from all the atoms. Prove that for a measure $\mu$ on $\mathscr{B}^1$ this new definition is equivalent to that given in Exercise 11 above provided we identify two sets which differ by a $\mathscr{P}$-null set.

**23.** Prove that if the p.m. $\mathscr{P}$ is atomless, then given any $\alpha$ in [0, 1] there exists a set $E \in \mathscr{F}$ with $\mathscr{P}(E) = \alpha$. [HINT: Prove first that there exists $E$ with "arbitrarily small" probability. A quick proof then follows from Zorn's lemma by considering a maximal collection of disjoint sets, the sum of whose probabilities does not exceed $\alpha$. But an elementary proof without using any maximality principle is also possible.]

***24.** A point $x$ is said to be in the support of a measure $\mu$ on $\mathscr{B}^n$ iff every open neighborhood of $x$ has strictly positive measure. The set of all such points is called the *support* of $\mu$. Prove that the support is a closed set whose complement is the maximal open set on which $\mu$ vanishes. Show that the support of a p.m. on $\mathscr{B}^1$ is the same as that of its d.f., defined in Exercise 6 of Sec. 1.2.

$^\star$**25.** Let $f$ be measurable with respect to $\mathscr{F}$, and $Z$ be contained in a null set. Define

$$\tilde{f} = \begin{cases} f & \text{on } Z^c, \\ K & \text{on } Z, \end{cases}$$

where $K$ is a constant. Then $\tilde{f}$ is measurable with respect to $\mathscr{F}$ provided that $(\Omega, \mathscr{F}, \mathscr{P})$ is complete. Show that the conclusion may be false otherwise.

# 3 Random variable. Expectation. Independence

## 3.1 General definitions

Let the probability space $(\Omega, \mathscr{F}, \mathscr{P})$ be given. $\mathscr{R}^1 = (-\infty, +\infty)$ the (finite) real line, $\mathscr{R}^* = [-\infty, +\infty]$ the extended real line, $\mathscr{B}^1 =$ the Euclidean Borel field on $\mathscr{R}^1$, $\mathscr{B}^* =$ the extended Borel field. A set in $\mathscr{B}^*$ is just a set in $\mathscr{B}$ possibly enlarged by one or both points $\pm\infty$.

DEFINITION OF A RANDOM VARIABLE. A real, extended-valued random variable is a function $X$ whose domain is a set $\Delta$ in $\mathscr{F}$ and whose range is contained in $\mathscr{R}^* = [-\infty, +\infty]$ such that for each $B$ in $\mathscr{B}^*$, we have

$$\{\omega: X(\omega) \in B\} \in \Delta \cap \mathscr{F} \tag{1}$$

where $\Delta \cap \mathscr{F}$ is the trace of $\mathscr{F}$ on $\Delta$. A complex-valued random variable is a function on a set $\Delta$ in $\mathscr{F}$ to the complex plane whose real and imaginary parts are both real, finite-valued random variables.

This definition in its generality is necessary for logical reasons in many applications, but for a discussion of basic properties *we may suppose* $\Delta = \Omega$ *and that* $X$ *is real and finite-valued with probability one*. This restricted meaning of a "random variable", abbreviated as "r.v.", will be understood in the book unless otherwise specified. The general case may be reduced to this one by considering the trace of $(\Omega, \mathscr{F}, \mathscr{P})$ on $\Delta$, or on the "domain of finiteness" $\Delta_0 = \{\omega: |X(\omega)| < \infty\}$, and taking real and imaginary parts.

Consider the "inverse mapping" $X^{-1}$ from $\mathscr{R}^1$ to $\Omega$, defined (as usual) as follows:

$$\forall A \subset \mathscr{R}^1: X^{-1}(A) = \{\omega: X(\omega) \in A\}.$$

Condition (1) then states that $X^{-1}$ carries members of $\mathscr{B}^1$ onto members of $\mathscr{F}$:

$$\forall B \in \mathscr{B}^1: X^{-1}(B) \in \mathscr{F}; \tag{2}$$

or in the briefest notation:

$$X^{-1}(\mathscr{B}^1) \subset \mathscr{F}.$$

Such a function is said to be *measurable (with respect to* $\mathscr{F}$). Thus, an r.v. is just a measurable function from $\Omega$ to $\mathscr{R}^1$ (or $\mathscr{R}^*$).

The next proposition, a standard exercise on inverse mapping, is essential.

**Theorem 3.1.1.** For any function $X$ from $\Omega$ to $\mathscr{R}^1$ (or $\mathscr{R}^*$), not necessarily an r.v., the inverse mapping $X^{-1}$ has the following properties:

$$X^{-1}(A^c) = (X^{-1}(A))^c.$$

$$X^{-1}\left(\bigcup_\alpha A_\alpha\right) = \bigcup_\alpha X^{-1}(A_\alpha),$$

$$X^{-1}\left(\bigcap_\alpha A_\alpha\right) = \bigcap_\alpha X^{-1}(A_\alpha).$$

where $\alpha$ ranges over an arbitrary index set, not necessarily countable.

**Theorem 3.1.2.** $X$ is an r.v. if and only if for each real number $x$, or each real number $x$ in a dense subset of $\mathscr{R}^1$, we have

$$\{\omega: X(\omega) \leq x\} \in \mathscr{F}.$$

PROOF. The preceding condition may be written as

$$\forall x: X^{-1}((-\infty, x]) \in \mathscr{F}. \tag{3}$$

Consider the collection $\mathscr{A}$ of all subsets $S$ of $\mathscr{R}^1$ for which $X^{-1}(S) \in \mathscr{F}$. From Theorem 3.1.1 and the defining properties of the Borel field $\mathscr{F}$, it follows that if $S \in \mathscr{A}$, then

$$X^{-1}(S^c) = (X^{-1}(S))^c \in \mathscr{F};$$

if $\forall j: S_j \in \mathscr{A}$, then

$$X^{-1}\left(\bigcup_j S_j\right) = \bigcup_j X^{-1}(S_j) \in \mathscr{F}.$$

Thus $S^c \in \mathscr{A}$ and $\bigcup_j S_j \in \mathscr{A}$ and consequently $\mathscr{A}$ is a B.F. This B.F. contains all intervals of the form $(-\infty, x]$, which generate $\mathscr{B}^1$ even if $x$ is restricted to a dense set, hence $\mathscr{A} \supset \mathscr{B}^1$, which means that $X^{-1}(B) \in \mathscr{F}$ for each $B \in \mathscr{B}^1$. Thus $X$ is an r.v. by definition. This proves the "if" part of the theorem; the "only if" part is trivial.

Since $\mathscr{P}(\cdot)$ is defined on $\mathscr{F}$, the probability of the set in (1) is defined and will be written as

$$\mathscr{P}\{X(\omega) \in B\} \quad \text{or} \quad \mathscr{P}\{X \in B\}.$$

The next theorem relates the p.m. $\mathscr{P}$ to a p.m. on $(\mathscr{R}^1, \mathscr{B}^1)$ as discussed in Sec. 2.2.

**Theorem 3.1.3.** Each r.v. on the probability space $(\Omega, \mathscr{F}, \mathscr{P})$ *induces* a probability space $(\mathscr{R}^1, \mathscr{B}^1, \mu)$ by means of the following correspondence:

$$\forall B \in \mathscr{B}^1: \quad \mu(B) = \mathscr{P}\{X^{-1}(B)) = \mathscr{P}\{X \in B\}. \tag{4}$$

PROOF. Clearly $\mu(B) \geq 0$. If the $B_n$'s are disjoint sets in $\mathscr{B}^1$, then the $X^{-1}(B_n)$'s are disjoint by Theorem 3.1.1. Hence

$$\mu\left(\bigcup_n B_n\right) = \mathscr{P}\left(X^{-1}\left(\bigcup_n B_n\right)\right) = \mathscr{P}\left(\bigcup_n X^{-1}(B_n)\right)$$
$$= \sum_n \mathscr{P}(X^{-1}(B_n)) = \sum_n \mu(B_n).$$

Finally $X^{-1}(\mathscr{R}^1) = \Omega$, hence $\mu(\mathscr{R}^1) = 1$. Thus $\mu$ is a p.m.

The collection of sets $\{X^{-1}(S), S \subset \mathscr{R}^1\}$ is a B.F. for any function $X$. If $X$ is a r.v. then the collection $\{X^{-1}(B), B \in \mathscr{B}^1\}$ is called *the B.F. generated by* $X$. It is the smallest Borel subfield of $\mathscr{F}$ which contains all sets of the form $\{\omega: X(\omega) \leq x\}$, where $x \in \mathscr{R}^1$. Thus (4) is a convenient way of representing the measure $\mathscr{P}$ when it is restricted to this subfield; symbolically we may write it as follows:

$$\mu = \mathscr{P} \circ X^{-1}.$$

This $\mu$ is called the "probability distribution measure" or p.m. of $X$, and its associated d.f. $F$ according to Theorem 2.2.4 will be called the d.f. of $X$.

Specifically, $F$ is given by

$$F(x) = \mu((-\infty, x]) = \mathscr{P}\{X \le x\}.$$

While the r.v. $X$ determines $\mu$ and therefore $F$, the converse is obviously false. A family of r.v.'s having the same distribution is said to be "identically distributed".

**Example 1.** Let $(\Omega, \mathscr{S})$ be a discrete sample space (see Example 1 of Sec. 2.2). Every numerically valued function is an r.v.

**Example 2.** $(\mathscr{U}, \mathscr{B}, m)$.

In this case an r.v. is by definition just a *Borel measurable* function. According to the usual definition, $f$ on $\mathscr{U}$ is Borel measurable iff $f^{-1}(\mathscr{B}^1) \subset \mathscr{B}$. In particular, the function $f$ given by $f(\omega) \equiv \omega$ is an r.v. The two r.v.'s $\omega$ and $1 - \omega$ are not identical but are identically distributed; in fact their common distribution is the underlying measure $m$.

**Example 3.** $(\mathscr{R}^1, \mathscr{B}^1, \mu)$.

The definition of a Borel measurable function is not affected, since no measure is involved; so any such function is an r.v., whatever the given p.m. $\mu$ may be. As in Example 2, there exists an r.v. with the underlying $\mu$ as its p.m.; see Exercise 3 below.

We proceed to produce new r.v.'s from given ones.

**Theorem 3.1.4.** If $X$ is an r.v., $f$ a Borel measurable function [on $(\mathscr{R}^1, \mathscr{B}^1)$], then $f(X)$ is an r.v.

PROOF. The quickest proof is as follows. Regarding the function $f(X)$ of $\omega$ as the "composite mapping":

$$f \circ X : \omega \to f(X(\omega)),$$

we have $(f \circ X)^{-1} = X^{-1} \circ f^{-1}$ and consequently

$$(f \circ X)^{-1}(\mathscr{B}^1) = X^{-1}(f^{-1}(\mathscr{B}^1)) \subset X^{-1}(\mathscr{B}^1) \subset \mathscr{F}.$$

The reader who is not familiar with operations of this kind is advised to spell out the proof above in the old-fashioned manner, which takes only a little longer.

We must now discuss the notion of a *random vector*. This is just a vector each of whose components is an r.v. It is sufficient to consider the case of two dimensions, since there is no essential difference in higher dimensions apart from complication in notation.

We recall first that in the 2-dimensional Euclidean space $\mathscr{R}^2$, or the plane, the Euclidean Borel field $\mathscr{B}^2$ is generated by rectangles of the form

$$\{(x, y): a < x \leq b, c < y \leq d\}.$$

*A fortiori*, it is also generated by *product sets* of the form

$$B_1 \times B_2 = \{(x, y): x \in B_1, y \in B_2\},$$

where $B_1$ and $B_2$ belong to $\mathscr{B}^1$. The collection of sets, each of which is a finite union of disjoint product sets, forms a field $\mathscr{B}_0^2$. A function from $\mathscr{R}^2$ into $\mathscr{R}^1$ is called a Borel measurable function (of two variables) iff $f^{-1}(\mathscr{B}^1) \subset \mathscr{B}^2$. Written out, this says that for each 1-dimensional Borel set $B$, viz., a member of $\mathscr{B}^1$, the set

$$\{(x, y): f(x, y) \in B\}$$

is a 2-dimensional Borel set, viz. a member of $\mathscr{B}^2$.

Now let $X$ and $Y$ be two r.v.'s on $(\Omega, \mathscr{F}, \mathscr{P})$. The random vector $(X, Y)$ *induces* a probability $\nu$ on $\mathscr{B}^2$ as follows:

$$\forall A \in \mathscr{B}^2: \nu(A) = \mathscr{P}\{(X, Y) \in A\}, \tag{5}$$

the right side being an abbreviation of $\mathscr{P}(\{\omega: (X(\omega), Y(\omega)) \in A\})$. This $\nu$ is called the (2-dimensional, probability) *distribution* or simply the p.m. of $(X, Y)$.

Let us also define, in imitation of $X^{-1}$, the inverse mapping $(X, Y)^{-1}$ by the following formula:

$$\forall A \in \mathscr{B}^2: (X, Y)^{-1}(A) = \{\omega: (X, Y) \in A\}.$$

This mapping has properties analogous to those of $X^{-1}$ given in Theorem 3.1.1, since the latter is actually true for a mapping of any two abstract spaces. We can now easily generalize Theorem 3.1.4.

**Theorem 3.1.5.** If $X$ and $Y$ are r.v.'s and $f$ is a Borel measurable function of two variables, then $f(X, Y)$ is an r.v.

PROOF.

$$[f \circ (X, Y)]^{-1}(\mathscr{B}^1) = (X, Y)^{-1} \circ f^{-1}(\mathscr{B}^1) \subset (X, Y)^{-1}(\mathscr{B}^2) \subset \mathscr{F}.$$

The last inclusion says the inverse mapping $(X, Y)^{-1}$ carries each 2-dimensional Borel set into a set in $\mathscr{F}$. This is proved as follows. If $A = B_1 \times B_2$, where $B_1 \in \mathscr{B}^1$, $B_2 \in \mathscr{B}^1$, then it is clear that

$$(X, Y)^{-1}(A) = X^{-1}(B_1) \cap Y^{-1}(B_2) \in \mathscr{F}$$

by (2). Now the collection of sets $A$ in $\mathscr{R}^2$ for which $(X, Y)^{-1}(A) \in \mathscr{F}$ forms a B.F. by the analogue of Theorem 3.1.1. It follows from what has just been shown that this B.F. contains $\mathscr{B}_0^2$ hence it must also contain $\mathscr{B}^2$. Hence each set in $\mathscr{B}^2$ belongs to the collection, as was to be proved.

Here are some important special cases of Theorems 3.1.4 and 3.1.5. Throughout the book we shall use the notation for numbers as well as functions:

$$x \vee y = \max(x, y), \quad x \wedge y = \min(x, y). \tag{6}$$

**Corollary.** If $X$ is an r.v. and $f$ is a continuous function on $\mathscr{R}^1$, then $f(X)$ is an r.v.; in particular $X^r$ for positive integer $r$, $|X|^r$ for positive real $r$, $e^{-\lambda X}$, $e^{itX}$ for real $\lambda$ and $t$, are all r.v.'s (the last being complex-valued). If $X$ and $Y$ are r.v.'s, then

$$X \vee Y, \quad X \wedge Y, \quad X + Y, \quad X - Y, \quad X \cdot Y, \quad X/Y$$

are r.v.'s, the last provided $Y$ does not vanish.

Generalization to a finite number of r.v.'s is immediate. Passing to an infinite sequence, let us state the following theorem, although its analogue in real function theory should be well known to the reader.

**Theorem 3.1.6.** If $\{X_j, j \geq 1\}$ is a sequence of r.v.'s, then

$$\inf_j X_j, \quad \sup_j X_j, \quad \liminf_j X_j, \quad \limsup_j X_j$$

are r.v.'s, not necessarily finite-valued with probability one though everywhere defined, and

$$\lim_{j \to \infty} X_j$$

is an r.v. on the set $\Delta$ on which there is either convergence or divergence to $\pm\infty$.

PROOF. To see, for example, that $\sup_j X_j$ is an r.v., we need only observe the relation

$$\forall x \in \mathscr{R}^1: \{\sup_j X_j \leq x\} = \bigcap_j \{X_j \leq x\}$$

and use Theorem 3.1.2. Since

$$\limsup_j X_j = \inf_n (\sup_{j \geq n} X_j),$$

and $\lim_{j\to\infty} X_j$ exists [and is finite] on the set where $\limsup_j X_j = \liminf_j X_j$ [and is finite], which belongs to $\mathscr{F}$, the rest follows.

Here already we see the necessity of the general definition of an r.v. given at the beginning of this section.

DEFINITION. An r.v. $X$ is called *discrete* (or *countably valued*) iff there is a countable set $B \subset \mathscr{R}^1$ such that $\mathscr{P}(X \in B) = 1$.

It is easy to see that $X$ is discrete if and only if its d.f. is. Perhaps it is worthwhile to point out that a discrete r.v. need not have a range that is discrete in the sense of Euclidean topology, even apart from a set of probability zero. Consider, for example, an r.v. with the d.f. in Example 2 of Sec. 1.1.

The following terminology and notation will be used throughout the book for an arbitrary set $\Omega$, not necessarily the sample space.

DEFINITION. For each $\Delta \subset \Omega$, the function $1_\Delta(\cdot)$ defined as follows:

$$\forall \omega \in \Omega: 1_\Delta(\omega) = \begin{cases} 1, & \text{if } \omega \in \Delta, \\ 0, & \text{if } \omega \in \Omega\backslash\Delta, \end{cases}$$

is called the *indicator (function)* of $\Delta$.

Clearly $1_\Delta$ is an r.v. if and only if $\Delta \in \mathscr{F}$.

A *countable partition* of $\Omega$ is a countable family of disjoint sets $\{\Lambda_j\}$, with $\Lambda_j \in \mathscr{F}$ for each $j$ and such that $\Omega = \bigcup_j \Lambda_j$. We have then

$$1 = 1_\Omega = \sum_j 1_{\Lambda_j}.$$

More generally, let $b_j$ be arbitrary real numbers, then the function $\varphi$ defined below:

$$\forall \omega \in \Omega: \varphi(\omega) = \sum_j b_j 1_{\Lambda_j}(\omega),$$

is a discrete r.v. We shall call $\varphi$ the r.v. *belonging to the weighted partition* $\{\Lambda_j; b_j\}$. Each discrete r.v. $X$ belongs to a certain partition. For let $\{b_j\}$ be the countable set in the definition of $X$ and let $\Lambda_j = \{\omega: X(\omega) = b_j\}$, then $X$ belongs to the weighted partition $\{\Lambda_j; b_j\}$. If $j$ ranges over a finite index set, the partition is called *finite* and the r.v. belonging to it *simple*.

## EXERCISES

**1.** Prove Theorem 3.1.1. For the "direct mapping" $X$, which of these properties of $X^{-1}$ holds?

**2.** If two r.v.'s are equal a.e., then they have the same p.m.

$\star$**3.** Given any p.m. $\mu$ on $(\mathscr{R}^1, \mathscr{B}^1)$, define an r.v. whose p.m. is $\mu$. Can this be done in an arbitrary probability space?

$\star$**4.** Let $\theta$ be uniformly distributed on [0,1]. For each d.f. $F$, define $G(y) = \sup\{x: F(x) \le y\}$. Then $G(\theta)$ has the d.f. $F$.

$\star$**5.** Suppose $X$ has the continuous d.f. $F$, then $F(X)$ has the uniform distribution on [0,1]. What if $F$ is not continuous?

**6.** Is the range of an r.v. necessarily Borel or Lebesgue measurable?

**7.** The sum, difference, product, or quotient (denominator nonvanishing) of the two discrete r.v.'s is discrete.

**8.** If $\Omega$ is discrete (countable), then every r.v. is discrete. Conversely, every r.v. in a probability space is discrete if and only if the p.m. is atomic. [HINT: Use Exercise 23 of Sec. 2.2.]

**9.** If $f$ is Borel measurable, and $X$ and $Y$ are identically distributed, then so are $f(X)$ and $f(Y)$.

**10.** Express the indicators of $\Lambda_1 \cup \Lambda_2$, $\Lambda_1 \cap \Lambda_2$, $\Lambda_1 \backslash \Lambda_2$, $\Lambda_1 \vartriangle \Lambda_2$, $\limsup \Lambda_n$, $\liminf \Lambda_n$ in terms of those of $\Lambda_1$, $\Lambda_2$, or $\Lambda_n$. [For the definitions of the limits see Sec. 4.2.]

$\star$**11.** Let $\mathscr{F}\{X\}$ be the minimal B.F. with respect to which $X$ is measurable. Show that $\Lambda \in \mathscr{F}\{X\}$ if and only if $\Lambda = X^{-1}(B)$ for some $B \in \mathscr{B}^1$. Is this $B$ unique? Can there be a set $A \notin \mathscr{B}^1$ such that $\Lambda = X^{-1}(A)$?

**12.** Generalize the assertion in Exercise 11 to a finite set of r.v.'s. [It is possible to generalize even to an arbitrary set of r.v.'s.]

## 3.2 Properties of mathematical expectation

The concept of "(mathematical) expectation" is the same as that of integration in the probability space with respect to the measure $\mathscr{P}$. The reader is supposed to have some acquaintance with this, at least in the particular case $(\mathscr{U}, \mathscr{B}, m)$ or $(\mathscr{R}^1, \mathscr{B}^1, m)$. [In the latter case, the measure not being finite, the theory of integration is slightly more complicated.] The general theory is not much different and will be briefly reviewed. The r.v.'s below will be tacitly assumed to be *finite everywhere* to avoid trivial complications.

For each positive discrete r.v. $X$ belonging to the weighted partition $\{\Lambda_j; b_j\}$, we define its *expectation* to be

$$\mathscr{E}(X) = \sum_j b_j \mathscr{P}\{\Lambda_j\}. \tag{1}$$

This is either a positive finite number or $+\infty$. It is trivial that if $X$ belongs to different partitions, the corresponding values given by (1) agree. Now let

$X$ be an arbitrary positive r.v. For any two positive integers $m$ and $n$, the set

$$\Lambda_{mn} = \left\{\omega: \frac{n}{2^m} \leq X(\omega) < \frac{n+1}{2^m}\right\}$$

belongs to $\mathscr{F}$. For each $m$, let $X_m$ denote the r.v. belonging to the weighted partition $\{\Lambda_{mn}; n/2^m\}$; thus $X_m = n/2^m$ if and only if $n/2^m \leq X < (n+1)/2^m$. It is easy to see that we have for each $m$:

$$\forall \omega: X_m(\omega) \leq X_{m+1}(\omega); \quad 0 \leq X(\omega) - X_m(\omega) < \frac{1}{2^m}.$$

Consequently there is monotone convergence:

$$\forall \omega: \lim_{m\to\infty} X_m(\omega) = X(\omega).$$

The expectation $X_m$ has just been defined; it is

$$\mathscr{E}(X_m) = \sum_{n=0}^{\infty} \frac{n}{2^m} \mathscr{P}\left\{\frac{n}{2^m} \leq X < \frac{n+1}{2^m}\right\}.$$

If for one value of $m$ we have $\mathscr{E}(X_m) = +\infty$, then we define $\mathscr{E}(X) = +\infty$; otherwise we define

$$\mathscr{E}(X) = \lim_{m\to\infty} \mathscr{E}(X_m),$$

the limit existing, finite or infinite, since $\mathscr{E}(X_m)$ is an increasing sequence of real numbers. It should be shown that when $X$ is discrete, this new definition agrees with the previous one.

For an arbitrary $X$, put as usual

$$(2) \qquad X = X^+ - X^- \quad \text{where} \quad X^+ = X \vee 0, \quad X^- = (-X) \vee 0.$$

Both $X^+$ and $X^-$ are positive r.v.'s, and so their expectations are defined. Unless both $\mathscr{E}(X^+)$ and $\mathscr{E}(X^-)$ are $+\infty$, we define

$$(3) \qquad \mathscr{E}(X) = \mathscr{E}(X^+) - \mathscr{E}(X^-)$$

with the usual convention regarding $\infty$. We say $X$ has a *finite* or *infinite expectation* (or *expected value*) according as $\mathscr{E}(X)$ is a finite number or $\pm\infty$. In the expected case we shall say that the expectation of $X$ *does not exist*. The expectation, when it exists, is also denoted by

$$\int_\Omega X(\omega)\mathscr{P}(d\omega).$$

More generally, for each $\Lambda$ in $\mathscr{F}$, we define

$$(4) \qquad \int_\Lambda X(\omega)\mathscr{P}(d\omega) = \mathscr{E}(X \cdot 1_\Lambda)$$

and call it "the integral of $X$ (with respect to $\mathscr{P}$) over the set $\Lambda$". We shall say that $X$ is *integrable with respect to* $\mathscr{P}$ *over* $\Lambda$ iff the integral above exists and is finite.

In the case of $(\mathscr{R}^1, \mathscr{B}^1, \mu)$, if we write $X = f$, $\omega = x$, the integral

$$\int_\Lambda X(\omega)\mathscr{P}(d\omega) = \int_\Lambda f(x)\mu(dx)$$

is just the ordinary Lebesgue–Stieltjes integral of $f$ with respect to $\mu$. If $F$ is the d.f. of $\mu$ and $\Lambda = (a, b]$, this is also written as

$$\int_{(a,b]} f(x)\, dF(x).$$

This classical notation is really an anachronism, originated in the days when a point function was more popular than a set function. The notation above must then amend to

$$\int_{a+0}^{b+0}, \int_{a-0}^{b+0}, \int_{a+0}^{b-0}, \int_{a-0}^{b-0}$$

to distinguish clearly between the four kinds of intervals $(a, b]$, $[a, b]$, $(a, b)$, $[a, b)$.

In the case of $(\mathscr{U}, \mathscr{B}, m)$, the integral reduces to the ordinary Lebesgue integral

$$\int_a^b f(x)m(dx) = \int_a^b f(x)\, dx.$$

Here $m$ is atomless, so the notation is adequate and there is no need to distinguish between the different kinds of intervals.

The general integral has the familiar properties of the Lebesgue integral on [0,1]. We list a few below for ready reference, some being easy consequences of others. As a general notation, the left member of (4) will be abbreviated to $\int_\Lambda X\, d\mathscr{P}$. In the following, $X$, $Y$ are r.v.'s; $a$, $b$ are constants; $\Lambda$ is a set in $\mathscr{F}$.

**(i)** *Absolute integrability.* $\int_\Lambda X\, d\mathscr{P}$ is finite if and only if

$$\int_\Lambda |X|\, d\mathscr{P} < \infty.$$

**(ii)** *Linearity.*

$$\int_\Lambda (aX + bY) d\mathscr{P} = a\int_\Lambda X\, d\mathscr{P} + b\int_\Lambda Y\, d\mathscr{P}$$

provided that the right side is meaningful, namely not $+\infty - \infty$ or $-\infty + \infty$.

**(iii)** *Additivity over sets*. If the $\Lambda_n$'s are disjoint, then

$$\int_{\cup_n \Lambda_n} X \, d\mathscr{P} = \sum_n \int_{\Lambda_n} X \, d\mathscr{P}.$$

**(iv)** *Positivity*. If $X \geq 0$ a.e. on $\Lambda$, then

$$\int_\Lambda X \, d\mathscr{P} \geq 0.$$

**(v)** *Monotonicity*. If $X_1 \leq X \leq X_2$ a.e. on $\Lambda$, then

$$\int_\Lambda X_1 \, d\mathscr{P} \leq \int_\Lambda X \, d\mathscr{P} \leq \int_\Lambda X_2 \, d\mathscr{P}.$$

**(vi)** *Mean value theorem*. If $a \leq X \leq b$ a.e. on $\Lambda$, then

$$a\mathscr{P}(\Lambda) \leq \int_\Lambda X \, d\mathscr{P} \leq b\mathscr{P}(\Lambda).$$

**(vii)** *Modulus inequality*.

$$\left| \int_\Lambda X \, d\mathscr{P} \right| \leq \int_\Lambda |X| \, d\mathscr{P}.$$

**(viii)** *Dominated convergence theorem*. If $\lim_{n\to\infty} X_n = X$ a.e. or merely in measure on $\Lambda$ and $\forall n$: $|X_n| \leq Y$ a.e. on $\Lambda$, with $\int_\Lambda Y \, d\mathscr{P} < \infty$, then

$$\lim_{n\to\infty} \int_\Lambda X_n \, d\mathscr{P} = \int_\Lambda X \, d\mathscr{P} = \int_\Lambda \lim_{n\to\infty} X_n \, d\mathscr{P}. \tag{5}$$

**(ix)** *Bounded convergence theorem*. If $\lim_{n\to\infty} X_n = X$ a.e. or merely in measure on $\Lambda$ and there exists a constant $M$ such that $\forall n$: $|X_n| \leq M$ a.e. on $\Lambda$, then (5) is true.

**(x)** *Monotone convergence theorem*. If $X_n \geq 0$ and $X_n \uparrow X$ a.e. on $\Lambda$, then (5) is again true provided that $+\infty$ is allowed as a value for either member. The condition "$X_n \geq 0$" may be weakened to: "$\mathscr{E}(X_n) > -\infty$ for some $n$".

**(xi)** *Integration term by term*. If

$$\sum_n \int_\Lambda |X_n| \, d\mathscr{P} < \infty,$$

then $\sum_n |X_n| < \infty$ a.e. on $\Lambda$ so that $\sum_n X_n$ converges a.e. on $\Lambda$ and

$$\int_\Lambda \sum_n X_n \, d\mathscr{P} = \sum_n \int_\Lambda X_n \, d\mathscr{P}.$$

**(xii)** *Fatou's lemma.* If $X_n \geq 0$ a.e. on $\Lambda$, then

$$\int_\Lambda (\underline{\lim}_{n\to\infty} X_n)\, d\mathscr{P} \leq \underline{\lim}_{n\to\infty} \int_\Lambda X_n\, d\mathscr{P}.$$

Let us prove the following useful theorem as an instructive example.

**Theorem 3.2.1.** We have

$$\sum_{n=1}^{\infty} \mathscr{P}(|X| \geq n) \leq \mathscr{E}(|X|) \leq 1 + \sum_{n=1}^{\infty} \mathscr{P}(|X| \geq n) \tag{6}$$

so that $\mathscr{E}(|X|) < \infty$ if and only if the series above converges.

PROOF. By the additivity property (iii), if $\Lambda_n = \{n \leq |X| < n+1\}$,

$$\mathscr{E}(|X|) = \sum_{n=0}^{\infty} \int_{\Lambda_n} |X|\, d\mathscr{P}.$$

Hence by the mean value theorem (vi) applied to each set $\Lambda_n$:

$$\sum_{n=0}^{\infty} n\mathscr{P}(\Lambda_n) \leq \mathscr{E}(|X|) \leq \sum_{n=0}^{\infty}(n+1)\mathscr{P}(\Lambda_n) = 1 + \sum_{n=0}^{\infty} n\mathscr{P}(\Lambda_n). \tag{7}$$

It remains to show

$$\sum_{n=0}^{\infty} n\mathscr{P}(\Lambda_n) = \sum_{n=0}^{\infty} \mathscr{P}(|X| \geq n), \tag{8}$$

finite or infinite. Now the partial sums of the series on the left may be rearranged (Abel's method of partial summation!) to yield, for $N \geq 1$,

$$\begin{aligned} &\sum_{n=0}^{N} n\{\mathscr{P}(|X| \geq n) - \mathscr{P}(|X| \geq n+1)\} \\ &= \sum_{n=1}^{N} \{n - (n-1)\}\mathscr{P}(|X| \geq n) - N\mathscr{P}(|X| \geq N+1) \\ &= \sum_{n=1}^{N} \mathscr{P}(|X| \geq n) - N\mathscr{P}(|X| \geq N+1). \end{aligned} \tag{9}$$

Thus we have

$$\sum_{n=1}^{N} n\mathscr{P}(\Lambda_n) \leq \sum_{n=1}^{N} \mathscr{P}(|X| \geq n) \leq \sum_{n=1}^{N} n\mathscr{P}(\Lambda_n) + N\mathscr{P}(|X| \geq N+1). \tag{10}$$

Another application of the mean value theorem gives

$$N\mathscr{P}(|X| \geq N+1) \leq \int_{(|X|\geq N+1)} |X|\, d\mathscr{P}.$$

Hence if $\mathscr{E}(|X|) < \infty$, then the last term in (10) converges to zero as $N \to \infty$ and (8) follows with both sides finite. On the other hand, if $\mathscr{E}(|X|) = \infty$, then the second term in (10) diverges with the first as $N \to \infty$, so that (8) is also true with both sides infinite.

**Corollary.** If $X$ takes only positive integer values, then

$$\mathscr{E}(X) = \sum_{n=1}^{\infty} \mathscr{P}(X \geq n).$$

EXERCISES

**1.** If $X \geq 0$ a.e. on $\Lambda$ and $\int_\Lambda X\, d\mathscr{P} = 0$, then $X = 0$ a.e. on $\Lambda$.

$^\star$**2.** If $\mathscr{E}(|X|) < \infty$ and $\lim_{n\to\infty} \mathscr{P}(\Lambda_n) = 0$, then $\lim_{n\to\infty} \int_{\Lambda_n} X\, d\mathscr{P} = 0$. In particular

$$\lim_{n\to\infty} \int_{\{|X|>n\}} X\, d\mathscr{P} = 0.$$

**3.** Let $X \geq 0$ and $\int_\Omega X\, d\mathscr{P} = A$, $0 < A < \infty$. Then the set function $\nu$ defined on $\mathscr{F}$ as follows:

$$\nu(\Lambda) = \frac{1}{A}\int_\Lambda X\, d\mathscr{P},$$

is a probability measure on $\mathscr{F}$.

**4.** Let $c$ be a fixed constant, $c > 0$. Then $\mathscr{E}(|X|) < \infty$ if and only if

$$\sum_{n=1}^{\infty} \mathscr{P}(|X| \geq cn) < \infty.$$

In particular, if the last series converges for one value of $c$, it converges for all values of $c$.

**5.** For any $r > 0$, $\mathscr{E}(|X|^r) < \infty$ if and only if

$$\sum_{n=1}^{\infty} n^{r-1}\mathscr{P}(|X| \geq n) < \infty.$$

**★6.** Suppose that $\sup_n |X_n| \le Y$ on $\Lambda$ with $\int_\Lambda Y \, d\mathscr{P} < \infty$. Deduce from Fatou's lemma:

$$\int_\Lambda (\overline{\lim_{n\to\infty}} X_n) \, d\mathscr{P} \ge \overline{\lim_{n\to\infty}} \int_\Lambda X_n \, d\mathscr{P}.$$

Show this is false if the condition involving $Y$ is omitted.

**★7.** Given the r.v. $X$ with finite $\mathscr{E}(X)$, and $\epsilon > 0$, there exists a simple r.v. $X_\epsilon$ (see the end of Sec. 3.1) such that

$$\mathscr{E}(|X - X_\epsilon|) < \epsilon.$$

Hence there exists a sequence of simple r.v.'s $X_m$ such that

$$\lim_{m\to\infty} \mathscr{E}(|X - X_m|) = 0.$$

We can choose $\{X_m\}$ so that $|X_m| \le |X|$ for all $m$.

**★8.** For any two sets $\Lambda_1$ and $\Lambda_2$ in $\mathscr{F}$, define

$$\rho(\Lambda_1, \Lambda_2) = \mathscr{P}(\Lambda_1 \vartriangle \Lambda_2);$$

then $\rho$ is a pseudo-metric in the space of sets in $\mathscr{F}$; call the resulting metric space $\mathbf{M}(\mathscr{F}, \mathscr{P})$. Prove that for each integrable r.v. $X$ the mapping of $\mathbf{M}(\mathscr{F}, \mathscr{P})$ to $\mathscr{R}^1$ given by $\Lambda \to \int_\Lambda X \, d\mathscr{P}$ is continuous. Similarly, the mappings on $\mathbf{M}(\mathscr{F}, \mathscr{P}) \times \mathbf{M}(\mathscr{F}, \mathscr{P})$ to $\mathbf{M}(\mathscr{F}, \mathscr{P})$ given by

$$(\Lambda_1, \Lambda_2) \to \Lambda_1 \cup \Lambda_2, \Lambda_1 \cap \Lambda_2, \Lambda_1 \backslash \Lambda_2, \Lambda_1 \vartriangle \Lambda_2$$

are all continuous. If (see Sec. 4.2 below)

$$\limsup_n \Lambda_n = \liminf_n \Lambda_n$$

modulo a null set, we denote the common equivalence class of these two sets by $\lim_n \Lambda_n$. Prove that in this case $\{\Lambda_n\}$ converges to $\lim_n \Lambda_n$ in the metric $\rho$. Deduce Exercise 2 above as a special case.

There is a basic relation between the abstract integral with respect to $\mathscr{P}$ over sets in $\mathscr{F}$ on the one hand, and the Lebesgue–Stieltjes integral with respect to $\mu$ over sets in $\mathscr{B}^1$ on the other, induced by each r.v. We give the version in one dimension first.

**Theorem 3.2.2.** Let $X$ on $(\Omega, \mathscr{F}, \mathscr{P})$ induce the probability space $(\mathscr{R}^1, \mathscr{B}^1, \mu)$ according to Theorem 3.1.3 and let $f$ be Borel measurable. Then we have

$$\int_\Omega f(X(\omega)) \mathscr{P}(d\omega) = \int_{\mathscr{R}^1} f(x) \mu(dx) \tag{11}$$

provided that either side exists.

PROOF. Let $B \in \mathscr{B}^1$, and $f = 1_B$, then the left side in (11) is $\mathscr{P}(X \in B)$ and the right side is $\mu(B)$. They are equal by the definition of $\mu$ in (4) of Sec. 3.1. Now by the linearity of both integrals in (11), it will also hold if

$$f = \sum_j b_j 1_{B_j}, \tag{12}$$

namely the r.v. on $(\mathscr{R}^1, \mathscr{B}^1, \mu)$ belonging to an arbitrary weighted partition $\{B_j; b_j\}$. For an arbitrary positive Borel measurable function $f$ we can define, as in the discussion of the abstract integral given above, a sequence $\{f_m, m \geq 1\}$ of the form (12) such that $f_m \uparrow f$ everywhere. For each of them we have

$$\int_\Omega f_m(X)\, d\mathscr{P} = \int_{\mathscr{R}^1} f_m\, d\mu; \tag{13}$$

hence, letting $m \to \infty$ and using the monotone convergence theorem, we obtain (11) whether the limits are finite or not. This proves the theorem for $f \geq 0$, and the general case follows in the usual way.

We shall need the generalization of the preceding theorem in several dimensions. No change is necessary except for notation, which we will give in two dimensions. Instead of the $\nu$ in (5) of Sec. 3.1, let us write the "mass element" as $\mu^2(dx, dy)$ so that

$$\nu(A) = \iint_A \mu^2(dx, dy).$$

**Theorem 3.2.3.** Let $(X, Y)$ on $(\Omega, \mathscr{F}, \mathscr{P})$ induce the probability space $(\mathscr{R}^2, \mathscr{B}^2, \mu^2)$ and let $f$ be a Borel measurable function of two variables. Then we have

$$\int_\Omega f(X(\omega), Y(\omega))\mathscr{P}(d\omega) = \iint_{\mathscr{R}^2} f(x, y)\mu^2(dx, dy). \tag{14}$$

Note that $f(X, Y)$ is an r.v. by Theorem 3.1.5.

As a consequence of Theorem 3.2.2, we have: if $\mu_X$ and $F_X$ denote, respectively, the p.m. and d.f. induced by $X$, then we have

$$\mathscr{E}(X) = \int_{\mathscr{R}^1} x\mu_X(dx) = \int_{-\infty}^{\infty} x\, dF_X(x);$$

and more generally

$$\mathscr{E}(f(X)) = \int_{\mathscr{R}^1} f(x)\mu_X(dx) = \int_{-\infty}^{\infty} f(x)\, dF_X(x) \tag{15}$$

with the usual proviso regarding existence and finiteness.

Another important application is as follows: let $\mu^2$ be as in Theorem 3.2.3 and take $f(x, y)$ to be $x + y$ there. We obtain

$$\mathscr{E}(X+Y) = \iint_{\mathscr{R}^2} (x+y)\mu^2(dx, dy) \tag{16}$$

$$= \iint_{\mathscr{R}^2} x\mu^2(dx, dy) + \iint_{\mathscr{R}^2} y\mu^2(dx, dy).$$

On the other hand, if we take $f(x, y)$ to be $x$ or $y$, respectively, we obtain

$$\mathscr{E}(X) = \iint_{\mathscr{R}^2} x\mu^2(dx, dy), \ \mathscr{E}(Y) = \iint_{\mathscr{R}^2} y\mu^2(dx, dy)$$

and consequently

$$\mathscr{E}(X+Y) = \mathscr{E}(X) + \mathscr{E}(Y). \tag{17}$$

This result is a case of the linearity of $\mathscr{E}$ given but not proved here; the proof above reduces this property in the general case of $(\Omega, \mathscr{F}, \mathscr{P})$ to the corresponding one in the special case $(\mathscr{R}^2, \mathscr{B}^2, \mu^2)$. Such a reduction is frequently useful when there are technical difficulties in the abstract treatment.

We end this section with a discussion of "moments".

Let $a$ be real, $r$ positive, then $\mathscr{E}(|X - a|^r)$ is called the *absolute moment of $X$ of order $r$, about $a$*. It may be $+\infty$; otherwise, and if $r$ is an integer, $\mathscr{E}((X - a)^r)$ is the corresponding *moment*. If $\mu$ and $F$ are, respectively, the p.m. and d.f. of $X$, then we have by Theorem 3.2.2:

$$\mathscr{E}(|X-a|^r) = \int_{\mathscr{R}^1} |x-a|^r \mu(dx) = \int_{-\infty}^{\infty} |x-a|^r \, dF(x),$$

$$\mathscr{E}((X-a)^r) = \int_{\mathscr{R}^1} (x-a)^r \mu(dx) = \int_{-\infty}^{\infty} (x-a)^r \, dF(x).$$

For $r = 1$, $a = 0$, this reduces to $\mathscr{E}(X)$, which is also called the *mean* of $X$. The moments about the mean are called *central moments*. That of order 2 is particularly important and is called the *variance*, var $(X)$; its positive square root the *standard deviation*. $\sigma(X)$:

$$\text{var}(X) = \sigma^2(X) = \mathscr{E}\{(X - \mathscr{E}(X))^2\} = \mathscr{E}(X^2) - \{\mathscr{E}(X)\}^2.$$

We note the inequality $\sigma^2(X) \le \mathscr{E}(X^2)$, which will be used a good deal in Chapter 5. For any positive number $p$, $X$ is said to belong to $L^p = L^p(\Omega, \mathscr{F}, \mathscr{P})$ iff $\mathscr{E}(|X|^p) < \infty$.

The well-known inequalities of Hölder and Minkowski (see, e.g., Natanson [3]) may be written as follows. Let $X$ and $Y$ be r.v.'s, $1 < p < \infty$ and $1/p + 1/q = 1$, then

$$|\mathscr{E}(XY)| \leq \mathscr{E}(|XY|) \leq \mathscr{E}(|X|^p)^{1/p}\mathscr{E}(|Y|^q)^{1/q}, \tag{18}$$

$$\{\mathscr{E}(|X+Y|^p)\}^{1/p} \leq \mathscr{E}(|X|^p)^{1/p} + \mathscr{E}(|Y|^p)^{1/p}. \tag{19}$$

If $Y \equiv 1$ in (18), we obtain

$$\mathscr{E}(|X|) \leq \mathscr{E}(|X|^p)^{1/p}; \tag{20}$$

for $p = 2$, (18) is called the Cauchy–Schwarz inequality. Replacing $|X|$ by $|X|^r$, where $0 < r < p$, and writing $r' = pr$ in (20) we obtain

$$\mathscr{E}(|X|^r)^{1/r} \leq \mathscr{E}(|X|^{r'})^{1/r'}, \quad 0 < r < r' < \infty. \tag{21}$$

The last will be referred to as the Liapounov inequality. It is a special case of the next inequality, of which we will sketch a proof.

**Jensen's inequality.** If $\varphi$ is a convex function on $\mathscr{R}^1$, and $X$ and $\varphi(X)$ are integrable r.v.'s, then

$$\varphi(\mathscr{E}(X)) \leq \mathscr{E}(\varphi(X)). \tag{22}$$

PROOF. Convexity means : for every positive $\lambda_1, \ldots, \lambda_n$ with sum 1 we have

$$\varphi\left(\sum_{j=1}^{n} \lambda_j y_j\right) \leq \sum_{j=1}^{n} \lambda_j \varphi(y_j). \tag{23}$$

This is known to imply the continuity of $\varphi$, so that $\varphi(X)$ is an r.v. We shall prove (22) for a simple r.v. and refer the general case to Theorem 9.1.4. Let then $X$ take the value $y_j$ with probability $\lambda_j$, $1 \leq j \leq n$. Then we have by definition (1):

$$\mathscr{E}(X) = \sum_{j=1}^{n} \lambda_j y_j, \quad \mathscr{E}(\varphi(X)) = \sum_{j=1}^{n} \lambda_j \varphi(y_j).$$

Thus (22) follows from (23).

Finally, we prove a famous inequality that is almost trivial but very useful.

**Chebyshev inequality.** If $\varphi$ is a strictly positive and increasing function on $(0, \infty)$, $\varphi(u) = \varphi(-u)$, and $X$ is an r.v. such that $\mathscr{E}\{\varphi(X)\} < \infty$, then for each $u > 0$:

$$\mathscr{P}\{|X| \geq u\} \leq \frac{\mathscr{E}\{\varphi(X)\}}{\varphi(u)}.$$

PROOF. We have by the mean value theorem:

$$\mathscr{E}\{\varphi(X)\} = \int_{\Omega} \varphi(X)d\mathscr{P} \geq \int_{\{|X|\geq u\}} \varphi(X)d\mathscr{P} \geq \varphi(u)\mathscr{P}\{|X| \geq u\}$$

from which the inequality follows.

The most familiar application is when $\varphi(u) = |u|^p$ for $0 < p < \infty$, so that the inequality yields an upper bound for the "tail" probability in terms of an absolute moment.

EXERCISES

**9.** Another proof of (14): verify it first for simple r.v.'s and then use Exercise 7 of Sec. 3.2.

**10.** Prove that if $0 \leq r < r'$ and $\mathscr{E}(|X|^{r'}) < \infty$, then $\mathscr{E}(|X|^r) < \infty$. Also that $\mathscr{E}(|X|^r) < \infty$ if and only if $\mathscr{E}(|X - a|^r) < \infty$ for every $a$.

*__11.__ If $\mathscr{E}(X^2) = 1$ and $\mathscr{E}(|X|) \geq a > 0$, then $\mathscr{P}\{|X| \geq \lambda a\} \geq (1 - \lambda)^2 a^2$ for $0 \leq \lambda \leq 1$.

*__12.__ If $X \geq 0$ and $Y \geq 0$, $p \geq 0$, then $\mathscr{E}\{(X + Y)^p\} \leq 2^p\{\mathscr{E}(X^p) + \mathscr{E}(Y^p)\}$. If $p > 1$, the factor $2^p$ may be replaced by $2^{p-1}$. If $0 \leq p \leq 1$, it may be replaced by 1.

*__13.__ If $X_j \geq 0$, then

$$\mathscr{E}\left\{\left(\sum_{j=1}^{n} X_j\right)^p\right\} \leq \quad \text{or} \quad \geq \sum_{j=1}^{n} \mathscr{E}(X_j^p)$$

according as $p \leq 1$ or $p \geq 1$.

*__14.__ If $p > 1$, we have

$$\left|\frac{1}{n}\sum_{j=1}^{n} X_j\right|^p \leq \frac{1}{n}\sum_{j=1}^{n} |X_j|^p$$

and so

$$\mathscr{E}\left\{\left|\frac{1}{n}\sum_{j=1}^{n} X_j\right|^p\right\} \leq \frac{1}{n}\sum_{j=1}^{n} \mathscr{E}(|X_j|^p);$$

we have also

$$\mathscr{E}\left\{\left|\frac{1}{n}\sum_{j=1}^{n} X_j\right|^p\right\} \leq \left\{\frac{1}{n}\sum_{j=1}^{n} \mathscr{E}(|X_j|^p)^{1/p}\right\}^p.$$

Compare the inequalities.

**15.** If $p > 0$, $\mathscr{E}(|X|^p) < \infty$, then $x^p \mathscr{P}\{|X| > x\} = o(1)$ as $x \to \infty$. Conversely, if $x^p \mathscr{P}\{|X| > x\} = o(1)$, then $\mathscr{E}(|X|^{p-\epsilon}) < \infty$ for $0 < \epsilon < p$.

$^\star$**16.** For any d.f. and any $a \geq 0$, we have

$$\int_{-\infty}^{\infty} [F(x+a) - F(x)]\, dx = a.$$

**17.** If $F$ is a d.f. such that $F(0-) = 0$, then

$$\int_0^{\infty} \{1 - F(x)\}\, dx = \int_0^{\infty} x\, dF(x) \leq +\infty.$$

Thus if $X$ is a positive r.v., then we have

$$\mathscr{E}(X) = \int_0^{\infty} \mathscr{P}\{X > x\}\, dx = \int_0^{\infty} \mathscr{P}\{X \geq x\}\, dx.$$

**18.** Prove that $\int_{-\infty}^{\infty} |x|\, dF(x) < \infty$ if and only if

$$\int_{-\infty}^{0} F(x)\, dx < \infty \quad \text{and} \quad \int_0^{\infty} [1 - F(x)]\, dx < \infty.$$

$^\star$**19.** If $\{X_n\}$ is a sequence of identically distributed r.v.'s with finite mean, then

$$\lim_n \frac{1}{n} \mathscr{E}\{\max_{1 \leq j \leq n} |X_j|\} = 0.$$

[HINT: Use Exercise 17 to express the mean of the maximum.]

**20.** For $r > 1$, we have

$$\int_0^{\infty} \frac{1}{u^r} \mathscr{E}(X \wedge u^r)\, du = \frac{r}{r-1} \mathscr{E}(X^{1/r}).$$

[HINT: By Exercise 17,

$$\mathscr{E}(X \wedge u^r) = \int_0^{u^r} \mathscr{P}(X > x)\, dx = \int_0^{u} \mathscr{P}(X^{1/r} > v) r v^{r-1}\, dv,$$

substitute and invert the order of the repeated integrations.]

## 3.3 Independence

We shall now introduce a fundamental new concept peculiar to the theory of probability, that of "(stochastic) independence".

DEFINITION OF INDEPENDENCE. The r.v.'s $\{X_j, 1 \le j \le n\}$ are said to be *(totally) independent* iff for any linear Borel sets $\{B_j, 1 \le j \le n\}$ we have

$$\mathscr{P}\left\{\bigcap_{j=1}^{n}(X_j \in B_j)\right\} = \prod_{j=1}^{n}\mathscr{P}(X_j \in B_j). \tag{1}$$

The r.v.'s of an infinite family are said to be independent iff those in every finite subfamily are. They are said to be *pairwise independent* iff every two of them are independent.

Note that (1) implies that the r.v.'s in every subset of $\{X_j, 1 \le j \le n\}$ are also independent, since we may take some of the $B_j$'s as $\mathscr{R}^1$. On the other hand, (1) is implied by the apparently weaker hypothesis: for every set of real numbers $\{x_j, 1 \le j \le n\}$:

$$\mathscr{P}\left\{\bigcap_{j=1}^{n}(X_j \le x_j)\right\} = \prod_{j=1}^{n}\mathscr{P}(X_j \le x_j). \tag{2}$$

The proof of the equivalence of (1) and (2) is left as an exercise. In terms of the p.m. $\mu^n$ induced by the random vector $(X_1, \dots, X_n)$ on $(\mathscr{R}^n, \mathscr{B}^n)$, and the p.m.'s $\{\mu_j, 1 \le j \le n\}$ induced by each $X_j$ on $(\mathscr{R}^1, \mathscr{B}^1)$, the relation (1) may be written as

$$\mu^n\left(\mathop{\times}_{j=1}^{n} B_j\right) = \prod_{j=1}^{n}\mu_j(B_j), \tag{3}$$

where $\times_{j=1}^{n} B_j$ is the product set $B_1 \times \cdots \times B_n$ discussed in Sec. 3.1. Finally, we may introduce the *n-dimensional distribution function* corresponding to $\mu^n$, which is defined by the left side of (2) or in alternative notation:

$$F(x_1, \dots, x_n) = \mathscr{P}\{X_j \le x_j, 1 \le j \le n\} = \mu^n\left(\mathop{\times}_{j=1}^{n}(-\infty, x_j]\right);$$

then (2) may be written as

$$F(x_1, \dots, x_n) = \prod_{j=1}^{n} F_j(x_j).$$

From now on when the probability space is fixed, a set in $\mathscr{F}$ will also be called an *event*. The events $\{E_j, 1 \le j \le n\}$ are said to be independent iff their indicators are independent; this is equivalent to: for any subset $\{j_1, \dots j_\ell\}$ of $\{1, \dots, n\}$, we have

$$\mathscr{P}\left\{\bigcap_{k=1}^{\ell} E_{j_k}\right\} = \prod_{k=1}^{\ell} \mathscr{P}(E_{j_k}). \tag{4}$$

**Theorem 3.3.1.** If $\{X_j, 1 \le j \le n\}$ are independent r.v.'s and $\{f_j, 1 \le j \le n\}$ are Borel measurable functions, then $\{f_j(X_j), 1 \le j \le n\}$ are independent r.v.'s.

PROOF. Let $A_j \in \mathscr{B}^1$, then $f_j^{-1}(A_j) \in \mathscr{B}^1$ by the definition of a Borel measurable function. By Theorem 3.1.1, we have

$$\bigcap_{j=1}^{n} \{f_j(X_j) \in A_j\} = \bigcap_{j=1}^{n} \{X_j \in f_j^{-1}(A_j)\}.$$

Hence we have

$$\mathscr{P}\left\{\bigcap_{j=1}^{n} [f_j(X_j) \in A_j]\right\} = \mathscr{P}\left\{\bigcap_{j=1}^{n} [X_j \in f_j^{-1}(A_j)]\right\} = \prod_{j=1}^{n} \mathscr{P}\{X_j \in f_j^{-1}(A_j)\}$$

$$= \prod_{j=1}^{n} \mathscr{P}\{f_j(X_j) \in A_j\}.$$

This being true for every choice of the $A_j$'s, the $f_j(X_j)$'s are independent by definition.

The proof of the next theorem is similar and is left as an exercise.

**Theorem 3.3.2.** Let $1 \le n_1 < n_2 < \cdots < n_k = n$; $f_1$ a Borel measurable function of $n_1$ variables, $f_2$ one of $n_2 - n_1$ variables, ..., $f_k$ one of $n_k - n_{k-1}$ variables. If $\{X_j, 1 \le j \le n\}$ are independent r.v.'s then the $k$ r.v.'s

$$f_1(X_1, \dots, X_{n_1}),\ f_2(X_{n_1+1}, \dots, X_{n_2}), \dots,\ f_k(X_{n_{k-1}+1}, \dots, X_{n_k})$$

are independent.

**Theorem 3.3.3.** If $X$ and $Y$ are independent and both have finite expectations, then

$$\mathscr{E}(XY) = \mathscr{E}(X)\mathscr{E}(Y). \tag{5}$$

PROOF. We give two proofs in detail of this important result to illustrate the methods. Cf. the two proofs of (14) in Sec. 3.2, one indicated there, and one in Exercise 9 of Sec. 3.2.

*First proof.* Suppose first that the two r.v.'s $X$ and $Y$ are both discrete belonging respectively to the weighted partitions $\{\Lambda_j; c_j\}$ and $\{M_k; d_k\}$ such that $\Lambda_j = \{X = c_j\}$, $M_k = \{Y = d_k\}$. Thus

$$\mathscr{E}(X) = \sum_j c_j \mathscr{P}(\Lambda_j), \qquad \mathscr{E}(Y) = \sum_k d_k \mathscr{P}(\mathrm{M}_k).$$

Now we have

$$\Omega = \left(\bigcup_j \Lambda_j\right) \cap \left(\bigcup_k \mathrm{M}_k\right) = \bigcup_{j,k} (\Lambda_j \mathrm{M}_k)$$

and

$$X(\omega)Y(\omega) = c_j d_k \quad \text{if} \quad \omega \in \Lambda_j \mathrm{M}_k.$$

Hence the r.v. $XY$ is discrete and belongs to the *superposed partitions* $\{\Lambda_j \mathrm{M}_k; c_j d_k\}$ with both the $j$ and the $k$ varying independently of each other. Since $X$ and $Y$ are independent, we have for every $j$ and $k$:

$$\mathscr{P}(\Lambda_j \mathrm{M}_k) = \mathscr{P}(X = c_j; Y = d_k) = \mathscr{P}(X = c_j)\mathscr{P}(Y = d_k) = \mathscr{P}(\Lambda_j)\mathscr{P}(\mathrm{M}_k);$$

and consequently by definition (1) of Sec. 3.2:

$$\begin{aligned}\mathscr{E}(XY) &= \sum_{j,k} c_j d_k \mathscr{P}(\Lambda_j \mathrm{M}_k) = \left\{\sum_j c_j \mathscr{P}(\Lambda_j)\right\}\left\{\sum_k d_k \mathscr{P}(\mathrm{M}_k)\right\} \\ &= \mathscr{E}(X)\mathscr{E}(Y).\end{aligned}$$

Thus (5) is true in this case.

Now let $X$ and $Y$ be arbitrary positive r.v.'s with finite expectations. Then, according to the discussion at the beginning of Sec. 3.2, there are discrete r.v.'s $X_m$ and $Y_m$ such that $\mathscr{E}(X_m) \uparrow \mathscr{E}(X)$ and $\mathscr{E}(Y_m) \uparrow \mathscr{E}(Y)$. Furthermore, for each $m$, $X_m$ and $Y_m$ are independent. Note that for the independence of discrete r.v.'s it is sufficient to verify the relation (1) when the $B_j$'s are their possible values and "$\in$" is replaced by "$=$" (why ?). Here we have

$$\begin{aligned}\mathscr{P}\left\{X_m = \frac{n}{2^m}; Y_m = \frac{n'}{2^m}\right\} &= \mathscr{P}\left\{\frac{n}{2^m} \le X < \frac{n+1}{2^m}; \frac{n'}{2^m} \le Y < \frac{n'+1}{2^m}\right\} \\ &= \mathscr{P}\left\{\frac{n}{2^m} \le X < \frac{n+1}{2^m}\right\}\mathscr{P}\left\{\frac{n'}{2^m} \le Y < \frac{n'+1}{2^m}\right\}\end{aligned}$$

$$= \mathscr{P}\left\{X_m = \frac{n}{2^m}\right\}\mathscr{P}\left\{Y_m = \frac{n'}{2^m}\right\}.$$

The independence of $X_m$ and $Y_m$ is also a consequence of Theorem 3.3.1, since $X_m = [2^m X]/2^m$, where $[X]$ denotes the greatest integer in $X$. Finally, it is clear that $X_m Y_m$ is increasing with $m$ and

$$0 \leq XY - X_m Y_m = X(Y - Y_m) + Y_m(X - X_m) \to 0.$$

Hence, by the monotone convergence theorem, we conclude that

$$\begin{aligned} \mathscr{E}(XY) &= \lim_{m\to\infty} \mathscr{E}(X_m Y_m) = \lim_{m\to\infty} \mathscr{E}(X_m)\mathscr{E}(Y_m) \\ &= \lim_{m\to\infty} \mathscr{E}(X_m) \lim_{m\to\infty} \mathscr{E}(Y_m) = \mathscr{E}(X)\mathscr{E}(Y). \end{aligned}$$

Thus (5) is true also in this case. For the general case, we use (2) and (3) of Sec. 3.2 and observe that the independence of $X$ and $Y$ implies that of $X^+$ and $Y^+$; $X^-$ and $Y^-$; and so on. This again can be seen directly or as a consequence of Theorem 3.3.1. Hence we have, under our finiteness hypothesis:

$$\begin{aligned} \mathscr{E}(XY) &= \mathscr{E}((X^+ - X^-)(Y^+ - Y^-)) \\ &= \mathscr{E}(X^+Y^+ - X^+Y^- - X^-Y^+ + X^-Y^-) \\ &= \mathscr{E}(X^+Y^+) - \mathscr{E}(X^+Y^-) - \mathscr{E}(X^-Y^+) + \mathscr{E}(X^-Y^-) \\ &= \mathscr{E}(X^+)\mathscr{E}(Y^+) - \mathscr{E}(X^+)\mathscr{E}(Y^-) - \mathscr{E}(X^-)\mathscr{E}(Y^+) + \mathscr{E}(X^-)\mathscr{E}(Y^-) \\ &= \{\mathscr{E}(X^+) - \mathscr{E}(X^-)\}\{\mathscr{E}(Y^+) - \mathscr{E}(Y^-)\} = \mathscr{E}(X)\mathscr{E}(Y). \end{aligned}$$

The first proof is completed.

*Second proof.* Consider the random vector $(X, Y)$ and let the p.m. induced by it be $\mu^2(dx, dy)$. Then we have by Theorem 3.2.3:

$$\mathscr{E}(XY) = \int_\Omega XY\, d\mathscr{P} = \iint_{\mathscr{R}^2} xy\, \mu^2(dx, dy)$$

By (3), the last integral is equal to

$$\int_{\mathscr{R}^1}\int_{\mathscr{R}^1} xy\, \mu_1(dx)\mu_2(dy) = \int_{\mathscr{R}^1} x\, \mu_1(dx) \int_{\mathscr{R}^1} y\, \mu_2(dx) = \mathscr{E}(X)\mathscr{E}(Y),$$

finishing the proof! Observe that we are using here a very simple form of Fubini's theorem (see below). Indeed, the second proof appears to be so much shorter only because we are relying on the theory of "product measure" $\mu^2 = \mu_1 \times \mu_2$ on $(\mathscr{R}^2, \mathscr{B}^2)$. This is another illustration of the method of reduction mentioned in connection with the proof of (17) in Sec. 3.2.

**Corollary.** If $\{X_j, 1 \le j \le n\}$ are independent r.v.'s with finite expectations, then

$$\mathscr{E}\left(\prod_{j=1}^{n} X_j\right) = \prod_{j=1}^{n} \mathscr{E}(X_j). \tag{6}$$

This follows at once by induction from (5), provided we observe that the two r.v.'s

$$\prod_{j=1}^{k} X_j \quad \text{and} \quad \prod_{j=k+1}^{n} X_j$$

are independent for each $k$, $1 \le k \le n-1$. A rigorous proof of this fact may be supplied by Theorem 3.3.2.

Do independent random variables exist? Here we can take the cue from the intuitive background of probability theory which not only has given rise historically to this branch of mathematical discipline, but remains a source of inspiration, inculcating a way of thinking peculiar to the discipline. It may be said that no one could have learned the subject properly without acquiring some feeling for the intuitive content of the concept of stochastic independence, and through it, certain degrees of dependence. Briefly then: events are determined by the *outcomes* of random *trials*. If an unbiased coin is tossed and the two possible outcomes are recorded as 0 and 1, this is an r.v., and it takes these two values with roughly the probabilities $\frac{1}{2}$ each. Repeated tossing will produce a sequence of outcomes. If now a die is cast, the outcome may be similarly represented by an r.v. taking the six values 1 to 6; again this may be repeated to produce a sequence. Next we may draw a card from a pack or a ball from an urn, or take a measurement of a physical quantity sampled from a given population, or make an observation of some fortuitous natural phenomenon, the outcomes in the last two cases being r.v.'s taking some rational values in terms of certain units; and so on. Now it is very easy to conceive of undertaking these various trials under conditions such that their respective outcomes do not appreciably affect each other; indeed it would take more imagination to conceive the opposite! In this circumstance, idealized, the trials are carried out "independently of one another" and the corresponding r.v.'s are "independent" according to definition. We have thus "constructed" sets of independent r.v.'s in varied contexts and with various distributions (although they are always discrete on realistic grounds), and the whole process can be continued indefinitely.

Can such a construction be made rigorous? We begin by an easy special case.

**Example 1.** Let $n \geq 2$ and $(\Omega_j, \mathcal{S}_j, \mathcal{P}_j)$ be $n$ discrete probability spaces. We define the *product space*

$$\Omega^n = \Omega_1 \times \cdots \times \Omega_n (n \text{ factors})$$

to be the space of all ordered $n$-tuples $\omega^n = (\omega_1, \dots, \omega_n)$, where each $\omega_j \in \Omega_j$. The *product B.F.* $\mathcal{S}^n$ is simply the collection of all subsets of $\Omega^n$, just as $\mathcal{S}_j$ is that for $\Omega_j$. Recall that (Example 1 of Sec. 2.2) the p.m. $\mathcal{P}_j$ is determined by its value for each point of $\Omega_j$. Since $\Omega^n$ is also a countable set, we may define a p.m. $\mathcal{P}^n$ on $\mathcal{S}^n$ by the following assignment:

$$(7) \qquad \mathcal{P}^n(\{\omega^n\}) = \prod_{j=1}^{n} \mathcal{P}_j(\{\omega_j\}),$$

namely, to the $n$-tuple $(\omega_1, \dots, \omega_n)$ the probability to be assigned is the *product* of the probabilities originally assigned to each component $\omega_j$ by $\mathcal{P}_j$. This p.m. will be called the *product measure* derived from the p.m.'s $\{\mathcal{P}_j, 1 \leq j \leq n\}$ and denoted by $\times_{j=1}^{n} \mathcal{P}_j$. It is trivial to verify that this is indeed a p.m. Furthermore, it has the following product property, extending its definition (7): if $S_j \in \mathcal{S}_j$, $1 \leq j \leq n$, then

$$(8) \qquad \mathcal{P}^n\left(\mathop{\times}_{j=1}^{n} S_j\right) = \prod_{j=1}^{n} \mathcal{P}_j(S_j).$$

To see this, we observe that the left side is, by definition, equal to

$$\sum_{\omega_1 \in S_1} \cdots \sum_{\omega_n \in S_n} \mathcal{P}^n(\{\omega_1, \dots, \omega_n\}) = \sum_{\omega_1 \in S_1} \cdots \sum_{\omega_n \in S_n} \prod_{j=1}^{n} \mathcal{P}_j(\{\omega_j\})$$

$$= \prod_{j=1}^{n} \left\{ \sum_{\omega_j \in S_j} \mathcal{P}_j(\{\omega_j\}) \right\} = \prod_{j=1}^{n} \mathcal{P}_j(S_j),$$

the second equation being a matter of simple algebra.

Now let $X_j$ be an r.v. (namely an arbitrary function) on $\Omega_j$; $B_j$ be an arbitrary Borel set; and $S_j = X_j^{-1}(B_j)$, namely:

$$S_j = \{\omega_j \in \Omega_j : X_j(\omega_j) \in B_j\}$$

so that $S_j \in \mathcal{S}_j$, We have then by (8):

$$(9) \qquad \mathcal{P}^n\left\{\mathop{\times}_{j=1}^{n} [X_j \in B_j]\right\} = \mathcal{P}^n\left\{\mathop{\times}_{j=1}^{n} S_j\right\} = \prod_{j=1}^{n} \mathcal{P}_j(S_j) = \prod_{j=1}^{n} \mathcal{P}_j\{X_j \in B_j\}.$$

To each function $X_j$ on $\Omega_j$ let correspond the function $\tilde{X}_j$ on $\Omega^n$ defined below, in which $\omega = (\omega_1, \dots, \omega_n)$ and each "coordinate" $\omega_j$ is regarded as a *function* of the point $\omega$:

$$\forall \omega \in \Omega^n : \tilde{X}_j(\omega) = X_j(\omega_j).$$

Then we have

$$\bigcap_{j=1}^{n}\{\omega: \tilde{X}_j(\omega) \in B_j\} = \bigtimes_{j=1}^{n} \{\omega_j: X_j(\omega_j) \in B_j\}$$

since

$$\{\omega: \tilde{X}_j(\omega) \in B_j\} = \Omega_1 \times \cdots \times \Omega_{j-1} \times \{\omega_j: X_j(\omega_j) \in B_j\} \times \Omega_{j+1} \times \cdots \times \Omega_n.$$

It follows from (9) that

$$\mathscr{P}^n\left\{\bigcap_{j=1}^{n}[\tilde{X}_j \in B_j]\right\} = \prod_{j=1}^{n} \mathscr{P}^n\{\tilde{X}_j \in B_j\}.$$

Therefore the r.v.'s $\{\tilde{X}_j, 1 \leq j \leq n\}$ are independent.

**Example 2.** Let $\mathscr{U}^n$ be the $n$-dimensional cube (immaterial whether it is closed or not):

$$\mathscr{U}^n = \{(x_1, \ldots, x_n): 0 \leq x_j \leq 1; 1 \leq j \leq n\}.$$

The trace on $\mathscr{U}^n$ of $(\mathscr{R}^n, \mathscr{B}^n, m^n)$, where $\mathscr{R}^n$ is the $n$-dimensional Euclidean space, $\mathscr{B}^n$ and $m^n$ the usual Borel field and measure, is a probability space. The p.m. $m^n$ on $\mathscr{U}^n$ is a product measure having the property analogous to (8). Let $\{f_j, 1 \leq j \leq n\}$ be $n$ Borel measurable functions of one variable, and

$$X_j((x_1, \ldots, x_n)) = f_j(x_j).$$

Then $\{X_j, 1 \leq j \leq n\}$ are independent r.v.'s. In particular if $f_j(x_j) \equiv x_j$, we obtain the $n$ *coordinate variables* in the cube. The reader may recall the term "independent variables" used in calculus, particularly for integration in several variables. The two usages have some accidental rapport.

**Example 3.** The point of Example 2 is that there is a ready-made product measure there. Now on $(\mathscr{R}^n, \mathscr{B}^n)$ it is possible to construct such a one based on given p.m.'s on $(\mathscr{R}^1, \mathscr{B}^1)$. Let these be $\{\mu_j, 1 \leq j \leq n\}$; we define $\mu^n$ for product sets, in analogy with (8), as follows:

$$\mu^n\left(\bigtimes_{j=1}^{n} B_j\right) = \prod_{j=1}^{n} \mu_j(B_j).$$

It remains to extend this definition to all of $\mathscr{B}^n$, or, more logically speaking, to prove that there exists a p.m. $\mu^n$ on $\mathscr{B}^n$ that has the above "product property". The situation is somewhat more complicated than in Example 1, just as Example 3 in Sec. 2.2 is more complicated than Example 1 there. Indeed, the required construction is exactly that of the corresponding Lebesgue–Stieltjes measure in $n$ dimensions. This will be subsumed in the next theorem. Assuming that it has been accomplished, then sets of $n$ independent r.v.'s can be defined just as in Example 2.

**Example 4.** Can we construct r.v.'s on the probability space $(\mathscr{U}, \mathscr{B}, m)$ itself, without going to a product space? Indeed we can, but only by *imbedding* a product structure in $\mathscr{U}$. The simplest case will now be described and we shall return to it in the next chapter.

For each real number in (0,1], consider its binary digital expansion

$$x = \cdot\epsilon_1\epsilon_2 \cdots \epsilon_n \cdots = \sum_{n=1}^{\infty} \frac{\epsilon_n}{2^n}, \qquad \text{each } \epsilon_n = 0 \quad \text{or} \quad 1. \tag{10}$$

This expansion is unique except when $x$ is of the form $m/2^n$; the set of such $x$ is countable and so of probability zero, hence whatever we decide to do with them will be immaterial for our purposes. For the sake of definiteness, let us agree that only expansions with infinitely many digits "1" are used. Now each digit $\epsilon_j$ of $x$ is a function of $x$ taking the values 0 and 1 on two Borel sets. Hence they are r.v.'s. Let $\{c_j, j \geq 1\}$ be a given sequence of 0's and 1's. Then the set

$$\{x : \epsilon_j(x) = c_j, 1 \leq j \leq n\} = \bigcap_{j=1}^{n} \{x : \epsilon_j(x) = c_j\}$$

is the set of numbers $x$ whose first $n$ digits are the given $c_j$'s, thus

$$x = \cdot c_1 c_2 \cdots c_n \epsilon_{n+1} \epsilon_{n+2} \cdots$$

with the digits from the $(n+1)$st on completely arbitrary. It is clear that this set is just an interval of length $1/2^n$, hence of probability $1/2^n$. On the other hand for each $j$, the set $\{x : \epsilon_j(x) = c_j\}$ has probability $\frac{1}{2}$ for a similar reason. We have therefore

$$\mathscr{P}\{\epsilon_j = c_j, 1 \leq j \leq n\} = \frac{1}{2^n} = \prod_{j=1}^{n} \left(\frac{1}{2}\right) = \prod_{j=1}^{n} \mathscr{P}\{\epsilon_j = c_j\}.$$

This being true for every choice of the $c_j$'s, the r.v.'s $\{\epsilon_j, j \geq 1\}$ are independent. Let $\{f_j, j \geq 1\}$ be arbitrary functions with domain the two points $\{0, 1\}$, then $\{f_j(\epsilon_j), j \geq 1\}$ are also independent r.v.'s.

This example seems extremely special, but easy extensions are at hand (see Exercises 13, 14, and 15 below).

We are now ready to state and prove the fundamental existence theorem of product measures.

**Theorem 3.3.4.** Let a finite or infinite sequence of p.m.'s $\{\mu_j\}$ on $(\mathscr{R}^1, \mathscr{B}^1)$, or equivalently their d.f.'s, be given. There exists a probability space $(\Omega, \mathscr{F}, \mathscr{P})$ and a sequence of independent r.v.'s $\{X_j\}$ defined on it such that for each $j$, $\mu_j$ is the p.m. of $X_j$.

PROOF. Without loss of generality we may suppose that the given sequence is infinite. (Why?) For each $n$, let $(\Omega_n, \mathscr{F}_n, \mathscr{P}_n)$ be a probability space in which there exists an r.v. $X_n$ with $\mu_n$ as its p.m. Indeed this is possible if we take $(\Omega_n, \mathscr{F}_n, \mathscr{P}_n)$ to be $(\mathscr{R}^1, \mathscr{B}^1, \mu_n)$ and $X_n$ to be the identical function of the sample point $x$ in $\mathscr{R}^1$, now to be written as $\omega_n$ (cf. Exercise 3 of Sec. 3.1).

Now define the *infinite product space*

$$\Omega = \mathop{\times}_{n=1}^{\infty} \Omega_n$$

on the collection of all "points" $\omega = \{\omega_1, \omega_2, \dots, \omega_n, \dots\}$, where for each $n$, $\omega_n$ is a point of $\Omega_n$. A subset $E$ of $\Omega$ will be called a "finite-product set" iff it is of the form

$$E = \mathop{\times}_{n=1}^{\infty} F_n, \tag{11}$$

where each $F_n \in \mathscr{F}_n$ and all but a finite number of these $F_n$'s are equal to the corresponding $\Omega_n$'s. Thus $\omega \in E$ if and only if $\omega_n \in F_n$, $n \geq 1$, but this is actually a restriction only for a finite number of values of $n$. Let the collection of subsets of $\Omega$, each of which is the union of a finite number of disjoint finite-product sets, be $\mathscr{F}_0$. It is easy to see that the collection $\mathscr{F}_0$ is closed with respect to complementation and pairwise intersection, hence it is a field. We shall take the $\mathscr{F}$ in the theorem to be the B.F. generated by $\mathscr{F}_0$. This $\mathscr{F}$ is called the *product B.F.* of the sequence $\{\mathscr{F}_n, n \geq 1\}$ and denoted by $\times_{n=1}^{\infty} \mathscr{F}_n$.

We define a set function $\mathscr{P}$ on $\mathscr{F}_0$ as follows. First, for each finite-product set such as the $E$ given in (11) we set

$$\mathscr{P}(E) = \prod_{n=1}^{\infty} \mathscr{P}_n(F_n), \tag{12}$$

where all but a finite number of the factors on the right side are equal to one. Next, if $E \in \mathscr{F}_0$ and

$$E = \bigcup_{k=1}^{n} E^{(k)},$$

where the $E^{(k)}$'s are disjoint finite-product sets, we put

$$\mathscr{P}(E) = \sum_{k=1}^{n} \mathscr{P}(E^{(k)}). \tag{13}$$

If a given set $E$ in $\mathscr{F}_0$ has two representations of the form above, then it is not difficult to see though it is tedious to explain (try drawing some pictures!) that the two definitions of $\mathscr{P}(E)$ agree. Hence the set function $\mathscr{P}$ is uniquely defined on $\mathscr{F}_0$; it is clearly positive with $\mathscr{P}(\Omega) = 1$, and it is finitely additive on $\mathscr{F}_0$ by definition. In order to verify countable additivity it is sufficient to verify the axiom of continuity, by the remark after Theorem 2.2.1. Proceeding by contraposition, suppose that there exist a $\delta > 0$ and a sequence of sets $\{C_n, n \geq 1\}$ in $\mathscr{F}_0$ such that for each $n$, we have $C_n \supset C_{n+1}$ and $\mathscr{P}(C_n) \geq \delta > 0$; we shall prove that $\bigcap_{n=1}^{\infty} C_n \neq \varnothing$. Note that each set $E$ in $\mathscr{F}_0$, as well as each finite-product set, is "determined" by a finite number of coordinates in the sense that there exists a finite integer $k$, depending only on $E$, such that if $\omega = (\omega_1, \omega_2, \dots)$ and $\omega' = (\omega'_1, \omega'_2, \dots)$ are two points of $\Omega$ with $\omega_j = \omega'_j$ for $1 \leq j \leq k$, then either both $\omega$ and $\omega'$ belong to $E$ or neither of them does. To simplify the notation and with no real loss of generality (why?) we may suppose that for each $n$, the set $C_n$ is determined by the first $n$ coordinates. Given $\omega_1^0$, for any subset $E$ of $\Omega$ let us write $(E \mid \omega_1^0)$ for $\Omega_1 \times E_1$, where $E_1$ is the set of points $(\omega_2, \omega_3, \dots)$ in $\times_{n=2}^{\infty} \Omega_n$ such that $(\omega_1^0, \omega_2, \omega_3, \dots) \in E$. If $\omega_1^0$ does not appear as first coordinate of any point in $E$, then $(E \mid \omega_1^0) = \varnothing$. Note that if $E \in \mathscr{F}_0$, then $(E \mid \omega_1^0) \in \mathscr{F}_0$ for each $\omega_1^0$. We claim that there exists an $\omega_1^0$ such that for every $n$, we have $\mathscr{P}((C_n \mid \omega_1^0)) \geq \delta/2$. To see this we begin with the equation

$$\mathscr{P}(C_n) = \int_{\Omega_1} \mathscr{P}((C_n \mid \omega_1))\mathscr{P}_1(d\omega_1). \tag{14}$$

This is trivial if $C_n$ is a finite-product set by definition (12), and follows by addition for any $C_n$ in $\mathscr{F}_0$. Now put $B_n = \{\omega_1 : \mathscr{P}((C_n \mid \omega_1)) \geq \delta/2\}$, a subset of $\Omega_1$; then it follows that

$$\delta \leq \int_{B_n} 1\mathscr{P}_1(d\omega_1) + \int_{B_n^c} \frac{\delta}{2}\mathscr{P}_1(d\omega_1)$$

and so $\mathscr{P}(B_n) \geq \delta/2$ for every $n \geq 1$. Since $B_n$ is decreasing with $C_n$, we have $\mathscr{P}_1(\bigcap_{n=1}^{\infty} B_n) \geq \delta/2$. Choose any $\omega_1^0$ in $\bigcap_{n=1}^{\infty} B_n$. Then $\mathscr{P}((C_n \mid \omega_1^0)) \geq \delta/2$. Repeating the argument for the set $(C_n \mid \omega_1^0)$, we see that there exists an $\omega_2^0$ such that for every $n$, $\mathscr{P}((C_n \mid \omega_1^0, \omega_2^0)) \geq \delta/4$, where $(C_n \mid \omega_1^0, \omega_2^0) = ((C_n \mid \omega_1^0) \mid \omega_2^0)$ is of the form $\Omega_1 \times \Omega_2 \times E_3$ and $E_3$ is the set $(\omega_3, \omega_4, \dots)$ in $\times_{n=3}^{\infty} \Omega_n$ such that $(\omega_1^0, \omega_2^0, \omega_3, \omega_4, \dots) \in C_n$; and so forth by induction. Thus for each $k \geq 1$, there exists $\omega_k^0$ such that

$$\forall n : \mathscr{P}((C_n \mid \omega_1^0, \dots, \omega_k^0)) \geq \frac{\delta}{2^k}.$$

Consider the point $\omega^0 = (\omega_1^0, \omega_2^0, \dots, \omega_n^0, \dots)$. Since $(C_k \mid \omega_1^0, \dots, \omega_k^0) \neq \varnothing$, there is a point in $C_k$ whose first $k$ coordinates are the same as those of $\omega^0$; since $C_k$ is determined by the first $k$ coordinates, it follows that $\omega^0 \in C_k$. This is true for all $k$, hence $\omega^0 \in \bigcap_{k=1}^{\infty} C_k$, as was to be shown.

We have thus proved that $\mathscr{P}$ as defined on $\mathscr{F}_0$ is a p.m. The next theorem, which is a generalization of the extension theorem discussed in Theorem 2.2.2 and whose proof can be found in Halmos [4], ensures that it can be extended to $\mathscr{F}$ as desired. This extension is called the *product measure* of the sequence $\{\mathscr{P}_n, n \geq 1\}$ and denoted by $\times_{n=1}^{\infty} \mathscr{P}_n$ with domain the product field $\times_{n=1}^{\infty} \mathscr{F}_n$.

**Theorem 3.3.5.** Let $\mathscr{F}_0$ be a field of subsets of an abstract space $\Omega$, and $\mathscr{P}$ a p.m. on $\mathscr{F}_0$. There exists a unique p.m. on the B.F. generated by $\mathscr{F}_0$ that agrees with $\mathscr{P}$ on $\mathscr{F}_0$.

The uniqueness is proved in Theorem 2.2.3.

Returning to Theorem 3.3.4, it remains to show that the r.v.'s $\{\omega_j\}$ are independent. For each $k \geq 2$, the independence of $\{\omega_j, 1 \leq j \leq k\}$ is a consequence of the definition in (12); this being so for every $k$, the independence of the infinite sequence follows by definition.

We now give a statement of Fubini's theorem for product measures, which has already been used above in special cases. It is sufficient to consider the case $n = 2$. Let $\Omega = \Omega_1 \times \Omega_2$, $\mathscr{F} = \mathscr{F}_1 \times \mathscr{F}_2$ and $\mathscr{P} = \mathscr{P}_1 \times \mathscr{P}_2$ be the product space, B.F., and measure respectively.

Let $(\overline{\mathscr{F}_1}, \overline{\mathscr{P}_1})$, $(\overline{\mathscr{F}_2}, \overline{\mathscr{P}_2})$ and $(\overline{\mathscr{F}_1 \times \mathscr{F}_2}, \overline{\mathscr{P}_1 \times \mathscr{P}_2})$ be the completions of $(\mathscr{F}_1, \mathscr{P}_1)$, $(\mathscr{F}_2, \mathscr{P}_2)$, and $(\mathscr{F}_1 \times \mathscr{F}_2, \mathscr{P}_1 \times \mathscr{P}_2)$, respectively, according to Theorem 2.2.5.

**Fubini's theorem.** Suppose that $f$ is measurable with respect to $\overline{\mathscr{F}_1 \times \mathscr{F}_2}$ and integrable with respect to $\overline{\mathscr{P}_1 \times \mathscr{P}_2}$. Then

**(i)** for each $\omega_1 \in \Omega_1 \backslash N_1$ where $N_1 \in \mathscr{F}_1$ and $\overline{\mathscr{P}_1}(N_1) = 0$, $f(\omega_1, \cdot)$ is measurable with respect to $\overline{\mathscr{F}_2}$ and integrable with respect to $\overline{\mathscr{P}_2}$;

**(ii)** the integral

$$\int_{\Omega_2} f(\cdot, \omega_2) \overline{\mathscr{P}_2}(d\omega_2)$$

is measurable with respect to $\overline{\mathscr{F}_1}$ and integrable with respect to $\overline{\mathscr{P}_1}$;

**(iii)** The following equation between a double and a repeated integral holds:

$$(15) \quad \iint_{\Omega_1 \times \Omega_2} f(\omega_1, \omega_2) \overline{\mathscr{P}_1 \times \mathscr{P}_2}(d\omega) = \int_{\Omega_1} \left[ \int_{\Omega_2} f(\omega_1, \omega_2) \overline{\mathscr{P}_2}(d\omega_2) \right] \overline{\mathscr{P}_1}(d\omega_1).$$

Furthermore, suppose $f$ is positive and measurable with respect to $\overline{\mathscr{F}_1 \times \mathscr{F}_2}$; then if either member of (15) exists finite or infinite, so does the other also and the equation holds.

Finally if we remove all the completion signs "—" in the hypotheses as well as the conclusions above, the resulting statement is true with the exceptional set $N_1$ in (i) empty, provided the word "integrable" in (i) be replaced by "has an integral."

The theorem remains true if the p.m.'s are replaced by $\sigma$-finite measures; see, e.g., Royden [5].

The reader should readily recognize the particular cases of the theorem with one or both of the factor measures discrete, so that the corresponding integral reduces to a sum. Thus it includes the familiar theorem on evaluating a double series by repeated summation.

We close this section by stating a generalization of Theorem 3.3.4 to the case where the finite-dimensional joint distributions of the sequence $\{X_j\}$ are arbitrarily given, subject only to mutual consistency. This result is a particular case of *Kolmogorov's extension theorem*, which is valid for an arbitrary family of r.v.'s.

Let $m$ and $n$ be integers: $1 \leq m < n$, and define $\pi_{mn}$ to be the "projection map" of $\mathscr{B}^m$ onto $\mathscr{B}^n$ given by

$$\forall B \in \mathscr{B}^m: \pi_{mn}(B) = \{(x_1, \dots, x_n): (x_1, \dots, x_m) \in B\}.$$

**Theorem 3.3.6.** For each $n \geq 1$, let $\mu^n$ be a p.m. on $(\mathscr{R}^n, \mathscr{B}^n)$ such that

$$\forall m < n: \mu^n \circ \pi_{mn} = \mu^m. \tag{16}$$

Then there exists a probability space $(\Omega, \mathscr{F}, \mathscr{P})$ and a sequence of r.v.'s $\{X_j\}$ on it such that for each $n$, $\mu^n$ is the $n$-dimensional p.m. of the vector

$$(X_1, \dots, X_n).$$

Indeed, the $\Omega$ and $\mathscr{F}$ may be taken exactly as in Theorem 3.3.4 to be the product space

$$\mathop{\times}_j \Omega_j \quad \text{and} \quad \mathop{\times}_j \mathscr{F}_j,$$

where $(\Omega_j, \mathscr{F}_j) = (\mathscr{R}^1, \mathscr{B}^1)$ for each $j$; only $\mathscr{P}$ is now more general. In terms of d.f.'s, the consistency condition (16) may be stated as follows. For each $m \geq 1$ and $(x_1, \dots, x_m) \in \mathscr{R}^m$, we have if $n > m$:

$$\lim_{\substack{x_{m+1} \to \infty \\ \cdots\cdots \\ x_n \to \infty}} F_n(x_1, \dots, x_m, x_{m+1}, \dots, x_n) = F_m(x_1, \dots, x_m).$$

For a proof of this first fundamental theorem in the theory of stochastic processes, see Kolmogorov [8].

EXERCISES

**1.** Show that two r.v.'s on $(\Omega, \mathscr{F})$ may be independent according to one p.m. $\mathscr{P}$ but not according to another!

$^\star$**2.** If $X_1$ and $X_2$ are independent r.v.'s each assuming the values $+1$ and $-1$ with probability $\frac{1}{2}$, then the three r.v.'s $\{X_1, X_2, X_1X_2\}$ are pairwise independent but not totally independent. Find an example of $n$ r.v.'s such that every $n-1$ of them are independent but not all of them. Find an example where the events $\Lambda_1$ and $\Lambda_2$ are independent, $\Lambda_1$ and $\Lambda_3$ are independent, but $\Lambda_1$ and $\Lambda_2 \cup \Lambda_3$ are not independent.

$^\star$**3.** If the events $\{E_\alpha, \alpha \in A\}$ are independent, then so are the events $\{F_\alpha, \alpha \in A\}$, where each $F_\alpha$ may be $E_\alpha$ or $E_a^c$; also if $\{A_\beta, \beta \in B\}$, where $B$ is an arbitrary index set, is a collection of disjoint countable subsets of $A$, then the events

$$\bigcup_{\alpha \in A_\beta} E_\alpha, \qquad \beta \in B,$$

are independent.

**4.** Fields or B.F.'s $\mathscr{F}_\alpha (\subset \mathscr{F})$ of any family are said to be independent iff any collection of events, one from each $\mathscr{F}_\alpha$, forms a set of independent events. Let $\mathscr{F}_\alpha^0$ be a field generating $\mathscr{F}_\alpha$. Prove that if the fields $\mathscr{F}_\alpha^0$ are independent, then so are the B.F.'s $\mathscr{F}_\alpha$. Is the same conclusion true if the fields are replaced by arbitrary generating sets? Prove, however, that the conditions (1) and (2) are equivalent. [HINT: Use Theorem 2.1.2.]

**5.** If $\{X_\alpha\}$ is a family of independent r.v.'s, then the B.F.'s generated by disjoint subfamilies are independent. [Theorem 3.3.2 is a corollary to this proposition.]

**6.** The r.v. $X$ is independent of itself if and only if it is constant with probability one. Can $X$ and $f(X)$ be independent where $f \in \mathscr{B}^1$?

**7.** If $\{E_j, 1 \le j < \infty\}$ are independent events then

$$\mathscr{P}\left(\bigcap_{j=1}^{\infty} E_j\right) = \prod_{j=1}^{\infty} \mathscr{P}(E_j),$$

where the infinite product is defined to be the obvious limit; similarly

$$\mathscr{P}\left(\bigcup_{j=1}^{\infty} E_j\right) = 1 - \prod_{j=1}^{\infty}(1 - \mathscr{P}(E_j)).$$

**8.** Let $\{X_j, 1 \le j \le n\}$ be independent with d.f.'s $\{F_j, 1 \le j \le n\}$. Find the d.f. of $\max_j X_j$ and $\min_j X_j$.

$^\star$**9.** If $X$ and $Y$ are independent and $\mathscr{E}(X)$ exists, then for any Borel set $B$, we have

$$\int_{\{Y \in B\}} X \, d\mathscr{P} = \mathscr{E}(X)\mathscr{P}(Y \in B).$$

$^\star$**10.** If $X$ and $Y$ are independent and for some $p > 0$: $\mathscr{E}(|X + Y|^p) < \infty$, then $\mathscr{E}(|X|^p) < \infty$ and $\mathscr{E}(|Y|^p) < \infty$.

**11.** If $X$ and $Y$ are independent, $\mathscr{E}(|X|^p) < \infty$ for some $p \geq 1$, and $\mathscr{E}(Y) = 0$, then $\mathscr{E}(|X + Y|^p) \geq \mathscr{E}(|X|^p)$. [This is a case of Theorem 9.3.2; but try a direct proof!]

**12.** The r.v.'s $\{\epsilon_j\}$ in Example 4 are related to the "Rademacher functions":

$$r_j(x) = \operatorname{sgn}(\sin 2^j \pi x).$$

What is the precise relation?

**13.** Generalize Example 4 by considering any $s$-ary expansion where $s \geq 2$ is an integer:

$$x = \sum_{n=1}^{\infty} \frac{\epsilon_n}{s^n}, \quad \text{where } \epsilon_n = 0, 1, \ldots, s - 1.$$

$^\star$**14.** Modify Example 4 so that to each $x$ in $[0, 1]$ there corresponds a sequence of independent and identically distributed r.v.'s $\{\epsilon_n, n \geq 1\}$, each taking the values 1 and 0 with probabilities $p$ and $1 - p$, respectively, where $0 < p < 1$. Such a sequence will be referred to as a "coin-tossing game (with probability $p$ for heads)"; when $p = \frac{1}{2}$ it is said to be "fair".

**15.** Modify Example 4 further so as to allow $\epsilon_n$ to take the values 1 and 0 with probabilities $p_n$ and $1 - p_n$, where $0 < p_n < 1$, but $p_n$ depends on $n$.

**16.** Generalize (14) to

$$\begin{aligned}
\mathscr{P}(C) &= \int_{C(1,\ldots,k)} \mathscr{P}((C \mid \omega_1, \ldots, \omega_k))\mathscr{P}_1(d\omega_1) \cdots \mathscr{P}_k(d\omega_k) \\
&= \int_{\Omega} \mathscr{P}((C \mid \omega_1, \ldots, \omega_k))\mathscr{P}(d\omega),
\end{aligned}$$

where $C(1, \ldots, k)$ is the set of $(\omega_1, \ldots, \omega_k)$ which appears as the first $k$ coordinates of at least one point in $C$ and in the second equation $\omega_1, \ldots, \omega_k$ are regarded as functions of $\omega$.

$^\star$**17.** For arbitrary events $\{E_j, 1 \leq j \leq n\}$, we have

$$\mathscr{P}\left(\bigcup_{j=1}^{n} E_j\right) \geq \sum_{j=1}^{n} \mathscr{P}(E_j) - \sum_{1 \leq j < k \leq n} \mathscr{P}(E_j E_k).$$

If $\forall n: \{E_j^{(n)}, 1 \le j \le n\}$ are independent events, and

$$\mathscr{P}\left(\bigcup_{j=1}^{n} E_j^{(n)}\right) \to 0 \quad \text{as} \quad n \to \infty,$$

then

$$\mathscr{P}\left(\bigcup_{j=1}^{n} E_j^{(n)}\right) \sim \sum_{j=1}^{n} \mathscr{P}(E_j^{(n)}).$$

**18.** Prove that $\overline{\mathscr{B}} \times \overline{\mathscr{B}} \neq \overline{\mathscr{B} \times \mathscr{B}}$, where $\overline{\mathscr{B}}$ is the completion of $\mathscr{B}$ with respect to the Lebesgue measure; similarly $\overline{\mathscr{B} \times \mathscr{B}}$.

**19.** If $f \in \mathscr{F}_1 \times \mathscr{F}_2$ and

$$\iint_{\Omega_1 \times \Omega_2} |f| \, d(\overline{\mathscr{P}_1 \times \mathscr{P}_2}) < \infty,$$

then

$$\int_{\Omega_1} \left[\int_{\Omega_2} f \, d\overline{\mathscr{P}}_2\right] d\overline{\mathscr{P}}_1 = \int_{\Omega_2} \left[\int_{\Omega_1} f \, d\overline{\mathscr{P}}_1\right] d\overline{\mathscr{P}}_2.$$

**20.** A typical application of Fubini's theorem is as follows. If $f$ is a Lebesgue measurable function of $(x, y)$ such that $f(x, y) = 0$ for each $x \in \mathscr{R}^1$ and $y \notin N_x$, where $m(N_x) = 0$ for each $x$, then we have also $f(x, y) = 0$ for each $y \notin N$ and $x \in N'_y$, where $m(N) = 0$ and $m(N'_y) = 0$ for each $y \notin N$.

# 4 Convergence concepts

## 4.1 Various modes of convergence

As numerical-valued functions, the convergence of a sequence of r.v.'s $\{X_n, n \geq 1\}$, to be denoted simply by $\{X_n\}$ below, is a well-defined concept. Here and hereafter the term "convergence" will be used to mean convergence to a *finite* limit. Thus it makes sense to say: for every $\omega \in \Delta$, where $\Delta \in \mathscr{F}$, the sequence $\{X_n(\omega)\}$ converges. The limit is then a finite-valued r.v. (see Theorem 3.1.6), say $X(\omega)$, defined on $\Delta$. If $\Omega = \Delta$, then we have "convergence every-where", but a more useful concept is the following one.

DEFINITION OF CONVERGENCE "ALMOST EVERYWHERE" (a.e.). The sequence of r.v. $\{X_n\}$ is said to converge almost everywhere [to the r.v. $X$] iff there exists a null set $\mathbf{N}$ such that

$$(1) \qquad \forall \omega \in \Omega \backslash \mathbf{N}: \lim_{n \to \infty} X_n(\omega) = X(\omega) \text{ finite.}$$

Recall that our convention stated at the beginning of Sec. 3.1 allows each r.v. a null set on which it may be $\pm\infty$. The union of all these sets being still a null set, it may be included in the set $\mathbf{N}$ in (1) without modifying the conclusion. This type of trivial consideration makes it possible, when dealing with a countable set of r.v.'s that are finite a.e., to regard them as finite everywhere.

The reader should learn at an early stage to reason with a single sample point $\omega_0$ and the corresponding sample sequence $\{X_n(\omega_0), n \geq 1\}$ as a numerical sequence, and translate the inferences on the latter into probability statements about sets of $\omega$. The following characterization of convergence a.e. is a good illustration of this method.

**Theorem 4.1.1.** The sequence $\{X_n\}$ converges a.e. to $X$ if and only if for every $\epsilon > 0$ we have

$$\lim_{m\to\infty} \mathscr{P}\{|X_n - X| \leq \epsilon \text{ for all } n \geq m\} = 1; \tag{2}$$

or equivalently

$$\lim_{m\to\infty} \mathscr{P}\{|X_n - X| > \epsilon \text{ for some } n \geq m\} = 0. \tag{2'}$$

PROOF. Suppose there is convergence a.e. and let $\Omega_0 = \Omega \backslash \mathbf{N}$ where $\mathbf{N}$ is as in (1). For $m \geq 1$ let us denote by $A_m(\epsilon)$ the event exhibited in (2), namely:

$$A_m(\epsilon) = \bigcap_{n=m}^{\infty} \{|X_n - X| \leq \epsilon\}. \tag{3}$$

Then $A_m(\epsilon)$ is increasing with $m$. For each $\omega_0$, the convergence of $\{X_n(\omega_0)\}$ to $X(\omega_0)$ implies that given any $\epsilon > 0$, there exists $m(\omega_0, \epsilon)$ such that

$$n \geq m(\omega_0, \epsilon) \Rightarrow |X_n(\omega_0) - X(\omega_0)| \leq \epsilon. \tag{4}$$

Hence each such $\omega_0$ belongs to some $A_m(\epsilon)$ and so $\Omega_0 \subset \bigcup_{m=1}^{\infty} A_m(\epsilon)$. It follows from the monotone convergence property of a measure that $\lim_{m\to\infty} \mathscr{P}(A_m(\epsilon)) = 1$, which is equivalent to (2).

Conversely, suppose (2) holds, then we see above that the set $A(\epsilon) = \bigcup_{m=1}^{\infty} A_m(\epsilon)$ has probability equal to one. For any $\omega_0 \in A(\epsilon)$, (4) is true for the given $\epsilon$. Let $\epsilon$ run through a sequence of values decreasing to zero, for instance $\{1/n\}$. Then the set

$$A = \bigcap_{n=1}^{\infty} A\left(\frac{1}{n}\right)$$

still has probability one since

$$\mathscr{P}(A) = \lim_{n} \mathscr{P}\left(A\left(\frac{1}{n}\right)\right).$$

If $\omega_0$ belongs to $A$, then (4) is true for all $\epsilon = 1/n$, hence for all $\epsilon > 0$ (why?). This means $\{X_n(\omega_0)\}$ converges to $X(\omega_0)$ for all $\omega_0$ in a set of probability one.

A weaker concept of convergence is of basic importance in probability theory.

DEFINITION OF CONVERGENCE "IN PROBABILITY" (in pr.). The sequence $\{X_n\}$ is said to converge in probability to $X$ iff for every $\epsilon > 0$ we have

$$\lim_{n\to\infty} \mathscr{P}\{|X_n - X| > \epsilon\} = 0. \tag{5}$$

Strictly speaking, the definition applies when all $X_n$ and $X$ are finite-valued. But we may extend it to r.v.'s that are finite a.e. either by agreeing to ignore a null set or by the logical convention that a formula must first be *defined* in order to be valid or invalid. Thus, for example, if $X_n(\omega) = +\infty$ and $X(\omega) = +\infty$ for some $\omega$, then $X_n(\omega) - X(\omega)$ is not defined and therefore such an $\omega$ cannot belong to the set $\{|X_n - X| > \epsilon\}$ figuring in (5).

Since (2′) clearly implies (5), we have the immediate consequence below.

**Theorem 4.1.2.** Convergence a.e. [to $X$] implies convergence in pr. [to $X$].

Sometimes we have to deal with questions of convergence when no limit is in evidence. For convergence a.e. this is immediately reducible to the numerical case where the *Cauchy criterion* is applicable. Specifically, $\{X_n\}$ converges a.e. iff there exists a null set $\mathbf{N}$ such that for every $\omega \in \Omega\backslash\mathbf{N}$ and every $\epsilon > 0$, there exists $m(\omega, \epsilon)$ such that

$$n' > n \geq m(\omega, \epsilon) \Rightarrow |X_n(\omega) - X_{n'}(\omega)| \leq \epsilon.$$

The following analogue of Theorem 4.1.1 is left to the reader.

**Theorem 4.1.3.** The sequence $\{X_n\}$ converges a.e. if and only if for every $\epsilon$ we have

$$\lim_{m\to\infty} \mathscr{P}\{|X_n - X_{n'}| > \epsilon \text{ for some } n' > n \geq m\} = 0. \tag{6}$$

For convergence in pr., the obvious analogue of (6) is

$$\lim_{\substack{n\to\infty \\ n'\to\infty}} \mathscr{P}\{|X_n - X_{n'}| > \epsilon\} = 0. \tag{7}$$

It can be shown (Exercise 6 of Sec. 4.2) that this implies the existence of a finite r.v. $X$ such that $X_n \to X$ in pr.

DEFINITION OF CONVERGENCE IN $L^p$, $0 < p < \infty$. The sequence $\{X_n\}$ is said to converge in $L^p$ to $X$ iff $X_n \in L^p$, $X \in L^p$ and

$$\lim_{n\to\infty} \mathscr{E}(|X_n - X|^p) = 0. \tag{8}$$

In all these definitions above, $X_n$ converges to $X$ if and only if $X_n - X$ converges to 0. Hence there is no loss of generality if we put $X \equiv 0$ in the discussion, provided that any hypothesis involved can be similarly reduced to this case. We say that $X$ is *dominated by* $Y$ if $|X| \le Y$ a.e., and that the sequence $\{X_n\}$ is dominated by $Y$ iff this is true for each $X_n$ with the same $Y$. We say that $X$ or $\{X_n\}$ is *uniformly bounded* iff the $Y$ above may be taken to be a constant.

**Theorem 4.1.4.** If $X_n$ converges to 0 in $L^p$, then it converges to 0 in pr. The converse is true provided that $\{X_n\}$ is dominated by some $Y$ that belongs to $L^p$.

*Remark.* If $X_n \to X$ in $L^p$, and $\{X_n\}$ is dominated by $Y$, then $\{X_n - X\}$ is dominated by $Y + |X|$, which is in $L^p$. Hence there is no loss of generality to assume $X \equiv 0$.

PROOF. By Chebyshev inequality with $\varphi(x) \equiv |x|^p$, we have

$$\mathscr{P}\{|X_n| \ge \epsilon\} \le \frac{\mathscr{E}(|X_n|^p)}{\epsilon^p}. \tag{9}$$

Letting $n \to \infty$, the right member $\to 0$ by hypothesis, hence so does the left, which is equivalent to (5) with $X = 0$. This proves the first assertion. If now $|X_n| \le Y$ a.e. with $E(Y^p) < \infty$, then we have

$$\mathscr{E}(|X_n|^p) = \int_{\{|X_n|<\epsilon\}} |X_n|^p \, d\mathscr{P} + \int_{\{X_n \ge \epsilon\}} |X_n|^p \, d\mathscr{P} \le \epsilon^p + \int_{\{|X_n|\ge\epsilon\}} Y^p \, d\mathscr{P}.$$

Since $\mathscr{P}\{|X_n| \ge \epsilon\} \to 0$, the last-written integral $\to 0$ by Exercise 2 in Sec. 3.2. Letting first $n \to \infty$ and then $\epsilon \to 0$, we obtain $\mathscr{E}(|X_n|^p) \to 0$; hence $X_n$ converges to 0 in $L^p$.

As a corollary, for a uniformly bounded sequence $\{X_n\}$ convergence in pr. and in $L^p$ are equivalent. The general result is as follows.

**Theorem 4.1.5.** $X_n \to 0$ in pr. if and only if

$$\mathscr{E}\left(\frac{|X_n|}{1 + |X_n|}\right) \to 0. \tag{10}$$

Furthermore, the functional $\rho(\cdot, \cdot)$ given by

$$\rho(X, Y) = \mathscr{E}\left(\frac{|X - Y|}{1 + |X - Y|}\right)$$

is a metric in the space of r.v.'s, provided that we identify r.v.'s that are equal a.e.

PROOF. If $\rho(X, Y) = 0$, then $\mathscr{E}(|X - Y|) = 0$, hence $X = Y$ a.e. by Exercise 1 of Sec. 3.2. To show that $\rho(\cdot, \cdot)$ is metric it is sufficient to show that

$$\mathscr{E}\left(\frac{|X-Y|}{1+|X-Y|}\right) \leq \mathscr{E}\left(\frac{|X|}{1+|X|}\right) + \mathscr{E}\left(\frac{|Y|}{1+|Y|}\right).$$

For this we need only verify that for every real $x$ and $y$:

$$\frac{|x+y|}{1+|x+y|} \leq \frac{|x|}{1+|x|} + \frac{|y|}{1+|y|}. \tag{11}$$

By symmetry we may suppose that $|y| \leq |x|$; then

$$\begin{aligned}\frac{|x+y|}{1+|x+y|} - \frac{|x|}{1+|x|} &= \frac{|x+y|-|x|}{(1+|x+y|)(1+|x|)} \\ &\leq \frac{||x+y|-|x||}{1+|x|} \leq \frac{|y|}{1+|y|}.\end{aligned} \tag{12}$$

For any $X$ the r.v. $|X|/(1+|X|)$ is bounded by 1, hence by the second part of Theorem 4.1.4 the first assertion of the theorem will follow if we show that $|X_n| \to 0$ in pr. if and only if $|X_n|/(1+|X_n|) \to 0$ in pr. But $|x| \leq \epsilon$ is equivalent to $|x|/(1+|x|) \leq \epsilon/(1+\epsilon)$; hence the proof is complete.

**Example 1.** Convergence in pr. does not imply convergence in $L^p$, and the latter does not imply convergence a.e.

Take the probability space $(\Omega, \mathscr{F}, \mathscr{P})$ to be $(\mathscr{U}, \mathscr{B}, m)$ as in Example 2 of Sec. 2.2. Let $\varphi_{k,j}$ be the indicator of the interval

$$\left(\frac{j-1}{k}, \frac{j}{k}\right], \qquad k \geq 1, 1 \leq j \leq k.$$

Order these functions lexicographically first according to $k$ increasing, and then for each $k$ according to $j$ increasing, into one sequence $\{X_n\}$ so that if $X_n = \varphi_{k_n j_n}$, then $k_n \to \infty$ as $n \to \infty$. Thus for each $p > 0$:

$$\mathscr{E}(X_n^p) = \frac{1}{k_n} \to 0,$$

and so $X_n \to 0$ in $L^p$. But for each $\omega$ and every $k$, there exists a $j$ such that $\varphi_{kj}(\omega) = 1$; hence there exist infinitely many values of $n$ such that $X_n(\omega) = 1$. Similarly there exist infinitely many values of $n$ such that $X_n(\omega) = 0$. It follows that the sequence $\{X_n(\omega)\}$ of 0's and 1's cannot converge for any $\omega$. In other words, the set on which $\{X_n\}$ converges is empty.

Now if we replace $\varphi_{kj}$ by $k^{1/p}\varphi_{kj}$, where $p > 0$, then $\mathscr{P}\{X_n > 0\} = 1/k_n \to 0$ so that $X_n \to 0$ in pr., but for each $n$, we have $\mathscr{E}(X_n{}^p) = 1$. Consequently $\lim_{n\to\infty} \mathscr{E}(|X_n - 0|^p) = 1$ and $X_n$ does not $\to 0$ in $L^p$.

**Example 2.** Convergence a.e. does not imply convergence in $L^p$.

In $(\mathscr{U}, \mathscr{B}, m)$ define

$$X_n(\omega) = \begin{cases} 2^n, & \text{if } \omega \in \left(0, \dfrac{1}{n}\right); \\ 0, & \text{otherwise.} \end{cases}$$

Then $\mathscr{E}(|X_n|^p) = 2^{np}/n \to +\infty$ for each $p > 0$, but $X_n \to 0$ everywhere.

The reader should not be misled by these somewhat "artificial" examples to think that such counterexamples are rare or abnormal. Natural examples abound in more advanced theory, such as that of stochastic processes, but they are not as simple as those discussed above. Here are some culled from later chapters, tersely stated. In the symmetrical Bernoullian random walk $\{S_n, n \geq 1\}$, let $\zeta_n = 1_{\{S_n=0\}}$. Then $\lim_{n\to\infty} \mathscr{E}(\zeta_n^p) = 0$ for every $p > 0$, but $\mathscr{P}\{\lim_{n\to\infty} \zeta_n \text{ exists}\} = 0$ because of intermittent return of $S_n$ to the origin (see Sec. 8.3). This kind of example can be formulated for any recurrent process such as a Brownian motion. On the other hand, if $\{\zeta_n, n \geq 1\}$ denotes the random walk above stopped at 1, then $\mathscr{E}(\zeta_n) = 0$ for all $n$ but $\mathscr{P}\{\lim_{n\to\infty} \zeta_n = 1\} = 1$. The same holds for the martingale that consists in tossing a fair coin and doubling the stakes until the first head (see Sec. 9.4). Another striking example is furnished by the simple Poisson process $\{N(t), t \geq 0\}$ (see Sect. 5.5). If $\zeta(t) = N(t)/t$, then $\mathscr{E}(\zeta(t)) = \lambda$ for all $t > 0$; but $\mathscr{P}\{\lim_{t\downarrow 0} \zeta(t) = 0\} = 1$ because for almost every $\omega$, $N(t, \omega) = 0$ for all sufficiently small values of $t$. The continuous parameter may be replaced by a sequence $t_n \downarrow 0$.

Finally, we mention another kind of convergence which is basic in functional analysis, but confine ourselves to $L^1$. The sequence of r.v.'s $\{X_n\}$ in $L^1$ is said to converge *weakly in* $L^1$ to $X$ iff for each bounded r.v. $Y$ we have

$$\lim_{n\to\infty} \mathscr{E}(X_n Y) = \mathscr{E}(XY), \text{ finite.}$$

It is easy to see that $X \in L^1$ and is unique up to equivalence by taking $Y = 1_{\{X \neq X'\}}$ if $X'$ is another candidate. Clearly convergence in $L^1$ defined above implies weak convergence; hence the former is sometimes referred to as "strong". On the other hand, Example 2 above shows that convergence a.e. does not imply weak convergence; whereas Exercises 19 and 20 below show that weak convergence does not imply convergence in pr. (nor even in distribution; see Sec. 4.4).

## EXERCISES

**1.** $X_n \to +\infty$ a.e. if and only if $\forall M > 0$: $\mathscr{P}\{X_n < M \text{ i.o.}\} = 0$.

**2.** If $0 \le X_n$, $X_n \le X \in L^1$ and $X_n \to X$ in pr., then $X_n \to X$ in $L^1$.

**3.** If $X_n \to X$, $Y_n \to Y$ both in pr., then $X_n \pm Y_n \to X \pm Y$, $X_n Y_n \to XY$, all in pr.

*__4.__ Let $f$ be a bounded uniformly continuous function in $\mathscr{R}^1$. Then $X_n \to 0$ in pr. implies $\mathscr{E}\{f(X_n)\} \to f(0)$. [Example:

$$f(x) = \frac{|X|}{1 + |X|}$$

as in Theorem 4.1.5.]

**5.** Convergence in $L^p$ implies that in $L^r$ for $r < p$.

**6.** If $X_n \to X$, $Y_n \to Y$, both in $L^p$, then $X_n \pm Y_n \to X \pm Y$ in $L^p$. If $X_n \to X$ in $L^p$ and $Y_n \to Y$ in $L^q$, where $p > 1$ and $1/p + 1/q = 1$, then $X_n Y_n \to XY$ in $L^1$.

**7.** If $X_n \to X$ in pr. and $X_n \to Y$ in pr., then $X = Y$ a.e.

**8.** If $X_n \to X$ a.e. and $\mu_n$ and $\mu$ are the p.m.'s of $X_n$ and $X$, it does not follow that $\mu_n(P) \to \mu(P)$ even for all intervals $P$.

**9.** Give an example in which $\mathscr{E}(X_n) \to 0$ but there does not exist any subsequence $\{n_k\} \to \infty$ such that $X_{n_k} \to 0$ in pr.

*__10.__ Let $f$ be a continuous function on $\mathscr{R}^1$. If $X_n \to X$ in pr., then $f(X_n) \to f(X)$ in pr. The result is false if $f$ is merely Borel measurable. [HINT: Truncate $f$ at $\pm A$ for large $A$.]

**11.** The extended-valued r.v. $X$ is said to be *bounded in pr.* iff for each $\epsilon > 0$, there exists a finite $M(\epsilon)$ such that $\mathscr{P}\{|X| \le M(\epsilon)\} \ge 1 - \epsilon$. Prove that $X$ is bounded in pr. if and only if it is finite a.e.

**12.** The sequence of extended-valued r.v. $\{X_n\}$ is said to be bounded in pr. iff $\sup_n |X_n|$ is bounded in pr.; $\{X_n\}$ is said to *diverge to $+\infty$ in pr.* iff for each $M > 0$ and $\epsilon > 0$ there exists a finite $n_0(M, \epsilon)$ such that if $n > n_0$, then $\mathscr{P}\{|X_n| > M\} > 1 - \epsilon$. Prove that if $\{X_n\}$ diverges to $+\infty$ in pr. and $\{Y_n\}$ is bounded in pr., then $\{X_n + Y_n\}$ diverges to $+\infty$ in pr.

**13.** If $\sup_n X_n = +\infty$ a.e., there need exist no subsequence $\{X_{n_k}\}$ that diverges to $+\infty$ in pr.

**14.** It is possible that for each $\omega$, $\overline{\lim}_n X_n(\omega) = +\infty$, but there does not exist a subsequence $\{n_k\}$ and a set $\Delta$ of positive probability such that $\lim_k X_{n_k}(\omega) = +\infty$ on $\Delta$. [HINT: On $(\mathscr{U}, \mathscr{B})$ define $X_n(\omega)$ according to the $n$th digit of $\omega$.]

***15.** Instead of the $\rho$ in Theorem 4.1.5 one may define other metrics as follows. Let $\rho_1(X, Y)$ be the infimum of all $\epsilon > 0$ such that

$$\mathscr{P}(|X - Y| > \epsilon) \leq \epsilon.$$

Let $\rho_2(X, Y)$ be the infimum of $\mathscr{P}\{|X - Y| > \epsilon\} + \epsilon$ over all $\epsilon > 0$. Prove that these are metrics and that convergence in pr. is equivalent to convergence according to either metric.

***16.** Convergence in pr. for arbitrary r.v.'s may be reduced to that of bounded r.v.'s by the transformation

$$X' = \arctan X.$$

In a space of uniformly bounded r.v.'s, convergence in pr. is equivalent to that in the metric $\rho_0(X, Y) = \mathscr{E}(|X - Y|)$; this reduces to the definition given in Exercise 8 of Sec. 3.2 when $X$ and $Y$ are indicators.

**17.** Unlike convergence in pr., convergence a.e. is not expressible by means of metric. [HINT: Metric convergence has the property that if $\rho(x_n, x) \nrightarrow 0$, then there exist $\epsilon > 0$ and $\{n_k\}$ such that $\rho(x_{n_k}, x) \geq \epsilon$ for every $k$.]

**18.** If $X_n \downarrow X$ a.s., each $X_n$ is integrable and $\inf_n \mathscr{E}(X_n) > -\infty$, then $X_n \to X$ in $L^1$.

**19.** Let $f_n(x) = 1 + \cos 2\pi nx$, $f(x) = 1$ in $[0, 1]$. Then for each $g \in L^1$ $[0, 1]$ we have

$$\int_0^1 f_n g\, dx \to \int_0^1 f g\, dx,$$

but $f_n$ does not converge to $f$ in measure. [HINT: This is just the Riemann–Lebesgue lemma in Fourier series, but we have made $f_n \geq 0$ to stress a point.]

**20.** Let $\{X_n\}$ be a sequence of independent r.v.'s with zero mean and unit variance. Prove that for any bounded r.v. $Y$ we have $\lim_{n\to\infty} \mathscr{E}(X_n Y) = 0$. [HINT: Consider $\mathscr{E}\{[Y - \sum_{k=1}^n \mathscr{E}(X_k Y)X_k]^2\}$ to get Bessel's inequality $\mathscr{E}(Y^2) \geq \sum_{k=1}^n \mathscr{E}(X_k Y)^2$. The stated result extends to case where the $X_n$'s are assumed only to be uniformly integrable (see Sec. 4.5) but now we must approximate $Y$ by a function of $(X_1, \ldots, X_m)$, cf. Theorem 8.1.1.]

## 4.2 Almost sure convergence; Borel–Cantelli lemma

An important concept in set theory is that of the "lim sup" and "lim inf" of a sequence of sets. These notions can be defined for subsets of an arbitrary space $\Omega$.

DEFINITION. Let $E_n$ be any sequence of subsets of $\Omega$; we define

$$\limsup_n E_n = \bigcap_{m=1}^{\infty} \bigcup_{n=m}^{\infty} E_n, \qquad \liminf_n E_n = \bigcup_{m=1}^{\infty} \bigcap_{n=m}^{\infty} E_n.$$

Let us observe at once that

$$\liminf_n E_n = (\limsup_n E_n^c)^c, \tag{1}$$

so that in a sense one of the two notions suffices, but it is convenient to employ both. The main properties of these sets will be given in the following two propositions.

**(i)** A point belongs to $\limsup_n E_n$ if and only if it belongs to infinitely many terms of the sequence $\{E_n, n \geq 1\}$. A point belongs to $\liminf_n E_n$ if and only if it belongs to all terms of the sequence from a certain term on. It follows in particular that both sets are independent of the enumeration of the $E_n$'s.

PROOF. We shall prove the first assertion. Since a point belongs to infinitely many of the $E_n$'s if and only if it does not belong to all $E_n^c$ from a certain value of $n$ on, the second assertion will follow from (1). Now if $\omega$ belongs to infinitely many of the $E_n$'s, then it belongs to

$$F_m = \bigcup_{n=m}^{\infty} E_n \qquad \text{for every } m;$$

hence it belongs to

$$\bigcap_{m=1}^{\infty} F_m = \limsup_n E_n.$$

Conversely, if $\omega$ belongs to $\bigcap_{m=1}^{\infty} F_m$, then $\omega \in F_m$ for every $m$. Were $\omega$ to belong to only a finite number of the $E_n$'s there would be an $m$ such that $\omega \notin E_n$ for $n \geq m$, so that

$$\omega \notin \bigcup_{n=m}^{\infty} E_n = F_m.$$

This contradiction proves that $\omega$ must belong to an infinite number of the $E_n$'s.

In more intuitive language: *the event* $\limsup_n E_n$ *occurs if and only if the events* $E_n$ *occur infinitely often*. Thus we may write

$$\mathscr{P}(\limsup_n E_n) = \mathscr{P}(E_n \text{ i.o.})$$

where the abbreviation "i.o." stands for "infinitely often". The advantage of such a notation is better shown if we consider, for example, the events "$|X_n| \geq \epsilon$" and the probability $\mathscr{P}\{|X_n| \geq \epsilon \text{ i.o.}\}$; see Theorem 4.2.3 below.

**(ii)** If each $E_n \in \mathscr{F}$, then we have

$$\mathscr{P}(\limsup_n E_n) = \lim_{m\to\infty} \mathscr{P}\left(\bigcup_{n=m}^{\infty} E_n\right); \tag{2}$$

$$\mathscr{P}(\liminf_n E_n) = \lim_{m\to\infty} \mathscr{P}\left(\bigcap_{n=m}^{\infty} E_n\right). \tag{3}$$

PROOF. Using the notation in the preceding proof, it is clear that $F_m$ decreases as $m$ increases. Hence by the monotone property of p.m.'s:

$$\mathscr{P}\left(\bigcap_{m=1}^{\infty} F_m\right) = \lim_{m\to\infty} \mathscr{P}(F_m),$$

which is (2); (3) is proved similarly or via (1).

**Theorem 4.2.1.** We have for arbitrary events $\{E_n\}$:

$$\sum_n \mathscr{P}(E_n) < \infty \Rightarrow \mathscr{P}(E_n \text{ i.o.}) = 0. \tag{4}$$

PROOF. By Boole's inequality for p.m.'s, we have

$$\mathscr{P}(F_m) \leq \sum_{n=m}^{\infty} \mathscr{P}(E_n).$$

Hence the hypothesis in (4) implies that $\mathscr{P}(F_m) \to 0$, and the conclusion in (4) now follows by (2).

As an illustration of the convenience of the new notions, we may restate Theorem 4.1.1 as follows. The intuitive content of condition (5) below is the point being stressed here.

**Theorem 4.2.2.** $X_n \to 0$ a.e. if and only if

$$\forall \epsilon > 0: \mathscr{P}\{|X_n| > \epsilon \text{ i.o.}\} = 0. \tag{5}$$

PROOF. Using the notation $A_m = \bigcap_{n=m}^{\infty}\{|X_n| \leq \epsilon\}$ as in (3) of Sec. 4.1 (with $X = 0$), we have

$$\{|X_n| > \epsilon \text{ i.o.}\} = \bigcap_{m=1}^{\infty} \bigcup_{n=m}^{\infty} \{|X_n| > \epsilon\} = \bigcap_{m=1}^{\infty} A_m^c.$$

According to Theorem 4.1.1, $X_n \to 0$ a.e. if and only if for each $\epsilon > 0$, $\mathscr{P}(A_m^c) \to 0$ as $m \to \infty$; since $A_m^c$ decreases as $m$ increases, this is equivalent to (5) as asserted.

**Theorem 4.2.3.** If $X_n \to X$ in pr., then there exists a sequence $\{n_k\}$ of integers increasing to infinity such that $X_{n_k} \to X$ a.e. Briefly stated: convergence in pr. implies convergence a.e. along a subsequence.

PROOF. We may suppose $X \equiv 0$ as explained before. Then the hypothesis may be written as

$$\forall k > 0: \qquad \lim_{n\to\infty} \mathscr{P}\left(|X_n| > \frac{1}{2^k}\right) = 0.$$

It follows that for each $k$ we can find $n_k$ such that

$$\mathscr{P}\left(|X_{n_k}| > \frac{1}{2^k}\right) \le \frac{1}{2^k}$$

and consequently

$$\sum_k \mathscr{P}\left(|X_{n_k}| > \frac{1}{2^k}\right) \le \sum_k \frac{1}{2^k} < \infty.$$

Having so chosen $\{n_k\}$, we let $E_k$ be the event "$|X_{n_k}| > 1/2^k$". Then we have by (4):

$$\mathscr{P}\left(|X_{n_k}| > \frac{1}{2^k} \text{ i.o.}\right) = 0.$$

[Note: Here the index involved in "i.o." is $k$; such ambiguity being harmless and expedient.] This implies $X_{n_k} \to 0$ a.e. (why?) finishing the proof.

## EXERCISES

**1.** Prove that

$$\mathscr{P}(\limsup_n E_n) \ge \overline{\lim_n} \mathscr{P}(E_n),$$

$$\mathscr{P}(\liminf_n E_n) \le \underline{\lim_n} \mathscr{P}(E_n).$$

**2.** Let $\{B_n\}$ be a countable collection of Borel sets in $\mathscr{U}$. If there exists a $\delta > 0$ such that $m(B_n) \ge \delta$ for every $n$, then there is at least one point in $\mathscr{U}$ that belongs to infinitely many $B_n$'s.

**3.** If $\{X_n\}$ converges to a finite limit a.e., then for any $\epsilon$ there exists $M(\epsilon) < \infty$ such that $\mathscr{P}\{\sup |X_n| \le M(\epsilon)\} \ge 1 - \epsilon$.

*4. For any sequence of r.v.'s $\{X_n\}$ there exists a sequence of constants $\{A_n\}$ such that $X_n/A_n \to 0$ a.e.

*5. If $\{X_n\}$ converges on the set $C$, then for any $\epsilon > 0$, there exists $C_0 \subset C$ with $\mathscr{P}(C \backslash C_0) < \epsilon$ such that $X_n$ converges uniformly in $C_0$. [This is Egorov's theorem. We may suppose $C = \Omega$ and the limit to be zero. Let $F_{mk} = \bigcap_{n=k}^{\infty} \{\omega: | X_n(\omega) \le 1/m\}$; then $\forall m$, $\exists k(m)$ such that

$$\mathscr{P}(F_{m,k(m)}) > 1 - \epsilon/2^m.$$

Take $C_0 = \bigcap_{m=1}^{\infty} F_{m,k(m)}$.]

**6.** Cauchy convergence of $\{X_n\}$ in pr. (or in $L^p$) implies the existence of an $X$ (finite a.e.), such that $X_n$ converges to $X$ in pr. (or in $L^p$). [HINT: Choose $n_k$ so that

$$\sum_k \mathscr{P}\left\{|X_{n_{k+1}} - X_{n_k}| > \frac{1}{2^k}\right\} < \infty;$$

cf. Theorem 4.2.3.]

*7. $\{X_n\}$ converges in pr. to $X$ if and only if every subsequence $\{X_{n_k}\}$ contains a further subsequence that converges a.e. to $X$. [HINT: Use Theorem 4.1.5.]

**8.** Let $\{X_n, n \ge 1\}$ be any sequence of functions on $\Omega$ to $\mathscr{R}^1$ and let $C$ denote the set of $\omega$ for which the numerical sequence $\{X_n(\omega), n \ge 1\}$ converges. Show that

$$C = \bigcap_{m=1}^{\infty} \bigcup_{n=1}^{\infty} \bigcap_{n'=n+1}^{\infty} \left\{|X_n - X_{n'}| \le \frac{1}{m}\right\} = \bigcap_{m=1}^{\infty} \bigcup_{n=1}^{\infty} \bigcap_{n'=n+1}^{\infty} \bigwedge(m, n, n')$$

where

$$\bigwedge(m, n, n') = \left\{\omega: \max_{n<j<k\le n'} |X_j(\omega) - X_k(\omega)| \le \frac{1}{m}\right\}.$$

Hence if the $X_n$'s are r.v.'s, we have

$$\mathscr{P}(C) = \lim_{m\to\infty} \lim_{n\to\infty} \lim_{n'\to\infty} \mathscr{P}\left(\bigwedge(m, n, n')\right).$$

**9.** As in Exercise 8 show that

$$\{\omega: \lim_{n\to\infty} X_n(\omega) = 0\} = \bigcap_{m=1}^{\infty} \bigcup_{k=1}^{\infty} \bigcap_{n=k}^{\infty} \left\{|X_n| \le \frac{1}{m}\right\}.$$

**10.** Suppose that for $a < b$, we have

$$\mathscr{P}\{X_n < a \text{ i.o.} \quad \text{and} \quad X_n > b \text{ i.o.}\} = 0;$$

then $\lim_{n\to\infty} X_n$ exists a.e. but it may be infinite. [HINT: Consider all pairs of rational numbers $(a, b)$ and take a union over them.]

Under the assumption of independence, Theorem 4.2.1 has a striking complement.

**Theorem 4.2.4.** If the events $\{E_n\}$ are independent, then

$$\sum_n \mathscr{P}(E_n) = \infty \Rightarrow \mathscr{P}(E_n \text{ i.o.}) = 1. \tag{6}$$

PROOF. By (3) we have

$$\mathscr{P}\{\liminf_n E_n^c\} = \lim_{m\to\infty} \mathscr{P}\left(\bigcap_{n=m}^{\infty} E_n^c\right). \tag{7}$$

The events $\{E_n^c\}$ are independent as well as $\{E_n\}$, by Exercise 3 of Sec. 3.3; hence we have if $m' > m$:

$$\mathscr{P}\left(\bigcap_{n=m}^{m'} E_n^c\right) = \prod_{n=m}^{m'} \mathscr{P}(E_n^c) = \prod_{n=m}^{m'} (1 - \mathscr{P}(E_n)).$$

Now for any $x \geq 0$, we have $1 - x \leq e^{-x}$; it follows that the last term above does not exceed

$$\prod_{n=m}^{m'} e^{-\mathscr{P}(E_n)} = \exp\left(-\sum_{n=m}^{m'} \mathscr{P}(E_n)\right).$$

Letting $m' \to \infty$, the right member above $\to 0$, since the series in the exponent $\to +\infty$ by hypothesis. It follows by monotone property that

$$\mathscr{P}\left(\bigcap_{n=m}^{\infty} E_n^c\right) = \lim_{m'\to\infty} \mathscr{P}\left(\bigcap_{n=m}^{m'} E_n^c\right) = 0.$$

Thus the right member in (7) is equal to 0, and consequently so is the left member in (7). This is equivalent to $\mathscr{P}(E_n \text{ i.o.}) = 1$ by (1).

Theorems 4.2.1 and 4.2.4 together will be referred to as *the Borel–Cantelli lemma*, the former the "convergence part" and the latter "the divergence part". The first is more useful since the events there may be completely arbitrary. The second has an extension to pairwise independent r.v.'s; although the result is of some interest, it is the method of proof to be given below that is more important. It is a useful technique in probability theory.

**Theorem 4.2.5.** The implication (6) remains true if the events $\{E_n\}$ are pairwise independent.

PROOF. Let $I_n$ denote the indicator of $E_n$, so that our present hypothesis becomes

$$\forall m \neq n: \qquad \mathscr{E}(I_m I_n) = \mathscr{E}(I_m)\mathscr{E}(I_n). \tag{8}$$

Consider the series of r.v.'s: $\sum_{n=1}^{\infty} I_n(\omega)$. It diverges to $+\infty$ if and only if an infinite number of its terms are equal to one, namely if $\omega$ belongs to an infinite number of the $E_n$'s. Hence the conclusion in (6) is equivalent to

$$\mathscr{P}\left\{\sum_{n=1}^{\infty} I_n = +\infty\right\} = 1. \tag{9}$$

What has been said so far is true for arbitrary $E_n$'s. Now the hypothesis in (6) may be written as

$$\sum_{n=1}^{\infty} \mathscr{E}(I_n) = +\infty.$$

Consider the partial sum $J_k = \sum_{n=1}^{k} I_n$. Using Chebyshev's inequality, we have for every $A > 0$:

$$\mathscr{P}\{|J_k - \mathscr{E}(J_k)| \leq A\sigma(J_k)\} \geq 1 - \frac{\sigma^2(J_k)}{A^2\sigma^2(J_k)} = 1 - \frac{1}{A^2}, \tag{10}$$

where $\sigma^2(J)$ denotes the variance of $J$. Writing

$$p_n = \mathscr{E}(I_n) = \mathscr{P}(E_n),$$

we may calculate $\sigma^2(J_k)$ by using (8), as follows:

$$\begin{aligned}
\mathscr{E}(J_k^2) &= \mathscr{E}\left(\sum_{n=1}^{k} I_n^2 + 2\sum_{1\leq m<n\leq k} I_m I_n\right) \\
&= \sum_{n=1}^{k} \mathscr{E}(I_n^2) + 2\sum_{1\leq m<n\leq k} \mathscr{E}(I_m)\mathscr{E}(I_n) \\
&= \sum_{n=1}^{k} \mathscr{E}(I_n)^2 + 2\sum_{1\leq m<n\leq k} \mathscr{E}(I_m)\mathscr{E}(I_n) + \sum_{n=1}^{k}\{\mathscr{E}(I_n) - \mathscr{E}(I_n)^2\} \\
&= \left(\sum_{n=1}^{k} p_n\right)^2 + \sum_{n=1}^{k}(p_n - p_n^2).
\end{aligned}$$

Hence

$$\sigma^2(J_k) = \mathscr{E}(J_k^2) - \mathscr{E}(J_k)^2 = \sum_{n=1}^{k}(p_n - p_n^2) \quad \left[= \sum_{n=1}^{k}\sigma^2(I_n)\right].$$

This calculation will turn out to be a particular case of a simple basic formula; see (6) of Sec. 5.1. Since $\sum_{n=1}^{k} p_n = \mathscr{E}(J_k) \to \infty$, it follows that

$$\sigma(J_k) \leq \mathscr{E}(J_k)^{1/2} = o(\mathscr{E}(J_k))$$

in the classical "$o$, $O$" notation of analysis. Hence if $k > k_0(A)$, (10) implies

$$\mathscr{P}\left\{J_k > \frac{1}{2}\mathscr{E}(J_k)\right\} \geq 1 - \frac{1}{A^2}$$

(where $\frac{1}{2}$ may be replaced by any constant $< 1$). Since $J_k$ increases with $k$ the inequality above holds *a fortiori* when the first $J_k$ there is replaced by $\lim_{k\to\infty} J_k$; after that we can let $k \to \infty$ in $\mathscr{E}(J_k)$ to obtain

$$\mathscr{P}\{\lim_{k\to\infty} J_k = +\infty\} \geq 1 - \frac{1}{A^2}.$$

Since the left member does not depend on $A$, and $A$ is arbitrary, this implies that $\lim_{k\to\infty} J_k = +\infty$ a.e., namely (9).

**Corollary.** If the events $\{E_n\}$ are pairwise independent, then

$$\mathscr{P}(\limsup_n E_n) = 0 \text{ or } 1$$

according as $\sum_n \mathscr{P}(E_n) < \infty$ or $= \infty$.

This is an example of a "zero-or-one" law to be discussed in Chapter 8, though it is not included in any of the general results there.

EXERCISES

Below $X_n$, $Y_n$ are r.v.'s, $E_n$ events.

**11.** Give a trivial example of dependent $\{E_n\}$ satisfying the hypothesis but not the conclusion of (6); give a less trivial example satisfying the hypothesis but with $\mathscr{P}(\limsup_n E_n) = 0$. [HINT: Let $E_n$ be the event that a real number in [0, 1] has its $n$-ary expansion begin with 0.]

***12.** Prove that the probability of convergence of a sequence of independent r.v.'s is equal to zero or one.

**13.** If $\{X_n\}$ is a sequence of independent and identically distributed r.v.'s not constant a.e., then $\mathscr{P}\{X_n \text{ converges}\} = 0$.

$^\star$**14.** If $\{X_n\}$ is a sequence of independent r.v.'s with d.f.'s $\{F_n\}$, then $\mathscr{P}\{\lim_n X_n = 0\} = 1$ if and only if $\forall_\epsilon > 0$: $\sum_n \{1 - F_n(\epsilon) + F_n(-\epsilon)\} < \infty$.

**15.** If $\sum_n \mathscr{P}(|X_n| > n) < \infty$, then

$$\limsup_n \frac{|X_n|}{n} \leq 1 \text{ a.e.}$$

$^\star$**16.** Strengthen Theorem 4.2.4 by proving that

$$\lim_{n\to\infty} \frac{J_n}{\mathscr{E}(J_n)} = 1 \text{ a.e.}$$

[HINT: Take a subsequence $\{k_n\}$ such that $\mathscr{E}(J_{n_k}) \sim k^2$; prove the result first for this subsequence by estimating $\mathscr{P}\{|J_k - \mathscr{E}(J_k)| > \delta\mathscr{E}(J_k)\}$; the general case follows because if $n_k \leq n < n_{k+1}$,

$$J_{n_k}/\mathscr{E}(J_{n_{k+1}}) \leq J_n/\mathscr{E}(J_n) \leq J_{n_{k+1}}/\mathscr{E}(J_{n_k}).$$

**17.** If $\mathscr{E}(X_n) = 1$ and $\mathscr{E}(X_n^2)$ is bounded in $n$, then

$$\mathscr{P}\{\overline{\lim_{n\to\infty}} X_n \geq 1\} > 0.$$

[This is due to Kochen and Stone. Truncate $X_n$ at $A$ to $Y_n$ with $A$ so large that $\mathscr{E}(Y_n) > 1 - \epsilon$ for all $n$; then apply Exercise 6 of Sec. 3.2.]

$^\star$**18.** Let $\{E_n\}$ be events and $\{I_n\}$ their indicators. Prove the inequality

$$\mathscr{P}\left(\bigcup_{k=1}^n E_k\right) \geq \left\{\mathscr{E}\left(\sum_{k=1}^n I_k\right)\right\}^2 \Big/ \mathscr{E}\left\{\left(\sum_{k=1}^n I_k\right)^2\right\}. \tag{11}$$

Deduce from this that if (i) $\sum_n \mathscr{P}(E_n) = \infty$ and (ii) there exists $c > 0$ such that we have

$$\forall m < n: \mathscr{P}(E_m E_n) \leq c\mathscr{P}(E_m)\mathscr{P}(E_{n-m});$$

then

$$\mathscr{P}\{\limsup_n E_n\} > 0.$$

**19.** If $\sum_n \mathscr{P}(E_n) = \infty$ and

$$\varliminf_n \left\{\sum_{j=1}^n \sum_{k=1}^n \mathscr{P}(E_j E_k)\right\} \Big/ \left\{\sum_{k=1}^n \mathscr{P}(E_k)\right\}^2 = 1,$$

then $\mathscr{P}\{\limsup_n E_n\} = 1$. [HINT: Use (11) above.]

**20.** Let $\{E_n\}$ be arbitrary events satisfying

$$\text{(i)} \lim_n \mathscr{P}(E_n) = 0, \qquad \text{(ii)} \sum_n \mathscr{P}(E_n E_{n+1}^c) < \infty;$$

then $\mathscr{P}\{\limsup_n E_n\} = 0$. [This is due to Barndorff–Nielsen.]

## 4.3 Vague convergence

If a sequence of r.v.'s $\{X_n\}$ tends to a limit, the corresponding sequence of p.m.'s $\{\mu_n\}$ ought to tend to a limit in some sense. Is it true that $\lim_n \mu_n(A)$ exists for all $A \in \mathscr{B}^1$ or at least for all intervals $A$? The answer is no from trivial examples.

**Example 1.** Let $X_n = c_n$ where the $c_n$'s are constants tending to zero. Then $X_n \to 0$ deterministically. For any interval $I$ such that $0 \notin \bar{I}$, where $\bar{I}$ is the closure of $I$, we have $\lim_n \mu_n(I) = 0 = \mu(I)$; for any interval such that $0 \in I^\circ$, where $I^\circ$ is the interior of $I$, we have $\lim_n \mu_n(I) = 1 = \mu(I)$. But if $\{c_n\}$ oscillates between strictly positive and strictly negative values, and $I = (a, 0)$ or $(0, b)$, where $a < 0 < b$, then $\mu_n(I)$ oscillates between 0 and 1, while $\mu(I) = 0$. On the other hand, if $I = (a, 0]$ or $[0, b)$, then $\mu_n(I)$ oscillates as before but $\mu(I) = 1$. Observe that $\{0\}$ is the sole atom of $\mu$ and it is the root of the trouble.

Instead of the point masses concentrated at $c_n$, we may consider, e.g., r.v.'s $\{X_n\}$ having uniform distributions over intervals $(c_n, c'_n)$ where $c_n < 0 < c'_n$ and $c_n \to 0$, $c'_n \to 0$. Then again $X_n \to 0$ a.e. but $\mu_n((a, 0))$ may not converge at all, or converge to any number between 0 and 1.

Next, even if $\{\mu_n\}$ does converge in some weaker sense, is the limit necessarily a p.m.? The answer is again no.

**Example 2.** Let $X_n = c_n$ where $c_n \to +\infty$. Then $X_n \to +\infty$ deterministically. According to our definition of a r.v., the constant $+\infty$ indeed qualifies. But for any finite interval $(a, b)$ we have $\lim_n \mu_n((a, b)) = 0$, so any limiting measure must also be identically zero. This example can be easily ramified; e.g. let $a_n \to -\infty$, $b_n \to +\infty$ and

$$X_n = \begin{cases} a_n & \text{with probability } \alpha, \\ 0 & \text{with probability } 1 - \alpha - \beta, \\ b_n & \text{with probability } \beta. \end{cases}$$

Then $X_n \to X$ where

$$X = \begin{cases} +\infty & \text{with probability } \alpha, \\ 0 & \text{with probability } 1 - \alpha - \beta, \\ -\infty & \text{with probability } \beta. \end{cases}$$

For any finite interval $(a, b)$ containing 0 we have

$$\lim_n \mu_n((a, b)) = \lim_n \mu_n(\{0\}) = 1 - \alpha - \beta.$$

In this situation it is said that masses of amount $\alpha$ and $\beta$ "have wandered off to $+\infty$ and $-\infty$ respectively." The remedy here is obvious: we should consider measures on the extended line $\mathscr{R}^* = [-\infty, +\infty]$, with possible atoms at $\{+\infty\}$ and $\{-\infty\}$. We leave this to the reader but proceed to give the appropriate definitions which take into account the two kinds of troubles discussed above.

DEFINITION. A measure $\mu$ on $(\mathscr{R}^1, \mathscr{B}^1)$ with $\mu(\mathscr{R}^1) \leq 1$ will be called a subprobability measure (s.p.m.).

DEFINITION OF VAGUE CONVERGENCE. A sequence $\{\mu_n, n \geq 1\}$ of s.p.m.'s is said to *converge vaguely* to an s.p.m. $\mu$ iff there exists a dense subset $D$ of $\mathscr{R}^1$ such that

$$\forall a \in D, b \in D, a < b: \qquad \mu_n((a, b]) \to \mu((a, b]). \tag{1}$$

This will be denoted by

$$\mu_n \overset{\to}{v} \mu \tag{2}$$

and $\mu$ is called the *vague limit* of $\{\mu_n\}$. We shall see that it is unique below.

For brevity's sake, we will write $\mu((a, b])$ as $\mu(a, b]$ below, and similarly for other kinds of intervals. An interval $(a, b)$ is called a *continuity interval* of $\mu$ iff neither $a$ nor $b$ is an atom of $\mu$; in other words iff $\mu(a, b) = \mu[a, b]$. As a notational convention, $\mu(a, b) = 0$ when $a > b$.

**Theorem 4.3.1.** Let $\{\mu_n\}$ and $\mu$ be s.p.m.'s. The following propositions are equivalent.

**(i)** For every finite interval $(a, b)$ and $\epsilon > 0$, there exists an $n_0(a, b, \epsilon)$ such that if $n \geq n_0$, then

$$\mu(a + \epsilon, b - \epsilon) - \epsilon \leq \mu_n(a, b) \leq \mu(a - \epsilon, b + \epsilon) + \epsilon. \tag{3}$$

Here and hereafter the first term is interpreted as 0 if $a + \epsilon > b - \epsilon$.

**(ii)** For every continuity interval $(a, b]$ of $\mu$, we have

$$\mu_n(a, b] \to \mu(a, b].$$

**(iii)** $\mu_n \overset{v}{\to} \mu$.

PROOF. To prove that (i) $\Rightarrow$ (ii), let $(a, b)$ be a continuity interval of $\mu$. It follows from the monotone property of a measure that

$$\lim_{\epsilon \downarrow 0} \mu(a + \epsilon, b - \epsilon) = \mu(a, b) = \mu[a, b] = \lim_{\epsilon \downarrow 0} \mu(a - \epsilon, b + \epsilon).$$

Letting $n \to \infty$, then $\epsilon \downarrow 0$ in (3), we have

$$\mu(a, b) \leq \underline{\lim}_n \mu_n(a, b) \leq \overline{\lim}_n \mu_n[a, b] \leq \mu[a, b] = \mu(a, b),$$

which proves (ii); indeed $(a, b]$ may be replaced by $(a, b)$ or $[a, b)$ or $[a, b]$ there. Next, since the set of atoms of $\mu$ is countable, the complementary set $D$ is certainly dense in $\mathscr{R}^1$. If $a \in D$, $b \in D$, then $(a, b)$ is a continuity interval of $\mu$. This proves (ii) $\Rightarrow$ (iii). Finally, suppose (iii) is true so that (1) holds. Given any $(a, b)$ and $\epsilon > 0$, there exist $a_1$, $a_2$, $b_1$, $b_2$ all in $D$ satisfying

$$a - \epsilon < a_1 < a < a_2 < a + \epsilon, \qquad b - \epsilon < b_1 < b < b_2 < b + \epsilon.$$

By (1), there exists $n_0$ such that if $n \geq n_0$, then

$$|\mu_n(a_i, b_j] - \mu(a_i, b_j]| \leq \epsilon$$

for $i = 1, 2$ and $j = 1, 2$. It follows that

$$\mu(a + \epsilon, b - \epsilon) - \epsilon \leq \mu(a_2, b_1] - \epsilon \leq \mu_n(a_2, b_1] \leq \mu_n(a, b) \leq \mu_n(a_1, b_2]$$
$$\leq \mu(a_1, b_2] + \epsilon \leq \mu(a - \epsilon, b + \epsilon) + \epsilon.$$

Thus (iii) $\Rightarrow$ (i). The theorem is proved.

As an immediate consequence, the *vague limit is unique*. More precisely, if besides (1) we have also

$$\forall a \in D', b \in D', a < b: \mu_n(a, b] \to \mu'(a, b],$$

then $\mu \equiv \mu'$. For let $A$ be the set of atoms of $\mu$ and of $\mu'$; then if $a \in A^c$, $b \in A^c$, we have by Theorem 4.3.1, (ii):

$$\mu(a, b] \leftarrow \mu_n(a, b] \to \mu'(a, b]$$

so that $\mu(a, b] = \mu'(a, b]$. Now $A^c$ is dense in $\mathscr{R}^1$, hence the two measures $\mu$ and $\mu'$ coincide on a set of intervals that is dense in $\mathscr{R}^1$ and must therefore be identical (see Corollary to Theorem 2.2.3).

Another consequence is: if $\mu_n \xrightarrow{v} \mu$ and $(a, b)$ is a continuity interval of $\mu$, then $\mu_n(I) \to \mu(I)$, where $I$ is any of the four kinds of intervals with endpoints $a$, $b$. For it is clear that we may replace $\mu_n(a, b)$ by $\mu_n[a, b]$ in (3). In particular, we have $\mu_n(\{a\}) \to 0$, $\mu_n(\{b\}) \to 0$.

The case of strict probability measures will now be treated.

**Theorem 4.3.2.** Let $\{\mu_n\}$ and $\mu$ be p.m.'s. Then (i), (ii), and (iii) in the preceding theorem are equivalent to the following "uniform" strengthening of (i).

**(i′)** For any $\delta > 0$ and $\epsilon > 0$, there exists $n_0(\delta, \epsilon)$ such that if $n \geq n_0$ then we have for *every* interval $(a, b)$, possibly infinite:

$$\mu(a + \delta, b - \delta) - \epsilon \leq \mu_n(a, b) \leq \mu(a - \delta, b + \delta) + \epsilon. \tag{4}$$

PROOF. The employment of two numbers $\delta$ and $\epsilon$ instead of one $\epsilon$ as in (3) is an apparent extension only (why?). Since (i′) $\Rightarrow$ (i), the preceding theorem yields at once (i′) $\Rightarrow$ (ii) $\Leftrightarrow$ (iii). It remains to prove (ii) $\Rightarrow$ (i′) when the $\mu_n$'s and $\mu$ are p.m.'s. Let $A$ denote the set of atoms of $\mu$. Then there exist an integer $\ell$ and $a_j \in A^c$, $1 \leq j \leq \ell$, satisfying:

$$a_j < a_{j+1} \leq a_j + \delta, \qquad 1 \leq j \leq \ell - 1;$$

and

$$\mu((a_1, a_\ell)^c) < \frac{\epsilon}{4}. \tag{5}$$

By (ii), there exist $n_0$ depending on $\epsilon$ and $\ell$ (and so on $\epsilon$ and $\delta$) such that if $n \geq n_0$ then

$$\sup_{1 \leq j \leq \ell - 1} |\mu(a_j, a_{j+1}] - \mu_n(a_j, a_{j+1}]| \leq \frac{\epsilon}{4\ell}. \tag{6}$$

It follows by additivity that

$$|\mu(a_1, a_\ell] - \mu_n(a_1, a_\ell]| < \frac{\epsilon}{4}$$

and consequently by (5):

$$\mu_n((a_1, a_\ell)^c) < \frac{\epsilon}{2}. \tag{7}$$

From (5) and (7) we see that by ignoring the part of $(a, b)$ outside $(a_1, a_\ell)$, we commit at most an error $<\epsilon/2$ in either one of the inequalities in (4). Thus it is sufficient to prove (4) with $\delta$ and $\epsilon/2$, assuming that $(a, b) \subset (a_1, a_\ell)$. Let then $a_j \leq a < a_{j+1}$ and $a_k \leq b < a_{k+1}$, where $0 \leq j \leq k \leq \ell - 1$. The desired inequalities follow from (6), since when $n \geq n_0$ we have

$$\begin{aligned}\mu_n(a + \delta, b - \delta) - \frac{\epsilon}{4} &\leq \mu_n(a_{j+1}, a_k) - \frac{\epsilon}{4} \leq \mu(a_{j+1}, a_k) \leq \mu(a, b) \\ &\leq \mu(a_j, a_{k+1}) \leq \mu_n(a_j, a_{k+1}) + \frac{\epsilon}{4} \\ &\leq \mu_n(a - \delta, b + \delta) + \frac{\epsilon}{4}.\end{aligned}$$

The set of all s.p.m.'s on $\mathscr{R}^1$ bears close analogy to the set of all real numbers in [0, 1]. Recall that the latter is *sequentially compact*, which means: Given any sequence of numbers in the set, there is a subsequence which converges, and the limit is also a number in the set. This is the fundamental Bolzano–Weierstrass theorem. We have the following analogue which states: The set of all s.p.m.'s is sequentially compact with respect to vague convergence. It is often referred to as "Helly's extraction (or selection) principle".

**Theorem 4.3.3.** Given any sequence of s.p.m.'s, there is a subsequence that converges vaguely to an s.p.m.

PROOF. Here it is convenient to consider the *subdistribution function* (s.d.f.) $F_n$ defined as follows:

$$\forall x: F_n(x) = \mu_n(-\infty, x].$$

If $\mu_n$ is a p.m., then $F_n$ is just its d.f. (see Sec. 2.2); in general $F_n$ is increasing, right continuous with $F_n(-\infty) = 0$ and $F_n(+\infty) = \mu_n(\mathscr{R}^1) \leq 1$.

Let $D$ be a countable dense set of $\mathscr{R}^1$, and let $\{r_k, k \geq 1\}$ be an enumeration of it. The sequence of numbers $\{F_n(r_1), n \geq 1\}$ is bounded, hence by the Bolzano–Weierstrass theorem there is a subsequence $\{F_{1k}, k \geq 1\}$ of the given sequence such that the limit

$$\lim_{k\to\infty} F_{1k}(r_1) = \ell_1$$

exists; clearly $0 \leq \ell_1 \leq 1$. Next, the sequence of numbers $\{F_{1k}(r_2), k \geq 1\}$ is bounded, hence there is a subsequence $\{F_{2k}, k \geq 1\}$ of $\{F_{1k}, k \geq 1\}$ such that

$$\lim_{k\to\infty} F_{2k}(r_2) = \ell_2$$

where $0 \leq \ell_2 \leq 1$. Since $\{F_{2k}\}$ is a subsequence of $\{F_{1k}\}$, it converges also at $r_1$ to $\ell_1$. Continuing, we obtain

$$\begin{array}{ll}
F_{11}, F_{12}, \ldots, F_{1k}, \ldots & \text{converging at } r_1; \\
F_{21}, F_{22}, \ldots, F_{2k}, \ldots & \text{converging at } r_1, r_2; \\
\cdots\cdots\cdots\cdots\cdots & \\
F_{j1}, F_{j2}, \ldots, F_{jk}, \ldots & \text{converging at } r_1, r_2, \ldots, r_j; \\
\cdots\cdots\cdots\cdots\cdots &
\end{array}$$

Now consider the diagonal sequence $\{F_{kk}, k \geq 1\}$. We assert that it converges at every $r_j$, $j \geq 1$. To see this let $r_j$ be given. Apart from the first $j-1$ terms, the sequence $\{F_{kk}, k \geq 1\}$ is a subsequence of $\{F_{jk}, k \geq 1\}$, which converges at $r_j$ and hence $\lim_{k\to\infty} F_{kk}(r_j) = \ell_j$, as desired.

We have thus proved the existence of an infinite subsequence $\{n_k\}$ and a function $G$ defined and increasing on $D$ such that

$$\forall r \in D: \lim_{k\to\infty} F_{n_k}(r) = G(r).$$

From $G$ we define a function $F$ on $\mathscr{R}^1$ as follows:

$$\forall x \in \mathscr{R}^1: F(x) = \inf_{x<r\in D} G(r).$$

By Sec. 1.1, (vii), $F$ is increasing and right continuous. Let $C$ denote the set of its points of continuity; $C$ is dense in $\mathscr{R}^1$ and we show that

$$\forall x \in C: \lim_{k\to\infty} F_{n_k}(x) = F(x). \tag{8}$$

For, let $x \in C$ and $\epsilon > 0$ be given, there exist $r$, $r'$, and $r''$ in $D$ such that $r < r' < x < r''$ and $F(r'') - F(r) \leq \epsilon$. Then we have

$$F(r) \leq G(r') \leq F(x) \leq G(r'') \leq F(r'') \leq F(r) + \epsilon;$$

and

$$\uparrow \qquad\qquad \uparrow$$

$$F_{n_k}(r') \leq F_{n_k}(x) \leq F_{n_k}(r'').$$

From these (8) follows, since $\epsilon$ is arbitrary.

To $F$ corresponds a (unique) s.p.m. $\mu$ such that

$$F(x) - F(-\infty) = \mu(-\infty, x]$$

as in Theorem 2.2.2. Now the relation (8) yields, upon taking differences:

$$\forall a \in C, b \in C, a < b: \lim_{k\to\infty} \mu_{n_k}(a, b] = \mu(a, b].$$

Thus $\mu_{n_k} \xrightarrow{v} \mu$, and the theorem is proved.

We say that $F_n$ converges vaguely to $F$ and write $F_n \xrightarrow{v} F$ for $\mu_n \xrightarrow{v} \mu$ where $\mu_n$ and $\mu$ are the s.p.m.'s corresponding to the s.d.f.'s $F_n$ and $F$.

The reader should be able to confirm the truth of the following proposition about real numbers. Let $\{x_n\}$ be a sequence of real numbers such that every subsequence that tends to a limit ($\pm\infty$ allowed) has the same value for the limit; then the whole sequence tends to this limit. In particular a bounded sequence such that every convergent subsequence has the same limit is convergent to this limit.

The next theorem generalizes this result to vague convergence of s.p.m.'s. It is not contained in the preceding proposition but can be reduced to it if we use the properties of vague convergence; see also Exercise 9 below.

**Theorem 4.3.4.** If every vaguely convergent subsequence of the sequence of s.p.m.'s $\{\mu_n\}$ converges to the same $\mu$, then $\mu_n \xrightarrow{v} \mu$.

PROOF. To prove the theorem by contraposition, suppose $\mu_n$ does not converge vaguely to $\mu$. Then by Theorem 4.3.1, (ii), there exists a continuity interval $(a, b)$ of $\mu$ such that $\mu_n(a, b)$ does not converge to $\mu(a, b)$. By the Bolzano–Weierstrass theorem there exists a subsequence $\{n_k\}$ tending to infinity such that the numbers $\mu_{n_k}(a, b)$ converge to a limit, say $L \neq \mu(a, b)$. By Theorem 4.3.3, the sequence $\{\mu_{n_k}, k \geq 1\}$ contains a subsequence, say $\{\mu_{n'_k}, k \geq 1\}$, which converges vaguely, hence to $\mu$ by hypothesis of the theorem. Hence again by Theorem 4.3.1, (ii), we have

$$\mu_{n'_k}(a, b) \to \mu(a, b).$$

But the left side also $\to L$, which is a contradiction.

EXERCISES

**1.** Perhaps the most logical approach to vague convergence is as follows. The sequence $\{\mu_n, n \geq 1\}$ of s.p.m.'s is said to converge vaguely iff there exists a dense subset $D$ of $\mathscr{R}^1$ such that for every $a \in D$, $b \in D$, $a < b$, the sequence $\{\mu_n(a, b), n \geq 1\}$ converges. The definition given before implies this, of course, but prove the converse.

**2.** Prove that if (1) is true, then there exists a dense set $D'$, such that $\mu_n(I) \to \mu(I)$ where $I$ may be any of the four intervals $(a, b)$, $(a, b]$, $[a, b)$, $[a, b]$ with $a \in D'$, $b \in D'$.

**3.** Can a sequence of absolutely continuous p.m.'s converge vaguely to a discrete p.m.? Can a sequence of discrete p.m.'s converge vaguely to an absolutely continuous p.m.?

**4.** If a sequence of p.m.'s converges vaguely to an atomless p.m., then the convergence is uniform for all intervals, finite or infinite. (This is due to Pólya.)

**5.** Let $\{f_n\}$ be a sequence of functions increasing in $\mathscr{R}^1$ and uniformly bounded there: $\sup_{n,x} |f_n(x)| \leq M < \infty$. Prove that there exists an increasing function $f$ on $\mathscr{R}^1$ and a subsequence $\{n_k\}$ such that $f_{n_k}(x) \to f(x)$ for every $x$. (This is a form of Theorem 4.3.3 frequently given; the insistence on "every $x$" requires an additional argument.)

**6.** Let $\{\mu_n\}$ be a sequence of finite measures on $\mathscr{B}^1$. It is said to converge vaguely to a measure $\mu$ iff (1) holds. The limit $\mu$ is not necessarily a finite measure. But if $\mu_n(\mathscr{R}^1)$ is bounded in $n$, then $\mu$ is finite.

**7.** If $\mathscr{P}_n$ is a sequence of p.m.'s on $(\Omega, \mathscr{F})$ such that $\mathscr{P}_n(E)$ converges for every $E \in \mathscr{F}$, then the limit is a p.m. $\mathscr{P}$. Furthermore, if $f$ is bounded and $\mathscr{F}$-measurable, then

$$\int_\Omega f \, d\mathscr{P}_n \to \int_\Omega f \, d\mathscr{P}.$$

(The first assertion is the Vitali–Hahn–Saks theorem and rather deep, but it can be proved by reducing it to a problem of summability; see A. Rényi, [24].

**8.** If $\mu_n$ and $\mu$ are p.m.'s and $\mu_n(E) \to \mu(E)$ for every open set $E$, then this is also true for every Borel set. [HINT: Use (7) of Sec. 2.2.]

**9.** Prove a convergence theorem in metric space that will include both Theorem 4.3.3 for p.m.'s and the analogue for real numbers given before the theorem. [HINT: Use Exercise 9 of Sec. 4.4.]

## 4.4 Continuation

We proceed to discuss another kind of criterion, which is becoming ever more popular in measure theory as well as functional analysis. This has to do with classes of continuous functions on $\mathscr{R}^1$.

$C_K$ = the class of continuous functions $f$ each vanishing outside a compact set $K(f)$;

$C_0$ = the class of continuous functions $f$ such that

$$\lim_{|x|\to\infty} f(x) = 0;$$

$C_B$ = the class of bounded continuous functions;

$C$ = the class of continuous functions.

We have $C_K \subset C_0 \subset C_B \subset C$. It is well known that $C_0$ is the closure of $C_K$ with respect to uniform convergence.

An arbitrary function $f$ defined on an arbitrary space is said to *have support in* a subset $S$ of the space iff it vanishes outside $S$. Thus if $f \in C_K$, then it has support in a certain compact set, hence also in a certain compact interval. *A step function on a finite or infinite interval* $(a, b)$ is one with support in it such that $f(x) = c_j$ for $x \in (a_j, a_{j+1})$ for $1 \le j \le \ell$, where $\ell$ is finite, $a = a_1 < \cdots < a_\ell = b$, and the $c_j$'s are arbitrary real numbers. It will be called $D$-valued iff all the $a_j$'s and $c_j$'s belong to a given set $D$. When the interval $(a, b)$ is $\mathscr{R}^1$, $f$ is called just a *step function*. Note that the values of $f$ at the points $a_j$ are left unspecified to allow for flexibility; frequently they are defined by right or left continuity. The following lemma is basic.

**Approximation Lemma.** Suppose that $f \in C_K$ has support in the compact interval $[a, b]$. Given any dense subset $A$ of $\mathscr{R}^1$ and $\epsilon > 0$, there exists an $A$-valued step function $f_\epsilon$ on $(a, b)$ such that

$$\sup_{x\in\mathscr{R}^1} |f(x) - f_\epsilon(x)| \le \epsilon. \tag{1}$$

If $f \in C_0$, the same is true if $(a, b)$ is replaced by $\mathscr{R}^1$.

This lemma becomes obvious as soon as its geometric meaning is grasped. In fact, for any $f$ in $C_K$, one may even require that either $f_\epsilon \le f$ or $f_\epsilon \ge f$.

The problem is then that of the approximation of the graph of a plane curve by inscribed or circumscribed polygons, as treated in elementary calculus. But let us remark that the lemma is also a particular case of the Stone–Weierstrass theorem (see, e.g., Rudin [2]) and should be so verified by the reader. Such a sledgehammer approach has its merit, as other kinds of approximation soon to be needed can also be subsumed under the same theorem. Indeed, the discussion in this section is meant in part to introduce some modern terminology to the relevant applications in probability theory. We can now state the following alternative criterion for vague convergence.

**Theorem 4.4.1.** Let $\{\mu_n\}$ and $\mu$ be s.p.m.'s. Then $\mu_n \xrightarrow{v} \mu$ if and only if

$$(2) \qquad \forall f \in C_K[\text{or } C_0]: \int_{\mathscr{R}^1} f(x)\mu_n(dx) \to \int_{\mathscr{R}^1} f(x)\mu(dx).$$

PROOF. Suppose $\mu_n \xrightarrow{v} \mu$; (2) is true by definition when $f$ is the indicator of $(a, b]$ for $a \in D$, $b \in D$, where $D$ is the set in (1) of Sec. 4.3. Hence by the linearity of integrals it is also true when $f$ is any $D$-valued step function. Now let $f \in C_0$ and $\epsilon > 0$; by the approximation lemma there exists a $D$-valued step function $f_\epsilon$ satisfying (1). We have

$$(3) \quad \left|\int f\,d\mu_n - \int f\,d\mu\right| \le \left|\int (f - f_\epsilon)\,d\mu_n\right| + \left|\int f_\epsilon\,d\mu_n - \int f_\epsilon\,d\mu\right| + \left|\int (f_\epsilon - f)\,d\mu\right|.$$

By the modulus inequality and mean value theorem for integrals (see Sec. 3.2), the first term on the right side above is bounded by

$$\int |f - f_\epsilon|\,d\mu_n \le \epsilon \int d\mu_n \le \epsilon;$$

similarly for the third term. The second term converges to zero as $n \to \infty$ because $f_\epsilon$ is a $D$-valued step function. Hence the left side of (3) is bounded by $2\epsilon$ as $n \to \infty$, and so converges to zero since $\epsilon$ is arbitrary.

Conversely, suppose (2) is true for $f \in C_K$. Let $A$ be the set of atoms of $\mu$ as in the proof of Theorem 4.3.2; we shall show that vague convergence holds with $D = A^c$. Let $g = 1_{(a,b]}$ be the indicator of $(a, b]$ where $a \in D$, $b \in D$. Then, given $\epsilon > 0$, there exists $\delta(\epsilon) > 0$ such that $a + \delta < b - \delta$, and such that $\mu(U) < \epsilon$ where

$$U = (a - \delta, a + \delta) \cup (b - \delta, b + \delta).$$

Now define $g_1$ to be the function that coincides with $g$ on $(-\infty, a] \cup [a + \delta, b - \delta] \cup [b, \infty)$ and that is linear in $(a, a + \delta)$ and in $(b - \delta, b)$; $g_2$ to be the function that coincides with $g$ on $(-\infty, a - \delta] \cup [a, b] \cup [b + \delta, \infty)$ and that is linear in $(a - \delta, a)$ and $(b, b + \delta)$. It is clear that $g_1 \leq g \leq g_2 \leq g_1 + 1$ and consequently

$$\int g_1 \, d\mu_n \leq \int g \, d\mu_n \leq \int g_2 \, d\mu_n, \tag{4}$$

$$\downarrow \qquad\qquad\qquad \downarrow$$

$$\int g_1 \, d\mu \leq \int g \, d\mu \leq \int g_2 \, d\mu. \tag{5}$$

Since $g_1 \in C_K$, $g_2 \in C_K$, it follows from (2) that the extreme terms in (4) converge to the corresponding terms in (5). Since

$$\int g_2 \, d\mu - \int g_1 \, d\mu \leq \int_U 1 \, d\mu = \mu(U) < \epsilon,$$

and $\epsilon$ is arbitrary, it follows that the middle term in (4) also converges to that in (5), proving the assertion.

**Corollary.** If $\{\mu_n\}$ is a sequence of s.p.m.'s such that for every $f \in C_K$,

$$\lim_n \int_{\mathscr{R}^1} f(x)\mu_n(dx)$$

exists, then $\{\mu_n\}$ converges vaguely.

For by Theorem 4.3.3 a subsequence converges vaguely, say to $\mu$. By Theorem 4.4.1, the limit above is equal to $\int_{\mathscr{R}^1} f(x)\mu(dx)$. This must then be the same for every vaguely convergent subsequence, according to the hypothesis of the corollary. The vague limit of every such sequence is therefore uniquely determined (why?) to be $\mu$, and the corollary follows from Theorem 4.3.4.

**Theorem 4.4.2.** Let $\{\mu_n\}$ and $\mu$ be p.m.'s. Then $\mu_n \xrightarrow{v} \mu$ if and only if

$$\forall f \in C_B: \int_{\mathscr{R}^1} f(x)\mu_n(dx) \to \int_{\mathscr{R}^1} f(x)\mu(dx). \tag{6}$$

PROOF. Suppose $\mu_n \xrightarrow{v} \mu$. Given $\epsilon > 0$, there exist $a$ and $b$ in $D$ such that

$$\mu((a, b]^c) = 1 - \mu((a, b]) < \epsilon. \tag{7}$$

It follows from vague convergence that there exists $n_0(\epsilon)$ such that if $n \geq n_0(\epsilon)$, then

$$\mu_n((a, b]^c) = 1 - \mu_n((a, b]) < \epsilon. \tag{8}$$

Let $f \in C_B$ be given and suppose that $|f| \leq M < \infty$. Consider the function $f_\epsilon$, which is equal to $f$ on $[a, b]$, to zero on $(-\infty, a - 1) \cup (b + 1, \infty)$, and which is linear in $[a-1, a)$ and in $(b, b+1]$. Then $f_\epsilon \in C_K$ and $|f - f_\epsilon| \leq 2M$. We have by Theorem 4.4.1

(9) $$\int_{\mathscr{R}^1} f_\epsilon \, d\mu_n \to \int_{\mathscr{R}^1} f_\epsilon \, d\mu.$$

On the other hand, we have

(10) $$\int_{\mathscr{R}^1} |f - f_\epsilon| \, d\mu_n \leq \int_{(a,b]^c} 2M \, d\mu_n \leq 2M\epsilon$$

by (8). A similar estimate holds with $\mu$ replacing $\mu_n$ above, by (7). Now the argument leading from (3) to (2) finishes the proof of (6) in the same way. This proves that $\mu_n \xrightarrow{v} \mu$ implies (6); the converse has already been proved in Theorem 4.4.1.

Theorems 4.3.3 and 4.3.4 deal with s.p.m.'s. Even if the given sequence $\{\mu_n\}$ consists only of strict p.m.'s, the sequential vague limit may not be so. This is the sense of Example 2 in Sec. 4.3. It is sometimes demanded that such a limit be a p.m. The following criterion is not deep, but applicable.

**Theorem 4.4.3.** Let a family of p.m.'s $\{\mu_\alpha, \alpha \in A\}$ be given on an arbitrary index set $A$. In order that every sequence of them contains a subsequence which converges vaguely to a p.m., it is necessary and sufficient that the following condition be satisfied: for any $\epsilon > 0$, there exists a finite interval $I$ such that

(11) $$\inf_{\alpha \in A} \mu_\alpha(I) > 1 - \epsilon.$$

PROOF. Suppose (11) holds. For any sequence $\{\mu_n\}$ from the family, there exists a subsequence $\{\mu'_n\}$ such that $\mu'_n \xrightarrow{v} \mu$. We show that $\mu$ is a p.m. Let $J$ be a continuity interval of $\mu$ which contains the $I$ in (11). Then

$$\mu(\mathscr{R}^1) \geq \mu(J) = \lim_n \mu'_n(J) \geq \overline{\lim_n} \mu'_n(I) \geq 1 - \epsilon.$$

Since $\epsilon$ is arbitrary, $\mu(\mathscr{R}^1) = 1$. Conversely, suppose the condition involving (11) is not satisfied, then there exists $\epsilon > 0$, a sequence of finite intervals $I_n$ increasing to $\mathscr{R}^1$, and a sequence $\{\mu_n\}$ from the family such that

$$\forall n: \mu_n(I_n) \leq 1 - \epsilon.$$

Let $\{\mu'_n\}$ and $\mu$ be as before and $J$ any continuity interval of $\mu$. Then $J \subset I_n$ for all sufficiently large $n$, so that

$$\mu(J) = \lim_n \mu'_n(J) \leq \frac{\lim}{n} \mu'_n(I_n) \leq 1 - \epsilon.$$

Thus $\mu(\mathscr{R}^1) \leq 1 - \epsilon$ and $\mu$ is not a p.m. The theorem is proved.

A family of p.m.'s satisfying the condition above involving (11) is said to be *tight*. The preceding theorem can be stated as follows: a family of p.m.'s is relatively compact if and only if it is tight. The word "relatively" purports that the limit need not belong to the family; the word "compact" is an abbreviation of "sequentially vaguely convergent to a strict p.m." Extension of the result to p.m.'s in more general topological spaces is straight-forward but plays an important role in the convergence of stochastic processes.

The new definition of vague convergence in Theorem 4.4.1 has the advantage over the older ones in that it can be at once carried over to measures in more general topological spaces. There is no substitute for "intervals" in such a space but the classes $C_K$, $C_0$ and $C_B$ are readily available. We will illustrate the general approach by indicating one more result in this direction.

Recall the notion of a *lower semicontinuous* function on $\mathscr{R}^1$ defined by:

$$\forall x \in \mathscr{R}^1 : f(x) \le \underline{\lim}_{\substack{y \to x \\ y \ne x}} f(y). \tag{12}$$

There are several equivalent definitions (see, e.g., Royden [5]) but the following characterization is most useful: $f$ is bounded and lower semicontinuous if and only if there exists a sequence of functions $f_k \in C_B$ which increases to $f$ everywhere, and we call $f$ *upper semicontinuous* iff $-f$ is lower semicontinuous. Usually $f$ is allowed to be extended-valued; but to avoid complications we will deal with bounded functions only and denote by $L$ and $U$ respectively the classes of bounded lower semicontinuous and bounded upper semicontinuous functions.

**Theorem 4.4.4.** If $\{\mu_n\}$ and $\mu$ are p.m.'s, then $\mu_n \xrightarrow{v} \mu$ if and only if one of the two conditions below is satisfied:

$$\forall f \in L : \underline{\lim}_n \int f(x)\mu_n(dx) \ge \int f(x)\mu(dx) \tag{13}$$

$$\forall g \in U : \overline{\lim}_n \int g(x)\mu_n(dx) \le \int g(x)\mu(dx).$$

PROOF. We begin by observing that the two conditions above are equivalent by putting $f = -g$. Now suppose $\mu_n \xrightarrow{v} \mu$ and let $f_k \in C_B$, $f_k \uparrow f$. Then we have

$$\underline{\lim}_n \int f(x)\mu_n(dx) \ge \lim_n \int f_k(x)\mu_n(dx) = \int f_k(x)\mu(dx) \tag{14}$$

by Theorem 4.4.2. Letting $k \to \infty$ the last integral above converges to $\int f(x)\mu(dx)$ by monotone convergence. This gives the first inequality in (13). Conversely, suppose the latter is satisfied and let $\varphi \in C_B$, then $\varphi$ belongs to both $L$ and $U$, so that

$$\int \varphi(x)\mu\,(dx) \le \varliminf_{n} \int \varphi(x)\mu_n\,(dx) \le \varlimsup_{n} \int \varphi(x)\mu_n\,(dx) \le \int \varphi(x)\mu\,(dx)$$

which proves

$$\lim_{n} \int \varphi(x)\mu_n\,(dx) = \int \varphi(x)\mu\,(dx).$$

Hence $\mu_n \xrightarrow{v} \mu$ by Theorem 4.4.2.

*Remark.* (13) remains true if, e.g., $f$ is lower semicontinuous, with $+\infty$ as a possible value but bounded below.

**Corollary.** The conditions in (13) may be replaced by the following:

$$\text{for every open } O\colon \varliminf_{n} \mu_n(O) \ge \mu(O);$$

$$\text{for every closed } C\colon \varlimsup_{n} \mu_n(C) \le \mu(C).$$

We leave this as an exercise.

Finally, we return to the connection between the convergence of r.v.'s and that of their distributions.

DEFINITION OF CONVERGENCE "IN DISTRIBUTION" (in dist.). A sequence of r.v.'s $\{X_n\}$ is said to converge in distribution to $F$ iff the sequence $\{F_n\}$ of corresponding d.f.'s converges vaguely to the d.f. $F$.

If $X$ is an r.v. that has the d.f. $F$, then by an abuse of language we shall also say that $\{X_n\}$ converges in dist. to $X$.

**Theorem 4.4.5.** Let $\{F_n\}$, $F$ be the d.f.'s of the r.v.'s $\{X_n\}$, $X$. If $X_n \to X$ in pr., then $F_n \xrightarrow{v} F$. More briefly stated, convergence in pr. implies convergence in dist.

PROOF. If $X_n \to X$ in pr., then for each $f \in C_K$, we have $f(X_n) \to f(X)$ in pr. as easily seen from the uniform continuity of $f$ (actually this is true for any continuous $f$, see Exercise 10 of Sec. 4.1). Since $f$ is bounded the convergence holds also in $L^1$ by Theorem 4.1.4. It follows that

$$\mathscr{E}\{f(X_n)\} \to \mathscr{E}\{f(X)\},$$

which is just the relation (2) in another guise, hence $\mu_n \xrightarrow{v} \mu$.

Convergence of r.v.'s in dist. is merely a convenient turn of speech; it does not have the usual properties associated with convergence. For instance, if $X_n \to X$ in dist. and $Y_n \to Y$ in dist., it does not follow by any means that $X_n + Y_n$ will converge in dist. to $X + Y$. This is in contrast to the true convergence concepts discussed before; cf. Exercises 3 and 6 of Sec. 4.1. But if $X_n$ and $Y_n$ are independent, then the preceding assertion is indeed true as a property of the convergence of convolutions of distributions (see Chapter 6). However, in the simple situation of the next theorem no independence assumption is needed. The result is useful in dealing with limit distributions in the presence of nuisance terms.

**Theorem 4.4.6.** If $X_n \to X$ in dist, and $Y_n \to 0$ in dist., then

(a) $X_n + Y_n \to X$ in dist.
(b) $X_n Y_n \to 0$ in dist.

PROOF. We begin with the remark that for any constant $c$, $Y_n \to c$ in dist. is equivalent to $Y_n \to c$ in pr. (Exercise 4 below). To prove (a), let $f \in C_K$, $|f| \leq M$. Since $f$ is uniformly continuous, given $\epsilon > 0$ there exists $\delta$ such that $|x - y| \leq \delta$ implies $|f(x) - f(y)| \leq \epsilon$. Hence we have

$$\begin{aligned}
&\mathscr{E}\{|f(X_n + Y_n) - f(X_n)|\} \\
&\quad \leq \epsilon \mathscr{P}\{|f(X_n + Y_n) - f(X_n)| \leq \epsilon\} \mid + 2M\mathscr{P}\{|f(X_n + Y_n) - f(X_n)| > \epsilon\} \\
&\quad \leq \epsilon + 2M\mathscr{P}\{|Y_n| > \delta\}.
\end{aligned}$$

The last-written probability tends to zero as $n \to \infty$; it follows that

$$\lim_{n\to\infty} \mathscr{E}\{f(X_n + Y_n)\} = \lim_{n\to\infty} \mathscr{E}\{f(X_n)\} = \mathscr{E}\{f(X)\}$$

by Theorem 4.4.1, and (a) follows by the same theorem.

To prove (b), for given $\epsilon > 0$ we choose $A_0$ so that both $\pm A_0$ are points of continuity of the d.f. of $X$, and so large that

$$\lim_{n\to\infty} \mathscr{P}\{|X_n| > A_0\} = \mathscr{P}\{|X| > A_0\} < \epsilon.$$

This means $\mathscr{P}\{|X_n| > A_0\} < \epsilon$ for $n > n_0(\epsilon)$. Furthermore we choose $A \geq A_0$ so that the same inequality holds also for $n \leq n_0(\epsilon)$. Now it is clear that

$$\mathscr{P}\{|X_n Y_n| > \epsilon\} \leq \mathscr{P}\{|X_n| > A\} + \mathscr{P}\left\{|Y_n| > \frac{\epsilon}{A}\right\} \leq \epsilon + \mathscr{P}\left\{|Y_n| > \frac{\epsilon}{A}\right\}.$$

The last-written probability tends to zero as $n \to \infty$, and (b) follows.

**Corollary.** If $X_n \to X$, $\alpha_n \to a$, $\beta_n \to b$, all in dist. where $a$ and $b$ are constants, then $\alpha_n X_n + \beta_n \to aX + b$ in dist.

EXERCISES

$^\star$**1.** Let $\mu_n$ and $\mu$ be p.m.'s such that $\mu_n \xrightarrow{v} \mu$. Show that the conclusion in (2) need not hold if (a) $f$ is bounded and Borel measurable and all $\mu_n$ and $\mu$ are absolutely continuous, or (b) $f$ is continuous except at one point and every $\mu_n$ is absolutely continuous. (To find even sharper counterexamples would not be too easy, in view of Exercise 10 of Sec. 4.5.)

**2.** Let $\mu_n \xrightarrow{v} \mu$ when the $\mu_n$'s are s.p.m.'s. Then for each $f \in C$ and each finite continuity interval $I$ we have $\int_I f\, d\mu_n \to \int_I f\, d\mu$.

$^\star$**3.** Let $\mu_n$ and $\mu$ be as in Exercise 1. If the $f_n$'s are bounded continuous functions converging uniformly to $f$, then $\int f_n\, d\mu_n \to \int f\, d\mu$.

$^\star$**4.** Give an example to show that convergence in dist. does not imply that in pr. However, show that convergence to the unit mass $\delta_a$ does imply that in pr. to the constant $a$.

**5.** A set $\{\mu_\alpha\}$ of p.m.'s is tight if and only if the corresponding d.f.'s $\{F_\alpha\}$ converge uniformly in $\alpha$ as $x \to -\infty$ and as $x \to +\infty$.

**6.** Let the r.v.'s $\{X_\alpha\}$ have the p.m.'s $\{\mu_\alpha\}$. If for some real $r > 0$, $\mathscr{E}\{|X_\alpha|^r\}$ is bounded in $\alpha$, then $\{\mu_\alpha\}$ is tight.

**7.** Prove the Corollary to Theorem 4.4.4.

**8.** If the r.v.'s $X$ and $Y$ satisfy

$$\mathscr{P}\{|X - Y| \geq \epsilon\} \leq \epsilon$$

for some $\epsilon$, then their d.f.'s $F$ and $G$ satisfying the inequalities:

$$\forall x \in \mathscr{R}^1 : F(x - \epsilon) - \epsilon \leq G(x) \leq F(x + \epsilon) + \epsilon. \tag{15}$$

Derive another proof of Theorem 4.4.5 from this.

$^\star$**9.** The *Lévy distance* of two s.d.f.'s $F$ and $G$ is defined to be the infimum of all $\epsilon > 0$ satisfying the inequalities in (15). Prove that this is indeed a metric in the space of s.d.f.'s, and that $F_n$ converges to $F$ in this metric if and only if $F_n \xrightarrow{v} F$ and $\int_{-\infty}^{\infty} dF_n \to \int_{-\infty}^{\infty} dF$.

**10.** Find two sequences of p.m.'s $\{\mu_n\}$ and $\{\nu_n\}$ such that

$$\forall f \in C_K\colon \int f\, d\mu_n - \int f\, d\nu_n \to 0;$$

but for *no* finite $(a, b)$ is it true that

$$\mu_n(a, b) - \nu_n(a, b) \to 0.$$

[HINT: Let $\mu_n = \delta_{r_n}$, $\nu_n = \delta_{s_n}$ and choose $\{r_n\}$, $\{s_n\}$ suitably.]

**11.** Let $\{\mu_n\}$ be a sequence of p.m.'s such that for each $f \in C_B$, the sequence $\int_{\mathscr{R}^1} f\, d\mu_n$ converges; then $\mu_n \xrightarrow{v} \mu$, where $\mu$ is a p.m. [HINT: If the hypothesis is strengthened to include every $f$ in $C$, and convergence of real numbers is interpreted as usual as convergence to a finite limit, the result is easy by taking an $f$ going to $\infty$. In general one may proceed by contradiction using an $f$ that oscillates at infinity.]

***12.** Let $F_n$ and $F$ be d.f.'s such that $F_n \xrightarrow{v} F$. Define $G_n(\theta)$ and $G(\theta)$ as in Exercise 4 of Sec. 3.1. Then $G_n(\theta) \to G(\theta)$ in a.e. [HINT: Do this first when $F_n$ and $F$ are continuous and strictly increasing. The general case is obtained by smoothing $F_n$ and $F$ by convoluting with a uniform distribution in $[-\delta, +\delta]$ and letting $\delta \downarrow 0$; see the end of Sec. 6.1 below.]

## 4.5 Uniform integrability; convergence of moments

The function $|x|^r$, $r > 0$, is in $C$ but not in $C_B$, hence Theorem 4.4.2 does not apply to it. Indeed, we have seen in Example 2 of Sec. 4.1 that even convergence a.e. does not imply convergence of any moment of order $r > 0$. For, given $r$, a slight modification of that example will yield $X_n \to X$ a.e., $\mathscr{E}(|X_n|^r) = 1$ but $\mathscr{E}(|X|^r) = 0$.

It is useful to have conditions to ensure the convergence of moments when $X_n$ converges a.e. We begin with a standard theorem in this direction from classical analysis.

**Theorem 4.5.1.** If $X_n \to X$ a.e., then for every $r > 0$:

$$\mathscr{E}(|X|^r) \leq \varliminf_{n\to\infty} \mathscr{E}(|X_n|^r). \tag{1}$$

If $X_n \to X$ in $L^r$, and $X \in L^r$, then $\mathscr{E}(|X_n|^r) \to \mathscr{E}(|X|^r)$.

PROOF. (1) is just a case of Fatou's lemma (see Sec. 3.2):

$$\int_\Omega |X|^r \, d\mathscr{P} = \int_\Omega \lim_n |X_n|^r \, d\mathscr{P} \leq \varliminf_n \int_\Omega |X_n|^r \, d\mathscr{P},$$

where $+\infty$ is allowed as a value for each member with the usual convention. In case of convergence in $L^r$, $r > 1$, we have by Minkowski's inequality (Sec. 3.2), since $X = X_n + (X - X_n) = X_n - (X_n - X)$:

$$\mathscr{E}(|X_n|^r)^{1/r} - \mathscr{E}(|X_n - X|^r)^{1/r} \leq \mathscr{E}(|X|^r)^{1/r} \leq \mathscr{E}(|X_n|^r)^{1/r} + \mathscr{E}(|X_n - X|^r)^{1/r}.$$

Letting $n \to \infty$ we obtain the second assertion of the theorem. For $0 < r \leq 1$, the inequality $|x + y|^r \leq |x|^r + |y|^r$ implies that

$$\mathscr{E}(|X_n|^r) - \mathscr{E}(|X - X_n|^r) \leq \mathscr{E}(|X|^r) \leq \mathscr{E}(|X_n|^r) + \mathscr{E}(|X - X_n|^r),$$

whence the same conclusion.

The next result should be compared with Theorem 4.1.4.

**Theorem 4.5.2.** If $\{X_n\}$ converges in dist. to $X$, and for some $p > 0$, $\sup_n \mathscr{E}\{|X_n|^p\} = M < \infty$, then for each $r < p$:

$$\text{(2)} \qquad \lim_{n\to\infty} \mathscr{E}(|X_n|^r) = \mathscr{E}(|X|^r) < \infty.$$

If $r$ is a positive integer, then we may replace $|X_n|^r$ and $|X|^r$ above by $X_n{}^r$ and $X^r$.

PROOF. We prove the second assertion since the first is similar. Let $F_n$, $F$ be the d.f.'s of $X_n$, $X$; then $F_n \xrightarrow{v} F$. For $A > 0$ define $f_A$ on $\mathscr{R}^1$ as follows:

$$\text{(3)} \qquad f_A(x) = \begin{cases} x^r, & \text{if } |x| \le A; \\ A^r, & \text{if } x > A; \\ (-A)^r, & \text{if } x < -A. \end{cases}$$

Then $f_A \in C_B$, hence by Theorem 4.4.4 the "truncated moments" converge:

$$\int_{-\infty}^{\infty} f_A(x)\, dF_n(x) \to \int_{-\infty}^{\infty} f_A(x)\, dF(x).$$

Next we have

$$\int_{-\infty}^{\infty} |f_A(x) - x^r|\, dF_n(x) \le \int_{|x|>A} |x|^r\, dF_n(x) = \int_{|X_n|>A} |X_n|^r\, d\mathscr{P}$$

$$\le \frac{1}{A^{p-r}} \int_{\Omega} |X_n|^p\, d\mathscr{P} \le \frac{M}{A^{p-r}}.$$

The last term does not depend on $n$, and converges to zero as $A \to \infty$. It follows that as $A \to \infty$, $\int_{-\infty}^{\infty} f_A\, dF_n$ converges uniformly in $n$ to $\int_{-\infty}^{\infty} x^r\, dF$. Hence by a standard theorem on the inversion of repeated limits, we have

$$\text{(4)} \qquad \int_{-\infty}^{\infty} x^r\, dF = \lim_{A\to\infty} \int_{-\infty}^{\infty} f_A\, dF = \lim_{A\to\infty} \lim_{n\to\infty} \int_{-\infty}^{\infty} f_A\, dF_n$$

$$= \lim_{n\to\infty} \lim_{A\to\infty} \int_{-\infty}^{\infty} f_A\, dF_n = \lim_{n\to\infty} \int_{-\infty}^{\infty} x^r\, dF_n.$$

We now introduce the concept of uniform integrability, which is of basic importance in this connection. It is also an essential hypothesis in certain convergence questions arising in the theory of martingales (to be treated in Chapter 9).

DEFINITION OF UNIFORM INTEGRABILITY. A family of r.v.'s $\{X_t\}$, $t \in T$, where $T$ is an arbitrary index set, is said to be *uniformly integrable* iff

$$\text{(5)} \qquad \lim_{A\to\infty} \int_{|X_t|>A} |X_t|\, d\mathscr{P} = 0$$

uniformly in $t \in T$.

**Theorem 4.5.3.** The family $\{X_t\}$ is uniformly integrable if and only if the following two conditions are satisfied:

(a) $\mathscr{E}(|X_t|)$ is bounded in $t \in T$;
(b) For every $\epsilon > 0$, there exists $\delta(\epsilon) > 0$ such that for any $E \in \mathscr{F}$:

$$\mathscr{P}(E) < \delta(\epsilon) \Rightarrow \int_E |X_t| \, d\mathscr{P} < \epsilon \text{ for every } t \in T.$$

PROOF. Clearly (5) implies (a). Next, let $E \in \mathscr{F}$ and write $E_t$ for the set $\{\omega : |X_t(\omega)| > A\}$. We have by the mean value theorem

$$\int_E |X_t| \, d\mathscr{P} = \left( \int_{E \cap E_t} + \int_{E \backslash E_t} \right) |X_t| \, d\mathscr{P} \leq \int_{E_t} |X_t| \, d\mathscr{P} + A\mathscr{P}(E).$$

Given $\epsilon > 0$, there exists $A = A(\epsilon)$ such that the last-written integral is less than $\epsilon/2$ for every $t$, by (5). Hence (b) will follow if we set $\delta = \epsilon/2A$. Thus (5) implies (b).

Conversely, suppose that (a) and (b) are true. Then by the Chebyshev inequality we have for every $t$,

$$\mathscr{P}\{|X_t| > A\} \leq \frac{\mathscr{E}(|X_t|)}{A} \leq \frac{M}{A},$$

where $M$ is the bound indicated in (a). Hence if $A > M/\delta$, then $\mathscr{P}(E_t) < \delta$ and we have by (b):

$$\int_{E_t} |X_t| \, d\mathscr{P} < \epsilon.$$

Thus (5) is true.

**Theorem 4.5.4.** Let $0 < r < \infty$, $X_n \in L^r$, and $X_n \to X$ in pr. Then the following three propositions are equivalent:

**(i)** $\{|X_n|^r\}$ is uniformly integrable;
**(ii)** $X_n \to X$ in $L^r$;
**(iii)** $\mathscr{E}(|X_n|^r) \to \mathscr{E}(|X|^r) < \infty$.

PROOF. Suppose (i) is true; since $X_n \to X$ in pr. by Theorem 4.2.3, there exists a subsequence $\{n_k\}$ such that $X_{n_k} \to X$ a.e. By Theorem 4.5.1 and (a) above, $X \in L^r$. The easy inequality

$$|X_n - X|^r \leq 2^r\{|X_n|^r + |X|^r\},$$

valid for all $r > 0$, together with Theorem 4.5.3 then implies that the sequence $\{|X_n - X|^r\}$ is also uniformly integrable. For each $\epsilon > 0$, we have

(6)
$$\int_\Omega |X_n - X|^r \, d\mathscr{P} = \int_{|X_n - X| > \epsilon} |X_n - X|^r \, d\mathscr{P} + \int_{|X_n - X| \le \epsilon} |X_n - X|^r \, d\mathscr{P}$$
$$\le \int_{|X_n - X| > \epsilon} |X_n - X|^r \, d\mathscr{P} + \epsilon^r.$$

Since $\mathscr{P}\{|X_n - X| > \epsilon\} \to 0$ as $n \to \infty$ by hypothesis, it follows from (b) above that the last written integral in (6) also tends to zero. This being true for every $\epsilon > 0$, (ii) follows.

Suppose (ii) is true, then we have (iii) by the second assertion of Theorem 4.5.1.

Finally suppose (iii) is true. To prove (i), let $A > 0$ and consider a function $f_A$ in $C_K$ satisfying the following conditions:

$$f_A(x) \begin{cases} = |x|^r & \text{for } |x|^r \le A; \\ \le |x|^r & \text{for } A < |x|^r \le A + 1; \\ = 0 & \text{for } |x|^r > A + 1; \end{cases}$$

cf. the proof of Theorem 4.4.1 for the construction of such a function. Hence we have

$$\varliminf_{n\to\infty} \int_{|X_n|^r \le A+1} |X_n|^r \, d\mathscr{P} \ge \lim_{n\to\infty} \mathscr{E}\{f_A(X_n)\} = \mathscr{E}\{f_A(X)\} \ge \int_{|X|^r \le A} |X|^r \, d\mathscr{P},$$

where the inequalities follow from the shape of $f_A$, while the limit relation in the middle as in the proof of Theorem 4.4.5. Subtracting from the limit relation in (iii), we obtain

$$\varlimsup_{n\to\infty} \int_{|X_n|^r > A+1} |X_n|^r \, d\mathscr{P} \le \int_{|X|^r > A} |X|^r \, d\mathscr{P}.$$

The last integral does not depend on $n$ and converges to zero as $A \to \infty$. This means: for any $\epsilon > 0$, there exists $A_0 = A_0(\epsilon)$ and $n_0 = n_0(A_0(\epsilon))$ such that we have

$$\sup_{n > n_0} \int_{|X_n|^r > A+1} |X_n|^r \, d\mathscr{P} < \epsilon$$

provided that $A > A_0$. Since each $|X_n|^r$ is integrable, there exists $A_1 = A_1(\epsilon)$ such that the supremum above may be taken over all $n \ge 1$ provided that $A > A_0 \vee A_1$. This establishes (i), and completes the proof of the theorem.

In the remainder of this section the term "moment" will be restricted to a moment of positive integral order. It is well known (see Exercise 5 of Sec. 6.6) that on $(\mathscr{U}, \mathscr{B})$ any p.m. or equivalently its d.f. is uniquely determined by its moments of all orders. Precisely, if $F_1$ and $F_2$ are two d.f.'s such that

$F_i(0) = 0$, $F_i(1) = 1$ for $i = 1, 2$; and

$$\forall n \geq 1: \int_0^1 x^n \, dF_1(x) = \int_0^1 x^n \, dF_2(x),$$

then $F_1 \equiv F_2$. The corresponding result is false in $(\mathscr{R}^1, \mathscr{B}^1)$ and a further condition on the moments is required to ensure uniqueness. The sufficient condition due to Carleman is as follows:

$$\sum_{r=1}^{\infty} \frac{1}{m_{2r}^{1/2r}} = +\infty,$$

where $m_r$ denotes the moment of order $r$. When a given sequence of numbers $\{m_r, r \geq 1\}$ uniquely determines a d.f. $F$ such that

$$m_r = \int_{-\infty}^{\infty} x^r \, dF(x), \tag{7}$$

we say that "the moment problem is determinate" for the sequence. Of course an arbitrary sequence of numbers need not be the sequence of moments for any d.f.; a necessary but far from sufficient condition, for example, is that the Liapounov inequality (Sec. 3.2) be satisfied. We shall not go into these questions here but shall content ourselves with the useful result below, which is often referred to as the "method of moments"; see also Theorem 6.4.5.

**Theorem 4.5.5.** Suppose there is a unique d.f. $F$ with the moments $\{m^{(r)}, r \geq 1\}$, all finite. Suppose that $\{F_n\}$ is a sequence of d.f.'s, each of which has all its moments finite:

$$m_n^{(r)} = \int_{-\infty}^{\infty} x^r \, dF_n.$$

Finally, suppose that for every $r \geq 1$:

$$\lim_{n \to \infty} m_n^{(r)} = m^{(r)}. \tag{8}$$

Then $F_n \xrightarrow{v} F$.

PROOF. Let $\mu_n$ be the p.m. corresponding to $F_n$. By Theorem 4.3.3 there exists a subsequence of $\{\mu_n\}$ that converges vaguely. Let $\{\mu_{n_k}\}$ be *any* subsequence converging vaguely to some $\mu$. We shall show that $\mu$ is indeed a p.m. with the d.f. $F$. By the Chebyshev inequality, we have for each $A > 0$:

$$\mu_{n_k}(-A, +A) \geq 1 - A^{-2} m_{n_k}^{(2)}.$$

Since $m_{n_k}^{(2)} \to m^{(2)} < \infty$, it follows that as $A \to \infty$, the left side converges uniformly in $k$ to one. Letting $A \to \infty$ along a sequence of points such that both $\pm A$ belong to the dense set $D$ involved in the definition of vague convergence, we obtain as in (4) above:

$$\begin{aligned}\mu(\mathscr{R}^1) &= \lim_{A\to\infty} \mu(-A, +A) = \lim_{A\to\infty} \lim_{k\to\infty} \mu_{n_k}(-A, +A) \\ &= \lim_{k\to\infty} \lim_{A\to\infty} \mu_{n_k}(-A, +A) = \lim_{k\to\infty} \mu_{n_k}(\mathscr{R}^1) = 1.\end{aligned}$$

Now for each $r$, let $p$ be the next larger even integer. We have

$$\int_{-\infty}^{\infty} x^p \, d\mu_{n_k} = m_{n_k}^{(p)} \to m^{(p)},$$

hence $m_{n_k}^{(p)}$ is bounded in $k$. It follows from Theorem 4.5.2 that

$$\int_{-\infty}^{\infty} x^r \, d\mu_{n_k} \to \int_{-\infty}^{\infty} x^r \, d\mu.$$

But the left side also converges to $m^{(r)}$ by (8). Hence by the uniqueness hypothesis $\mu$ is the p.m. determined by $F$. We have therefore proved that every vaguely convergent subsequence of $\{\mu_n\}$, or equivalently $\{F_n\}$, has the same limit $\mu$, or equivalently $F$. Hence the theorem follows from Theorem 4.3.4.

EXERCISES

**1.** If $\sup_n |X_n| \in L^p$ and $X_n \to X$ a.e., then $X \in L^p$ and $X_n \to X$ in $L^p$.

**2.** If $\{X_n\}$ is dominated by some $Y$ in $L^p$, and converges in dist. to $X$, then $\mathscr{E}(|X_n|^p) \to \mathscr{E}(|X|^p)$.

**3.** If $X_n \to X$ in dist., and $f \in C$, then $f(X_n) \to f(X)$ in dist.

***4.** Exercise 3 may be reduced to Exercise 10 of Sec. 4.1 as follows. Let $F_n$, $1 \le n \le \infty$, be d.f.'s such that $F_n \xrightarrow{v} F$. Let $\theta$ be uniformly distributed on $[0, 1]$ and put $X_n = F_n^{-1}(\theta)$, $1 \le n \le \infty$, where $F_n^{-1}(y) = \sup\{x\colon F_n(x) \le y\}$ (cf. Exercise 4 of Sec. 3.1). Then $X_n$ has d.f. $F_n$ and $X_n \to X_\infty$ in pr.

**5.** Find the moments of the *normal* d.f. $\Phi$ and the *positive normal* d.f. $\Phi_+$ below:

$$\Phi(x) = \frac{1}{\sqrt{2\pi}} \int_{-\infty}^{\infty} e^{-y^2/2} \, dy, \qquad \Phi_+(x) = \begin{cases} \sqrt{\dfrac{2}{\pi}} \int_0^x e^{-y^2/2} \, dy, & \text{if } x \ge 0; \\ 0, & \text{if } x < 0. \end{cases}$$

Show that in either case the moments satisfy Carleman's condition.

**6.** If $\{X_t\}$ and $\{Y_t\}$ are uniformly integrable, then so is $\{X_t + Y_1\}$ and $\{X_t + Y_t\}$.

**7.** If $\{X_n\}$ is dominated by some $Y$ in $L^1$ or if it is identically distributed with finite mean, then it is uniformly integrable.

★**8.** If $\sup_n \mathscr{E}(|X_n|^p) < \infty$ for some $p > 1$, then $\{X_n\}$ is uniformly integrable.

**9.** If $\{X_n\}$ is uniformly integrable, then the sequence

$$\left\{\frac{1}{n}\sum_{j=1}^{n} X_j, n \geq 1\right\}$$

is uniformly integrable.

★**10.** Suppose the distributions of $\{X_n, 1 \leq n \leq \infty\}$ are absolutely continuous with densities $\{g_n\}$ such that $g_n \to g_\infty$ in Lebesgue measure. Then $g_n \to g_\infty$ in $L^1(-\infty, \infty)$, and consequently for every bounded Borel measurable function $f$ we have $\mathscr{E}\{f(X_n)\} \to \mathscr{E}\{f(X_\infty)\}$. [HINT: $\int (g_\infty - g_n)^+ \, dx = \int (g_\infty - g_n)^- \, dx$ and $(g_\infty - g_n)^+ \leq g_\infty$; use dominated convergence.]

# 5 Law of large numbers. Random series

## 5.1 Simple limit theorems

The various concepts of Chapter 4 will be applied to the so-called "law of large numbers" — a famous name in the theory of probability. This has to do with partial sums

$$S_n = \sum_{j=1}^{n} X_j$$

of a sequence of r.v.'s. In the most classical formulation, the "weak" or the "strong" law of large numbers is said to hold for the sequence according as

$$\frac{S_n - \mathscr{E}(S_n)}{n} \to 0 \tag{1}$$

in pr. or a.e. This, of course, presupposes the finiteness of $\mathscr{E}(S_n)$. A natural generalization is as follows:

$$\frac{S_n - a_n}{b_n} \to 0,$$

where $\{a_n\}$ is a sequence of real numbers and $\{b_n\}$ a sequence of positive numbers tending to infinity. We shall present several stages of the development, even though they overlap each other to some extent, for it is just as important to learn the basic techniques as the results themselves.

The simplest cases follow from Theorems 4.1.4 and 4.2.3, according to which if $Z_n$ is any sequence of r.v.'s, then $\mathscr{E}(Z_n^2) \to 0$ implies that $Z_n \to 0$ in pr. and $Z_{n_k} \to 0$ a.e. for a subsequence $\{n_k\}$. Applied to $Z_n = S_n/n$, the first assertion becomes

$$\mathscr{E}(S_n^2) = o(n^2) \Rightarrow \frac{S_n}{n} \to 0 \text{ in pr.} \tag{2}$$

Now we can calculate $\mathscr{E}(S_n^2)$ more explicitly as follows:

$$\begin{aligned} \mathscr{E}(S_n^2) &= \mathscr{E}\left(\left(\sum_{j=1}^{n} X_j\right)^2\right) = \mathscr{E}\left(\sum_{j=1}^{n} X_j^2 + 2\sum_{1\le j<k\le n} X_j X_k\right) \\ &= \sum_{j=1}^{n} \mathscr{E}(X_j^2) + 2\sum_{1\le j<k\le n} \mathscr{E}(X_j X_k). \end{aligned} \tag{3}$$

Observe that there are $n^2$ terms above, so that even if all of them are bounded by a fixed constant, only $\mathscr{E}(S_n^2) = O(n^2)$ will result, which falls critically short of the hypothesis in (2). The idea then is to introduce certain assumptions to cause enough cancellation among the "mixed terms" in (3). A salient feature of probability theory and its applications is that such assumptions are not only permissible but realistic. We begin with the simplest of its kind.

DEFINITION. Two r.v.'s $X$ and $Y$ are said to be *uncorrelated* iff both have finite second moments and

$$\mathscr{E}(XY) = \mathscr{E}(X)\mathscr{E}(Y). \tag{4}$$

They are said to be *orthogonal* iff (4) is replaced by

$$\mathscr{E}(XY) = 0. \tag{5}$$

The r.v.'s of any family are said to be uncorrelated [orthogonal] iff every two of them are.

Note that (4) is equivalent to

$$\mathscr{E}\{(X - \mathscr{E}(X))(Y - \mathscr{E}(Y))\} = 0,$$

which reduces to (5) when $\mathscr{E}(X) = \mathscr{E}(Y) = 0$. The requirement of finite second moments seems unnecessary, but it does ensure the finiteness of $\mathscr{E}(XY)$ (Cauchy–Schwarz inequality!) as well as that of $\mathscr{E}(X)$ and $\mathscr{E}(Y)$, and without it the definitions are hardly useful. Finally, it is obvious that pairwise independence implies uncorrelatedness, provided second moments are finite.

If $\{X_n\}$ is a sequence of uncorrelated r.v.'s, then the sequence $\{X_n - \mathscr{E}(X_n)\}$ is orthogonal, and for the latter (3) reduces to the fundamental relation below:

$$\sigma^2(S_n)^- = \sum_{j=1}^{n} \sigma^2(X_j), \tag{6}$$

which may be called the "additivity of the variance". Conversely, the validity of (6) for $n = 2$ implies that $X_1$ and $X_2$ are uncorrelated. There are only $n$ terms on the right side of (6), hence if these are bounded by a fixed constant we have now $\sigma^2(S_n) = O(n) = o(n^2)$. Thus (2) becomes applicable, and we have proved the following result.

**Theorem 5.1.1.** If the $X_j$'s are uncorrelated and their second moments have a common bound, then (1) is true in $L^2$ and hence also in pr.

This simple theorem is actually due to Chebyshev, who invented his famous inequalities for its proof. The next result, due to Rajchman (1932), strengthens the conclusion by proving convergence a.e. This result is interesting by virtue of its simplicity, and serves well to introduce an important method, that of taking subsequences.

**Theorem 5.1.2.** Under the same hypotheses as in Theorem 5.1.1, (1) holds also a.e.

PROOF. Without loss of generality we may suppose that $\mathscr{E}(X_j) = 0$ for each $j$, so that the $X_j$'s are orthogonal. We have by (6):

$$\mathscr{E}(S_n^2) \le Mn,$$

where $M$ is a bound for the second moments. It follows by Chebyshev's inequality that for each $\epsilon > 0$ we have

$$\mathscr{P}\{|S_n| > n\epsilon\} \le \frac{Mn}{n^2\epsilon^2} = \frac{M}{n\epsilon^2}.$$

If we sum this over $n$, the resulting series on the right diverges. However, if we confine ourselves to the subsequence $\{n^2\}$, then

$$\sum_n \mathscr{P}\{|S_{n^2}| > n^2\epsilon\} = \sum_n \frac{M}{n^2\epsilon^2} < \infty.$$

Hence by Theorem 4.2.1 (Borel–Cantelli) we have

$$\mathscr{P}\{|S_{n^2}| > n^2\epsilon \text{ i.o.}\} = 0; \tag{7}$$

and consequently by Theorem 4.2.2

(8) $$\frac{S_{n^2}}{n^2} \to 0 \quad \text{a.e.}$$

We have thus proved the desired result for a subsequence; and the "method of subsequences" aims in general at extending to the whole sequence a result proved (relatively easily) for a subsequence. In the present case we must show that $S_k$ does not differ enough from the nearest $S_{n^2}$ to make any real difference.

Put for each $n \geq 1$:

$$D_n = \max_{n^2 \leq k < (n+1)^2} |S_k - S_{n^2}|.$$

Then we have

$$\mathscr{E}(D_n^2) \leq 2n\,\mathscr{E}(|S_{(n+1)^2} - S_{n^2}|^2) = 2n \sum_{j=n^2+1}^{(n+1)^2} \sigma^2(X_j) \leq 4n^2 M$$

and consequently by Chebyshev's inequality

$$\mathscr{P}\{D_n > n^2\epsilon\} \leq \frac{4M}{\epsilon^2 n^2}.$$

It follows as before that

(9) $$\frac{D_n}{n^2} \to 0 \quad \text{a.e.}$$

Now it is clear that (8) and (9) together imply (1), since

$$\frac{|S_k|}{k} \leq \frac{|S_{n^2}| + D_n}{n^2}$$

for $n^2 \leq k < (n+1)^2$. The theorem is proved.

The hypotheses of Theorems 5.1.1 and 5.1.2 are certainly satisfied for a sequence of independent r.v.'s that are uniformly bounded or that are identically distributed with a finite second moment. The most celebrated, as well as the very first case of the strong law of large numbers, due to Borel (1909), is formulated in terms of the so-called "normal numbers." Let each real number in [0, 1] be expanded in the usual decimal system:

(10) $$\omega = \cdot x_1 x_2 \ldots x_n \ldots .$$

Except for the countable set of terminating decimals, for which there are two distinct expansions, this representation is unique. Fix a $k: 0 \leq k \leq 9$, and let

$\nu_k^{(n)}(\omega)$ denote the number of digits among the first $n$ digits of $\omega$ that are equal to $k$. Then $\nu_k^{(n)}(\omega)/n$ is the relative frequency of the digit $k$ in the first $n$ places, and the limit, if existing:

$$\lim_{n\to\infty} \frac{\nu_k^{(n)}(\omega)}{n} = \varphi_k(\omega), \tag{11}$$

may be called the frequency of $k$ in $\omega$. The number $\omega$ is called *simply normal* (*to the scale* 10) iff this limit exists for each $k$ and is equal to 1/10. Intuitively all ten possibilities should be equally likely for each digit of a number picked "at random". On the other hand, one can write down "at random" any number of numbers that are "abnormal" according to the definition given, such as $\cdot 1111\ldots$, while it is a relatively difficult matter to name even one normal number in the sense of Exercise 5 below. It turns out that the number

$$\cdot 12345678910111213\ldots,$$

which is obtained by writing down in succession all the natural numbers in the decimal system, is a normal number to the scale 10 even in the stringent definition of Exercise 5 below, but the proof is not so easy. As for determining whether certain well-known numbers such as $e - 2$ or $\pi - 3$ are normal, the problem seems beyond the reach of our present capability for mathematics. In spite of these difficulties, Borel's theorem below asserts that in a perfectly precise sense almost every number *is* normal. Furthermore, this striking proposition is merely a very particular case of Theorem 5.1.2 above.

**Theorem 5.1.3.** Except for a Borel set of measure zero, every number in [0, 1] is simply normal.

PROOF. Consider the probability space $(\mathscr{U}, \mathscr{B}, m)$ in Example 2 of Sec. 2.2. Let $Z$ be the subset of the form $m/10^n$ for integers $n \geq 1$, $m \geq 1$, then $m(Z) = 0$. If $\omega \in \mathscr{U}\backslash Z$, then it has a unique decimal expansion; if $\omega \in Z$, it has two such expansions, but we agree to use the "terminating" one for the sake of definiteness. Thus we have

$$\omega = \cdot \xi_1 \xi_2 \ldots \xi_n \ldots,$$

where for each $n \geq 1$, $\xi_n(\cdot)$ is a Borel measurable function of $\omega$. Just as in Example 4 of Sec. 3.3, the sequence $\{\xi_n, n \geq 1\}$ is a sequence of independent r.v.'s with

$$\mathscr{P}\{\xi_n = k\} = \tfrac{1}{10}, \quad k = 0, 1, \ldots, 9.$$

Indeed according to Theorem 5.1.2 we need only verify that the $\xi_n$'s are uncorrelated, which is a very simple matter. For a fixed $k$ we define the r.v. $X_n$ to be the indicator of the set $\{\omega: \xi_n(\omega) = k\}$, then $\mathscr{E}(X_n) = 1/10$, $\mathscr{E}({X_n}^2) = 1/10$, and

$$\frac{1}{n}\sum_{j=1}^{n} X_j(\omega)$$

is the relative frequency of the digit $k$ in the first $n$ places of the decimal for $\omega$. According to Theorem 5.1.2, we have then

$$\frac{S_n}{n} \to \frac{1}{10} \quad \text{a.e.}$$

Hence in the notation of (11), we have $\mathscr{P}\{\varphi_k = 1/10\} = 1$ for each $k$ and consequently also

$$\mathscr{P}\left\{\bigcap_{k=0}^{9}\left[\varphi_k = \frac{1}{10}\right]\right\} = 1,$$

which means that the set of normal numbers has Borel measure one. Theorem 5.1.3 is proved.

The preceding theorem makes a deep impression (at least on the older generation!) because it interprets a general proposition in probability theory at a most classical and fundamental level. If we use the intuitive language of probability such as coin-tossing, the result sounds almost trite. For it merely says that if an unbiased coin is tossed indefinitely, the limiting frequency of "heads" will be equal to $\frac{1}{2}$ — that is, its *a priori* probability. A mathematician who is unacquainted with and therefore skeptical of probability theory tends to regard the last statement as either "obvious" or "unprovable", but he can scarcely question the authenticity of Borel's theorem about ordinary decimals. As a matter of fact, the proof given above, essentially Borel's own, is a lot easier than a straightforward measure-theoretic version, deprived of the intuitive content [see, e.g., Hardy and Wright, *An introduction to the theory of numbers*, 3rd. ed., Oxford University Press, Inc., New York, 1954].

## EXERCISES

**1.** For any sequence of r.v.'s $\{X_n\}$, if $\mathscr{E}(X_n^2) \to 0$, then (1) is true in pr. but not necessarily a.e.

$^\star$**2.** Theorem 5.1.2 may be sharpened as follows: under the same hypotheses we have $S_n/n^\alpha \to 0$ a.e. for any $\alpha > \frac{3}{4}$.

**3.** Theorem 5.1.2 remains true if the hypothesis of bounded second moments is weakened to: $\sigma^2(X_n) = O(n^\theta)$ where $0 \le \theta < \frac{1}{2}$. Various combinations of Exercises 2 and 3 are possible.

$^\star$**4.** If $\{X_n\}$ are independent r.v.'s such that the fourth moments $\mathscr{E}(X_n^4)$ have a common bound, then (1) is true a.e. [This is Cantelli's strong law of large numbers. Without using Theorem 5.1.2 we may operate with $\mathscr{E}(S_n^4/n^4)$ as we did with $\mathscr{E}(S_n^2/n^2)$. Note that the full strength of independence is not needed.]

**5.** We may strengthen the definition of a normal number by considering *blocks* of digits. Let $r \geq 1$, and consider the successive overlapping blocks of $r$ consecutive digits in a decimal; there are $n - r + 1$ such blocks in the first $n$ places. Let $\nu^{(n)}(\omega)$ denote the number of such blocks that are identical with a given one; for example, if $r = 5$, the given block may be "21212". Prove that for a.e. $\omega$, we have for every $r$:

$$\lim_{n\to\infty} \frac{\nu^{(n)}(\omega)}{n} = \frac{1}{10^r}$$

[HINT: Reduce the problem to disjoint blocks, which are independent.]

***6.** The above definition may be further strengthened if we consider different scales of expansion. A real number in [0, 1] is said to be *completely normal* iff the relative frequency of each block of length $r$ in the scale $s$ tends to the limit $1/s^r$ for every $s$ and $r$. Prove that almost every number in [0, 1] is completely normal.

**7.** Let $\alpha$ be completely normal. Show that by looking at the expansion of $\alpha$ in some scale we can rediscover the complete works of Shakespeare from end to end without a single misprint or interruption. [This is Borel's paradox.]

***8.** Let $X$ be an arbitrary r.v. with an absolutely continuous distribution. Prove that with probability one the fractional part of $X$ is a normal number. [HINT: Let $N$ be the set of normal numbers and consider $\mathscr{P}\{X - [X] \in N\}$.]

**9.** Prove that the set of real numbers in [0, 1] whose decimal expansions do not contain the digit 2 is of measure zero. Deduce from this the existence of two sets $A$ and $B$ both of measure zero such that every real number is representable as a sum $a + b$ with $a \in A$, $b \in B$.

***10.** Is the sum of two normal numbers, modulo 1, normal? Is the product? [HINT: Consider the differences between a fixed abnormal number and all normal numbers: this is a set of probability one.]

## 5.2 Weak law of large numbers

The law of large numbers in the form (1) of Sec. 5.1 involves only the first moment, but so far we have operated with the second. In order to drop any assumption on the second moment, we need a new device, that of "equivalent sequences", due to Khintchine (1894–1959).

DEFINITION. Two sequences of r.v.'s $\{X_n\}$ and $\{Y_n\}$ are said to be *equivalent* iff

$$\sum_n \mathscr{P}\{X_n \neq Y_n\} < \infty. \tag{1}$$

In practice, an equivalent sequence is obtained by "truncating" in various ways, as we shall see presently.

**Theorem 5.2.1.** If $\{X_n\}$ and $\{Y_n\}$ are equivalent, then

$$\sum_n (X_n - Y_n) \quad \text{converges a.e.}$$

Furthermore if $a_n \uparrow \infty$, then

$$\frac{1}{a_n}\sum_{j=1}^{n}(X_j - Y_j) \to 0 \quad \text{a.e.} \tag{2}$$

PROOF. By the Borel–Cantelli lemma, (1) implies that

$$\mathscr{P}\{X_n \neq Y_n \text{ i.o.}\} = 0.$$

This means that there exists a null set N with the following property: if $\omega \in \Omega\backslash\text{N}$, then there exists $n_0(\omega)$ such that

$$n \geq n_0(\omega) \Rightarrow X_n(\omega) = Y_n(\omega).$$

Thus for such an $\omega$, the two numerical sequences $\{X_n(\omega)\}$ and $\{Y_n(\omega)\}$ differ only in a finite number of terms (how many depending on $\omega$). In other words, the series

$$\sum_n (X_n(\omega) - Y_n(\omega))$$

consists of zeros from a certain point on. Both assertions of the theorem are trivial consequences of this fact.

**Corollary.** With probability one, the expression

$$\sum_n X_n \quad \text{or} \quad \frac{1}{a_n}\sum_{j=1}^{n} X_j$$

converges, diverges to $+\infty$ or $-\infty$, or fluctuates in the same way as

$$\sum_n Y_n \quad \text{or} \quad \frac{1}{a_n}\sum_{j=1}^{n} Y_j,$$

respectively. In particular, if

$$\frac{1}{a_n}\sum_{j=1}^{n} X_j$$

converges to $X$ in pr., then so does

$$\frac{1}{a_n}\sum_{j=1}^{n} Y_j.$$

To prove the last assertion of the corollary, observe that by Theorem 4.1.2 the relation (2) holds also in pr. Hence if

$$\frac{1}{a_n}\sum_{j=1}^{n} X_j \to X \quad \text{in pr.},$$

then we have

$$\frac{1}{a_n}\sum_{j=1}^{n} Y_j = \frac{1}{a_n}\sum_{j=1}^{n} X_j + \frac{1}{a_n}\sum_{j=1}^{n}(Y_j - X_j) \to X + 0 = X \quad \text{in pr.}$$

(see Exercise 3 of Sec. 4.1).

The next law of large numbers is due to Khintchine. Under the stronger hypothesis of total independence, it will be proved again by an entirely different method in Chapter 6.

**Theorem 5.2.2.** Let $\{X_n\}$ be pairwise independent and identically distributed r.v.'s with finite mean $m$. Then we have

$$\frac{S_n}{n} \to m \quad \text{in pr.} \tag{3}$$

PROOF. Let the common d.f. be $F$ so that

$$m = \mathscr{E}(X_n) = \int_{-\infty}^{\infty} x\,dF(x), \qquad \mathscr{E}(|X_n|) = \int_{-\infty}^{\infty} |x|\,dF(x) < \infty.$$

By Theorem 3.2.1 the finiteness of $\mathscr{E}(|X_1|)$ is equivalent to

$$\sum_n \mathscr{P}(|X_1| > n) < \infty.$$

Hence we have, since the $X_n$'s have the same distribution:

$$\sum_n \mathscr{P}(|X_n| > n) < \infty. \tag{4}$$

We introduce a sequence of r.v.'s $\{Y_n\}$ by "truncating at $n$":

$$Y_n(\omega) = \begin{cases} X_n(\omega), & \text{if } |X_n(\omega)| \le n; \\ 0, & \text{if } |X_n(\omega)| > n. \end{cases}$$

This is equivalent to $\{X_n\}$ by (4), since $\mathscr{P}(|X_n| > n) = \mathscr{P}(X_n \neq Y_n)$. Let

$$T_n = \sum_{j=1}^{n} Y_j.$$

By the corollary above, (3) will follow if (and only if) we can prove $T_n/n \to m$ in pr. Now the $Y_n$'s are also pairwise independent by Theorem 3.3.1 (applied to each pair), hence they are uncorrelated, since each, being bounded, has a finite second moment. Let us calculate $\sigma^2(T_n)$; we have by (6) of Sec. 5.1,

$$\sigma^2(T_n) = \sum_{j=1}^{n} \sigma^2(Y_j) \le \sum_{j=1}^{n} \mathscr{E}(Y_j^2) = \sum_{j=1}^{n} \int_{|x| \le j} x^2 \, dF(x).$$

The crudest estimate of the last term yields

$$\sum_{j=1}^{n} \int_{|x| \le j} x^2 \, dF(x) \le \sum_{j=1}^{n} j \int_{|x| \le j} |x| \, dF(x) \le \frac{n(n+1)}{2} \int_{-\infty}^{\infty} |x| \, dF(x),$$

which is $O(n^2)$, but not $o(n^2)$ as required by (2) of Sec. 5.1. To improve on it, let $\{a_n\}$ be a sequence of integers such that $0 < a_n < n$, $a_n \to \infty$ but $a_n = o(n)$. We have

$$\begin{aligned}
\sum_{j=1}^{n} \int_{|x| \le j} x^2 \, dF(x) &= \sum_{j \le a_n} + \sum_{a_n < j \le n} \\
&\le \sum_{j \le a_n} a_n \int_{|x| \le a_n} |x| \, dF(x) + \sum_{a_n < j \le n} a_n \int_{|x| \le a_n} |x| \, dF(x) \\
&\quad + \sum_{a_n < j \le n} n \int_{a_n < |x| \le n} |x| \, dF(x) \\
&\le n a_n \int_{-\infty}^{\infty} |x| \, dF(x) + n^2 \int_{|x| > a_n} |x| \, dF(x).
\end{aligned}$$

The first term is $O(na_n) = o(n^2)$; and the second is $n^2 o(1) = o(n^2)$, since the set $\{x: |x| > a_n\}$ decreases to the empty set and so the last-written integral above converges to zero. We have thus proved that $\sigma^2(T_n) = o(n^2)$ and consequently, by (2) of Sec. 5.1,

$$\frac{T_n - \mathscr{E}(T_n)}{n} = \frac{1}{n}\sum_{j=1}^{n}\{Y_j - \mathscr{E}(Y_j)\} \to 0 \quad \text{in pr.}$$

Now it is clear that as $n \to \infty$, $\mathscr{E}(Y_n) \to \mathscr{E}(X) = m$; hence also

$$\frac{1}{n}\sum_{j=1}^{n}\mathscr{E}(Y_j) \to m.$$

It follows that

$$\frac{T_n}{n} = \frac{1}{n}\sum_{j=1}^{n} Y_j \to m \quad \text{in pr.,}$$

as was to be proved.

For totally independent r.v.'s, necessary and sufficient conditions for the weak law of large numbers in the most general formulation, due to Kolmogorov and Feller, are known. The sufficiency of the following criterion is easily proved, but we omit the proof of its necessity (cf. Gnedenko and Kolmogorov [12]).

**Theorem 5.2.3.** Let $\{X_n\}$ be a sequence of independent r.v.'s with d.f.'s $\{F_n\}$; and $S_n = \sum_{j=1}^{n} X_j$. Let $\{b_n\}$ be a given sequence of real numbers increasing to $+\infty$.

Suppose that we have

**(i)** $\sum_{j=1}^{n} \int_{|x|>b_n} dF_j(x) = o(1)$,

**(ii)** $\dfrac{1}{b_n^2} \sum_{j=1}^{n} \int_{|x|\le b_n} x^2\, dF_j(x) = o(1)$;

then if we put

$$a_n = \sum_{j=1}^{n} \int_{|x|\le b_n} x\, dF_j(x), \tag{5}$$

we have

$$\frac{1}{b_n}(S_n - a_n) \to 0 \quad \text{in pr.} \tag{6}$$

Next suppose that the $F_n$'s have the property that there exists a $\lambda > 0$ such that

$$\forall n: F_n(0) \ge \lambda, \quad 1 - F_n(0-) \ge \lambda. \tag{7}$$

Then if (6) holds for the given $\{b_n\}$ and any sequence of real numbers $\{a_n\}$, the conditions (i) and (ii) must hold.

*Remark.* Condition (7) may be written as

$$\mathscr{P}\{X_n \le 0\} \ge \lambda, \quad \mathscr{P}\{X_n \ge 0\} \ge \lambda;$$

when $\lambda = \frac{1}{2}$ this means that 0 is a *median* for each $X_n$; see Exercise 9 below. In general it ensures that none of the distribution is too far *off center*, and it is certainly satisfied if all $F_n$ are the same; see also Exercise 11 below.

It is possible to replace the $a_n$ in (5) by

$$\sum_{j=1}^{n} \int_{|x| \le b_j} x \, dF_j(x)$$

and maintain (6); see Exercise 8 below.

PROOF OF SUFFICIENCY. Define for each $n \ge 1$ and $1 \le j \le n$:

$$Y_{n,j} = \begin{cases} X_j, & \text{if } |X_j| \le b_n; \\ 0, & \text{if } |X_j| > b_n; \end{cases}$$

and write

$$T_n = \sum_{j=1}^{n} Y_{n,j}.$$

Then condition (i) may be written as

$$\sum_{j=1}^{n} \mathscr{P}\{Y_{n,j} \ne X_j\} = o(1);$$

and it follows from Boole's inequality that

$$\mathscr{P}\{T_n \ne S_n\} \le \mathscr{P}\left\{\bigcup_{j=1}^{n} (Y_{n,j} \ne X_j)\right\} \le \sum_{j=1}^{n} \mathscr{P}\{Y_{n,j} \ne X_j\} = o(1). \tag{8}$$

Next, condition (ii) may be written as

$$\sum_{j=1}^{n} \mathscr{E}\left(\left(\frac{Y_{n,j}}{b_n}\right)^2\right) = o(1);$$

from which it follows, since $\{Y_{n,j}, 1 \le j \le n\}$ are independent r.v.'s:

$$\sigma^2\left(\frac{T_n}{b_n}\right) = \sum_{j=1}^{n} \sigma^2\left(\frac{Y_{n,j}}{b_n}\right) \le \sum_{j=1}^{n} \mathscr{E}\left(\left(\frac{Y_{n,j}}{b_n}\right)^2\right) = o(1).$$

Hence as in (2) of Sec. 5.1,

$$\frac{T_n - \mathscr{E}(T_n)}{b_n} \to 0 \quad \text{in pr.} \tag{9}$$

It is clear (why?) that (8) and (9) together imply

$$\frac{S_n - \mathscr{E}(T_n)}{b_n} \to 0 \quad \text{in pr.}$$

Since

$$\mathscr{E}(T_n) = \sum_{j=1}^{n} \mathscr{E}(Y_{n,j}) = \sum_{j=1}^{n} \int_{|x| \le b_n} x \, dF_j(x) = a_n,$$

(6) is proved.

As an application of Theorem 5.2.3 we give an example where the weak but not the strong law of large numbers holds.

**Example.** Let $\{X_n\}$ be independent r.v.'s with a common d.f. $F$ such that

$$\mathscr{P}\{X_1 = n\} = \mathscr{P}\{X_1 = -n\} = \frac{c}{n^2 \log n}, \qquad n = 3, 4, \ldots,$$

where $c$ is the constant

$$\frac{1}{2}\left(\sum_{n=3}^{\infty} \frac{1}{n^2 \log n}\right)^{-1}.$$

We have then, for large values of $n$,

$$n \int_{|x|>n} dF(x) = n \sum_{k>n} \frac{c}{k^2 \log k} \sim \frac{c}{\log n},$$

$$\frac{1}{n^2} \cdot n \cdot \int_{|x| \le n} x^2 \, dF(x) = \frac{1}{n} \sum_{k=3}^{n} \frac{ck^2}{k^2 \log k} \sim \frac{c}{\log n}.$$

Thus conditions (i) and (ii) are satisfied with $b_n = n$; and we have $a_n = 0$ by (5). Hence $S_n/n \to 0$ in pr. in spite of the fact that $\mathscr{E}(|X_1|) = +\infty$. On the other hand, we have

$$\mathscr{P}\{|X_1| > n\} \sim \frac{c}{n \log n},$$

so that, since $X_1$ and $X_n$ have the same d.f.,

$$\sum_n \mathscr{P}\{|X_n| > n\} = \sum_n \mathscr{P}\{|X_1| > n\} = \infty.$$

Hence by Theorem 4.2.4 (Borel–Cantelli),

$$\mathscr{P}\{|X_n| > n \text{ i.o.}\} = 1.$$

But $|S_n - S_{n-1}| = |X_n| > n$ implies $|S_n| > n/2$ or $|S_{n-1}| > n/2$; it follows that

$$\mathscr{P}\left\{|S_n| > \frac{n}{2} \text{ i.o.}\right\} = 1,$$

and so it is certainly false that $S_n/n \to 0$ a.e. However, we can prove more. For any $A > 0$, the same argument as before yields

$$\mathscr{P}\{|X_n| > An \text{ i.o.}\} = 1$$

and consequently

$$\mathscr{P}\left\{|S_n| > \frac{An}{2} \text{ i.o.}\right\} = 1.$$

This means that for each $A$ there is a null set $Z(A)$ such that if $\omega \in \Omega \backslash Z(A)$, then

$$\varlimsup_{n\to\infty} \frac{S_n(\omega)}{n} \geq \frac{A}{2}. \tag{10}$$

Let $Z = \bigcup_{m=1}^{\infty} Z(m)$; then $Z$ is still a null set, and if $\omega \in \Omega\backslash Z$, (10) is true for every $A$, and therefore the upper limit is $+\infty$. Since $X$ is "symmetric" in the obvious sense, it follows that

$$\varliminf_{n\to\infty} \frac{S_n}{n} = -\infty, \quad \varlimsup_{n\to\infty} \frac{S_n}{n} = +\infty \quad \text{a.e.}$$

## EXERCISES

$$S_n = \sum_{j=1}^{n} X_j.$$

**1.** For any sequence of r.v.'s $\{X_n\}$, and any $p \geq 1$:

$$X_n \to 0 \text{ a.e.} \Rightarrow \frac{S_n}{n} \to 0 \text{ a.e.},$$

$$X_n \to 0 \text{ in } L^p \Rightarrow \frac{S_n}{n} \to 0 \text{ in } L^p.$$

The second result is false for $p < 1$.

**2.** Even for a sequence of independent r.v.'s $\{X_n\}$,

$$X_n \to 0 \text{ in pr.} \not\Rightarrow \frac{S_n}{n} \to 0 \text{ in pr.}$$

[HINT: Let $X_n$ take the values $2^n$ and 0 with probabilities $n^{-1}$ and $1 - n^{-1}$.]

**3.** For any sequence $\{X_n\}$:

$$\frac{S_n}{n} \to 0 \text{ in pr.} \Rightarrow \frac{X_n}{n} \to 0 \text{ in pr.}$$

More generally, this is true if $n$ is replaced by $b_n$, where $b_{n+1}/b_n \to 1$.

$^\star$**4.** For any $\delta > 0$, we have

$$\lim_{n\to\infty} \sum_{|k-np|>n\delta} \binom{n}{k} p^k(1-p)^{n-k} = 0$$

uniformly in $p: 0 < p < 1$.

$^\star$**5.** Let $\mathscr{P}(X_1 = 2^n) = 1/2^n$, $n \geq 1$; and let $\{X_n, n \geq 1\}$ be independent and identically distributed. Show that the weak law of large numbers does not hold for $b_n = n$; namely, with this choice of $b_n$ no sequence $\{a_n\}$ exists for which (6) is true. [This is the St. Petersburg paradox, in which you win $2^n$ if it takes $n$ tosses of a coin to obtain a head. What would you consider as a fair entry fee? and what is your mathematical expectation?]

$^\star$**6.** Show on the contrary that a weak law of large numbers does hold for $b_n = n \log n$ and find the corresponding $a_n$. [HINT: Apply Theorem 5.2.3.]

**7.** Conditions (i) and (ii) in Theorem 5.2.3 imply that for any $\delta > 0$,

$$\sum_{j=1}^{n} \int_{|x|>\delta b_n} dF_j(x) = o(1)$$

and that $a_n = o(\sqrt{n} b_n)$.

**8.** They also imply that

$$\frac{1}{b_n} \sum_{j=1}^{n} \int_{b_j<|x|\leq b_n} x\, dF_j(x) = o(1).$$

[HINT: Use the first part of Exercise 7 and divide the interval of integration $b_j < |x| \leq b_n$ into parts of the form $\lambda^k < |x| \leq \lambda^{k+1}$ with $\lambda > 1$.]

**9.** A median of the r.v. $X$ is any number $\alpha$ such that

$$\mathscr{P}\{X \leq \alpha\} \geq \tfrac{1}{2}, \quad \mathscr{P}\{X \geq \alpha\} \geq \tfrac{1}{2}.$$

Show that such a number always exists but need not be unique.

$^\star$**10.** Let $\{X_n, 1 \leq n \leq \infty\}$ be arbitrary r.v.'s and for each $n$ let $m_n$ be a median of $X_n$. Prove that if $X_n \to X_\infty$ in pr. and $m_\infty$ is unique, then $m_n \to m_\infty$. Furthermore, if there exists any sequence of real numbers $\{c_n\}$ such that $X_n - c_n \to 0$ in pr., then $X_n - m_n \to 0$ in pr.

**11.** Derive the following form of the weak law of large numbers from Theorem 5.2.3. Let $\{b_n\}$ be as in Theorem 5.2.3 and put $X_n = 2b_n$ for $n \geq 1$. Then there exists $\{a_n\}$ for which (6) holds but condition (i) does not.

**12.** Theorem 5.2.2 may be slightly generalized as follows. Let $\{X_n\}$ be pairwise independent with a common d.f. $F$ such that

$$\text{(i)} \int_{|x|\leq n} x\,dF(x) = o(1), \qquad \text{(ii)}\ n\int_{|x|>n} dF(x) = o(1);$$

then $S_n/n \to 0$ in pr.

**13.** Let $\{X_n\}$ be a sequence of identically distributed strictly positive random variables. For any $\varphi$ such that $\varphi(n)/n \to 0$ as $n \to \infty$, show that $\mathscr{P}\{S_n > \varphi(n) \text{ i.o.}\} = 1$, and so $S_n \to \infty$ a.e. [HINT: Let $N_n$ denote the number of $k \leq n$ such that $X_k \leq \varphi(n)/n$. Use Chebyshev's inequality to estimate $\mathscr{P}\{N_n > n/2\}$ and so conclude $\mathscr{P}\{S_n > \varphi(n)/2\} \geq 1 - 2F(\varphi(n)/n)$. This problem was proposed as a teaser and the rather unexpected solution was given by Kesten.]

**14.** Let $\{b_n\}$ be as in Theorem 5.2.3. and put $X_n = 2b_n$ for $n \geq 1$. Then there exists $\{a_n\}$ for which (6) holds, but condition (i) does not hold. Thus condition (7) cannot be omitted.

## 5.3 Convergence of series

If the terms of an infinite series are independent r.v.'s, then it will be shown in Sec. 8.1 that the probability of its convergence is either zero or one. Here we shall establish a concrete criterion for the latter alternative. Not only is the result a complete answer to the question of convergence of independent r.v.'s, but it yields also a satisfactory form of the strong law of large numbers. This theorem is due to Kolmogorov (1929). We begin with his two remarkable inequalities. The first is also very useful elsewhere; the second may be circumvented (see Exercises 3 to 5 below), but it is given here in Kolmogorov's original form as an example of true virtuosity.

**Theorem 5.3.1.** Let $\{X_n\}$ be independent r.v.'s such that

$$\forall n: \mathscr{E}(X_n) = 0, \qquad \mathscr{E}(X_n^2) = \sigma^2(X_n) < \infty.$$

Then we have for every $\epsilon > 0$:

$$\mathscr{P}\{\max_{1\leq j\leq n} |S_j| > \epsilon\} \leq \frac{\sigma^2(S_n)}{\epsilon^2}. \tag{1}$$

*Remark.* If we replace the $\max_{1\leq j\leq n} |S_j|$ in the formula by $|S_n|$, this becomes a simple case of Chebyshev's inequality, of which it is thus an essential improvement.

PROOF. Fix $\epsilon > 0$. For any $\omega$ in the set

$$\Lambda = \{\omega: \max_{1 \le j \le n} |S_j(\omega)| > \epsilon\},$$

let us define

$$\nu(\omega) = \min\{j: 1 \le j \le n, |S_j(\omega)| > \epsilon\}.$$

Clearly $\nu$ is an r.v. with domain $\Lambda$. Put

$$\Lambda_k = \{\omega: \nu(\omega) = k\} = \{\omega: \max_{1 \le j \le k-1} |S_j(\omega)| \le \epsilon, |S_k(\omega)| > \epsilon\},$$

where for $k = 1$, $\max_{1 \le j \le 0} |S_j(\omega)|$ is taken to be zero. Thus $\nu$ is the "first time" that the indicated maximum exceeds $\epsilon$, and $\Lambda_k$ is the event that this occurs "for the first time at the $k$th step". The $\Lambda_k$'s are disjoint and we have

$$\Lambda = \bigcup_{k=1}^{n} \Lambda_k.$$

It follows that

$$\int_\Lambda S_n^2 \, d\mathscr{P} = \sum_{k=1}^{n} \int_{\Lambda_k} S_n^2 \, d\mathscr{P} = \sum_{k=1}^{n} \int_{\Lambda_k} [S_k + (S_n - S_k)]^2 \, d\mathscr{P} \tag{2}$$

$$= \sum_{k=1}^{n} \int_{\Lambda_k} [S_k^2 + 2S_k(S_n - S_k) + (S_n - S_k)^2] \, d\mathscr{P}.$$

Let $\varphi_k$ denote the indicator of $\Lambda_k$, then the two r.v.'s $\varphi_k S_k$ and $S_n - S_k$ are independent by Theorem 3.3.2, and consequently (see Exercise 9 of Sec. 3.3)

$$\int_{\Lambda_k} S_k(S_n - S_k) \, d\mathscr{P} = \int_\Omega (\varphi_k S_k)(S_n - S_k) \, d\mathscr{P}$$

$$= \int_\Omega \varphi_k S_k \, d\mathscr{P} \int_\Omega (S_n - S_k) \, d\mathscr{P} = 0,$$

since the last-written integral is

$$\mathscr{E}(S_n - S_k) = \sum_{j=k+1}^{n} \mathscr{E}(X_j) = 0.$$

Using this in (2), we obtain

$$\sigma^2(S_n) = \int_\Omega S_n^2 \, d\mathscr{P} \ge \int_\Lambda S_n^2 \, d\mathscr{P} \ge \sum_{k=1}^{n} \int_{\Lambda_k} S_k^2 \, d\mathscr{P}$$

$$\ge \epsilon^2 \sum_{k=1}^{n} \mathscr{P}(\Lambda_k) = \epsilon^2 \mathscr{P}(\Lambda),$$

where the last inequality is by the mean value theorem, since $|S_k| > \epsilon$ on $\Lambda_k$ by definition. The theorem now follows upon dividing the inequality above by $\epsilon^2$.

**Theorem 5.3.2.** Let $\{X_n\}$ be independent r.v.'s with finite means and suppose that there exists an $A$ such that

$$\forall n: |X_n - \mathscr{E}(X_n)| \le A < \infty. \tag{3}$$

Then for every $\epsilon > 0$ we have

$$\mathscr{P}\{\max_{1\le j\le n} |S_j| \le \epsilon\} \le \frac{(2A+4\epsilon)^2}{\sigma^2(S_n)}. \tag{4}$$

PROOF. Let $\mathrm{M}_0 = \Omega$, and for $1 \le k \le n$:

$$\mathrm{M}_k = \{\omega: \max_{1\le j\le k} |S_j| \le \epsilon\},$$

$$\Delta_k = \mathrm{M}_{k-1} - \mathrm{M}_k.$$

We may suppose that $\mathscr{P}(\mathrm{M}_n) > 0$, for otherwise (4) is trivial. Furthermore, let $S'_0 = 0$ and for $k \ge 1$,

$$X'_k = X_k - \mathscr{E}(X_k), \quad S'_k = \sum_{j=1}^{k} X'_j.$$

Define numbers $a_k$, $0 \le k \le n$, as follows:

$$a_k = \frac{1}{\mathscr{P}(\mathrm{M}_k)} \int_{\mathrm{M}_k} S'_k \, d\mathscr{P},$$

so that

$$\int_{\mathrm{M}_k} (S'_k - a_k)\, d\mathscr{P} = 0. \tag{5}$$

Now we write

$$\int_{\mathrm{M}_{k+1}} (S'_{k+1} - a_{k+1})^2 \, d\mathscr{P} = \int_{\mathrm{M}_k} (S'_k - a_k + a_k - a_{k+1} + X'_{k+1})^2 \, d\mathscr{P} \tag{6}$$
$$- \int_{\Delta_{k+1}} (S'_k - a_k + a_k - a_{k+1} + X'_{k+1})^2 \, d\mathscr{P}$$

and denote the two integrals on the right by $I_1$ and $I_2$, respectively. Using the definition of $\mathrm{M}_k$ and (3), we have

$$|S'_k - a_k| = \left| S_k - \mathscr{E}(S_k) - \frac{1}{\mathscr{P}(\mathrm{M}_k)} \int_{\mathrm{M}_k} [S_k - \mathscr{E}(S_k)] \, d\mathscr{P} \right|$$
$$= \left| S_k - \frac{1}{\mathscr{P}(\mathrm{M}_k)} \int_{\mathrm{M}_k} S_k \, d\mathscr{P} \right| \le |S_k| + \epsilon;$$

$$
\begin{aligned}
|a_k - a_{k+1}| = \Bigg| \frac{1}{\mathscr{P}(\mathrm{M}_k)} \int_{\mathrm{M}_k} S_k \, d\mathscr{P} - \frac{1}{\mathscr{P}(\mathrm{M}_{k+1})} \int_{\mathrm{M}_{k+1}} S_k \, d\mathscr{P} \\
- \frac{1}{\mathscr{P}(\mathrm{M}_{k+1})} \int_{\mathrm{M}_{k+1}} X'_{k+1} \, d\mathscr{P} \Bigg| \le 2\epsilon + A.
\end{aligned} \tag{7}
$$

It follows that, since $|S_k| \le \epsilon$ on $\Delta_{k+1}$,

$$I_2 \le \int_{\Delta_{k+1}} (|S_k| + \epsilon + 2\epsilon + A + A)^2 \, d\mathscr{P} \le (4\epsilon + 2A)^2 \mathscr{P}(\Delta_{k+1}).$$

On the other hand, we have

$$
\begin{aligned}
I_1 = \int_{\mathrm{M}_k} \{ & (S'_k - a_k)^2 + (a_k - a_{k+1})^2 + X'^2_{k+1} + 2(S'_k - a_k)(a_k - a_{k+1}) \\
& + 2(S'_k - a_k)X'_{k+1} + 2(a_k - a_{k+1})X'_{k+1} \} \, d\mathscr{P}.
\end{aligned}
$$

The integrals of the last three terms all vanish by (5) and independence, hence

$$
\begin{aligned}
I_1 &\ge \int_{\mathrm{M}_k} (S'_k - a_k)^2 \, d\mathscr{P} + \int_{\mathrm{M}_k} X'^2_{k+1} \, d\mathscr{P} \\
&= \int_{\mathrm{M}_k} (S'_k - a_k)^2 \, d\mathscr{P} + \mathscr{P}(\mathrm{M}_k)\sigma^2(X_{k+1}).
\end{aligned}
$$

Substituting into (6), and using $\mathrm{M}_k \supset \mathrm{M}_n$, we obtain for $0 \le k \le n - 1$:

$$
\begin{aligned}
\int_{\mathrm{M}_{k+1}} & (S'_{k+1} - a_{k+1})^2 \, d\mathscr{P} - \int_{\mathrm{M}_k} (S'_k - a_k)^2 \, d\mathscr{P} \\
& \ge \mathscr{P}(\mathrm{M}_n)\sigma^2(X_{k+1}) - (4\epsilon + 2A)^2 \mathscr{P}(\Delta_{k+1}).
\end{aligned}
$$

Summing over $k$ and using (7) again for $k = n$:

$$
\begin{aligned}
4\epsilon^2 \mathscr{P}(\mathrm{M}_n) &\ge \int_{\mathrm{M}_n} (|S_n| + \epsilon)^2 \, d\mathscr{P} \ge \int_{\mathrm{M}_n} (S'_n - a_n)^2 \, d\mathscr{P} \\
&\ge \mathscr{P}(\mathrm{M}_n) \sum_{j=1}^{n} \sigma^2(X_j) - (4\epsilon + 2A)^2 \mathscr{P}(\Omega \backslash \mathrm{M}_n),
\end{aligned}
$$

hence

$$(2A + 4\epsilon)^2 \ge \mathscr{P}(\mathrm{M}_n) \sum_{j=1}^{n} \sigma^2(X_j),$$

which is (4).

We can now prove the "three series theorem" of Kolmogorov (1929).

**Theorem 5.3.3.** Let $\{X_n\}$ be independent r.v.'s and define for a fixed constant $A > 0$:

$$Y_n(\omega) = \begin{cases} X_n(\omega), & \text{if } |X_n(\omega)| \leq A; \\ 0, & \text{if } |X_n(\omega)| > A. \end{cases}$$

Then the series $\sum_n X_n$ converges a.e. if and only if the following three series all converge:

**(i)** $\sum_n \mathscr{P}\{|X_n| > A\} = \sum_n \mathscr{P}\{X_n \neq Y_n\}$,

**(ii)** $\sum_n \mathscr{E}(Y_n)$,

**(iii)** $\sum_n \sigma^2(Y_n)$.

PROOF. Suppose that the three series converge. Applying Theorem 5.3.1 to the sequence $\{Y_n - \mathscr{E}(Y_n)\}$, we have for every $m \geq 1$:

$$\mathscr{P}\left\{\max_{n \leq k \leq n'} \left|\sum_{j=n}^{k} \{Y_j - \mathscr{E}(Y_j)\}\right| \leq \frac{1}{m}\right\} \geq 1 - m^2 \sum_{j=n}^{n'} \sigma^2(Y_j).$$

If we denote the probability on the left by $\mathscr{P}(m, n, n')$, it follows from the convergence of (iii) that for each $m$:

$$\lim_{n \to \infty} \lim_{n' \to \infty} \mathscr{P}(m, n, n') = 1.$$

This means that the tail of $\sum_n \{Y_n - \mathscr{E}(Y_n)\}$ converges to zero a.e., so that the series converges a.e. Since (ii) converges, so does $\sum_n Y_n$. Since (i) converges, $\{X_n\}$ and $\{Y_n\}$ are equivalent sequences; hence $\sum_n X_n$ also converges a.e. by Theorem 5.2.1. We have therefore proved the "if" part of the theorem.

Conversely, suppose that $\sum_n X_n$ converges a.e. Then for each $A > 0$:

$$\mathscr{P}(|X_n| > A \text{ i.o.}) = 0.$$

It follows from the Borel–Cantelli lemma that the series (i) must converge. Hence as before $\sum_n Y_n$ also converges a.e. But since $|Y_n - \mathscr{E}(Y_n)| \leq 2A$, we have by Theorem 5.3.2

$$\mathscr{P}\left\{\max_{n \leq k \leq n'} \left|\sum_{j=n}^{k} Y_j\right| \leq 1\right\} \leq \frac{(4A + 4)^2}{\sum_{j=n}^{n'} \sigma^2(Y_j)}.$$

Were the series (iii) to diverge, the probability above would tend to zero as $n' \to \infty$ for each $n$, hence the tail of $\sum_n Y_n$ almost surely would not be bounded by 1, so the series could not converge. This contradiction proves that (iii) must converge. Finally, consider the series $\sum_n \{Y_n - \mathscr{E}(Y_n)\}$ and apply the proven part of the theorem to it. We have $\mathscr{P}\{|Y_n - \mathscr{E}(Y_n)| > 2A\} = 0$ and $\mathscr{E}(Y_n - \mathscr{E}(Y_n)) = 0$ so that the two series corresponding to (i) and (ii) with $2A$ for $A$ vanish identically, while (iii) has just been shown to converge. It follows from the sufficiency of the criterion that $\sum_n \{Y_n - \mathscr{E}(Y_n)\}$ converges a.e., and so by equivalence the same is true of $\sum_n \{X_n - \mathscr{E}(Y_n)\}$. Since $\sum_n X_n$ also converges a.e. by hypothesis, we conclude by subtraction that the series (ii) converges. This completes the proof of the theorem.

The convergence of a series of r.v.'s is, of course, by definition the same as its partial sums, in any sense of convergence discussed in Chapter 4. For series of independent terms we have, however, the following theorem due to Paul Lévy.

**Theorem 5.3.4.** If $\{X_n\}$ is a sequence of independent r.v.'s, then the convergence of the series $\sum_n X_n$ in pr. is equivalent to its convergence a.e.

PROOF. By Theorem 4.1.2, it is sufficient to prove that convergence of $\sum_n X_n$ in pr. implies its convergence a.e. Suppose the former; then, given $\epsilon: 0 < \epsilon < 1$, there exists $m_0$ such that if $n > m > m_0$, we have

$$\mathscr{P}\{|S_{m,n}| > \epsilon\} < \epsilon, \tag{8}$$

where

$$S_{m,n} = \sum_{j=m+1}^{n} X_j.$$

It is obvious that for $m < k \leq n$ we have

$$\bigcup_{k=m+1}^{n} \{\max_{m<j\leq k-1} |S_{m,j}| \leq 2\epsilon; |S_{m,k}| > 2\epsilon; |S_{k,n}| \leq \epsilon\} \subset \{|S_{m,n}| > \epsilon\} \tag{9}$$

where the sets in the union are disjoint. Going to probabilities and using independence, we obtain

$$\sum_{k=m+1}^{n} \mathscr{P}\{\max_{m<j\leq k-1} |S_{m,j}| \leq 2\epsilon; |S_{m,k}| > 2\epsilon\}\mathscr{P}\{|S_{k,n}| \leq \epsilon\} \leq \mathscr{P}\{|S_{m,n}| > \epsilon\}.$$

If we suppress the factors $\mathscr{P}\{|S_{k,n}| \leq \epsilon\}$, then the sum is equal to

$$\mathscr{P}\{\max_{m<j\leq n} |S_{m,j}| > 2\epsilon\}$$

(cf. the beginning of the proof of Theorem 5.3.1). It follows that

$$\mathscr{P}\{\max_{m<j\leq n} |S_{m,j}| > 2\epsilon\} \min_{m<k\leq n} \mathscr{P}\{|S_{k,n}| \leq \epsilon\} \leq \mathscr{P}\{|S_{m,n}| > \epsilon\}.$$

This inequality is due to Ottaviani. By (8), the second factor on the left exceeds $1 - \epsilon$, hence if $m > m_0$,

$$(10) \qquad \mathscr{P}\{\max_{m<j\leq n} |S_{m,j}| > 2\epsilon\} \leq \frac{1}{1-\epsilon}\mathscr{P}\{|S_{m,n}| > \epsilon\} < \frac{\epsilon}{1-\epsilon}.$$

Letting $n \to \infty$, then $m \to \infty$, and finally $\epsilon \to 0$ through a sequence of values, we see that the triple limit of the first probability in (10) is equal to zero. This proves the convergence a.e. of $\sum_n X_n$ by Exercise 8 of Sec. 4.2.

It is worthwhile to observe the underlying similarity between the inequalities (10) and (1). Both give a "probability upper bound" for the maximum of a number of summands in terms of the last summand as in (10), or its variance as in (1). The same principle will be used again later.

Let us remark also that even the convergence of $S_n$ in dist. is equivalent to that in pr. or a.e. as just proved; see Theorem 9.5.5.

We give some examples of convergence of series.

**Example.** $\sum_n \pm 1/n$.

This is meant to be the "harmonic series" with a random choice of signs in each term, the choices being totally independent and equally likely to be $+$ or $-$ in each case. More precisely, it is the series

$$\sum_n \frac{X_n}{n},$$

where $\{X_n, n \geq 1\}$ is a sequence of independent, identically distributed r.v.'s taking the values $\pm 1$ with probability $\frac{1}{2}$ each.

We may take $A = 1$ in Theorem 5.3.3 so that the two series (i) and (ii) vanish identically. Since $\sigma^2(X_n) = \sigma^2(Y_n) = 1/n^2$, the series (iii) converges. Hence, $\sum_n \pm 1/n$ converges a.e. by the criterion above. The same conclusion applies to $\sum_n \pm 1/n^\theta$ if $\frac{1}{2} < \theta \leq 1$. Clearly there is no absolute convergence. For $0 \leq \theta \leq \frac{1}{2}$, the probability of convergence is zero by the same criterion and by Theorem 8.1.2 below.

EXERCISES

**1.** Theorem 5.3.1 has the following "one-sided" analogue. Under the same hypotheses, we have

$$\mathscr{P}\{\max_{1\leq j\leq n} S_j \geq \epsilon\} \leq \frac{\sigma^2(S_n)}{\epsilon^2 + \sigma^2(S_n)}.$$

[This is due to A. W. Marshall.]

★**2.** Let $\{X_n\}$ be independent and identically distributed with mean 0 and variance 1. Then we have for every $x$:

$$\mathscr{P}\{\max_{1 \le j \le n} S_j \ge x\} \le 2\mathscr{P}\{S_n \ge x - \sqrt{2n}\}.$$

[HINT: Let

$$\Lambda_k = \{\max_{1 \le j < k} S_j < x; S_k \ge x\}$$

then $\sum_{k=1}^{n} \mathscr{P}\{\Lambda_k; S_n - S_k \ge -\sqrt{2n}\} \le \mathscr{P}\{S_n \ge x - \sqrt{2n}\}$.]

**3.** Theorem 5.3.2 has the following companion, which is easier to prove. Under the joint hypotheses in Theorems 5.3.1 and 5.3.2, we have

$$\mathscr{P}\{\max_{1 \le j \le n} |S_j| \le \epsilon\} \le \frac{(A+\epsilon)^2}{\sigma^2(S_n)}.$$

**4.** Let $\{X_n, X'_n, n \ge 1\}$ be independent r.v.'s such that $X_n$ and $X'_n$ have the same distribution. Suppose further that all these r.v.'s are bounded by the same constant $A$. Then

$$\sum_n (X_n - X'_n)$$

converges a.e. if and only if

$$\sum_n \sigma^2(X_n) < \infty.$$

Use Exercise 3 to prove this without recourse to Theorem 5.3.3, and so finish the converse part of Theorem 5.3.3.

★**5.** But neither Theorem 5.3.2 nor the alternative indicated in the preceding exercise is necessary; what we need is merely the following result, which is an easy consequence of a general theorem in Chapter 7. Let $\{X_n\}$ be a sequence of independent and uniformly bounded r.v.'s with $\sigma^2(S_n) \to +\infty$. Then for every $A > 0$ we have

$$\lim_{n \to \infty} \mathscr{P}\{|S_n| \le A\} = 0.$$

Show that this is sufficient to finish the proof of Theorem 5.3.3.

★**6.** The following analogue of the inequalities of Kolmogorov and Ottaviani is due to P. Lévy. Let $S_n$ be the sum of $n$ independent r.v.'s and $S_n^0 = S_n - m_0(S_n)$, where $m_0(S_n)$ is a median of $S_n$. Then we have

$$\mathscr{P}\{\max_{1 \le j \le n} |S_j^0| > \epsilon\} \le 3\mathscr{P}\left\{|S_n^0| > \frac{\epsilon}{2}\right\}.$$

[HINT: Try "4" in place of "3" on the right.]

**7.** For arbitrary $\{X_n\}$, if

$$\sum_n \mathscr{E}(|X_n|) < \infty,$$

then $\sum_n X_n$ converges absolutely a.e.

**8.** Let $\{X_n\}$, where $n = 0, \pm 1, \pm 2, \dots$, be independent and identically distributed according to the normal distribution $\Phi$ (see Exercise 5 of Sec. 4.5). Then the series of complex-valued r.v.'s

$$xX_0 + \sum_{n=1}^{\infty} \frac{e^{inx}X_n}{in} + \sum_{n=1}^{\infty} \frac{e^{-inx}X_{-n}}{-in},$$

where $i = \sqrt{-1}$ and $x$ is real, converges a.e. and uniformly in $x$. (This is Wiener's representation of the Brownian motion process.)

$^{\star}$**9.** Let $\{X_n\}$ be independent and identically distributed, taking the values 0 and 2 with probability $\frac{1}{2}$ each; then

$$\sum_{n=1}^{\infty} \frac{X_n}{3^n}$$

converges a.e. Prove that the limit has the Cantor d.f. discussed in Sec. 1.3. Do Exercise 11 in that section again; it is easier now.

$^{\star}$**10.** If $\sum_n \pm X_n$ converges a.e. for all choices of $\pm 1$, where the $X_n$'s are arbitrary r.v.'s, then $\sum_n {X_n}^2$ converges a.e. [HINT: Consider $\sum_n r_n(t)X_n(\omega)$ where the $r_n$'s are coin-tossing r.v.'s and apply Fubini's theorem to the space of $(t, \omega)$.]

## 5.4 Strong law of large numbers

To return to the strong law of large numbers, the link is furnished by the following lemma on "summability".

**Kronecker's lemma.** Let $\{x_k\}$ be a sequence of real numbers, $\{a_k\}$ a sequence of numbers $>0$ and $\uparrow \infty$. Then

$$\sum_n \frac{x_n}{a_n} < \text{converges} \Rightarrow \frac{1}{a_n} \sum_{j=1}^{n} x_j \to 0.$$

PROOF. For $1 \le n \le \infty$ let

$$b_n = \sum_{j=1}^{n} \frac{x_j}{a_j}.$$

If we also write $a_0 = 0$, $b_0 = 0$, we have

$$x_n = a_n(b_n - b_{n-1})$$

and

$$\frac{1}{a_n}\sum_{j=1}^{n} x_j = \frac{1}{a_n}\sum_{j=1}^{n} a_j(b_j - b_{j-1}) = b_n - \frac{1}{a_n}\sum_{j=0}^{n-1} b_j(a_{j+1} - a_j)$$

(Abel's method of partial summation). Since $a_{j+1} - a_j \geq 0$,

$$\frac{1}{a_n}\sum_{j=0}^{n-1}(a_{j+1} - a_j) = 1,$$

and $b_n \to b_\infty$, we have

$$\frac{1}{a_n}\sum_{j=1}^{n} x_j \to b_\infty - b_\infty = 0.$$

The lemma is proved.

Now let $\varphi$ be a positive, even, and continuous function on $\mathscr{R}^1$ such that as $|x|$ increases,

$$\frac{\varphi(x)}{|x|} \uparrow, \quad \frac{\varphi(x)}{x^2} \downarrow. \tag{1}$$

**Theorem 5.4.1.** Let $\{X_n\}$ be a sequence of independent r.v.'s with $\mathscr{E}(X_n) = 0$ for every $n$; and $0 < a_n \uparrow \infty$. If $\varphi$ satisfies the conditions above and

$$\sum_n \frac{\mathscr{E}(\varphi(X_n))}{\varphi(a_n)} < \infty, \tag{2}$$

then

$$\sum_n \frac{X_n}{a_n} \text{ converges a.e.} \tag{3}$$

PROOF. Denote the d.f. of $X_n$ by $F_n$. Define for each $n$:

$$Y_n(\omega) = \begin{cases} X_n(\omega), & \text{if } |X_n(\omega)| \leq a_n, \\ 0, & \text{if } |X_n(\omega)| > a_n. \end{cases} \tag{4}$$

Then

$$\sum_n \mathscr{E}\left(\frac{Y_n^2}{a_n^2}\right) = \sum_n \int_{|x| \leq a_n} \frac{x^2}{a_n^2}\, dF_n(x).$$

By the second hypothesis in (1), we have

$$\frac{x^2}{a_n^2} \leq \frac{\varphi(x)}{\varphi(a_n)} \quad \text{for } |x| \leq a_n.$$

It follows that

$$\sum_n \sigma^2\left(\frac{Y_n}{a_n}\right) \leq \sum_n \mathscr{E}\left(\frac{Y_n^2}{a_n^2}\right) \leq \sum_n \int_{|x|\leq a_n} \frac{\varphi(x)}{\varphi(a_n)}\, dF_n(x)$$

$$\leq \sum_n \frac{\mathscr{E}(\varphi(X_n))}{\varphi(a_n)} < \infty.$$

Thus, for the r.v.'s $\{Y_n - \mathscr{E}(Y_n)\}/a_n$, the series (iii) in Theorem 5.3.3 converges, while the two other series vanish for $A = 2$, since $|Y_n - \mathscr{E}(Y_n)| \leq 2a_n$; hence

(5) $$\sum_n \frac{1}{a_n}\{Y_n - \mathscr{E}(Y_n)\} \quad \text{converges a.e.}$$

Next we have

$$\sum_n \frac{|\mathscr{E}(Y_n)|}{a_n} = \sum_n \frac{1}{a_n}\left|\int_{|x|\leq a_n} x\, dF_n(x)\right| = \sum_n \frac{1}{a_n}\left|\int_{|x|>a_n} x\, dF_n(x)\right|$$

$$\leq \sum_n \int_{|x|>a_n} \frac{|x|}{a_n}\, dF_n(x),$$

where the second equation follows from $\int_{-\infty}^{\infty} x\, dF_n(x) = 0$. By the first hypothesis in (1), we have

$$\frac{|x|}{a_n} \leq \frac{\varphi(x)}{\varphi(a_n)} \quad \text{for } |x| > a_n.$$

It follows that

$$\sum_n \frac{|\mathscr{E}(Y_n)|}{a_n} \leq \sum_n \int_{|x|>a_n} \frac{\varphi(x)}{\varphi(a_n)}\, dF_n(x) \leq \sum_n \frac{\mathscr{E}(\varphi(X_n))}{\varphi(a_n)} < \infty.$$

This and (5) imply that $\sum_n (Y_n/a_n)$ converges a.e. Finally, since $\varphi \uparrow$, we have

$$\sum_n \mathscr{P}\{X_n \neq Y_n\} = \sum_n \int_{|x|>a_n} dF_n(x) \leq \sum_n \int_{|x|>a_n} \frac{\varphi(x)}{\varphi(a_n)}\, dF_n(x)$$

$$\leq \sum_n \frac{\mathscr{E}(\varphi(X_n))}{\varphi(a_n)} < \infty.$$

Thus, $\{X_n\}$ and $\{Y_n\}$ are equivalent sequences and (3) follows.

Applying Kronecker's lemma to (3) for each $\omega$ in a set of probability one, we obtain the next result.

**Corollary.** Under the hypotheses of the theorem, we have

$$\frac{1}{a_n}\sum_{j=1}^{n} X_j \to 0 \quad \text{a.e.} \tag{6}$$

*Particular cases.* **(i)** Let $\varphi(x) = |x|^p$, $1 \le p \le 2$; $a_n = n$. Then we have

$$\sum_n \frac{1}{n^p}\mathscr{E}(|X_n|^p) < \infty \Rightarrow \frac{1}{n}\sum_{j=1}^{n} X_j \to 0 \quad \text{a.e.} \tag{7}$$

For $p = 2$, this is due to Kolmogorov; for $1 \le p < 2$, it is due to Marcinkiewicz and Zygmund.

**(ii)** Suppose for some $\delta$, $0 < \delta \le 1$ and $M < \infty$ we have

$$\forall n: \mathscr{E}(|X_n|^{1+\delta}) \le M.$$

Then the hypothesis in (7) is clearly satisfied with $p = 1 + \delta$. This case is due to Markov. Cantelli's theorem under total independence (Exercise 4 of Sec. 5.1) is a special case.

**(iii)** By proper choice of $\{a_n\}$, we can considerably sharpen the conclusion (6). Suppose

$$\forall n: \sigma^2(X_n) = \sigma_n^2 < \infty, \quad \sigma^2(S_n) = s_n^2 = \sum_{j=1}^{n}\sigma_j^2 \to \infty.$$

Choose $\varphi(x) = x^2$ and $a_n = s_n(\log s_n)^{(1/2)+\epsilon}$, $\epsilon > 0$, in the corollary to Theorem 5.4.1. Then

$$\sum_n \frac{\mathscr{E}(X_n^2)}{a_n^2} = \sum_n \frac{\sigma_n^2}{s_n^2(\log s_n)^{1+2\epsilon}} < \infty$$

by Dini's theorem, and consequently

$$\frac{S_n}{s_n(\log s_n)^{(1/2)+\epsilon}} \to 0 \quad \text{a.e.}$$

In case all $\sigma_n^2 = 1$ so that $s_n^2 = n$, the above ratio has a denominator that is close to $n^{1/2}$. Later we shall see that $n^{1/2}$ is a critical order of magnitude for $S_n$.

We now come to the strong version of Theorem 5.2.2 in the totally independent case. This result is also due to Kolmogorov.

**Theorem 5.4.2.** Let $\{X_n\}$ be a sequence of independent and identically distributed r.v.'s. Then we have

$$\mathscr{E}(|X_1|) < \infty \Rightarrow \frac{S_n}{n} \to \mathscr{E}(X_1) \quad \text{a.e.,} \tag{8}$$

$$\mathscr{E}(|X_1|) = \infty \Rightarrow \varlimsup_{n\to\infty} \frac{|S_n|}{n} = +\infty \quad \text{a.e.} \tag{9}$$

PROOF. To prove (8) define $\{Y_n\}$ as in (4) with $a_n = n$. Since

$$\sum_n \mathscr{P}\{X_n \neq Y_n\} = \sum_n \mathscr{P}\{|X_n| > n\} = \sum_n \mathscr{P}\{|X_1| > n\} < \infty$$

by Theorem 3.2.1, $\{X_n\}$ and $\{Y_n\}$ are equivalent sequences. Let us apply (7) to $\{Y_n - \mathscr{E}(Y_n)\}$, with $\varphi(x) = x^2$. We have

$$\sum_n \frac{\sigma^2(Y_n)}{n^2} \le \sum_n \frac{\mathscr{E}(Y_n^2)}{n^2} = \sum_n \frac{1}{n^2} \int_{|x|\le n} x^2\, dF(x). \tag{10}$$

We are obliged to estimate the last written second moment in terms of the first moment, since this is the only one assumed in the hypothesis. The standard technique is to split the interval of integration and then invert the repeated summation, as follows:

$$\begin{aligned}
&\sum_{n=1}^{\infty} \frac{1}{n^2} \sum_{j=1}^{n} \int_{j-1<|x|\le j} x^2\, dF(x) \\
&= \sum_{j=1}^{\infty} \int_{j-1<|x|\le j} x^2\, dF(x) \sum_{n=j}^{\infty} \frac{1}{n^2} \\
&\le \sum_{j=1}^{\infty} j \int_{j-1<|x|\le j} |x|\, dF(x) \cdot \frac{C}{j} \le C \sum_{j=1}^{\infty} \cdot \int_{j-1\le|x|\le j} |x|\, dF(x) \\
&= C\mathscr{E}(|X_1|) < \infty.
\end{aligned}$$

In the above we have used the elementary estimate $\sum_{n=j}^{\infty} n^{-2} \le Cj^{-1}$ for some constant $C$ and all $j \ge 1$. Thus the first sum in (10) converges, and we conclude by (7) that

$$\frac{1}{n} \sum_{j=1}^{n} \{Y_j - \mathscr{E}(Y_j)\} \to 0 \quad \text{a.e.}$$

Clearly $\mathscr{E}(Y_n) \to \mathscr{E}(X_1)$ as $n \to \infty$; hence also

$$\frac{1}{n}\sum_{j=1}^{n} \mathscr{E}(Y_j) \to \mathscr{E}(X_1),$$

and consequently

$$\frac{1}{n}\sum_{j=1}^{n} Y_j \to \mathscr{E}(X_1) \quad \text{a.e.}$$

By Theorem 5.2.1, the left side above may be replaced by $(1/n)\sum_{j=1}^{n} X_j$, proving (8).

To prove (9), we note that $\mathscr{E}(|X_1|) = \infty$ implies $\mathscr{E}(|X_1|/A) = \infty$ for each $A > 0$ and hence, by Theorem 3.2.1,

$$\sum_n \mathscr{P}(|X_1| > An) = +\infty.$$

Since the r.v.'s are identically distributed, it follows that

$$\sum_n \mathscr{P}(|X_n| > An) = +\infty.$$

Now the argument in the example at the end of Sec. 5.2 may be repeated without any change to establish the conclusion of (9).

Let us remark that the first part of the preceding theorem is a special case of G. D. Birkhoff's ergodic theorem, but it was discovered a little earlier and the proof is substantially simpler.

N. Etemadi proved an unexpected generalization of Theorem 5.4.2: (8) is true when the (total) independence of $\{X_n\}$ is weakened to pairwise independence (*An elementary proof of the strong law of large numbers*, Z. Wahrscheinlichkeitstheorie **55** (1981), 119–122).

Here is an interesting extension of the law of large numbers when the mean is infinite, due to Feller (1946).

**Theorem 5.4.3.** Let $\{X_n\}$ be as in Theorem 5.4.2 with $\mathscr{E}(|X_1|) = \infty$. Let $\{a_n\}$ be a sequence of positive numbers satisfying the condition $a_n/n \uparrow$. Then we have

$$\overline{\lim_n} \frac{|S_n|}{a_n} = 0 \quad \text{a.e.,} \quad \text{or} = \infty \quad \text{a.e.} \tag{11}$$

according as

$$\sum_n \mathscr{P}\{|X_n| \geq a_n\} = \sum_n \int_{|x| \geq a_n} dF(x) < \infty, \qquad \text{or} = \infty. \tag{12}$$

PROOF. Writing

$$\int_{|x| \geq a_n} dF(x) = \sum_{k=n}^{\infty} \int_{a_k \leq |x| < a_{k+1}} dF(x),$$

substituting into (12) and rearranging the double series, we see that the series in (12) converges if and only if

$$\sum_k k \int_{a_{k-1} \le |x| < a_k} dF(x) < \infty. \tag{13}$$

Assuming, this is the case, we put

$$\mu_n = \int_{|x|<a_n} x\, dF(x);$$

$$Y_n = \begin{cases} X_n - \mu_n & \text{if } |X_n| < a_n, \\ -\mu_n & \text{if } |X_n| \ge a_n. \end{cases}$$

Thus $\mathscr{E}(Y_n) = 0$. We have by (12),

$$\sum_n \mathscr{P}\{Y_n \ne X_n - \mu_n\} < \infty. \tag{14}$$

Next, with $a_0 = 0$:

$$\begin{aligned} \sum_n \mathscr{E}\left(\frac{Y_n^2}{a_n^2}\right) &\le \sum_n \frac{1}{a_n^2} \int_{|x|<a_n} x^2\, dF(x) \\ &= \sum_{n=1}^{\infty} \frac{1}{a_n^2} \sum_{k=1}^{n} \int_{a_{k-1} \le |x| < a_k} x^2\, dF(x) \\ &\le \sum_{k=1}^{\infty} \int_{a_{k-1} \le |x| < a_k} dF(x) a_k^2 \sum_{n=k}^{\infty} \frac{1}{a_n^2}. \end{aligned}$$

Since $a_n/n \ge a_k/k$ for $n \ge k$, we have

$$\sum_{n=k}^{\infty} \frac{1}{a_n^2} \le \frac{k^2}{a_k^2} \sum_{n=k}^{\infty} \frac{1}{n^2} \le \frac{2k}{a_k^2},$$

and so

$$\sum_n \mathscr{E}\left(\frac{Y_n^2}{a_n^2}\right) \le \sum_{k=1}^{\infty} 2k \int_{a_{k-1} \le |x| < a_k} dF(x) < \infty$$

by (13). Hence $\sum Y_n/a_n$ converges (absolutely) a.e. by Theorem 5.4.1, and so by Kronecker's lemma:

$$\frac{1}{a_n} \sum_{k=1}^{n} Y_k \to 0 \quad \text{a.e.} \tag{15}$$

We now estimate the quantity

$$\frac{1}{a_n}\sum_{k=1}^{n}\mu_k = \frac{1}{a_n}\sum_{k=1}^{n}\int_{|x|<a_k} x\,dF(x) \tag{16}$$

as $n \to \infty$. Clearly for any $N < n$, it is bounded in absolute value by

$$\frac{n}{a_n}\left(a_N + \int_{a_N<|x|<a_n} |x|\,dF(x)\right). \tag{17}$$

Since $\mathscr{E}(|X_1|) = \infty$, the series in (12) cannot converge if $a_n/n$ remains bounded (see Exercise 4 of Sec. 3.2). Hence for fixed $N$ the term $(n/a_n)a_N$ in (17) tends to 0 as $n \to \infty$. The rest is bounded by

$$\frac{n}{a_n}\sum_{j=N+1}^{n} a_j \int_{a_{j-1}\le|x|<a_j} dF(x) \le \sum_{j=N+1}^{n} j \int_{a_{j-1}\le|x|<a_j} dF(x) \tag{18}$$

because $na_j/a_n \le j$ for $j \le n$. We may now as well replace the $n$ in the right-hand member above by $\infty$; as $N \to \infty$, it tends to 0 as the remainder of the convergent series in (13). Thus the quantity in (16) tends to 0 as $n \to \infty$; combine this with (14) and (15), we obtain the first alternative in (11).

The second alternative is proved in much the same way as in Theorem 5.4.2 and is left as an exercise. Note that when $a_n = n$ it reduces to (9) above.

**Corollary.** Under the conditions of the theorem, we have

$$\mathscr{P}\{|S_n| \ge a_n \text{ i.o.}\} = \mathscr{P}\{|X_n| \ge a_n \text{ i.o.}\}. \tag{19}$$

This follows because the second probability above is equal to 0 or 1 according as the series in (12) converges or diverges, by the Borel–Cantelli lemma. The result is remarkable in suggesting that the $n$th partial sum and the $n$th individual term of the sequence $\{X_n\}$ have comparable growth in a certain sense. This is in contrast to the situation when $\mathscr{E}(|X_1|) < \infty$ and (19) is false for $a_n = n$ (see Exercise 12 below).

## EXERCISES

The $X_n$'s are independent throughout; in Exercises 1, 6, 8, 9, and 12 they are also identically distributed; $S_n = \sum_{j=1}^{n} X_j$.

***1.** If $\mathscr{E}(X_1^+) = +\infty$, $\mathscr{E}(X_1^-) < \infty$, then $S_n/n \to +\infty$ a.e.

***2.** There is a complement to Theorem 5.4.1 as follows. Let $\{a_n\}$ and $\varphi$ be as there except that the conditions in (1) are replaced by the condition that $\varphi(x) \uparrow$ and $\varphi(x)/|x| \downarrow$. Then again (2) implies (3).

**3.** Let $\{X_n\}$ be independent and identically distributed r.v.'s such that $\mathscr{E}(|X_1|^p) < \infty$ for some $p: 0 < p < 2$; in case $p > 1$, we assume also that $\mathscr{E}(X_1) = 0$. Then $S_n n^{-(1/p)-\epsilon} \to 0$ a.e. For $p = 1$ the result is weaker than Theorem 5.4.2.

**4.** Both Theorem 5.4.1 and its complement in Exercise 2 above are "best possible" in the following sense. Let $\{a_n\}$ and $\varphi$ be as before and suppose that $b_n > 0$,

$$\sum_n \frac{b_n}{\varphi(a_n)} = \infty.$$

Then there exists a sequence of independent and identically distributed r.v.'s $\{X_n\}$ such that $\mathscr{E}(X_n) = 0$, $\mathscr{E}(\varphi(X_n)) = b_n$, and

$$\mathscr{P}\left\{\sum_n \frac{X_n}{a_n} \text{ converges}\right\} = 0.$$

[HINT: Make $\sum_n \mathscr{P}\{|X_n| \geq a_n\} = \infty$ by letting each $X_n$ take two or three values only according as $b_n/\varphi(a_n) \leq 1$ or $>1$.]

**5.** Let $X_n$ take the values $\pm n^\theta$ with probability $\frac{1}{2}$ each. If $0 \leq \theta < \frac{1}{2}$, then $S_n/n \to 0$ a.e. What if $\theta \geq \frac{1}{2}$? [HINT: To answer the question, use Theorem 5.2.3 or Exercise 12 of Sec. 5.2; an alternative method is to consider the characteristic function of $S_n/n$ (see Chapter 6).]

**6.** Let $\mathscr{E}(X_1) = 0$ and $\{c_n\}$ be a bounded sequence of real numbers. Then

$$\frac{1}{n}\sum_{j=1}^{n} c_j X_j \to 0 \quad \text{a.e.}$$

[HINT: Truncate $X_n$ at $n$ and proceed as in Theorem 5.4.2.]

**7.** We have $S_n/n \to 0$ a.e. if and only if the following two conditions are satisfied:

**(i)** $S_n/n \to 0$ in pr.,

**(ii)** $S_{2^n}/2^n \to 0$ a.e.;

an alternative set of conditions is (i) and

(ii$'$) $\forall \epsilon > 0: \sum_n \mathscr{P}(|S_{2^{n+1}} - S_{2^n}| > 2^n \epsilon) < \infty$.

***8.** If $\mathscr{E}(|X_1|) < \infty$, then the sequence $\{S_n/n\}$ is uniformly integrable and $S_n/n \to \mathscr{E}(X_1)$ in $L^1$ as well as a.e.

**9.** Construct an example where $\mathscr{E}(X_1^+) = \mathscr{E}(X_1^-) = +\infty$ and $S_n/n \to +\infty$ a.e. [HINT: Let $0 < \alpha < \beta < 1$ and take a d.f. $F$ such that $1 - F(x) \sim x^{-\alpha}$ as $x \to \infty$, and $\int_{-\infty}^0 |x|^\beta \, dF(x) < \infty$. Show that

$$\sum_n \mathscr{P}\{\max_{1 \le j \le n} X_j^+ \le n^{1/\alpha'}\} < \infty$$

for every $\alpha' > \alpha$ and use Exercise 3 for $\sum_{j=1}^n X_j^-$. This example is due to Derman and Robbins. Necessary and sufficient condition for $S_n/n \to +\infty$ has been given recently by K. B. Erickson.

**10.** Suppose there exist an $\alpha$, $0 < \alpha < 2$, $\alpha \neq 1$, and two constants $A_1$ and $A_2$ such that

$$\forall n, \forall x > 0: \quad \frac{A_1}{x^\alpha} \le \mathscr{P}\{|X_n| > x\} \le \frac{A_2}{x^\alpha}.$$

If $\alpha > 1$, suppose also that $\mathscr{E}(X_n) = 0$ for each $n$. Then for any sequence $\{a_n\}$ increasing to infinity, we have

$$\mathscr{P}\{|S_n| > a_n \text{ i.o.}\} = \begin{Bmatrix} 0 \\ 1 \end{Bmatrix} \quad \text{if } \sum_n \frac{1}{a_n^\alpha} \begin{Bmatrix} < \\ = \end{Bmatrix} \infty.$$

[This result, due to P. Lévy and Marcinkiewicz, was stated with a superfluous condition on $\{a_n\}$. Proceed as in Theorem 5.3.3 but truncate $X_n$ at $a_n$; direct estimates are easy.]

**11.** Prove the second alternative in Theorem 5.4.3.

**12.** If $\mathscr{E}(X_1) \neq 0$, then $\max_{1 \le k \le n} |X_k|/|S_n| \to 0$ a.e. [HINT: $|X_n|/n \to 0$ a.e.]

**13.** Under the assumptions in Theorem 5.4.2, if $S_n/n$ converges a.e. then $\mathscr{E}(|X_1|) < \infty$. [Hint: $X_n/n$ converges to 0 a.e., hence $\mathscr{P}\{|X_n| > n \text{ i.o.}\} = 0$; use Theorem 4.2.4 to get $\sum_n \mathscr{P}\{|X_1| > n\} < \infty$.]

## 5.5 Applications

The law of large numbers has numerous applications in all parts of probability theory and in other related fields such as combinatorial analysis and statistics. We shall illustrate this by two examples involving certain important new concepts.

The first deals with so-called "empiric distributions" in sampling theory. Let $\{X_n, n \ge 1\}$ be a sequence of independent, identically distributed r.v.'s with the common d.f. $F$. This is sometimes referred to as the "underlying" or "theoretical distribution" and is regarded as "unknown" in statistical lingo. For each $\omega$, the values $X_n(\omega)$ are called "samples" or "observed values", and the idea is to get some information on $F$ by looking at the samples. For each

$n$, and each $\omega \in \Omega$, let the $n$ real numbers $\{X_j(\omega), 1 \leq j \leq n\}$ be arranged in increasing order as

$$(1) \qquad Y_{n1}(\omega) \leq Y_{n2}(\omega) \leq \cdots \leq Y_{nn}(\omega).$$

Now define a discrete d.f. $F_n(\cdot, \omega)$ as follows:

$$F_n(x, \omega) = 0, \quad \text{if } x < Y_{n1}(\omega),$$

$$F_n(x, \omega) = \frac{k}{n}, \quad \text{if } Y_{nk}(\omega) \leq x < Y_{n,k+1}(\omega), \ 1 \leq k \leq n-1,$$

$$F_n(x, \omega) = 1, \quad \text{if } x \geq Y_{nn}(\omega).$$

In other words, for each $x$, $nF_n(x, \omega)$ is the number of values of $j$, $1 \leq j \leq n$, for which $X_j(\omega) \leq x$; or again $F_n(x, \omega)$ is the observed frequency of sample values not exceeding $x$. The function $F_n(\cdot, \omega)$ is called the *empiric distribution function based on n samples from F*.

For each $x$, $F_n(x, \cdot)$ is an r.v., as we can easily check. Let us introduce also the indicator r.v.'s $\{\xi_j(x), j \geq 1\}$ as follows:

$$\xi_j(x, \omega) = \begin{cases} 1 & \text{if } X_j(\omega) \leq x, \\ 0 & \text{if } X_j(\omega) > x. \end{cases}$$

We have then

$$F_n(x, \omega) = \frac{1}{n} \sum_{j=1}^{n} \xi_j(x, \omega).$$

For each $x$, the sequence $\{\xi_j(x)\}$ is totally independent since $\{X_j\}$ is, by Theorem 3.3.1. Furthermore they have the common "Bernoullian distribution", taking the values 1 and 0 with probabilities $p$ and $q = 1 - p$, where

$$p = F(x), \quad q = 1 - F(x);$$

thus $\mathscr{E}(\xi_j(x)) = F(x)$. The strong law of large numbers in the form Theorem 5.1.2 or 5.4.2 applies, and we conclude that

$$(2) \qquad F_n(x, \omega) \to F(x) \quad \text{a.e.}$$

Matters end here if we are interested only in a particular value of $x$, or a finite number of values, but since both members in (2) contain the parameter $x$, which ranges over the whole real line, how much better it would be to make a global statement about the *functions* $F_n(\cdot, \omega)$ and $F(\cdot)$. We shall do this in the theorem below. Observe first the precise meaning of (2): for each $x$, there exists a null set $N(x)$ such that (2) holds for $\omega \in \Omega \backslash N(x)$. It follows that (2) also holds simultaneously for all $x$ in any given countable set $Q$, such as the set of rational numbers, for $\omega \in \Omega \backslash \mathrm{N}$, where

$$\mathrm{N} = \bigcup_{x \in Q} \mathrm{N}(x)$$

is again a null set. Hence by the definition of vague convergence in Sec. 4.4, we can already assert that

$$F_n(\cdot, \omega) \xrightarrow{v} F(\cdot) \quad \text{for a.e. } \omega.$$

This will be further strengthened in two ways: convergence for all $x$ and uniformity. The result is due to Glivenko and Cantelli.

**Theorem 5.5.1.** We have as $n \to \infty$

$$\sup_{-\infty < x < \infty} |F_n(x, \omega) - F(x)| \to 0 \quad \text{a.e.}$$

PROOF. Let $J$ be the countable set of jumps of $F$. For each $x \in J$, define

$$\eta_j(x, \omega) = \begin{cases} 1, & \text{if } X_j(\omega) = x; \\ 0, & \text{if } X_j(\omega) \neq x. \end{cases}$$

Then for $x \in J$:

$$F_n(x+, \omega) - F_n(x-, \omega) = \frac{1}{n} \sum_{j=1}^{n} \eta_j(x, \omega),$$

and it follows as before that there exists a null set $\mathrm{N}(x)$ such that if $\omega \in \Omega \backslash \mathrm{N}(x)$, then

$$F_n(x+, \omega) - F_n(x-, \omega) \to F(x+) - F(x-). \tag{3}$$

Now let $\mathrm{N}_1 = \bigcup_{x \in Q \cup J} \mathrm{N}(x)$, then $\mathrm{N}_1$ is a null set, and if $\omega \in \Omega \backslash \mathrm{N}_1$, then (3) holds for every $x \in J$ and we have also

$$F_n(x, \omega) \to F(x) \tag{4}$$

for every $x \in Q$. Hence the theorem will follow from the following analytical result.

**Lemma.** Let $F_n$ and $F$ be (right continuous) d.f.'s, $Q$ and $J$ as before. Suppose that we have

$$\forall x \in Q: F_n(x) \to F(x);$$

$$\forall x \in J: F_n(x) - F_n(x-) \to F(x) - F(x-).$$

Then $F_n$ converges uniformly to $F$ in $\mathscr{R}^1$.

PROOF. Suppose the contrary, then there exist $\epsilon > 0$, a sequence $\{n_k\}$ of integers tending to infinity, and a sequence $\{x_k\}$ in $\mathscr{R}^1$ such that for all $k$:

$$(5) \qquad |F_{n_k}(x_k) - F(x_k)| \geq \epsilon > 0.$$

This is clearly impossible if $x_k \to +\infty$ or $x_k \to -\infty$. Excluding these cases, we may suppose by taking a subsequence that $x_k \to \xi \in \mathscr{R}^1$. Now consider four possible cases and the respective inequalities below, valid for all sufficiently large $k$, where $r_1 \in Q$, $r_2 \in Q$, $r_1 < \xi < r_2$.

*Case 1.* $x_k \uparrow \xi, x_k < \xi$ :

$$\begin{aligned}\epsilon &\leq F_{n_k}(x_k) - F(x_k) \leq F_{n_k}(\xi-) - F(r_1) \\ &\leq F_{n_k}(\xi-) - F_{n_k}(\xi) + F_{n_k}(r_2) - F(r_2) + F(r_2) - F(r_1).\end{aligned}$$

*Case 2.* $x_k \uparrow \xi, x_k < \xi$ :

$$\begin{aligned}\epsilon &\leq F(x_k) - F_{n_k}(x_k) \leq F(\xi-) - F_{n_k}(r_1) \\ &= F(\xi-) - F(r_1) + F(r_1) - F_{n_k}(r_1).\end{aligned}$$

*Case 3.* $x_k \downarrow \xi, x_k \geq \xi$ :

$$\begin{aligned}\epsilon &\leq F(x_k) - F_{n_k}(x_k) \leq F(r_2) - F_{n_k}(\xi) \\ &\leq F(r_2) - F(r_1) + F(r_1) - F_{n_k}(r_1) + F_{n_k}(\xi-) - F_{n_k}(\xi).\end{aligned}$$

*Case 4.* $x_k \downarrow \xi, x_k \geq \xi$ :

$$\begin{aligned}\epsilon &\leq F_{n_k}(x_k) - F(x_k) \leq F_{n_k}(r_2) - F(\xi) \\ &= F_{n_k}(r_2) - F_{n_k}(r_1) + F_{n_k}(r_1) - F(r_1) + F(r_1) - F(\xi).\end{aligned}$$

In each case let first $k \to \infty$, then $r_1 \uparrow \xi, r_2 \downarrow \xi$; then the last member of each chain of inequalities does not exceed a quantity which tends to 0 and a contradiction is obtained.

*Remark.* The reader will do well to observe the way the proof above is arranged. Having chosen a set of $\omega$ with probability one, for each fixed $\omega$ in this set we reason with the corresponding sample functions $F_n(\cdot, \omega)$ and $F(\cdot, \omega)$ without further intervention of probability. Such a procedure is standard in the theory of stochastic processes.

Our next application is to renewal theory. Let $\{X_n, n \geq 1\}$ again be a sequence of independent and identically distributed r.v.'s. We shall further assume that they are positive, although this hypothesis can be dropped, and that they are not identically zero a.e. It follows that the common mean is strictly positive but may be $+\infty$. Now the successive r.v.'s are interpreted as "lifespans" of certain objects undergoing a process of renewal, or the "return periods" of certain recurrent phenomena. Typical examples are the ages of a succession of living beings and the durations of a sequence of services. This raises theoretical as well as practical questions such as: given an epoch in time, how many renewals have there been before it? how long ago was the last renewal? how soon will the next be?

Let us consider the first question. Given the epoch $t \geq 0$, let $N(t, \omega)$ be the number of renewals up to and including the time $t$. It is clear that we have

$$\{\omega: N(t, \omega) = n\} = \{\omega: S_n(\omega) \leq t < S_{n+1}(\omega)\}, \tag{6}$$

valid for $n \geq 0$, provided $S_0 \equiv 0$. Summing over $n \leq m - 1$, we obtain

$$\{\omega: N(t, \omega) < m\} = \{\omega: S_m(\omega) > t\}. \tag{7}$$

This shows in particular that for each $t > 0$, $N(t) = N(t, \cdot)$ is a discrete r.v. whose range is the set of all natural numbers. The family of r.v.'s $\{N(t)\}$ indexed by $t \in [0, \infty)$ may be called a *renewal process*. If the common distribution $F$ of the $X_n$'s is the exponential $F(x) = 1 - e^{-\lambda x}$, $x \geq 0$; where $\lambda > 0$, then $\{N(t), t \geq 0\}$ is just the *simple Poisson process with parameter* $\lambda$.

Let us prove first that

$$\lim_{t \to \infty} N(t) = +\infty \quad \text{a.e.}, \tag{8}$$

namely that the total number of renewals becomes infinite with time. This is almost obvious, but the proof follows. Since $N(t, \omega)$ increases with $t$, the limit in (8) certainly exists for every $\omega$. Were it finite on a set of strictly positive probability, there would exist an integer $M$ such that

$$\mathscr{P}\{\sup_{0 \leq t < \infty} N(t, \omega) < M\} > 0.$$

This implies by (7) that

$$\mathscr{P}\{S_M(\omega) = +\infty\} > 0,$$

which is impossible. (Only because we have laid down the convention long ago that an r.v. such as $X_1$ should be finite-valued unless otherwise specified.)

Next let us write

$$0 < m = \mathscr{E}(X_1) \leq +\infty, \tag{9}$$

and suppose for the moment that $m < +\infty$. Then, according to the strong law of large numbers (Theorem 5.4.2), $S_n/n \to m$ a.e. Specifically, there exists a null set $Z_1$ such that

$$\forall \omega \in \Omega \backslash Z_1: \lim_{n \to \infty} \frac{X_1(\omega) + \cdots + X_n(\omega)}{n} = m.$$

We have just proved that there exists a null set $Z_2$ such that

$$\forall \omega \in \Omega \backslash Z_2: \lim_{t \to \infty} N(t, \omega) = +\infty.$$

Now for each fixed $\omega_0$, if the numerical sequence $\{a_n(\omega_0), n \geq 1\}$ converges to a finite (or infinite) limit $m$ and at the same time the numerical function $\{N(t, \omega_0), 0 \leq t < \infty\}$ tends to $+\infty$ as $t \to +\infty$, then the very definition of a limit implies that the numerical function $\{a_{N(t,\omega_0)}(\omega_0), 0 \leq t < \infty\}$ converges to the limit $m$ as $t \to +\infty$. Applying this trivial but fundamental observation to

$$a_n = \frac{1}{n} \sum_{j=1}^{n} X_j$$

for each $\omega$ in $\Omega \backslash (Z_1 \cup Z_2)$, we conclude that

$$\lim_{t \to \infty} \frac{S_{N(t,\omega)}(\omega)}{N(t, \omega)} = m \quad \text{a.e.} \tag{10}$$

By the definition of $N(t, \omega)$, the numerator on the left side should be close to $t$; this will be confirmed and strengthened in the following theorem.

**Theorem 5.5.2.** We have

$$\lim_{t \to \infty} \frac{N(t)}{t} = \frac{1}{m} \quad \text{a.e.} \tag{11}$$

and

$$\lim_{t \to \infty} \frac{\mathscr{E}\{N(t)\}}{t} = \frac{1}{m};$$

both being true even if $m = +\infty$, provided we take $1/m$ to be 0 in that case.

PROOF. It follows from (6) that for every $\omega$:

$$S_{N(t,\omega)}(\omega) \leq t < S_{N(t,\omega)+1}(\omega)$$

and consequently, as soon as $t$ is large enough to make $N(t, \omega) > 0$,

$$\frac{S_{N(t,\omega)}(\omega)}{N(t, \omega)} \leq \frac{t}{N(t, \omega)} < \frac{S_{N(t,\omega)+1}(\omega)}{N(t, \omega) + 1} \frac{N(t, \omega) + 1}{N(t, \omega)}.$$

Letting $t \to \infty$ and using (8) and (10) (together with Exercise 1 of Sec. 5.4 in case $m = +\infty$), we conclude that (11) is true.

The deduction of (12) from (11) is more tricky than might have been thought. Since $X_n$ is not zero a.e., there exists $\delta > 0$ such that

$$\forall n: \mathscr{P}\{X_n \geq \delta\} = p > 0.$$

Define

$$X_n'(\omega) = \begin{cases} \delta, & \text{if } X_n(\omega) \geq \delta; \\ 0, & \text{if } X_n(\omega) < \delta; \end{cases}$$

and let $S_n'$ and $N'(t)$ be the corresponding quantities for the sequence $\{X_n', n \geq 1\}$. It is obvious that $S_n' \leq S_n$ and $N'(t) \geq N(t)$ for each $t$. Since the r.v.'s $\{X_n'/\delta\}$ are independent with a Bernoullian distribution, elementary computations (see Exercise 7 below) show that

$$\mathscr{E}\{N'(t)^2\} = O\left(\frac{t^2}{\delta^2}\right) \quad \text{as } t \to \infty.$$

Hence we have, $\delta$ being fixed,

$$\mathscr{E}\left\{\left(\frac{N(t)}{t}\right)^2\right\} \leq \mathscr{E}\left\{\left(\frac{N'(t)}{t}\right)^2\right\} = O(1).$$

Since (11) implies the convergence of $N(t)/t$ in distribution to $\delta_{1/m}$, an application of Theorem 4.5.2 with $X_n = N(n)/n$ and $p = 2$ yields (12) with $t$ replaced by $n$ in (12), from which (12) itself follows at once.

**Corollary.** For each $t$, $\mathscr{E}\{N(t)\} < \infty$.

An interesting relation suggested by (10) is that $\mathscr{E}\{S_{N(t)}\}$ should be close to $m\mathscr{E}\{N(t)\}$ when $t$ is large. The precise result is as follows:

$$\mathscr{E}\{X_1 + \cdots + X_{N(t)+1}\} = \mathscr{E}\{X_1\}\mathscr{E}\{N(t) + 1\}.$$

This is a striking generalization of the additivity of expectations when the number of terms as well as the summands is "random." This follows from the following more general result, known as "Wald's equation".

**Theorem 5.5.3.** Let $\{X_n, n \geq 1\}$ be a sequence of independent and identically distributed r.v.'s with finite mean. For $k \geq 1$ let $\mathscr{F}_k$, $1 \leq k < \infty$, be the Borel field generated by $\{X_j, 1 \leq j \leq k\}$. Suppose that $N$ is an r.v. taking positive integer values such that

$$\forall k \geq 1: \{N \leq k\} \in \mathscr{F}_k, \tag{13}$$

and $\mathscr{E}(N) < \infty$. Then we have

$$\mathscr{E}(S_N) = \mathscr{E}(X_1)\mathscr{E}(N).$$

PROOF. Since $S_0 = 0$ as usual, we have

$$(14)\quad \mathscr{E}(S_N) = \int_\Omega S_N \, d\mathscr{P} = \sum_{k=1}^\infty \int_{(N=k)} S_k \, d\mathscr{P} = \sum_{k=1}^\infty \sum_{j=1}^k \int_{\{N=k\}} X_j \, d\mathscr{P}$$

$$= \sum_{j=1}^\infty \sum_{k=j}^\infty \int_{\{N=k\}} X_j \, d\mathscr{P} = \sum_{j=1}^\infty \int_{\{N\geq j\}} X_j \, d\mathscr{P}$$

$$= \sum_{j=1}^\infty \left\{ \mathscr{E}(X_j) - \int_{\{N\leq j-1\}} X_j \, d\mathscr{P} \right\}.$$

Now the set $\{N \leq j-1\}$ and the r.v. $X_j$ are independent, hence the last written integral is equal to $\mathscr{E}(X_j)\mathscr{P}\{N \leq j-1\}$. Substituting this into the above, we obtain

$$\mathscr{E}(S_N) = \sum_{j=1}^\infty \mathscr{E}(X_j)\mathscr{P}\{N \geq j\} = \mathscr{E}(X_1)\sum_{j=1}^\infty \mathscr{P}\{N \geq j\} = \mathscr{E}(X_1)\mathscr{E}(N),$$

the last equation by the corollary to Theorem 3.2.1.

It remains to justify the interchange of summations in (14), which is essential here. This is done by replacing $X_j$ with $|X_j|$ and obtaining as before that the repeated sum is equal to $\mathscr{E}(|X_1|)\mathscr{E}(N) < \infty$.

We leave it to the reader to verify condition (13) for the r.v. $N(t) + 1$ above. Such an r.v. will be called "optional" in Chapter 8 and will play an important role there.

Our last example is a noted triumph of the ideas of probability theory applied to classical analysis. It is S. Bernstein's proof of Weierstrass' theorem on the approximation of continuous functions by polynomials.

**Theorem 5.5.4.** Let $f$ be a continuous function on [0, 1], and define the Bernstein polynomials $\{p_n\}$ as follows:

$$(15)\qquad p_n(x) = \sum_{k=0}^n f\left(\frac{k}{n}\right)\binom{n}{k} x^k (1-x)^{n-k}.$$

Then $p_n$ converges uniformly to $f$ in [0, 1].

PROOF. For each $x$, consider a sequence of independent Bernoullian r.v.'s $\{X_n, n \geq 1\}$ with success probability $x$, namely:

$$X_n = \begin{cases} 1 & \text{with probability } x, \\ 0 & \text{with probability } 1-x; \end{cases}$$

and let $S_n = \sum_{k=1}^{n} X_k$ as usual. We know from elementary probability theory that

$$\mathscr{P}\{S_n = k\} = \binom{n}{k} x^k (1-x)^{n-k}, \quad 0 \le k \le n,$$

so that

$$p_n(x) = \mathscr{E}\left\{ f\left(\frac{S_n}{n}\right)\right\}.$$

We know from the law of large numbers that $S_n/n \to x$ with probability one, but it is sufficient to have convergence in probability, which is Bernoulli's weak law of large numbers. Since $f$ is uniformly continuous in [0, 1], it follows as in the proof of Theorem 4.4.5 that

$$\mathscr{E}\left\{ f\left(\frac{S_n}{n}\right)\right\} \to \mathscr{E}\{f(x)\} = f(x).$$

We have therefore proved the convergence of $p_n(x)$ to $f(x)$ for each $x$. It remains to check the uniformity. Now we have for any $\delta > 0$:

$$\begin{aligned}
(16) \qquad |p_n(x) - f(x)| &\le \mathscr{E}\left\{ \left| f\left(\frac{S_n}{n}\right) - f(x)\right| \right\} \\
&= \mathscr{E}\left\{ \left| f\left(\frac{S_n}{n}\right) - f(x)\right| ; \left|\frac{S_n}{n} - x\right| > \delta \right\} \\
&\quad + \mathscr{E}\left\{ \left| f\left(\frac{S_n}{n}\right) - f(x)\right| ; \left|\frac{S_n}{n} - x\right| \le \delta \right\},
\end{aligned}$$

where we have written $\mathscr{E}\{Y; \Lambda\}$ for $\int_\Lambda Y \, d\mathscr{P}$. Given $\epsilon > 0$, there exists $\delta(\epsilon)$ such that

$$|x - y| \le \delta \Rightarrow |f(x) - f(y)| \le \epsilon/2.$$

With this choice of $\delta$ the last term in (16) is bounded by $\epsilon/2$. The preceding term is clearly bounded by

$$2\|f\| P\left\{ \left|\frac{S_n}{n} - x\right| > \delta \right\}.$$

Now we have by Chebyshev's inequality, since $\mathscr{E}(S_n) = nx$, $\sigma^2(S_n) = nx(1 - x)$, and $x(1-x) \le \frac{1}{4}$ for $0 \le x \le 1$:

$$\mathscr{P}\left\{ \left|\frac{S_n}{n} - x\right| > \delta \right\} \le \frac{1}{\delta^2}\sigma^2\left(\frac{S_n}{n}\right) = \frac{nx(1-x)}{\delta^2 n^2} \le \frac{1}{4\delta^2 n}.$$

This is nothing but Chebyshev's proof of Bernoulli's theorem. Hence if $n \ge \|f\|/\delta^2\epsilon$, we get $|p_n(x) - f(x)| \le \epsilon$ in (16). This proves the uniformity.

One should remark not only on the lucidity of the above derivation but also the meaningful construction of the approximating polynomials. Similar methods can be used to establish a number of well-known analytical results with relative ease; see Exercise 11 below for another example.

EXERCISES

$\{X_n\}$ is a sequence of independent and identically distributed r.v.'s;

$$S_n = \sum_{j=1}^{n} X_j.$$

**1.** Show that equality can hold somewhere in (1) with strictly positive probability if and only if the discrete part of $F$ does not vanish.

***2.** Let $F_n$ and $F$ be as in Theorem 5.5.1; then the distribution of

$$\sup_{-\infty<x<\infty} |F_n(x, \omega) - F(x)|$$

is the same for all continuous $F$. [HINT: Consider $F(X)$, where $X$ has the d.f. $F$.]

**3.** Find the distribution of $Y_{nk}$, $1 \leq k \leq n$, in (1). [These r.v.'s are called *order statistics*.]

***4.** Let $S_n$ and $N(t)$ be as in Theorem 5.5.2. Show that

$$\mathscr{E}\{N(t)\} = \sum_{n=1}^{\infty} \mathscr{P}\{S_n \leq t\}.$$

This remains true if $X_1$ takes both positive and negative values.

**5.** If $\mathscr{E}(X_1) > 0$, then

$$\lim_{t\to-\infty} \mathscr{P}\left\{\bigcup_{n=1}^{\infty} [S_n \leq t]\right\} = 0.$$

***6.** For each $t > 0$, define

$$\nu(t, \omega) = \min\{n: |S_n(\omega)| > t\}$$

if such an $n$ exists, or $+\infty$ if not. If $\mathscr{P}(X_1 \neq 0) > 0$, then for every $t > 0$ and $r > 0$ we have $\mathscr{P}\{\nu(t) > n\} \leq \lambda^n$ for some $\lambda < 1$ and all large $n$; consequently $\mathscr{E}\{\nu(t)^r\} < \infty$. This implies the corollary of Theorem 5.5.2 without recourse to the law of large numbers. [This is Charles Stein's theorem.]

***7.** Consider the special case of renewal where the r.v.'s are Bernoullian taking the values 1 and 0 with probabilities $p$ and $1 - p$, where $0 < p < 1$.

Find explicitly the d.f. of $\nu(0)$ as defined in Exercise 6, and hence of $\nu(t)$ for every $t > 0$. Find $\mathscr{E}\{\nu(t)\}$ and $\mathscr{E}\{\nu(t)^2\}$. Relate $\nu(t, \omega)$ to the $N(t, \omega)$ in Theorem 5.5.2 and so calculate $\mathscr{E}\{N(t)\}$ and $\mathscr{E}\{N(t)^2\}$.

**8.** Theorem 5.5.3 remains true if $\mathscr{E}(X_1)$ is defined, possibly $+\infty$ or $-\infty$.

***9.** In Exercise 7, find the d.f. of $X_{\nu(t)}$ for a given $t$. $\mathscr{E}\{X_{\nu(t)}\}$ is the mean lifespan of the object living at the epoch $t$; should it not be the same as $\mathscr{E}\{X_1\}$, the mean lifespan of the given species? [This is one of the best examples of the use or misuse of intuition in probability theory.]

**10.** Let $\tau$ be a positive integer-valued r.v. that is independent of the $X_n$'s. Suppose that both $\tau$ and $X_1$ have finite second moments, then

$$\sigma^2(S_\tau) = \mathscr{E}(\tau)\sigma^2(X_1) + \sigma^2(\tau)(\mathscr{E}(X_1))^2.$$

***11.** Let $f$ be continuous and belong to $L^r(0, \infty)$ for some $r > 1$, and

$$g(\lambda) = \int_0^\infty e^{-\lambda t} f(t)\, dt.$$

Then

$$f(x) = \lim_{n\to\infty} \frac{(-1)^{n-1}}{(n-1)!} \left(\frac{n}{x}\right)^n g^{(n-1)}\left(\frac{n}{x}\right),$$

where $g^{(n-1)}$ is the $(n-1)$st derivative of $g$, uniformly in every finite interval. [HINT: Let $\lambda > 0$, $\mathscr{P}\{X_1(\lambda) \leq t\} = 1 - e^{-\lambda t}$. Then

$$\mathscr{E}\{f(S_n(\lambda))\} = [(-1)^{n-1}/(n-1)!]\lambda^n g^{(n-1)}(\lambda)$$

and $S_n(n/x) \to x$ in pr. This is a somewhat easier version of Widder's inversion formula for Laplace transforms.]

**12.** Let $\mathscr{P}\{X_1 = k\} = p_k$, $1 \leq k \leq \ell$, $\sum_{k=1}^{\ell} p_k = 1$. Let $N(n, \omega)$ be the number of values of $j$, $1 \leq j \leq n$, for which $X_j = k$ and

$$\prod(n, \omega) = \prod_{k=1}^{\ell} p_k^{N(n,\omega)}.$$

Prove that

$$\lim_{n\to\infty} \frac{1}{n} \log \prod(n, \omega) \quad \text{exists a.e.}$$

and find the limit. [This is from information theory.]

#### Bibliographical Note

Borel's theorem on normal numbers, as well as the Borel–Cantelli lemma in Secs. 4.2–4.3, is contained in

Émile Borel, *Sur les probabilités dénombrables et leurs applications arithmétiques*, Rend. Circ. Mat. Palermo **26** (1909), 247–271.

This pioneering paper, despite some serious gaps (see Fréchet [9] for comments), is well worth reading for its historical interest. In Borel's *Jubilé Selecta* (Gauthier-Villars, 1940), it is followed by a commentary by Paul Lévy with a bibliography on later developments.

Every serious student of probability theory should read:

A. N. Kolmogoroff, *Über die Summen durch den Zufall bestimmten unabhängiger Grössen*, Math. Annalen **99** (1928), 309–319; *Bermerkungen*, **102** (1929), 484–488.

This contains Theorems 5.3.1 to 5.3.3 as well as the original version of Theorem 5.2.3.

For all convergence questions regarding sums of independent r.v.'s, the deepest study is given in Chapter 6 of Lévy's book [11]. After three decades, this book remains a source of inspiration.

Theorem 5.5.2 is taken from

J. L. Doob, *Renewal theory from the point of view of probability*, Trans. Am. Math. Soc. **63** (1942), 422–438.

Feller's book [13], both volumes, contains an introduction to renewal theory as well as some of its latest developments.

# 6 Characteristic function

## 6.1 General properties; convolutions

An important tool in the study of r.v.'s and their p.m.'s or d.f.'s is the *characteristic function* (ch.f.). For any r.v. $X$ with the p.m. $\mu$ and d.f. $F$, this is defined to be the function $f$ on $\mathscr{R}^1$ as follows, $\forall t \in \mathscr{R}^1$:

$$(1) \quad f(t) = \mathscr{E}(e^{itX}) = \int_{\Omega} e^{itX(\omega)} \mathscr{P}(d\omega) = \int_{\mathscr{R}^1} e^{itx} \mu(dx) = \int_{-\infty}^{\infty} e^{itx}\, dF(x).$$

The equality of the third and fourth terms above is a consequence of Theorem 3.32.2, while the rest is by definition and notation. We remind the reader that the last term in (1) is *defined* to be the one preceding it, where the one-to-one correspondence between $\mu$ and $F$ is discussed in Sec. 2.2. We shall use both of them below. Let us also point out, since our general discussion of integrals has been confined to the real domain, that $f$ is a complex-valued function of the real variable $t$, whose real and imaginary parts are given respectively by

$$\mathbf{R}f(t) = \int \cos xt \mu(dx), \quad \mathbf{I}f(t) = \int \sin xt \mu(dx).$$

Here and hereafter, integrals without an indicated domain of integration are taken over $\mathscr{R}^1$.

Clearly, the ch.f. is a creature associated with $\mu$ or $F$, not with $X$, but the first equation in (1) will prove very convenient for us, see e.g. (iii) and (v) below. In analysis, the ch.f. is known as the Fourier–Stieltjes transform of $\mu$ or $F$. It can also be defined over a wider class of $\mu$ or $F$ and, furthermore, be considered as a function of a complex variable $t$, under certain conditions that ensure the existence of the integrals in (1). This extension is important in some applications, but we will not need it here except in Theorem 6.6.5. As specified above, it is always well defined (for all real $t$) and has the following simple properties.

**(i)** $\forall t \in \mathscr{R}^1$:

$$|f(t)| \leq 1 = f(0); \quad f(-t) = \overline{f(t)},$$

where $\bar{z}$ denotes the conjugate complex of $z$.

**(ii)** $f$ is uniformly continuous in $\mathscr{R}^1$.

To see this, we write for real $t$ and $h$:

$$f(t+h) - f(t) = \int (e^{i(t+h)x} - e^{itx})\mu(dx),$$

$$|f(t+h) - f(t)| \leq \int |e^{itx}||e^{ihx} - 1|\mu(dx) = \int |e^{ihx} - 1|\mu(dx).$$

The last integrand is bounded by 2 and tends to 0 as $h \to 0$, for each $x$. Hence, the integral converges to 0 by bounded convergence. Since it does not involve $t$, the convergence is surely uniform with respect to $t$.

**(iii)** If we write $f_X$ for the ch.f. of $X$, then for any real numbers $a$ and $b$, we have

$$f_{aX+b}(t) = f_X(at)e^{itb},$$
$$f_{-X}(t) = \overline{f_X(t)}.$$

This is easily seen from the first equation in (1), for

$$\mathscr{E}(e^{it(aX+b)}) = \mathscr{E}(e^{i(ta)X} \cdot e^{itb}) = \mathscr{E}(e^{i(ta)X})e^{itb}.$$

**(iv)** If $\{f_n, n \geq 1\}$ are ch.f.'s, $\lambda_n \geq 0$, $\sum_{n=1}^{\infty} \lambda_n = 1$, then

$$\sum_{n=1}^{\infty} \lambda_n f_n$$

is a ch.f. Briefly: a convex combination of ch.f.'s is a ch.f.

For if $\{\mu_n, n \geq 1\}$ are the corresponding p.m.'s, then $\sum_{n=1}^{\infty} \lambda_n \mu_n$ is a p.m. whose ch.f. is $\sum_{n=1}^{\infty} \lambda_n f_n$.

(**v**) If $\{f_j, 1 \leq j \leq n\}$ are ch.f.'s, then

$$\prod_{j=1}^{n} f_j$$

is a ch.f.

By Theorem 3.3.4, there exist independent r.v.'s $\{X_j, 1 \leq j \leq n\}$ with probability distributions $\{\mu_j, 1 \leq j \leq n\}$, where $\mu_j$ is as in (iv). Letting

$$S_n = \sum_{j=1}^{n} X_j,$$

we have by the corollary to Theorem 3.3.3:

$$\mathscr{E}(e^{itS_n}) = \mathscr{E}\left(\prod_{j=1}^{n} e^{itX_j}\right) = \prod_{j=1}^{n} \mathscr{E}(e^{itX_j}) = \prod_{j=1}^{n} f_j(t);$$

or in the notation of (iii):

$$f_{S_n} = \prod_{j=1}^{n} f_{X_j}. \tag{2}$$

(For an extension to an infinite number of $f_j$'s see Exercise 4 below.)

The ch.f. of $S_n$ being so neatly expressed in terms of the ch.f.'s of the summands, we may wonder about the d.f. of $S_n$. We need the following definitions.

DEFINITION. The *convolution* of two d.f.'s $F_1$ and $F_2$ is defined to be the d.f. $F$ such that

$$\forall x \in \mathscr{R}^1: F(x) = \int_{-\infty}^{\infty} F_1(x - y)\, dF_2(y), \tag{3}$$

and written as

$$F = F_1 * F_2.$$

It is easy to verify that $F$ is indeed a d.f. The other basic properties of convolution are consequences of the following theorem.

**Theorem 6.1.1.** Let $X_1$ and $X_2$ be independent r.v.'s with d.f.'s $F_1$ and $F_2$, respectively. Then $X_1 + X_2$ has the d.f. $F_1 * F_2$.

PROOF. We wish to show that

$$\forall x: \mathscr{P}\{X_1 + X_2 \leq x\} = (F_1 * F_2)(x). \tag{4}$$

For this purpose we define a function $f$ of $(x_1, x_2)$ as follows, for fixed $x$:

$$f(x_1, x_2) = \begin{cases} 1, & \text{if } x_1 + x_2 \leq x; \\ 0, & \text{otherwise.} \end{cases}$$

$f$ is a Borel measurable function of two variables. By Theorem 3.3.3 and using the notation of the second proof of Theorem 3.3.3, we have

$$\begin{aligned}
\int_\Omega f(X_1, X_2)\, d\mathscr{P} &= \iint_{\mathscr{R}^2} f(x_1, x_2) \mu^2(dx_1, dx_2) \\
&= \int_{\mathscr{R}^1} \mu_2(dx_2) \int_{\mathscr{R}^1} f(x_1, x_2) \mu_1(dx_1) \\
&= \int_{\mathscr{R}^1} \mu_2(dx_2) \int_{(-\infty, x - x_2]} \mu_1(dx_1) \\
&= \int_{-\infty}^{\infty} dF_2(x_2) F_1(x - x_2).
\end{aligned}$$

This reduces to (4). The second equation above, evaluating the double integral by an iterated one, is an application of Fubini's theorem (see Sec. 3.3).

**Corollary.** The binary operation of convolution $*$ is commutative and associative.

For the corresponding binary operation of addition of independent r.v.'s has these two properties.

DEFINITION. The convolution of two probability density functions $p_1$ and $p_2$ is defined to be the probability density function $p$ such that

$$\forall x \in \mathscr{R}^1: p(x) = \int_{-\infty}^{\infty} p_1(x - y) p_2(y)\, dy, \tag{5}$$

and written as

$$p = p_1 * p_2.$$

We leave it to the reader to verify that $p$ is indeed a density, but we will spell out the following connection.

**Theorem 6.1.2.** The convolution of two absolutely continuous d.f.'s with densities $p_1$ and $p_2$ is absolutely continuous with density $p_1 * p_2$.

PROOF. We have by Fubini's theorem:

$$\begin{aligned}\int_{-\infty}^{x} p(u)\,du &= \int_{-\infty}^{x} du \int_{-\infty}^{\infty} p_1(u-v)p_2(v)\,dv \\ &= \int_{-\infty}^{\infty} \left[\int_{-\infty}^{x} p_1(u-v)\,du\right] p_2(v)\,dv \\ &= \int_{-\infty}^{\infty} F_1(x-v)p_2(v)\,dv \\ &= \int_{-\infty}^{\infty} F_1(x-v)\,dF_2(v) = (F_1 * F_2)(x).\end{aligned}$$

This shows that $p$ is a density of $F_1 * F_2$.

What is the p.m., to be denoted by $\mu_1 * \mu_2$, that corresponds to $F_1 * F_2$? For arbitrary subsets $A$ and $B$ of $\mathscr{R}^1$, we denote their vector sum and difference by $A + B$ and $A - B$, respectively:

$$(6) \qquad A \pm B = \{x \pm y : x \in A,\ y \in B\};$$

and write $x \pm B$ for $\{x\} \pm B$, $-B$ for $0 - B$. There should be no danger of confusing $A - B$ with $A \backslash B$.

**Theorem 6.1.3.** For each $B \in \mathscr{B}$, we have

$$(7) \qquad (\mu_1 * \mu_2)(B) = \int_{\mathscr{R}^1} \mu_1(B - y)\mu_2(dy).$$

For each Borel measurable function $g$ that is integrable with respect to $\mu_1 * \mu_2$, we have

$$(8) \qquad \int_{\mathscr{R}^1} g(u)(\mu_1 * \mu_2)(du) = \iint_{\mathscr{R}^1 \mathscr{R}^1} g(x+y)\mu_1(dx)\mu_2(dy).$$

PROOF. It is easy to verify that the set function $(\mu_1 * \mu_2)(\cdot)$ defined by (7) is a p.m. To show that its d.f. is $F_1 * F_2$, we need only verify that its value for $B = (-\infty, x]$ is given by the $F(x)$ defined in (3). This is obvious, since the right side of (7) then becomes

$$\int_{\mathscr{R}^1} F_1(x-y)\mu_2(dy) = \int_{-\infty}^{\infty} F_1(x-y)\,dF_2(y).$$

Now let $g$ be the indicator of the set $B$, then for each $y$, the function $g_y$ defined by $g_y(x) = g(x+y)$ is the indicator of the set $B - y$. Hence

$$\int_{\mathscr{R}^1} g(x+y)\mu_1(dx) = \mu_1(B - y)$$

and, substituting into the right side of (8), we see that it reduces to (7) in this case. The general case is proved in the usual way by first considering simple functions $g$ and then passing to the limit for integrable functions.

As an instructive example, let us calculate the ch.f. of the convolution $\mu_1 * \mu_2$. We have by (8)

$$\int e^{itu}(\mu_1 * \mu_2)(du) = \iint e^{ity}e^{itx}\mu_1(dx)\mu_2(dy)$$

$$= \int e^{itx}\mu_1(dx)\int e^{ity}\mu_2(dy).$$

This is as it should be by (v), since the first term above is the ch.f. of $X + Y$, where $X$ and $Y$ are independent with $\mu_1$ and $\mu_2$ as p.m.'s. Let us restate the results, after an obvious induction, as follows.

**Theorem 6.1.4.** Addition of (a finite number of) independent r.v.'s corresponds to convolution of their d.f.'s and multiplication of their ch.f.'s.

**Corollary.** If $f$ is a ch.f., then so is $|f|^2$.

To prove the corollary, let $X$ have the ch.f. $f$. Then there exists on some $\Omega$ (why?) an r.v. $Y$ independent of $X$ and having the same d.f., and so also the same ch.f. $f$. The ch.f. of $X - Y$ is

$$\mathscr{E}(e^{it(X-Y)}) = \mathscr{E}(e^{itX})\mathscr{E}(e^{-itY}) = f(t)f(-t) = |f(t)|^2.$$

The technique of considering $X - Y$ and $|f|^2$ instead of $X$ and $f$ will be used below and referred to as "symmetrization" (see the end of Sec. 6.2). This is often expedient, since a real and particularly a positive-valued ch.f. such as $|f|^2$ is easier to handle than a general one.

Let us list a few well-known ch.f.'s together with their d.f.'s or p.d.'s (probability densities), the last being given in the interval outside of which they vanish.

(1) Point mass at $a$:

$$\text{d.f. } \delta_a; \quad \text{ch.f. } e^{iat}.$$

(2) Symmetric Bernoullian distribution with mass $\frac{1}{2}$ each at $+1$ and $-1$:

$$\text{d.f. } \tfrac{1}{2}(\delta_1 + \delta_{-1}); \quad \text{ch.f. } \cos t.$$

(3) Bernoullian distribution with "success probability" $p$, and $q = 1 - p$:

$$\text{d.f. } q\delta_0 + p\delta_1; \quad \text{ch.f. } q + pe^{it} = 1 + p(e^{it} - 1).$$

(4) Binomial distribution for $n$ trials with success probability $p$:

$$\text{d.f. } \sum_{k=0}^{n} \binom{n}{k} p^k q^{n-k} \delta_k; \quad \text{ch.f. } (q + pe^{it})^n.$$

(5) Geometric distribution with success probability $p$:

$$\text{d.f. } \sum_{n=0}^{\infty} q^n p \delta_n; \quad \text{ch.f. } p(1 - qe^{it})^{-1}.$$

(6) Poisson distribution with (mean) parameter $\lambda$:

$$\text{d.f. } \sum_{n=0}^{\infty} e^{-\lambda} \frac{\lambda^n}{n!} \delta_n; \quad \text{ch.f. } e^{\lambda(e^{it}-1)}.$$

(7) Exponential distribution with mean $\lambda^{-1}$:

$$\text{p.d. } \lambda e^{-\lambda x} \text{ in } [0, \infty); \quad \text{ch.f. } (1 - \lambda^{-1} it)^{-1}.$$

(8) Uniform distribution in $[-a, +a]$:

$$\text{p.d. } \frac{1}{2a} \text{ in } [-a, a]; \quad \text{ch.f. } \frac{\sin at}{at} (= 1 \text{ for } t = 0).$$

(9) Triangular distribution in $[-a, a]$:

$$\text{p.d. } \frac{a - |x|}{a^2} \text{ in } [-a, a]; \quad \text{ch.f. } \frac{2(1 - \cos at)}{a^2 t^2} = \left( \frac{\sin \frac{at}{2}}{\frac{at}{2}} \right)^2.$$

(10) Reciprocal of (9):

$$\text{p.d. } \frac{1 - \cos ax}{\pi a x^2} \text{ in } (-\infty, \infty); \quad \text{ch.f. } \left(1 - \frac{|t|}{a}\right) \vee 0.$$

(11) Normal distribution $N(m, \sigma^2)$ with mean $m$ and variance $\sigma^2$:

$$\text{p.d. } \frac{1}{\sqrt{2\pi}\sigma} \exp\left[-\frac{(x - m)^2}{2\sigma^2}\right] \text{ in } (-\infty, \infty);$$

$$\text{ch.f. } \exp\left(imt - \frac{\sigma^2 t^2}{2}\right).$$

Unit normal distribution $N(0, 1) = \Phi$ with mean 0 and variance 1:

$$\text{p.d. } \frac{1}{\sqrt{2\pi}} e^{-x^2/2} \text{ in } (-\infty, \infty); \quad \text{ch.f. } e^{-t^2/2}.$$

(12) Cauchy distribution with parameter $a > 0$:

$$\text{p.d. } \frac{a}{\pi(a^2 + x^2)} \text{ in } (-\infty, \infty); \quad \text{ch.f. } e^{-a|t|}.$$

Convolution is a smoothing operation widely used in mathematical analysis, for instance in the proof of Theorem 6.5.2 below. Convolution with the normal kernel is particularly effective, as illustrated below.

Let $n_\delta$ be the density of the normal distribution $N(0, \delta^2)$, namely

$$n_\delta(x) = \frac{1}{\sqrt{2\pi}\delta} \exp\left(-\frac{x^2}{2\delta^2}\right), \quad -\infty < x < \infty.$$

For any bounded measurable function $f$ on $\mathscr{R}^1$, put

$$(9) \quad f_\delta(x) = (f * n_\delta)(x) = \int_{-\infty}^{\infty} f(x-y)n_\delta(y)\,dy = \int_{-\infty}^{\infty} n_\delta(x-y)f(y)\,dy.$$

It will be seen below that the integrals above converge. Let $C_B^\infty$ denote the class of functions on $\mathscr{R}^1$ which have bounded derivatives of all orders; $C_U$ the class of bounded and uniformly continuous functions on $\mathscr{R}^1$.

**Theorem 6.1.5.** For each $\delta > 0$, we have $f_\delta \in C_B^\infty$. Furthermore if $f \in C_U$, then $f_\delta \to f$ uniformly in $\mathscr{R}^1$.

PROOF. It is easily verified that $n_\delta \in C_B^\infty$. Moreover its $k$th derivative $n_\delta^{(k)}$ is dominated by $c_{k,\delta}n_{2\delta}$ where $c_{k,\delta}$ is a constant depending only on $k$ and $\delta$ so that

$$\left|\int_{-\infty}^{\infty} n_\delta^{(k)}(x-y)f(y)\,dy\right| \le c_{k,\delta}||f|| \int_{-\infty}^{\infty} n_{2\delta}(x-y)\,dy = c_{k,\delta}||f||.$$

Thus the first assertion follows by differentiation under the integral of the last term in (9), which is justified by elementary rules of calculus. The second assertion is proved by standard estimation as follows, for any $\eta > 0$:

$$\begin{aligned} |f(x) - f_\delta(x)| &\le \int_{-\infty}^{\infty} |f(x) - f(x-y)|n_\delta(y)\,dy \\ &\le \sup_{|y|\le\eta} |f(x) - f(x-y)| + 2||f|| \int_{|y|>\eta} n_\delta(y)\,dy. \end{aligned}$$

Here is the probability idea involved. If $f$ is integrable over $\mathscr{R}^1$, we may think of $f$ as the density of a r.v. $X$, and $n_\delta$ as that of an independent normal r.v. $Y_\delta$. Then $f_\delta$ is the density of $X + Y_\delta$ by Theorem 6.1.2. As $\delta \downarrow 0$, $Y_\delta$ converges to 0 in probability and so $X + Y_\delta$ converges to $X$ likewise, hence also in distribution by Theorem 4.4.5. This makes it plausible that the densities will also converge under certain analytical conditions.

As a corollary, we have shown that the class $C_B^\infty$ is *dense* in the class $C_U$ with respect to the uniform topology on $\mathscr{R}^1$. This is a basic result in the theory of "generalized functions" or "Schwartz distributions". By way of application, we state the following strengthening of Theorem 4.4.1.

**Theorem 6.1.6.** If $\{\mu_n\}$ and $\mu$ are s.p.m.'s such that

$$\forall f \in C_B^\infty : \int_{\mathscr{R}^1} f(x)\mu_n(dx) \to \int_{\mathscr{R}^1} f(x)\mu(dx),$$

then $\mu_n \xrightarrow{v} \mu$.

This is an immediate consequence of Theorem 4.4.1, and Theorem 6.1.5, if we observe that $C_0 \subset C_U$. The reduction of the class of "test functions" from $C_0$ to $C_B^x$ is often expedient, as in Lindeberg's method for proving central limit theorems; see Sec. 7.1 below.

EXERCISES

**1.** . If $f$ is a ch.f., and $G$ a d.f. with $G(0-) = 0$, then the following functions are all ch.f.'s:

$$\int_0^1 f(ut)\,du, \quad \int_0^\infty f(ut)e^{-u}\,du, \quad \int_0^\infty e^{-|t|u}\,dG(u),$$

$$\int_0^\infty e^{-t^2u}\,dG(u), \quad \int_0^\infty f(ut)\,dG(u).$$

***2.** Let $f(u, t)$ be a function on $(-\infty, \infty) \times (-\infty, \infty)$ such that for each $u$, $f(u, \cdot)$ is a ch.f. and for each $t$, $f(\cdot, t)$ is a continuous function; then

$$\int_{-\infty}^\infty f(u, t)\,dG(u)$$

is a ch.f. for any d.f. $G$. In particular, if $f$ is a ch.f. such that $\lim_{t\to\infty} f(t)$ exists and $G$ a d.f. with $G(0-) = 0$, then

$$\int_0^\infty f\left(\frac{t}{u}\right)\,dG(u) \quad \text{is a ch.f.}$$

**3.** Find the d.f. with the following ch.f.'s ($\alpha > 0$, $\beta > 0$):

$$\frac{\alpha^2}{\alpha^2 + t^2}, \quad \frac{1}{(1 - \alpha i t)^\beta}, \quad \frac{1}{(1 + \alpha\beta - \alpha\beta e^{it})^{1/\beta}}.$$

[HINT: The second and third steps correspond respectively to the *gamma* and *Pólya distributions*.]

**4.** Let $S_n$ be as in (v) and suppose that $S_n \to S_\infty$ in pr. Prove that

$$\prod_{j=1}^{\infty} f_j(t)$$

converges in the sense of infinite product for each $t$ and is the ch.f. of $S_\infty$.

**5.** If $F_1$ and $F_2$ are d.f.'s such that

$$F_1 = \sum_j b_j \delta_{a_j}$$

and $F_2$ has density $p$, show that $F_1 * F_2$ has a density and find it.

***6.** Prove that the convolution of two discrete d.f.'s is discrete; that of a continuous d.f. with any d.f. is continuous; that of an absolutely continuous d.f. with any d.f. is absolutely continuous.

**7.** The convolution of two discrete distributions with exactly $m$ and $n$ atoms, respectively, has at least $m + n - 1$ and at most $mn$ atoms.

**8.** Show that the family of normal (Cauchy, Poisson) distributions is closed with respect to convolution in the sense that the convolution of any two in the family with arbitrary parameters is another in the family with some parameter(s).

**9.** Find the $n$th iterated convolution of an exponential distribution.

***10.** Let $\{X_j, j \geq 1\}$ be a sequence of independent r.v.'s having the common exponential distribution with mean $1/\lambda$, $\lambda > 0$. For given $x > 0$ let $\nu$ be the maximum of $n$ such that $S_n \leq x$, where $S_0 = 0$, $S_n = \sum_{j=1}^{n} X_j$ as usual. Prove that the r.v. $\nu$ has the Poisson distribution with mean $\lambda x$. See Sec. 5.5 for an interpretation by renewal theory.

**11.** Let $X$ have the normal distribution $\Phi$. Find the d.f., p.d., and ch.f. of $X^2$.

**12.** Let $\{X_j, 1 \leq j \leq n\}$ be independent r.v.'s each having the d.f. $\Phi$. Find the ch.f. of

$$\sum_{j=1}^{n} X_j^2$$

and show that the corresponding p.d. is $2^{-n/2}\Gamma(n/2)^{-1}x^{(n/2)-1}e^{-x/2}$ in $(0, \infty)$. This is called in statistics the "$\chi^2$ distribution with $n$ degrees of freedom".

**13.** For any ch.f. $f$ we have for every $t$:

$$\mathbf{R}[1 - f(t)] \geq \tfrac{1}{4}\mathbf{R}[1 - f(2t)].$$

**14.** Find an example of two r.v.'s $X$ and $Y$ with the same p.m. $\mu$ that are not independent but such that $X + Y$ has the p.m. $\mu * \mu$. [HINT: Take $X = Y$ and use ch.f.]

**★15.** For a d.f. $F$ and $h \geq 0$, define

$$Q_F(h) = \sup_x [F(x+h) - F(x-)];$$

$Q_F$ is called the *Lévy concentration function* of $F$. Prove that the sup above is attained, and if $G$ is also a d.f., we have

$$\forall h > 0: Q_{F*G}(h) \leq Q_F(h) \wedge Q_G(h).$$

**16.** If $0 < h\lambda \leq 2\pi$, then there is an absolute constant $A$ such that

$$Q_F(h) \leq \frac{A}{\lambda} \int_0^\lambda |f(t)|\, dt,$$

where $f$ is the ch.f. of $F$. [HINT: Use Exercise 2 of Sec. 6.2 below.]

**17.** Let $F$ be a symmetric d.f., with ch.f. $f \geq 0$ then

$$\varphi_F(h) = \int_{-\infty}^{\infty} \frac{h^2}{h^2 + x^2}\, dF(x) = h \int_0^\infty e^{-ht} f(t)\, dt$$

is a sort of average concentration function. Prove that if $G$ is also a d.f. with ch.f. $g \geq 0$, then we have $\forall h > 0$:

$$\varphi_{F*G}(h) \leq \varphi_F(h) \wedge \varphi_G(h);$$
$$1 - \varphi_{F*G}(h) \leq [1 - \varphi_F(h)] + [1 - \varphi_G(h)].$$

**★18.** Let the support of the p.m. $\mu$ on $\mathscr{R}^1$ be denoted by supp $\mu$. Prove that

$$\text{supp } (\mu * \nu) = \text{closure of supp } \mu + \text{supp } \nu;$$
$$\text{supp } (\mu_1 * \mu_2 * \cdots) = \text{closure of (supp } \mu_1 + \text{supp } \mu_2 + \cdots)$$

where "+" denotes vector sum.

## 6.2 Uniqueness and inversion

To study the deeper properties of Fourier–Stieltjes transforms, we shall need certain "Dirichlet integrals". We begin with three basic formulas, where "sgn $\alpha$" denotes 1, 0 or $-1$, according as $\alpha > 0$, $= 0$, or $< 0$.

(1)
$$\forall y \geq 0: 0 \leq (\text{sgn } \alpha) \int_0^y \frac{\sin \alpha x}{x}\, dx \leq \int_0^\pi \frac{\sin x}{x}\, dx.$$

(2)
$$\int_0^\infty \frac{\sin \alpha x}{x}\, dx = \frac{\pi}{2} \text{ sgn } \alpha.$$

(3) $$\int_0^\infty \frac{1-\cos\alpha x}{x^2}\,dx = \frac{\pi}{2}|\alpha|.$$

The substitution $\alpha x = u$ shows at once that it is sufficient to prove all three formulas for $\alpha = 1$. The inequality (1) is proved by partitioning the interval $[0, \infty)$ with positive multiples of $\pi$ so as to convert the integral into a series of alternating signs and decreasing moduli. The integral in (2) is a standard exercise in contour integration, as is also that in (3). However, we shall indicate the following neat heuristic calculations, leaving the justifications, which are not difficult, as exercises.

$$\int_0^\infty \frac{\sin x}{x}\,dx = \int_0^\infty \sin x\left[\int_0^\infty e^{-xu}\,du\right]dx = \int_0^\infty\left[\int_0^\infty e^{-xu}\sin x\,dx\right]du$$
$$= \int_0^\infty \frac{du}{1+u^2} = \frac{\pi}{2};$$
$$\int_0^\infty \frac{1-\cos x}{x^2}\,dx = \int_0^\infty \frac{1}{x^2}\left[\int_0^x \sin u\,du\right]dx = \int_0^\infty \sin u\left[\int_u^\infty \frac{dx}{x^2}\right]du$$
$$= \int_0^\infty \frac{\sin u}{u}\,du = \frac{\pi}{2}.$$

We are ready to answer the question: given a ch.f. $f$, how can we find the corresponding d.f. $F$ or p.m. $\mu$? The formula for doing this, called the *inversion formula*, is of theoretical importance, since it will establish a one-to-one correspondence between the class of d.f.'s or p.m.'s and the class of ch.f.'s (see, however, Exercise 12 below). It is somewhat complicated in its most general form, but special cases or variants of it can actually be employed to derive certain properties of a d.f. or p.m. from its ch.f.; see, e.g., (14) and (15) of Sec. 6.4.

**Theorem 6.2.1.** If $x_1 < x_2$, then we have

(4) $$\mu((x_1, x_2)) + \tfrac{1}{2}\mu(\{x_1\}) + \tfrac{1}{2}\mu(\{x_2\})$$
$$= \lim_{T\to\infty}\frac{1}{2\pi}\int_{-T}^{T}\frac{e^{-itx_1}-e^{-itx_2}}{it}f(t)\,dt$$

(the integrand being defined by continuity at $t = 0$).

PROOF. Observe first that the integrand above is bounded by $|x_1 - x_2|$ everywhere and is $O(|t|^{-1})$ as $|t| \to \infty$; yet we cannot assert that the "infinite integral" $\int_{-\infty}^{\infty}$ exists (in the Lebesgue sense). Indeed, it does not in general (see Exercise 9 below). The fact that the indicated limit, the so-called *Cauchy limit*, does exist is part of the assertion.

We shall prove (4) by actually substituting the definition of $f$ into (4) and carrying out the integrations. We have

$$
(5) \qquad \frac{1}{2\pi}\int_{-T}^{T}\frac{e^{-itx_1}-e^{-itx_2}}{it}\left[\int_{-\infty}^{\infty}e^{itx}\mu(dx)\right]dt
$$

$$
=\int_{-\infty}^{\infty}\left[\int_{-T}^{T}\frac{e^{it(x-x_1)}-e^{it(x-x_2)}}{2\pi it}\,dt\right]\mu(dx).
$$

Leaving aside for the moment the justification of the interchange of the iterated integral, let us denote the quantity in square brackets above by $I(T,x,x_1,x_2)$. Trivial simplification yields

$$
I(T,x,x_1,x_2)=\frac{1}{\pi}\int_0^T\frac{\sin t(x-x_1)}{t}\,dt-\frac{1}{\pi}\int_0^T\frac{\sin t(x-x_2)}{t}\,dt.
$$

It follows from (2) that

$$
\lim_{T\to\infty}I(T,x,x_1,x_2)=\begin{cases}-\frac{1}{2}-(-\frac{1}{2})=0 & \text{for } x<x_1,\\ 0-(-\frac{1}{2})=\frac{1}{2} & \text{for } x=x_1,\\ \frac{1}{2}-(-\frac{1}{2})=1 & \text{for } x_1<x<x_2,\\ \frac{1}{2}-0=\frac{1}{2} & \text{for } x=x_2,\\ \frac{1}{2}-\frac{1}{2}=0 & \text{for } x>x_2.\end{cases}
$$

Furthermore, $I$ is bounded in $T$ by (1). Hence we may let $T\to\infty$ under the integral sign in the right member of (5) by bounded convergence, since

$$
|I(T,x,x_1,x_2)|\le\frac{2}{\pi}\int_0^{\pi}\frac{\sin x}{x}\,dx
$$

by (1). The result is

$$
\left\{\int_{(-\infty,x_1)}0+\int_{\{x_1\}}\frac{1}{2}+\int_{(x_1,x_2)}1+\int_{\{x_2\}}\frac{1}{2}+\int_{(x_2,\infty)}0\right\}\mu(dx)
$$

$$
=\tfrac{1}{2}\mu(\{x_1\})+\mu((x_1,x_2))+\tfrac{1}{2}\mu(\{x_2\}).
$$

This proves the theorem. For the justification mentioned above, we invoke Fubini's theorem and observe that

$$
\left|\frac{e^{it(x-x_1)}-e^{it(x-x_2)}}{it}\right|=\left|\int_{x_1}^{x_2}e^{-itu}\,du\right|\le|x_1-x_2|,
$$

where the integral is taken along the real axis, and

$$
\int_{\mathscr{R}^1}\int_{-T}^{T}|x_1-x_2|\,dt\,\mu(dx)\le 2T|x_1-x_2|<\infty,
$$

so that the integrand on the right of (5) is dominated by a finitely integrable function with respect to the finite product measure $dt \cdot \mu(dx)$ on $[-T, +T] \times \mathscr{R}^1$. This suffices.

*Remark.* If $x_1$ and $x_2$ are points of continuity of $F$, the left side of (4) is $F(x_2) - F(x_1)$.

The following result is often referred to as the "uniqueness theorem" for the "determining" $\mu$ or $F$ (see also Exercise 12 below).

**Theorem 6.2.2.** If two p.m.'s or d.f.'s have the same ch.f., then they are the same.

PROOF. If neither $x_1$ nor $x_2$ is an atom of $\mu$, the inversion formula (4) shows that the value of $\mu$ on the interval $(x_1, x_2)$ is determined by its ch.f. It follows that two p.m.'s having the same ch.f. agree on each interval whose endpoints are not atoms for either measure. Since each p.m. has only a countable set of atoms, points of $\mathscr{R}^1$ that are not atoms for either measure form a dense set. Thus the two p.m.'s agree on a dense set of intervals, and therefore they are identical by the corollary to Theorem 2.2.3.

We give next an important particular case of Theorem 6.2.1.

**Theorem 6.2.3.** If $f \in L^1(-\infty, +\infty)$, then $F$ is continuously differentiable, and we have

$$F'(x) = \frac{1}{2\pi} \int_{-\infty}^{\infty} e^{-ixt} f(t)\, dt. \tag{6}$$

PROOF. Applying (4) for $x_2 = x$ and $x_1 = x - h$ with $h > 0$ and using $F$ instead of $\mu$, we have

$$\frac{F(x) + F(x-)}{2} - \frac{F(x-h) + F(x-h-)}{2} = \frac{1}{2\pi} \int_{-\infty}^{\infty} \frac{e^{ith} - 1}{it} e^{-itx} f(t)\, dt.$$

Here the infinite integral exists by the hypothesis on $f$, since the integrand above is dominated by $|hf(t)|$. Hence we may let $h \to 0$ under the integral sign by dominated convergence and conclude that the left side is 0. Thus, $F$ is left continuous and so continuous in $\mathscr{R}^1$. Now we can write

$$\frac{F(x) - F(x-h)}{h} = \frac{1}{2\pi} \int_{-\infty}^{\infty} \frac{e^{ith} - 1}{ith} e^{-itx} f(t)\, dt.$$

The same argument as before shows that the limit exists as $h \to 0$. Hence $F$ has a left-hand derivative at $x$ equal to the right member of (6), the latter being clearly continuous [cf. Proposition (ii) of Sec. 6.1]. Similarly, $F$ has a right-hand derivative given by the same formula. Actually it is known for a continuous function that, if one of the four "derivates" exists and is continuous at a point, then the function is continuously differentiable there (see, e.g., Titchmarsh, *The theory of functions*, 2nd ed., Oxford Univ. Press, New York, 1939, p. 355).

The derivative $F'$ being continuous, we have (why?)

$$\forall x\colon F(x) = \int_{-\infty}^{x} F'(u)\,du.$$

Thus $F'$ is a probability density function. We may now state Theorem 6.2.3 in a more symmetric form familiar in the theory of Fourier integrals.

**Corollary.** If $f \in L^1$, then $p \in L^1$, where

$$p(x) = \frac{1}{2\pi}\int_{-\infty}^{\infty} e^{-ixt} f(t)\,dt,$$

and

$$f(t) = \int_{-\infty}^{\infty} e^{itx} p(x)\,dx.$$

The next two theorems yield information on the atoms of $\mu$ by means of $f$ and are given here as illustrations of the method of "harmonic analysis".

**Theorem 6.2.4.** For each $x_0$, we have

$$\lim_{T\to\infty} \frac{1}{2T}\int_{-T}^{T} e^{-itx_0} f(t)\,dt = \mu(\{x_0\}). \tag{7}$$

PROOF. Proceeding as in the proof of Theorem 6.2.1, we obtain for the integral average on the left side of (7):

$$\int_{\mathscr{R}^1-\{x_0\}} \frac{\sin T(x-x_0)}{T(x-x_0)}\mu(dx) + \int_{\{x_0\}} 1\mu(dx). \tag{8}$$

The integrand of the first integral above is bounded by 1 and tends to 0 as $T \to \infty$ everywhere in the domain of integration; hence the integral converges to 0 by bounded convergence. The second term is simply the right member of (7).

**Theorem 6.2.5.** We have

$$\lim_{T\to\infty} \frac{1}{2T}\int_{-T}^{T} |f(t)|^2\,dt = \sum_{x\in\mathscr{R}^1} \mu(\{x\})^2. \tag{9}$$

PROOF. Since the set of atoms is countable, all but a countable number of terms in the sum above vanish, making the sum meaningful with a value bounded by 1. Formula (9) can be established directly in the manner of (5) and (7), but the following proof is more illuminating. As noted in the proof of the corollary to Theorem 6.1.4, $|f|^2$ is the ch.f. of the r.v. $X - Y$ there, whose distribution is $\mu * \mu'$, where $\mu'(B) = \mu(-B)$ for each $B \in \mathscr{B}$. Applying Theorem 6.2.4 with $x_0 = 0$, we see that the left member of (9) is equal to

$$(\mu * \mu')(\{0\}).$$

By (7) of Sec. 6.1, the latter may be evaluated as

$$\int_{\mathscr{R}^1} \mu'(\{-y\})\mu(dy) = \sum_{y \in \mathscr{R}^1} \mu(\{y\})\mu(\{y\}),$$

since the integrand above is zero unless $-y$ is an atom of $\mu'$, which is the case if and only if $y$ is an atom of $\mu$. This gives the right member of (9). The reader may prefer to carry out the argument above using the r.v.'s $X$ and $-Y$ in the proof of the Corollary to Theorem 6.1.4.

**Corollary.** $\mu$ is atomless ($F$ is continuous) if and only if the limit in the left member of (9) is zero.

This criterion is occasionally practicable.

DEFINITION. The r.v. $X$ is called *symmetric* iff $X$ and $-X$ have the same distribution.

For such an r.v., the distribution $\mu$ has the following property:

$$\forall B \in \mathscr{B}: \mu(B) = \mu(-B).$$

Such a p.m. may be called symmetric; an equivalent condition on its d.f. $F$ is as follows:

$$\forall x \in \mathscr{R}^1: F(x) = 1 - F(-x-),$$

(the awkwardness of using d.f. being obvious here).

**Theorem 6.2.6.** $X$ or $\mu$ is symmetric if and only if its ch.f. is real-valued (for all $t$).

PROOF. If $X$ and $-X$ have the same distribution, they must "determine" the same ch.f. Hence, by (iii) of Sec. 6.1, we have

$$f(t) = \overline{f(t)}$$

and so $f$ is real-valued. Conversely, if $f$ is real-valued, the same argument shows that $X$ and $-X$ must have the same ch.f. Hence, they have the same distribution by the uniqueness theorem (Theorem 6.2.2).

## EXERCISES

$f$ is the ch.f. of $F$ below.

**1.** Show that

$$\int_0^\infty \left(\frac{\sin x}{x}\right)^2 dx = \frac{\pi}{2}.$$

$\star$**2.** Show that for each $T > 0$:

$$\frac{1}{\pi}\int_{-\infty}^{\infty} \frac{(1-\cos Tx)\cos tx}{x^2}\,dx = (T - |t|) \vee 0.$$

Deduce from this that for each $T > 0$, the function of $t$ given by

$$\left(1 - \frac{|t|}{T}\right) \vee 0$$

is a ch.f. Next, show that as a particular case of Theorem 6.2.3,

$$\frac{1-\cos Tx}{x^2} = \frac{1}{2}\int_{-T}^{T} (T - |t|)e^{itx}\,dt.$$

Finally, derive the following particularly useful relation (a case of Parseval's relation in Fourier analysis), for arbitrary $a$ and $T > 0$:

$$\int_{-\infty}^{\infty} \frac{1-\cos T(x-a)}{[T(x-a)]^2}\,dF(x) = \frac{1}{2}\frac{1}{T^2}\int_{-T}^{T} (T - |t|)e^{-ita} f(t)\,dt.$$

$\star$**3.** Prove that for each $\alpha > 0$:

$$\int_0^\alpha [F(x+u) - F(x-u)]\,du = \frac{1}{\pi}\int_{-\infty}^{\infty} \frac{1-\cos\alpha t}{t^2} e^{-itx} f(t)\,dt.$$

As a sort of reciprocal, we have

$$\frac{1}{2}\int_0^\alpha du \int_{-u}^{u} f(t)\,dt = \int_{-\infty}^{\infty} \frac{1-\cos\alpha x}{x^2}\,dF(x).$$

**4.** If $f(t)/t \in L^1(-\infty, \infty)$, then for each $\alpha > 0$ such that $\pm\alpha$ are points of continuity of $F$, we have

$$F(\alpha) - F(-\alpha) = \frac{1}{\pi}\int_{-\infty}^{\infty} \frac{\sin\alpha t}{t} f(t)\,dt.$$

**5.** What is the special case of the inversion formula when $f \equiv 1$? Deduce also the following special cases, where $\alpha > 0$:

$$\frac{1}{\pi}\int_{-\infty}^{\infty} \frac{\sin \alpha t \sin t}{t^2}\, dt = \alpha \wedge 1,$$

$$\frac{1}{\pi}\int_{-\infty}^{\infty} \frac{\sin \alpha t (\sin t)^2}{t^3}\, dt = \alpha - \frac{\alpha^2}{4} \text{ for } \alpha \le 2; 1 \text{ for } \alpha > 2.$$

**6.** For each $n \ge 0$, we have

$$\frac{1}{\pi}\int_{-\infty}^{\infty} \left(\frac{\sin t}{t}\right)^{n+2} dt = \int_0^2 \int_0^u \varphi_n(t)\, dt\, du,$$

where $\varphi_1 = \frac{1}{2} 1_{[-1,1]}$ and $\varphi_n = \varphi_{n-1} * \varphi_1$ for $n \ge 2$.

***7.** If $F$ is absolutely continuous, then $\lim_{|t|\to\infty} f(t) = 0$. Hence, if the absolutely continuous part of $F$ does not vanish, then $\overline{\lim}_{t\to\infty} |f(t)| < 1$. If $F$ is purely discontinuous, then $\overline{\lim}_{t\to\infty} f(t) = 1$. [The first assertion is the Riemann–Lebesgue lemma; prove it first when $F$ has a density that is a simple function, then approximate. The third assertion is an easy part of the observation that such an $f$ is "almost periodic".]

**8.** Prove that for $0 < r < 2$ we have

$$\int_{-\infty}^{\infty} |x|^r\, dF(x) = C(r)\int_{-\infty}^{\infty} \frac{1 - \mathbf{R} f(t)}{|t|^{r+1}}\, dt$$

where

$$C(r) = \left(\int_{-\infty}^{\infty} \frac{1 - \cos u}{|u|^{r+1}}\, du\right)^{-1} = \frac{\Gamma(r+1)}{\pi} \sin \frac{r\pi}{2},$$

thus $C(1) = 1/\pi$. [HINT:

$$|x|^r = C(r)\int_{-\infty}^{\infty} \frac{1 - \cos xt}{|t|^{r+1}}\, dt.]$$

***9.** Give a trivial example where the right member of (4) cannot be replaced by the Lebesgue integral

$$\frac{1}{2\pi}\int_{-\infty}^{\infty} \mathbf{R}\frac{e^{-itx_1} - e^{-itx_2}}{it} f(t)\, dt.$$

But it can always be replaced by the *improper Riemann integral*:

$$\lim_{\substack{T_1\to-\infty \\ T_2\to+\infty}} \frac{1}{2\pi}\int_{T_1}^{T_2} \mathbf{R}\frac{e^{-itx_1} - e^{-itx_2}}{it} f(t)\, dt.$$

**10.** Prove the following form of the inversion formula (due to Gil-Palaez):

$$\frac{1}{2}\{F(x+)+F(x-)\} = \frac{1}{2} + \lim_{\substack{\delta\downarrow 0\\ T\uparrow\infty}} \int_{\delta}^{T} \frac{e^{itx}f(-t) - e^{-itx}f(t)}{2\pi it}\,dt.$$

[HINT: Use the method of proof of Theorem 6.2.1 rather than the result.]

**11.** Theorem 6.2.3 has an analogue in $L^2$. If the ch.f. $f$ of $F$ belongs to $L^2$, then $F$ is absolutely continuous. [HINT: By Plancherel's theorem, there exists $\varphi \in L^2$ such that

$$\int_0^x \varphi(u)\,du = \frac{1}{\sqrt{2\pi}} \int_0^\infty \frac{e^{-itx}-1}{-it} f(t)\,dt.$$

Now use the inversion formula to show that

$$F(x) - F(0) = \frac{1}{\sqrt{2\pi}} \int_0^x \varphi(u)\,du.]$$

***12.** Prove Theorem 6.2.2 by the Stone–Weierstrass theorem. [HINT: Cf. Theorem 6.6.2 below, but beware of the differences. Approximate uniformly $g_1$ and $g_2$ in the proof of Theorem 4.4.3 by a periodic function with "arbitrarily large" period.]

**13.** The uniqueness theorem holds as well for signed measures [or functions of bounded variations]. Precisely, if each $\mu_i$, $i = 1, 2$, is the difference of two finite measures such that

$$\forall t\colon \int e^{itx}\mu_1(dx) = \int e^{itx}\mu_2(dx),$$

then $\mu_1 \equiv \mu_2$.

**14.** There is a deeper supplement to the inversion formula (4) or Exercise 10 above, due to B. Rosén. Under the condition

$$\int_{-\infty}^{\infty} (1 + \log|x|)\,dF(x) < \infty,$$

the improper Riemann integral in Exercise 10 may be replaced by a Lebesgue integral. [HINT: It is a matter of proving the existence of the latter. Since

$$\int_{-\infty}^{\infty} dF(y) \int_0^N \left|\frac{\sin(x-y)t}{t}\right| dt \le \int_{-\infty}^{\infty} dF(y)\{1 + \log(1 + N|x-y|)\} < \infty,$$

we have

$$\int_{-\infty}^{\infty} dF(y) \int_0^N \frac{\sin(x-y)t}{t}\,dt = \int_0^N \frac{dt}{t} \int_{-\infty}^{\infty} \sin(x-y)t\,dF(y).$$

For fixed $x$, we have

$$\int_{y\neq x} dF(y) \left| \int_N^\infty \frac{\sin(x-y)t}{t}\, dt \right| \le C_1 \int_{|x-y|\ge 1/N} \frac{dF(y)}{N|x-y|}$$

$$+ C_2 \int_{0<|x-y|<1/N} dF(y),$$

both integrals on the right converging to 0 as $N \to \infty$.]

## 6.3 Convergence theorems

For purposes of probability theory the most fundamental property of the ch.f. is given in the two propositions below, which will be referred to jointly as the *convergence theorem*, due to P. Lévy and H. Cramér. Many applications will be given in this and the following chapter. We begin with the easier half.

**Theorem 6.3.1.** Let $\{\mu_n, 1 \le n \le \infty\}$ be p.m.'s on $\mathscr{R}^1$ with ch.f.'s $\{f_n, 1 \le n \le \infty\}$. If $\mu_n$ converges vaguely to $\mu_\infty$, then $f_n$ converges to $f_\infty$ uniformly in every finite interval. We shall write this symbolically as

$$(1) \qquad \mu_n \xrightarrow{v} \mu_\infty \Rightarrow f_n \xrightarrow{u} f_\infty.$$

Furthermore, the family $\{f_n\}$ is equicontinuous on $\mathscr{R}^1$.

PROOF. Since $e^{itx}$ is a bounded continuous function on $\mathscr{R}^1$, although complex-valued, Theorem 4.4.2 applies to its real and imaginary parts and yields (1) at once, apart from the asserted uniformity. Now for every $t$ and $h$, we have, as in (ii) of Sec. 6.1:

$$|f_n(t+h) - f_n(t)| \le \int |e^{ihx} - 1| \mu_n(dx) \le \int_{|x|\le A} |hx| \mu_n(dx)$$

$$+ 2\int_{|x|>A} \mu_n(dx) \le |h|A + 2\int_{|x|>A} \mu(dx) + \epsilon$$

for any $\epsilon > 0$, suitable $A$ and $n \ge n_0(A, \epsilon)$. The equicontinuity of $\{f_n\}$ follows. This and the pointwise convergence $f_n \to f_\infty$ imply $f_n \xrightarrow{u} f_\infty$ by a simple compactness argument (the "3$\epsilon$ argument") left to the reader.

**Theorem 6.3.2.** Let $\{\mu_n, 1 \le n < \infty\}$ be p.m.'s on $\mathscr{R}^1$ with ch.f.'s $\{f_n, 1 \le n < \infty\}$. Suppose that

(a) $f_n$ converges everywhere in $\mathscr{R}^1$ and defines the limit function $f_\infty$;
(b) $f_\infty$ is continuous at $t = 0$.

Then we have

($\alpha$) $\mu_n \xrightarrow{v} \mu_\infty$, where $\mu_\infty$ is a p.m.;
($\beta$) $f_\infty$ is the ch.f. of $\mu_\infty$.

PROOF. Let us first relax the conditions (a) and (b) to require only convergence of $f_n$ in a neighborhood $(-\delta_0, \delta_0)$ of $t = 0$ and the continuity of the limit function $f$ (defined only in this neighborhood) at $t = 0$. We shall prove that any vaguely convergent subsequence of $\{\mu_n\}$ converges to a p.m. $\mu$. For this we use the following lemma, which illustrates a useful technique for obtaining estimates on a p.m. from its ch.f.

**Lemma.** For each $A > 0$, we have

$$\mu([-2A, 2A]) \geq A \left| \int_{-A^{-1}}^{A^{-1}} f(t)\, dt \right| - 1. \tag{2}$$

PROOF OF THE LEMMA. By (8) of Sec. 6.2, we have

$$\frac{1}{2T} \int_{-T}^{T} f(t)\, dt = \int_{-\infty}^{\infty} \frac{\sin Tx}{Tx} \mu(dx). \tag{3}$$

Since the integrand on the right side is bounded by 1 for all $x$ (it is defined to be 1 at $x = 0$), and by $|Tx|^{-1} \leq (2TA)^{-1}$ for $|x| > 2A$, the integral is bounded by

$$\mu([-2A, 2A]) + \frac{1}{2TA}\{1 - \mu([-2A, 2A])\}$$
$$= \left(1 - \frac{1}{2TA}\right) \mu([-2A, 2A]) + \frac{1}{2TA}.$$

Putting $T = A^{-1}$ in (3), we obtain

$$\left| \frac{A}{2} \int_{-A^{-1}}^{A^{-1}} f(t)\, dt \right| \leq \frac{1}{2}\mu([-2A, 2A]) + \frac{1}{2},$$

which reduces to (2). The lemma is proved.

Now for each $\delta$, $0 < \delta < \delta_0$, we have

$$\left| \frac{1}{2\delta} \int_{-\delta}^{\delta} f_n(t)\, dt \right| \geq \left| \frac{1}{2\delta} \int_{-\delta}^{\delta} f(t)\, dt \right| - \frac{1}{2\delta} \int_{-\delta}^{\delta} |f_n(t) - f(t)|\, dt. \tag{4}$$

The first term on the right side tends to 1 as $\delta \downarrow 0$, since $f(0) = 1$ and $f$ is continuous at 0; for fixed $\delta$ the second term tends to 0 as $n \to \infty$, by bounded convergence since $|f_n - f| \leq 2$. It follows that for any given $\epsilon > 0$, there exist $\delta = \delta(\epsilon) < \delta_0$ and $n_0 = n_0(\epsilon)$ such that if $n \geq n_0$, then the left member of (4) has a value not less than $1 - \epsilon$. Hence by (2)

$$\mu_n([-2\delta^{-1}, 2\delta^{-1}]) \geq 2(1-\epsilon) - 1 \geq 1 - 2\epsilon. \tag{5}$$

Let $\{\mu_{n_k}\}$ be a vaguely convergent subsequence of $\{\mu_n\}$, which always exists by Theorem 4.3.3; and let the vague limit be $\mu$, which is always an s.p.m. For each $\delta$ satisfying the conditions above, and such that neither $-2\delta^{-1}$ nor $2\delta^{-1}$ is an atom of $\mu$, we have by the property of vague convergence and (5):

$$\begin{aligned}\mu(\mathscr{R}^1) &\geq \mu([-2\delta^{-1}, 2\delta^{-1}]) \\ &= \lim_{n\to\infty} \mu_n([-2\delta^{-1}, 2\delta^{-1}]) \geq 1 - 2\epsilon.\end{aligned}$$

Since $\epsilon$ is arbitrary, we conclude that $\mu$ is a p.m., as was to be shown.

Let $f$ be the ch.f. of $\mu$. Then, by the preceding theorem, $f_{n_k} \to f$ everywhere; hence under the original hypothesis (a) we have $f = f_\infty$. Thus every vague limit $\mu$ considered above has the same ch.f. and therefore by the uniqueness theorem is the same p.m. Rename it $\mu_\infty$ so that $\mu_\infty$ is the p.m. having the ch.f. $f_\infty$. Then by Theorem 4.3.4 we have $\mu_n \xrightarrow{v} \mu_\infty$. Both assertions ($\alpha$) and ($\beta$) are proved.

As a particular case of the above theorem: if $\{\mu_n, 1 \leq n \leq \infty\}$ and $\{f_n, 1 \leq n \leq \infty\}$ are corresponding p.m.'s and ch.f.'s, then the converse of (1) is also true, namely:

$$\mu_n \xrightarrow{v} \mu_\infty \Leftrightarrow f_n \xrightarrow{u} f_\infty. \tag{6}$$

This is an elegant statement, but it lacks the full strength of Theorem 6.3.2, which lies in concluding that $f_\infty$ is a ch.f. from more easily verifiable conditions, rather than assuming it. We shall see presently how important this is.

Let us examine some cases of inapplicability of Theorems 6.3.1 and 6.3.2.

**Example 1.** Let $\mu_n$ have mass $\frac{1}{2}$ at 0 and mass $\frac{1}{2}$ at $n$. Then $\mu_n \to \mu_\infty$, where $\mu_\infty$ has mass $\frac{1}{2}$ at 0 and is not a p.m. We have

$$f_n(t) = \tfrac{1}{2} + \tfrac{1}{2}e^{int},$$

which does not converge as $n \to \infty$, except when $t$ is equal to a multiple of $2\pi$.

**Example 2.** Let $\mu_n$ be the uniform distribution $[-n, n]$. Then $\mu_n \to \mu_\infty$, where $\mu_\infty$ is identically zero. We have

$$f_n(t) = \begin{cases} \dfrac{\sin nt}{nt}, & \text{if } t \neq 0; \\ 1, & \text{if } t = 0; \end{cases}$$

and

$$f_n(t) \to f(t) = \begin{cases} 0, & \text{if } t \neq 0; \\ 1, & \text{if } t = 0. \end{cases}$$

Thus, condition (a) is satisfied but (b) is not.

Later we shall see that (a) cannot be relaxed to read: $f_n(t)$ converges in $|t| \le T$ for some fixed $T$ (Exercise 9 of Sec. 6.5).

The convergence theorem above settles the question of vague convergence of p.m.'s to a p.m. What about just vague convergence without restriction on the limit? Recalling Theorem 4.4.3, this suggests first that we replace the integrand $e^{itx}$ in the ch.f. $f$ by a function in $C_0$. Secondly, going over the last part of the proof of Theorem 6.3.2, we see that the choice should be made so as to determine uniquely an s.p.m. (see Sec. 4.3). Now the Fourier–Stieltjes transform of an s.p.m. is well defined and the inversion formula remains valid, so that there is unique correspondence just as in the case of a p.m. Thus a natural choice of $g$ is given by an "indefinite integral" of a ch.f., as follows:

$$(7) \qquad g(u) = \int_0^u \left[ \int_{\mathscr{R}^1} e^{itx} \mu(dx) \right] dt = \int_{\mathscr{R}^1} \frac{e^{iux} - 1}{ix} \mu(dx).$$

Let us call $g$ the *integrated characteristic function* of the s.p.m. $\mu$. We are thus led to the following companion of (6), the details of the proof being left as an exercise.

**Theorem 6.3.3.** A sequence of s.p.m.'s $\{\mu_n, 1 \le n < \infty\}$ converges (to $\mu_\infty$) if and only if the corresponding sequence of integrated ch.f.'s $\{g_n\}$ converges (to the integrated ch.f. of $\mu_\infty$).

Another question concerning (6) arises naturally. We know that vague convergence for p.m.'s is metric (Exercise 9 of Sec. 4.4); let the metric be denoted by $\langle \cdot, \cdot \rangle_1$. Uniform convergence on compacts (viz., in finite intervals) for uniformly bounded subsets of $C_B(\mathscr{R}^1)$ is also metric, with the metric denoted by $\langle \cdot, \cdot \rangle_2$, defined as follows:

$$\langle f, g \rangle_2 = \sup_{t \in \mathscr{R}^1} \frac{|f(t) - g(t)|}{1 + t^2}.$$

It is easy to verify that this is a metric on $C_B$ and that convergence in this metric is equivalent to uniform convergence on compacts; clearly the denominator $1 + t^2$ may be replaced by any function continuous on $\mathscr{R}^1$, bounded below by a strictly positive constant, and tending to $+\infty$ as $|t| \to \infty$. Since there is a one-to-one correspondence between ch.f.'s and p.m.'s, we may transfer the metric $\langle \cdot, \cdot \rangle_2$ to the latter by setting

$$\langle \mu, \nu \rangle_2 = \langle f_\mu, f_\nu \rangle_2$$

in obvious notation. Now the relation (6) may be restated as follows.

**Theorem 6.3.4.** The topologies induced by the two metrics $\langle\ \rangle_1$ and $\langle\ \rangle_2$ on the space of p.m.'s on $\mathscr{R}^1$ are equivalent.

This means that for each $\mu$ and given $\epsilon > 0$, there exists $\delta(\mu, \epsilon)$ such that:

$$\langle \mu, \nu \rangle_1 \leq \delta(\mu, \epsilon) \Rightarrow \langle \mu, \nu \rangle_2 \leq \epsilon,$$

$$\langle \mu, \nu \rangle_2 \leq \delta(\mu, \epsilon) \Rightarrow \langle \mu, \nu \rangle_1 \leq \epsilon.$$

Theorem 6.3.4 needs no new proof, since it is merely a paraphrasing of (6) in new words. However, it is important to notice the dependence of $\delta$ on $\mu$ (as well as $\epsilon$) above. The sharper statement without this dependence, which would mean the equivalence of the *uniform structures* induced by the two metrics, is false with a vengeance; see Exercises 10 and 11 below (Exercises 3 and 4 are also relevant).

EXERCISES

**1.** Prove the uniform convergence of $f_n$ in Theorem 6.3.1 by an integration by parts of $\int e^{itx}\, dF_n(x)$.

***2.** Instead of using the Lemma in the second part of Theorem 6.3.2, prove that $\mu$ is a p.m. by integrating the inversion formula, as in Exercise 3 of Sec. 6.2. (Integration is a smoothing operation and a standard technique in taming improper integrals: cf. the proof of the second part of Theorem 6.5.2 below.)

**3.** Let $F$ be a given absolutely continuous d.f. and let $F_n$ be a sequence of step functions with equally spaced steps that converge to $F$ uniformly in $\mathscr{R}^1$. Show that for the corresponding ch.f.'s we have

$$\forall n: \sup_{t \in \mathscr{R}^1} |f(t) - f_n(t)| = 1.$$

**4.** Let $F_n, G_n$ be d.f.'s with ch.f.'s $f_n$ and $g_n$. If $f_n - g_n \to 0$ a.e., then for each $f \in C_K$ we have $\int f\, dF_n - \int f\, dG_n \to 0$ (see Exercise 10 of Sec. 4.4). This does not imply the Lévy distance $\langle F_n, G_n \rangle_1 \to 0$; find a counterexample. [HINT: Use Exercise 3 of Sec. 6.2 and proceed as in Theorem 4.3.4.]

**5.** Let $F$ be a discrete d.f. with points of jump $\{a_j, j \geq 1\}$ and sizes of jump $\{b_j, j \geq 1\}$. Consider the approximating s.d.f.'s $F_n$ with the same jumps but restricted to $j \leq n$. Show that $F_n \xrightarrow{v} F$.

$^\star$**6.** If the sequence of ch.f.'s $\{f_n\}$ converges uniformly in a neighborhood of the origin, then $\{f_n\}$ is equicontinuous, and there exists a subsequence that converges to a ch.f. [HINT: Use Ascoli-Arzela's theorem.]

**7.** If $F_n \xrightarrow{v} F$ and $G_n \xrightarrow{v} G$, then $F_n * G_n \xrightarrow{v} F * G$. [A proof of this simple result without the use of ch.f.'s would be tedious.]

$^\star$**8.** Interpret the remarkable trigonometric identity

$$\frac{\sin t}{t} = \prod_{n=1}^{\infty} \cos \frac{t}{2^n}$$

in terms of ch.f.'s, and hence by addition of independent r.v.'s. (This is an example of Exercise 4 of Sec. 6.1.)

**9.** Rewrite the preceding formula as

$$\frac{\sin t}{t} = \left( \prod_{k=1}^{\infty} \cos \frac{t}{2^{2k-1}} \right) \left( \prod_{k=1}^{\infty} \cos \frac{t}{2^{2k}} \right).$$

Prove that either factor on the right is the ch.f. of a singular distribution. Thus the convolution of two such may be absolutely continuous. [HINT: Use the same r.v.'s as for the Cantor distribution in Exercise 9 of Sec. 5.3.]

**10.** Using the strong law of large numbers, prove that the convolution of two Cantor d.f.'s is still singular. [HINT: Inspect the frequency of the digits in the sum of the corresponding random series; see Exercise 9 of Sec. 5.3.]

$^\star$**11.** Let $F_n$, $G_n$ be the d.f.'s of $\mu_n$, $\nu_n$, and $f_n$, $g_n$ their ch.f.'s. Even if $\sup_{x \in \mathscr{R}^1} |F_n(x) - G_n(x)| \to 0$, it does not follow that $\langle f_n, g_n \rangle_2 \to 0$; indeed it may happen that $\langle f_n, g_n \rangle_2 = 1$ for every $n$. [HINT: Take two step functions "out of phase".]

**12.** In the notation of Exercise 11, even if $\sup_{t \in \mathscr{R}^1} |f_n(t) - g_n(t)| \to 0$, it does not follow that $\langle F_n, G_n \rangle \to 0$; indeed it may $\to 1$. [HINT: Let $f$ be any ch.f. vanishing outside $(-1, 1)$, $f_j(t) = e^{-in_j t} f(m_j t)$, $g_j(t) = e^{in_j t} f(m_j t)$, and $F_j$, $G_j$ be the corresponding d.f.'s. Note that if $m_j n_j^{-1} \to 0$, then $F_j(x) \to 1$, $G_j(x) \to 0$ for every $x$, and that $f_j - g_j$ vanishes outside $(-m_j^{-1}, m_j^{-1})$ and is $0(\sin n_j t)$ near $t = 0$. If $m_j = 2^{j^2}$ and $n_j = j m_j$ then $\sum_j (f_j - g_j)$ is uniformly bounded in $t$: for $n_{k+1}^{-1} < t \leq n_k^{-1}$ consider $j > k$, $j = k$, $j < k$ separately. Let

$$f_n^* = n^{-1} \sum_{j=1}^{n} f_j, \quad g_n^* = n^{-1} \sum_{j=1}^{n} g_j,$$

then $\sup |f_n^* - g_n^*| = O(n^{-1})$ while $F_n^* - G_n^* \to 0$. This example is due to Katznelson, rivaling an older one due to Dyson, which is as follows. For

$b > a > 0$, let

$$F(x) = \frac{1}{2}\frac{\log\left(\dfrac{x^2+b^2}{x^2+a^2}\right)}{a\log\left(\dfrac{b}{a}\right)}$$

for $x < 0$ and $= 1$ for $x > 0$; $G(x) = 1 - F(-x)$. Then,

$$f(t) - g(t) = -\pi i \frac{t}{|t|}\frac{[e^{-a|t|} - e^{-b|t|}]}{\log\left(\dfrac{b}{a}\right)}.$$

If $a$ is large, then $\langle F, G\rangle$ is near 1. If $b/a$ is large, then $\langle f, g\rangle$ is near 0.]

## 6.4 Simple applications

A common type of application of Theorem 6.3.2 depends on power-series expansions of the ch.f., for which we need its derivatives.

**Theorem 6.4.1.** If the d.f. has a finite absolute moment of positive integral order $k$, then its ch.f. has a bounded continuous derivative of order $k$ given by

$$(1) \qquad f^{(k)}(t) = \int_{-\infty}^{\infty} (ix)^k e^{itx}\, dF(x).$$

Conversely, if $f$ has a finite derivative of even order $k$ at $t = 0$, then $F$ has a finite moment of order $k$.

PROOF. For $k = 1$, the first assertion follows from the formula:

$$\frac{f(t+h) - f(t)}{h} = \int_{-\infty}^{\infty} \frac{e^{i(t+h)x} - e^{itx}}{h}\, dF(x).$$

An elementary inequality already used in the proof of Theorem 6.2.1 shows that the integrand above is dominated by $|x|$. Hence if $\int |x|\, dF(x) < \infty$, we may let $h \to 0$ under the integral sign and obtain (1). Uniform continuity of the integral as a function of $t$ is proved as in (ii) of Sec. 6.1. The case of a general $k$ follows easily by induction.

To prove the second assertion, let $k = 2$ and suppose that $f''(0)$ exists and is finite. We have

$$f''(0) = \lim_{h\to 0} \frac{f(h) - 2f(0) + f(-h)}{h^2}$$

$$= \lim_{h\to 0} \int \frac{e^{ihx} - 2 + e^{-ihx}}{h^2} \, dF(x)$$

$$(2) \qquad = -2 \lim_{h\to 0} \int \frac{1-\cos hx}{h^2} \, dF(x).$$

As $h \to 0$, we have by Fatou's lemma,

$$\int x^2 \, dF(x) = 2 \int \lim_{h\to 0} \frac{1-\cos hx}{h^2} \, dF(x) \le \underline{\lim}_{h\to 0} 2 \int \frac{1-\cos hx}{h^2} \, dF(x)$$
$$= -f''(0).$$

Thus $F$ has a finite second moment, and the validity of (1) for $k = 2$ now follows from the first assertion of the theorem.

The general case can again be reduced to this by induction, as follows. Suppose the second assertion of the theorem is true for $2k - 2$, and that $f^{(2k)}(0)$ is finite. Then $f^{(2k-2)}(t)$ exists and is continuous in the neighborhood of $t = 0$, and by the induction hypothesis we have in particular

$$(-1)^{k-1} \int x^{2k-2} \, dF(x) = f^{(2k-2)}(0).$$

Put $G(x) = \int_{-\infty}^{x} y^{2k-2} \, dF(y)$ for every $x$, then $G(\cdot)/G(\infty)$ is a d.f. with the ch.f.

$$\psi(t) = \frac{1}{G(\infty)} \int e^{itx} x^{2k-2} \, dF(x) = \frac{(-1)^{k-1} f^{(2k-2)}(t)}{G(\infty)}.$$

Hence $\psi''$ exists, and by the case $k = 2$ proved above, we have

$$-\psi^{(2)}(0) = \frac{1}{G(\infty)} \int x^2 \, dG(x) = \frac{1}{G(\infty)} \int x^{2k} \, dF(x)$$

Upon cancelling $G(\infty)$, we obtain

$$(-1)^k f^{(2k)}(0) = \int x^{2k} \, dF(x),$$

which proves the finiteness of the $2k$th moment. The argument above fails if $G(\infty) = 0$, but then we have (why?) $F = \delta_0$, $f = 1$, and the theorem is trivial.

Although the next theorem is an immediate corollary to the preceding one, it is so important as to deserve prominent mention.

**Theorem 6.4.2.** If $F$ has a finite absolute moment of order $k$, $k$ an integer $\ge 1$, then $f$ has the following expansion in the neighborhood of $t = 0$:

$$(3) \qquad f(t) = \sum_{j=0}^{k} \frac{i^j}{j!} m^{(j)} t^j + o(|t|^k),$$

$$(3') \qquad f(t) = \sum_{j=0}^{k-1} \frac{i^j}{j!} m^{(j)} t^j + \frac{\theta_k}{k!} \mu^{(k)} |t|^k;$$

where $m^{(j)}$ is the moment of order $j$, $\mu^{(k)}$ is the absolute moment of order $k$, and $|\theta_k| \leq 1$.

PROOF. According to a theorem in calculus (see, e.g., Hardy [1], p. 290]), if $f$ has a finite $k$th derivative at the point $t = 0$, then the Taylor expansion below is valid:

$$(4) \qquad f(t) = \sum_{j=0}^{k} \frac{f^{(j)}(0)}{j!} t^j + o(|t|^k).$$

If $f$ has a finite $k$th derivative in the neighborhood of $t = 0$, then

$$(4') \qquad f(t) = \sum_{j=0}^{k-1} \frac{f^{(j)}(0)}{j!} t^j + \frac{f^{(k)}(\theta t)}{k!} t^k, \quad |\theta| \leq 1.$$

Since the absolute moment of order $j$ is finite for $1 \leq j \leq k$, and

$$f^{(j)}(0) = i^j m^{(j)}, \quad |f^{(k)}(\theta t)| \leq \mu^{(k)}$$

from (1), we obtain (3) from (4), and (3′) from (4′).

It should be remarked that the form of Taylor expansion given in (4) is not always given in textbooks as cited above, but rather under stronger assumptions, such as "$f$ has a finite $k$th derivative in the neighborhood of 0". [For even $k$ this stronger condition is actually implied by the weaker one stated in the proof above, owing to Theorem 6.4.1.] The reader is advised to learn the sharper result in calculus, which incidentally also yields a quick proof of the first equation in (2). Observe that (3) implies (3′) if the last term in (3′) is replaced by the more ambiguous $O(|t|^k)$, but not as it stands, since the constant in "$O$" may depend on the function $f$ and not just on $\mu^{(k)}$.

By way of illustrating the power of the method of ch.f.'s without the encumbrance of technicalities, although anticipating more elaborate developments in the next chapter, we shall apply at once the results above to prove two classical limit theorems: the weak law of large numbers (cf. Theorem 5.2.2), and the *central limit theorem* in the identically distributed and finite variance case. We begin with an elementary lemma from calculus, stated here for the sake of clarity.

**Lemma.** If the complex numbers $c_n$ have the limit $c$, then

$$(5) \qquad \lim_{n\to\infty} \left(1 + \frac{c_n}{n}\right)^n = e^c.$$

(For real $c_n$'s this remains valid for $c = +\infty$.)

Now let $\{X_n, n \geq 1\}$ be a sequence of independent r.v.'s with the common d.f. $F$, and $S_n = \sum_{j=1}^n X_j$, as in Chapter 5.

**Theorem 6.4.3.** If $F$ has a finite mean $m$, then

$$\frac{S_n}{n} \to m \quad \text{in pr.}$$

PROOF. Since convergence to the constant $m$ is equivalent to that in dist. to $\delta_m$ (Exercise 4 of Sec. 4.4), it is sufficient by Theorem 6.3.2 to prove that the ch.f. of $S_n/n$ converges to $e^{imt}$ (which is continuous). Now, by (2) of Sec. 6.1 we have

$$E(e^{it(S_n/n)}) = E(e^{i(t/n)S_n}) = \left[f\left(\frac{t}{n}\right)\right]^n.$$

By Theorem 6.4.2, the last term above may be written as

$$\left(1 + im\frac{t}{n} + o\left(\frac{t}{n}\right)\right)^n$$

for fixed $t$ and $n \to \infty$. It follows from (5) that this converges to $e^{imt}$ as desired.

**Theorem 6.4.4.** If $F$ has mean $m$ and finite variance $\sigma^2 > 0$, then

$$\frac{S_n - mn}{\sigma\sqrt{n}} \to \Phi \quad \text{in dist.}$$

where $\Phi$ is the normal distribution with mean 0 and variance 1.

PROOF. We may suppose $m = 0$ by considering the r.v.'s $X_j - m$, whose second moment is $\sigma^2$. As in the preceding proof, we have

$$\begin{aligned}
\mathscr{E}\left(\exp\left(it\frac{S_n}{\sigma\sqrt{n}}\right)\right) &= f\left(\frac{t}{\sigma\sqrt{n}}\right)^n \\
&= \left\{1 + \frac{i^2\sigma^2}{2}\left(\frac{t}{\sigma\sqrt{n}}\right)^2 + o\left(\frac{|t|}{\sigma\sqrt{n}}\right)^2\right\}^n \\
&= \left\{1 - \frac{t^2}{2n} + o\left(\frac{t^2}{n}\right)\right\}^n \to e^{-t^2/2}.
\end{aligned}$$

The limit being the ch.f. of $\Phi$, the proof is ended.

The convergence theorem for ch.f.'s may be used to complete the method of moments in Theorem 4.5.5, yielding a result not very far from what ensues from that theorem coupled with Carleman's condition mentioned there.

**Theorem 6.4.5.** In the notation of Theorem 4.5.5, if (8) there holds together with the following condition:

$$\forall t \in \mathscr{R}^1: \lim_{k\to\infty} \frac{m^{(k)}t^k}{k!} = 0, \tag{6}$$

then $F_n \xrightarrow{v} F$.

PROOF. Let $f_n$ be the ch.f. of $F_n$. For fixed $t$ and an odd $k$ we have by the Taylor expansion for $e^{itx}$ with a remainder term:

$$f_n(t) = \int e^{itx}\, dF_n(x) = \int \left\{ \sum_{j=0}^{k} \frac{(itx)^j}{j!} + \theta \frac{|itx|^{k+1}}{(k+1)!} \right\} dF_n(x)$$

$$= \sum_{j=0}^{k} \frac{(it)^j}{j!} m_n^{(j)} + \theta \frac{m_n^{(k+1)} t^{k+1}}{(k+1)!},$$

where $\theta$ denotes a "generic" complex number of modulus $\leq 1$, not necessarily the same in different appearances. (The use of such a symbol, to be further indulged in the next chapter, avoids the necessity of transposing terms and taking moduli of long expressions. It is extremely convenient, but one must occasionally watch the dependence of $\theta$ on various quantities involved.) It follows that

$$f_n(t) - f(t) = \sum_{j=0}^{k} \frac{(it)^j}{j!} (m_n^{(j)} - m^{(j)}) + \frac{\theta t^{k+1}}{(k+1)!} (m_n^{(k+1)} + m^{(k+1)}). \tag{7}$$

Given $\epsilon > 0$, by condition (6) there exists an odd $k = k(\epsilon)$ such that for the fixed $t$ we have

$$\frac{(2m^{(k+1)} + 1)t^{k+1}}{(k+1)!} \leq \frac{\epsilon}{2}. \tag{8}$$

Since we have fixed $k$, there exists $n_0 = n_0(\epsilon)$ such that if $n \geq n_0$, then

$$m_n^{(k+1)} \leq m^{(k+1)} + 1,$$

and moreover,

$$\max_{1 \leq j \leq k} |m_n^{(j)} - m^{(j)}| \leq e^{-|t|} \frac{\epsilon}{2}.$$

Then the right side of (7) will not exceed in modulus:

$$\sum_{j=0}^{k} \frac{|t|^j}{j!} e^{-|t|} \frac{\epsilon}{2} + \frac{t^{k+1}(2m^{(k+1)} + 1)}{(k+1)!} \leq \epsilon.$$

Hence $f_n(t) \to f(t)$ for each $t$, and since $f$ is a ch.f., the hypotheses of Theorem 6.3.2 are satisfied and so $F_n \xrightarrow{v} F$.

As another kind of application of the convergence theorem in which a limiting process is implicit rather than explicit, let us prove the following characterization of the normal distribution.

**Theorem 6.4.6.** Let $X$ and $Y$ be independent, identically distributed r.v.'s with mean 0 and variance 1. If $X + Y$ and $X - Y$ are independent then the common distribution of $X$ and $Y$ is $\Phi$.

PROOF. Let $f$ be the ch.f., then by (1), $f'(0) = 0$, $f''(0) = -1$. The ch.f. of $X + Y$ is $f(t)^2$ and that of $X - Y$ is $f(t)f(-t)$. Since these two r.v.'s are independent, the ch.f. $f(2t)$ of their sum $2X$ must satisfy the following relation:

$$f(2t) = f(t)^3 f(-t). \tag{9}$$

It follows from (9) that the function $f$ never vanishes. For if it did at $t_0$, then it would also at $t_0/2$, and so by induction at $t_0/2^n$ for every $n \geq 1$. This is impossible, since $\lim_{n\to\infty} f(t_0/2^n) = f(0) = 1$. Setting for every $t$:

$$\rho(t) = \frac{f(t)}{f(-t)},$$

we obtain

$$\rho(2t) = \rho(t)^2. \tag{10}$$

Hence we have by iteration, for each $t$:

$$\rho(t) = \rho\left(\frac{t}{2^n}\right)^{2^n} = \left\{1 + o\left(\frac{t}{2^n}\right)\right\}^{2^n} \to 1$$

by Theorem 6.4.2 and the previous lemma. Thus $\rho(t) \equiv 1$, $f(t) \equiv f(-t)$, and (9) becomes

$$f(2t) = f(t)^4. \tag{11}$$

Repeating the argument above with (11), we have

$$f(t) = f\left(\frac{t}{2^n}\right)^{4^n} = \left\{1 - \frac{1}{2}\left(\frac{t}{2^n}\right)^2 + o\left[\left(\frac{t}{2^n}\right)^2\right]\right\}^{4^n} \to e^{-t^2/2}.$$

This proves the theorem without the use of "logarithms" (see Sec. 7.6).

EXERCISES

$^\star$**1.** If $f$ is the ch.f. of $X$, and

$$\lim_{t\downarrow 0} \frac{f(t) - 1}{t^2} = \frac{-\Sigma^2}{2} > -\infty,$$

then $\mathscr{E}(X) = 0$ and $\mathscr{E}(X^2) = \sigma^2$. In particular, if $f(t) = 1 + o(t^2)$ as $t \to 0$, then $f \equiv 1$.

★**2.** Let $\{X_n\}$ be independent, identically distributed with mean 0 and variance $\sigma^2$, $0 \le \sigma^2 \le \infty$. Prove that

$$\lim_{n\to\infty} \mathscr{E}\left(\frac{|S_n|}{\sqrt{n}}\right) = 2 \lim_{n\to\infty} \mathscr{E}\left(\frac{S_n^+}{\sqrt{n}}\right) = \sqrt{\frac{2}{\pi}}\sigma$$

[If we assume only $\mathscr{P}\{X_1 \ne 0\} > 0$, $\mathscr{E}(|X_1|) < \infty$ and $\mathscr{E}(X_1) = 0$, then we have $\mathscr{E}(|S_n|) \ge C\sqrt{n}$ for some constant $C$ and all $n$; this is known as Hornich's inequality.] [HINT: In case $\sigma^2 = \infty$, if $\underline{\lim}_n \mathscr{E}(| S_n/\sqrt{n}) < \infty$, then there exists $\{n_k\}$ such that $S_n/\sqrt{n_k}$ converges in distribution; use an extension of Exercise 1 to show $|f(t/\sqrt{n})|^{2n} \to 0$. This is due to P. Matthews.]

**3.** Let $\mathscr{P}\{X = k\} = p_k$, $1 \le k \le \ell < \infty$, $\sum_{k=1}^{\ell} p_k = 1$. The sum $S_n$ of $n$ independent r.v.'s having the same distribution as $X$ is said to have a *multinomial distribution*. Define it explicitly. Prove that $[S_n - \mathscr{E}(S_n)]/\sigma(S_n)$ converges to $\Phi$ in distribution as $n \to \infty$, provided that $\sigma(X) > 0$.

★**4.** Let $X_n$ have the binomial distribution with parameter $(n, p_n)$, and suppose that $n p_n \to \lambda \ge 0$. Prove that $X_n$ converges in dist. to the Poisson d.f. with parameter $\lambda$. (In the old days this was called *the law of small numbers*.)

**5.** Let $X_\lambda$ have the Poisson distribution with parameter $\lambda$. Prove that $[X_\lambda - \lambda]/\lambda^{1/2}$ converges in dist. to $\Phi$ as $\lambda \to \infty$.

★**6.** Prove that in Theorem 6.4.4, $S_n/\sigma\sqrt{n}$ does not converge in probability. [HINT: Consider $S_n/\sigma\sqrt{n}$ and $S_{2n}/\sigma\sqrt{2n}$.]

**7.** Let $f$ be the ch.f. of the d.f. $F$. Suppose that as $t \to 0$,

$$f(t) - 1 = O(|t|^\alpha),$$

where $0 < \alpha \le 2$, then as $A \to \infty$,

$$\int_{|x|>A} dF(x) = O(A^{-\alpha}).$$

[HINT: Integrate $\int_{|x|>A} (1 - \cos tx)\, dF(x) \le Ct^\alpha$ over $t$ in $(0, A)$.]

**8.** If $0 < \alpha < 1$ and $\int |x|^\alpha \, dF(x) < \infty$, then $f(t) - 1 = o(|t|^\alpha)$ as $t \to 0$. For $1 \le \alpha < 2$ the same result is true under the additional assumption that $\int x \, dF(x) = 0$. [HINT: The case $1 \le \alpha < 2$ is harder. Consider the real and imaginary parts of $f(t) - 1$ separately and write the latter as

$$\int_{|x|\le\epsilon/t} \sin tx \, dF(x) + \int_{|x|>\epsilon/|t|} \sin tx \, dF(x).$$

The second is bounded by $(|t|/\epsilon)^\alpha \int_{|x|>\epsilon/|t|} |x|^\alpha \, dF(x) = o(|t|^\alpha)$ for fixed $\epsilon$. In the first integral use $\sin tx = tx + O(|tx|^3)$,

$$\int_{|x|\le\epsilon/|t|} tx \, dF(x) = t \int_{|x|>\epsilon/|t|} x \, dF(x),$$

$$\int_{|x|\le\epsilon/|t|} |tx|^3 \, dF(x) \le \epsilon^{3-\alpha} \int_{-\infty}^{\infty} |tx|^\alpha \, dF(x).]$$

**9.** Suppose that $e^{-c|t|^\alpha}$, where $c \ge 0$, $0 < \alpha \le 2$, is a ch.f. (Theorem 6.5.4 below). Let $\{X_j, j \ge 1\}$ be independent and identically distributed r.v.'s with a common ch.f. of the form

$$1 - \beta|t|^\alpha + o(|t|^\alpha)$$

as $t \to 0$. Determine the constants $b$ and $\theta$ so that the ch.f. of $S_n/bn^\theta$ converges to $e^{-|t|^\alpha}$.

**10.** Suppose $F$ satisfies the condition that for every $\eta > 0$ such that as $A \to \infty$,

$$\int_{|x|>A} dF(x) = O(e^{-\eta A}).$$

Then all moments of $F$ are finite, and condition (6) in Theorem 6.4.5 is satisfied.

**11.** Let $X$ and $Y$ be independent with the common d.f. $F$ of mean 0 and variance 1. Suppose that $(X+Y)/\sqrt{2}$ also has the d.f. $F$. Then $F \equiv \Phi$. [HINT: Imitate Theorem 6.4.5.]

***12.** Let $\{X_j, j \ge 1\}$ be independent, identically distributed r.v.'s with mean 0 and variance 1. Prove that both

$$\frac{\sum_{j=1}^n X_j}{\sqrt{\sum_{j=1}^n X_j^2}} \quad \text{and} \quad \frac{\sqrt{n}\sum_{j=1}^n X_j}{\sum_{j=1}^n X_j^2}$$

converge in dist. to $\Phi$. [Hint: Use the law of large numbers.]

**13.** The converse part of Theorem 6.4.1 is false for an odd $k$. *Example.* $F$ is a discrete symmetric d.f. with mass $C/n^2 \log n$ for integers $n \ge 3$, where $O$ is the appropriate constant, and $k = 1$. [HINT: It is well known that the series

$$\sum_n \frac{\sin nt}{n \log n}$$

converges uniformly in $t$.]

We end this section with a discussion of a special class of distributions and their ch.f.'s that is of both theoretical and practical importance.

A distribution or p.m. on $\mathscr{R}^1$ is said to be of the *lattice type* iff its support is in an arithmetical progression — that is, it is completely atomic (i.e., discrete) with all the atoms located at points of the form $\{a + jd\}$, where $a$ is real, $d > 0$, and $j$ ranges over a certain nonempty set of integers. The corresponding *lattice d.f.* is of form:

$$F(x) = \sum_{j=-\infty}^{\infty} p_j \delta_{a+jd}(x),$$

where $p_j \geq 0$ and $\sum_{j=-\infty}^{\infty} p_j = 1$. Its ch.f. is

$$f(t) = e^{ait} \sum_{j=-\infty}^{\infty} p_j e^{jdit}, \tag{12}$$

which is an absolutely convergent *Fourier series*. Note that the degenerate d.f. $\delta_a$ with ch.f. $e^{ait}$ is a particular case. We have the following characterization.

**Theorem 6.4.7.** A ch.f. is that of a lattice distribution if and only if there exists a $t_0 \neq 0$ such that $|f(t_0)| = 1$.

PROOF. The "only if" part is trivial, since it is obvious from (12) that $|f|$ is periodic of period $2\pi/d$. To prove the "if" part, let $f(t_0) = e^{i\theta_0}$, where $\theta_0$ is real; then we have

$$1 = e^{-i\theta_0} f(t_0) = \int e^{i(t_0 x - \theta_0)} \mu(dx)$$

and consequently, taking real parts and transposing:

$$0 = \int [1 - \cos(t_0 x - \theta_0)] \mu(dx). \tag{13}$$

The integrand is positive everywhere and vanishes if and only if for some integer $j$,

$$x = \frac{\theta_0}{t_0} + j\left(\frac{2\pi}{t_0}\right).$$

It follows that the support of $\mu$ must be contained in the set of $x$ of this form in order that equation (13) may hold, for the integral of a strictly positive function over a set of strictly positive measure is strictly positive. The theorem is therefore proved, with $a = \theta_0/t_0$ and $d = 2\pi/t_0$ in the definition of a lattice distribution.

It should be observed that neither "$a$" nor "$d$" is uniquely determined above; for if we take, e.g., $a' = a + d'$ and $d'$ a divisor of $d$, then the support of $\mu$ is also contained in the arithmetical progression $\{a' + jd'\}$. However, unless $\mu$ is degenerate, there is a unique *maximum* $d$, which is called the "span" of the lattice distribution. The following corollary is easy.

**Corollary.** Unless $|f| \equiv 1$, there is a smallest $t_0 > 0$ such that $f(t_0) = 1$. The span is then $2\pi/t_0$.

Of particular interest is the "integer lattice" when $a = 0$, $d = 1$; that is, when the support of $\mu$ is a set of integers at least two of which differ by 1. We have seen many examples of these. The ch.f. $f$ of such an r.v. $X$ has period $2\pi$, and the following simple inversion formula holds, for each integer $j$:

$$\mathscr{P}(X = j) = p_j = \frac{1}{2\pi} \int_{-\pi}^{\pi} f(t) e^{-jit}\, dt, \tag{14}$$

where the range of integration may be replaced by any interval of length $2\pi$. This, of course, is nothing but the well-known formula for the "Fourier coefficients" of $f$. If $\{X_k\}$ is a sequence of independent, identically distributed r.v.'s with the ch.f. $f$, then $S_n = \sum^n X_k$ has the ch.f. $(f)^n$, and the inversion formula above yields:

$$\mathscr{P}(S_n = j) = \frac{1}{2\pi} \int_{-\pi}^{\pi} [f(t)]^n e^{-jit}\, dt. \tag{15}$$

This may be used to advantage to obtain estimates for $S_n$ (see Exercises 24 to 26 below).

EXERCISES

$f$ or $f_n$ is a ch.f. below.

**14.** If $|f(t)| = 1$, $|f(t')| = 1$ and $t/t'$ is an irrational number, then $f$ is degenerate. If for a sequence $\{t_k\}$ of nonvanishing constants tending to 0 we have $|f(t_k)| = 1$, then $f$ is degenerate.

$^\star$**15.** If $|f_n(t)| \to 1$ for every $t$ as $n \to \infty$, and $F_n$ is the d.f. corresponding to $f_n$, then there exist constants $a_n$ such that $F_n(x + a_n) \xrightarrow{v} \delta_0$. [HINT: Symmetrize and take $a_n$ to be a median of $F_n$.]

$^\star$**16.** Suppose $b_n > 0$ and $|f(b_n t)|$ converges everywhere to a ch.f. that is not identically 1, then $b_n$ converges to a finite and strictly positive limit. [HINT: Show that it is impossible that a subsequence of $b_n$ converges to 0 or to $+\infty$, or that two subsequences converge to different finite limits.]

$^\star$**17.** Suppose $c_n$ is real and that $e^{c_n it}$ converges to a limit for every $t$ in a set of strictly positive Lebesgue measure. Then $c_n$ converges to a finite limit. [HINT: Proceed as in Exercise 16, and integrate over $t$. Beware of any argument using "logarithms", as given in some textbooks, but see Exercise 12 of Sec. 7.6 later.]

***18.** Let $f$ and $g$ be two nondegenerate ch.f.'s. Suppose that there exist real constants $a_n$ and $b_n > 0$ such that for every $t$:

$$f_n(t) \to f(t) \quad \text{and} \quad e^{ita_n/b_n} f_n\left(\frac{t}{b_n}\right) \to g(t).$$

Then $a_n \to a$, $b_n \to b$, where $a$ is finite, $0 < b < \infty$, and $g(t) = e^{ita/b} f(t/b)$. [HINT: Use Exercises 16 and 17.]

***19.** Reformulate Exercise 18 in terms of d.f.'s and deduce the following consequence. Let $F_n$ be a sequence of d.f.'s $a_n$, $a_n'$ real constants, $b_n > 0$, $b_n' > 0$. If

$$F_n(b_n x + a_n) \xrightarrow{v} F(x) \quad \text{and} \quad F_n(b_n' x + a_n') \xrightarrow{v} F(x),$$

where $F$ is a nondegenerate d.f., then

$$\frac{b_n}{b_n'} \to 1 \quad \text{and} \quad \frac{a_n - a_n'}{b_n} \to 0.$$

[Two d.f.'s $F$ and $G$ such that $G(x) = F(bx + a)$ for every $x$, where $b > 0$ and $a$ is real, are said to be of the same "type". The two preceding exercises deal with the convergence of types.]

**20.** Show by using (14) that $|\cos t|$ is not a ch.f. Thus the modulus of a ch.f. need not be a ch.f., although the squared modulus always is.

**21.** The span of an integer lattice distribution is the greatest common divisor of the set of all differences between points of jump.

**22.** Let $f(s, t)$ be the ch.f. of a 2-dimensional p.m. $\nu$. If $|f(s_0, t_0)| = 1$ for some $(s_0, t_0) \neq (0, 0)$, what can one say about the support of $\nu$?

***23.** If $\{X_n\}$ is a sequence of independent and identically distributed r.v.'s, then there does not exist a sequence of constants $\{c_n\}$ such that $\sum_n (X_n - c_n)$ converges a.e., unless the common d.f. is degenerate.

In Exercises 24 to 26, let $S_n = \sum_{j=1}^n X_j$, where the $X_j'$s are independent r.v.'s with a common d.f. $F$ of the integer lattice type with span 1, and taking both $>0$ and $<0$ values.

***24.** If $\int x \, dF(x) = 0$, $\int x^2 \, dF(x) = \sigma^2$, then for each integer $j$:

$$n^{1/2} \mathscr{P}\{S_n = j\} \to \frac{1}{\sigma\sqrt{2\pi}}.$$

[HINT: Proceed as in Theorem 6.4.4, but use (15).]

**25.** If $F \neq \delta_0$, then there exists a constant $A$ such that for every $j$:

$$\mathscr{P}\{S_n = j\} \leq An^{-1/2}.$$

[HINT: Use a special case of Exercise 27 below.]

**26.** If $F$ is symmetric and $\int |x|\, dF(x) < \infty$, then

$$n\mathscr{P}\{S_n = j\} \to \infty.$$

[HINT: $1 - f(t) = o(|t|)$ as $t \to 0$.]

**27.** If $f$ is any nondegenerate ch. $f$, then there exist constants $A > 0$ and $\delta > 0$ such that

$$|f(t)| \leq 1 - At^2 \quad \text{for} \quad |t| \leq \delta.$$

[HINT: Reduce to the case where the d.f. has zero mean and finite variance by translating and truncating.]

**28.** Let $Q_n$ be the concentration function of $S_n = \sum_{j=1}^n X_j$, where the $X_j$'s are independent r.v.'s having a common nondegenerate d.f. $F$. Then for every $h > 0$,

$$Q_n(h) \leq An^{-1/2}$$

[HINT: Use Exercise 27 above and Exercise 16 of Sec. 6.1. This result is due to Lévy and Doeblin, but the proof is due to Rosén.]

In Exercises 29 to 35, $\mu$ or $\mu_k$ is a p.m. on $\mathscr{U} = (0, 1]$.

**29.** Define for each $n$:

$$f_\mu(n) = \int_{\mathscr{U}} e^{2\pi inx} \mu(dx).$$

Prove by Weierstrass's approximation theorem (by trigonometrical polynomials) that if $f_{\mu_1}(n) = f_{\mu_2}(n)$ for every $n \geq 1$, then $\mu_1 \equiv \mu_2$. The conclusion becomes false if $\mathscr{U}$ is replaced by [0, 1].

**30.** Establish the inversion formula expressing $\mu$ in terms of the $f_\mu(n)$'s. Deduce again the uniqueness result in Exercise 29. [HINT: Find the Fourier series of the indicator function of an interval contained in $\mathscr{U}$.]

**31.** Prove that $|f_\mu(n)| = 1$ if and only if $\mu$ has its support in the set $\{\theta_0 + jn^{-1}, 0 \leq j \leq n-1\}$ for some $\theta_0$ in $(0, n^{-1}]$.

$^\star$**32.** $\mu$ is *equidistributed* on the set $\{jn^{-1}, 0 \leq j \leq n-1\}$ if and only if $f_{\mu(j)} = 0$ or 1 according to $j \nmid n$ or $j \mid n$.

$^\star$**33.** $\mu_k \xrightarrow{v} \mu$ if and only if $f_{\mu_k}(\cdot) \to f_\mu(\cdot)$ everywhere.

**34.** Suppose that the space $\mathscr{U}$ is replaced by its closure [0, 1] and the two points 0 and 1 are identified; in other words, suppose $\mathscr{U}$ is regarded as the circumference $\mathscr{U}$ of a circle; then we can improve the result in Exercise 33 as follows. If there exists a function $g$ on the integers such that $f_{\mu_k}(\cdot) \to g(\cdot)$ everywhere, then there exists a p.m. $\mu$ on $\mathscr{U}$ such that $g = f_\mu$ and $\mu_k \xrightarrow{v} \mu$ on $\mathscr{U}$.

**35.** Random variables defined on $\mathscr{U}$ are also referred to as defined "modulo 1". The theory of addition of independent r.v.'s on $\mathscr{U}$ is somewhat simpler than on $\mathscr{R}^1$, as exemplified by the following theorem. Let $\{X_j, j \geq 1\}$ be independent and identically distributed r.v.'s on $\mathscr{U}$ and let $S_k = \sum_{j=1}^k X_j$. Then there are only two possibilities for the asymptotic distributions of $S_k$. Either there exist a constant $c$ and an integer $n \geq 1$ such that $S_k - kc$ converges in dist. to the equidistribution on $\{jn^{-1}, 0 \leq j \leq n-1\}$; or $S_k$ converges in dist. to the uniform distribution on $\mathscr{U}$. [HINT: Consider the possible limits of $(f_\mu(n))^k$ as $k \to \infty$, for each $n$.]

## 6.5 Representation theorems

A ch.f. is defined to be the Fourier–Stieltjes transform of a p.m. Can it be characterized by some other properties? This question arises because frequently such a function presents itself in a natural fashion while the underlying measure is hidden and has to be recognized. Several answers are known, but the following characterization, due to Bochner and Herglotz, is most useful. It plays a basic role in harmonic analysis and in the theory of second-order stationary processes.

A complex-valued function $f$ defined on $\mathscr{R}^1$ is called *positive definite* iff for any finite set of real numbers $t_j$ and complex numbers $z_j$ (with conjugate complex $\bar{z}_j$), $1 \leq j \leq n$, we have

$$\sum_{j=1}^{n}\sum_{k=1}^{n} f(t_j - t_k) z_j \bar{z}_k \geq 0. \tag{1}$$

Let us deduce at once some elementary properties of such a function.

**Theorem 6.5.1.** If $f$ is positive definite, then for each $t \in \mathscr{R}^1$:

$$f(-t) = \overline{f(t)}, \quad |f(t)| \leq f(0).$$

If $f$ is continuous at $t = 0$, then it is uniformly continuous in $\mathscr{R}^1$. In this case, we have for every continuous complex-valued function $\zeta$ on $\mathscr{R}^1$ and every $T > 0$:

$$\int_0^T \int_0^T f(s-t)\zeta(s)\overline{\zeta(t)}\, ds\, dt \geq 0. \tag{2}$$

PROOF. Taking $n = 1$, $t_1 = 0$, $z_1 = 1$ in (1), we see that

$$f(0) \geq 0.$$

Taking $n = 2$, $t_1 = 0$, $t_2 = t$, $z_1 = z_2 = 1$, we have

$$2f(0) + f(t) + f(-t) \geq 0;$$

changing $z_2$ to $i$, we have

$$f(0) + f(t)i - f(-t)i + f(0) \geq 0.$$

Hence $f(t) + f(-t)$ is real and $f(t) - f(-t)$ is pure imaginary, which imply that $\overline{f(t)} = f(-t)$. Changing $z_1$ to $f(t)$ and $z_2$ to $-|f(t)|$, we obtain

$$2f(0)|f(t)|^2 - 2|f(t)|^3 \geq 0.$$

Hence $f(0) \geq |f(t)|$, whether $|f(t)| = 0$ or $\geq 0$. Now, if $f(0) = 0$, then $f(\cdot) \equiv 0$; otherwise we may suppose that $f(0) = 1$. If we then take $n = 3$, $t_1 = 0$, $t_2 = t$, $t_3 = t + h$, a well-known result in positive definite quadratic forms implies that the determinant below must have a positive value:

$$\begin{vmatrix} f(0) & f(-t) & f(-t-h) \\ f(t) & f(0) & f(-h) \\ f(t+h) & f(h) & f(0) \end{vmatrix}$$
$$= 1 - |f(t)|^2 - |f(t+h)|^2 - |f(h)|^2 + 2\mathbf{R}\{f(t)f(h)\overline{f(t+h)}\} \geq 0.$$

It follows that

$$\begin{aligned} |f(t) - f(t+h)|^2 &= |f(t)|^2 + |f(t+h)|^2 - 2\mathbf{R}\{f(t)\overline{f(t+h)}\} \\ &\leq 1 - |f(h)|^2 + 2\mathbf{R}\{f(t)\overline{f(t+h)}[f(h) - 1]\} \\ &\leq 1 - |f(h)|^2 + 2|1 - f(h)| \leq 4|1 - f(h)|. \end{aligned}$$

Thus the "modulus of continuity" at each $t$ is bounded by twice the square root of that at 0; and so continuity at 0 implies uniform continuity everywhere. Finally, the integrand of the double integral in (2) being continuous, the integral is the limit of Riemann sums, hence it is positive because these sums are by (1).

**Theorem 6.5.2.** $f$ is a ch.f. if and only if it is positive definite and continuous at 0 with $f(0) = 1$.

*Remark.* It follows trivially that $f$ is the Fourier–Stieltjes transform

$$\int_{\mathscr{R}^1} e^{itx} \nu(dx)$$

of a finite measure $\nu$ if and only if it is positive definite and finite continuous at 0; then $\nu(\mathscr{R}^1) = f(0)$.

PROOF. If $f$ is the ch.f. of the p.m. $\mu$, then we need only verify that it is positive definite. This is immediate, since the left member of (1) is then

$$\int \sum_{j=1}^{n} \sum_{k=1}^{n} e^{it_j^x} z_j \overline{e^{it_k^x} z_k} \mu(dx) = \int \left| \sum_{j=1}^{n} e^{it_j^x} z_j \right|^2 \mu(dx) \geq 0.$$

Conversely, if $f$ is positive definite and continuous at 0, then by Theorem 6.5.1, (2) holds for $\zeta(t) = e^{-itx}$. Thus

$$\frac{1}{2\pi T} \int_0^T \int_0^T f(s-t) e^{-i(s-t)x} \, ds \, dt \geq 0. \tag{3}$$

Denote the left member by $p_T(x)$; by a change of variable and partial integration, we have

$$p_T(x) = \frac{1}{2\pi} \int_{-T}^{T} \left(1 - \frac{|t|}{T}\right) f(t) e^{-itx} \, dt. \tag{4}$$

Now observe that for $\alpha > 0$,

$$\frac{1}{\alpha} \int_0^\alpha d\beta \int_{-\beta}^{\beta} e^{-itx} \, dx = \frac{1}{\alpha} \int_0^\alpha \frac{2 \sin \beta t}{t} \, d\beta = \frac{2(1 - \cos \alpha t)}{\alpha t^2}$$

(where at $t = 0$ the limit value is meant, similarly later); it follows that

$$\begin{aligned} \frac{1}{\alpha} \int_0^\alpha d\beta \int_{-\beta}^{\beta} p_T(x) \, dx &= \frac{1}{\pi} \int_{-T}^{T} \left(1 - \frac{|t|}{T}\right) f(t) \frac{1 - \cos \alpha t}{\alpha t^2} \, dt \\ &= \frac{1}{\pi} \int_{-\infty}^{\infty} f_T(t) \frac{1 - \cos \alpha t}{\alpha t^2} \, dt \\ &= \frac{1}{\pi} \int_{-\infty}^{\infty} f_T\left(\frac{t}{\alpha}\right) \frac{1 - \cos t}{t^2} \, dt, \end{aligned}$$

where

$$f_T(t) = \begin{cases} \left(1 - \dfrac{|t|}{T}\right) f(t), & \text{if } |t| \leq T; \\ 0, & \text{if } |t| > T. \end{cases} \tag{5}$$

Note that this is the product of $f$ by a ch.f. (see Exercise 2 of Sec. 6.2) and corresponds to the smoothing of the would-be density which does not necessarily exist.

Since $|f_T(t)| \le |f(t)| \le 1$ by Theorem 6.5.1, and $(1 - \cos t)/t^2$ belongs to $L^1(-\infty, \infty)$, we have by dominated convergence:

$$(6) \quad \lim_{\alpha\to\infty} \frac{1}{\alpha}\int_0^\alpha d\beta \int_{-\beta}^{\beta} p_T(x)\,dx = \frac{1}{\pi}\int_{-\infty}^{\infty} \lim_{\alpha\to\infty} f_T\left(\frac{t}{\alpha}\right)\frac{1-\cos t}{t^2}\,dt$$
$$= \frac{1}{\pi}\int_{-\infty}^{\infty} \frac{1-\cos t}{t^2}\,dt = 1.$$

Here the second equation follows from the continuity of $f_T$ at 0 with $f_T(0) = 1$, and the third equation from formula (3) of Sec. 6.2. Since $p_T \ge 0$, the integral $\int_{-\beta}^{\beta} p_T(x)\,dx$ is increasing in $\beta$, hence the existence of the limit of its "integral average" in (6) implies that of the plain limit as $\beta \to \infty$, namely:

$$(7) \quad \int_{-\infty}^{\infty} p_T(x)\,dx = \lim_{\beta\to\infty}\int_{-\beta}^{\beta} p_T(x)\,dx = 1.$$

Therefore $p_T$ is a probability density function. Returning to (3) and observing that for real $\tau$:

$$\int_{-\beta}^{\beta} e^{i\tau x}e^{-itx}\,dx = \frac{2\sin\beta(\tau - t)}{\tau - t},$$

we obtain, similarly to (4):

$$\frac{1}{\alpha}\int_0^\alpha d\beta\int_{-\beta}^{\beta} e^{i\tau x}p_T(x)\,dx = \frac{1}{\pi}\int_{-\infty}^{\infty} f_T(t)\frac{1-\cos\alpha(\tau - t)}{\alpha(\tau - t)^2}\,dt$$
$$= \frac{1}{\pi}\int_{-\infty}^{\infty} f_T\left(\tau - \frac{t}{\alpha}\right)\frac{1-\cos t}{t^2}\,dt.$$

Note that the last two integrals are in reality over *finite* intervals. Letting $\alpha \to \infty$, we obtain by bounded convergence as before:

$$(8) \quad \int_{-\infty}^{\infty} e^{i\tau x}p_T(x)\,dx = f_T(\tau),$$

the integral on the left existing by (7). Since equation (8) is valid for each $\tau$, we have proved that $f_T$ is the ch.f. of the density function $p_T$. Finally, since $f_T(\tau) \to f(\tau)$ as $T \to \infty$ for every $\tau$, and $f$ is by hypothesis continuous at $\tau = 0$, Theorem 6.3.2 yields the desired conclusion that $f$ is a ch.f.

As a typical application of the preceding theorem, consider a family of (real-valued) r.v.'s $\{X_t, t \in \mathscr{R}_+\}$, where $\mathscr{R}_+ = [0, \infty)$, satisfying the following conditions, for every $s$ and $t$ in $\mathscr{R}_+$:

(i) $\mathscr{E}(X_t^2) = 1$;
(ii) there exists a function $r(\cdot)$ on $\mathscr{R}^1$ such that $\mathscr{E}(X_s X_t) = r(s-t)$;
(iii) $\lim_{t \downarrow 0} \mathscr{E}((X_0 - X_t)^2) = 0$.

A family satisfying (i) and (ii) is called a *second-order stationary process* or stationary process *in the wide sense*; and condition (iii) means that the process is *continuous in* $L^2(\Omega, \mathscr{F}, \mathscr{P})$.

For every finite set of $t_j$ and $z_j$ as in the definitions of positive definiteness, we have

$$0 \le \mathscr{E}\left\{\left|\sum_{j=1}^{n} X_{t_j} z_j\right|^2\right\} = \sum_{j=1}^{n}\sum_{k=1}^{n} \mathscr{E}(X_{t_j} X_{t_k}) z_j \bar{z}_k = \sum_{j=1}^{n}\sum_{k=1}^{n} r(t_j - t_k) z_j \bar{z}_k.$$

Thus $r$ is a positive definite function. Next, we have

$$r(0) - r(t) = \mathscr{E}(X_0(X_0 - X_t)),$$

hence by the Cauchy–Schwarz inequality,

$$|r(t) - r(0)|^2 \le \mathscr{E}(X_0^2)\mathscr{E}((X_0 - X_t)^2).$$

It follows that $r$ is continuous at 0, with $r(0) = 1$. Hence, by Theorem 6.5.2, $r$ is the ch.f. of a uniquely determined p.m. $R$:

$$r(t) = \int_{\mathscr{R}^1} e^{itx} R(dx).$$

This $R$ is called the *spectral distribution* of the process and is essential in its further analysis.

Theorem 6.5.2 is not practical in recognizing special ch.f.'s or in constructing them. In contrast, there is a sufficient condition due to Pólya that is very easy to apply.

**Theorem 6.5.3.** Let $f$ on $\mathscr{R}^1$ satisfy the following conditions, for each $t$:

$$f(0) = 1, \quad f(t) \ge 0, \quad f(t) = f(-t), \tag{9}$$

$f$ is decreasing and continuous convex in $\mathscr{R}_+ = [0, \infty)$. Then $f$ is a ch.f.

PROOF. Without loss of generality we may suppose that

$$f(\infty) = \lim_{t \to \infty} f(t) = 0;$$

otherwise we consider $[f(t) - f(\infty)]/[f(0) - f(\infty)]$ unless $f(\infty) = 1$, in which case $f \equiv 1$. It is well known that a convex function $f$ has right-hand

and left-hand derivatives everywhere that are equal except on a countable set, that $f$ is the integral of either one of them, say the right-hand one, which will be denoted simply by $f'$, and that $f'$ is increasing. Under the conditions of the theorem it is also clear that $f$ is decreasing and $f'$ is negative in $\mathscr{R}_+$. Now consider the $f_T$ as defined in (5) above, and observe that

$$-f_T'(t) = \begin{cases} -\left(1 - \dfrac{t}{T}\right) f'(t) + \dfrac{1}{T} f(t), & \text{if } 0 < t < T; \\ 0, & \text{if } t \geq T. \end{cases}$$

Thus $-f_T'$ is positive and decreasing in $\mathscr{R}_+$. We have for each $x \neq 0$:

$$\int_{-\infty}^{\infty} e^{-itx} f_T(t)\, dt = 2 \int_0^{\infty} \cos tx f_T(t)\, dt = \frac{2}{x} \int_0^{\infty} \sin tx(-f_T'(t))\, dt$$

$$= \frac{2}{x} \sum_{k=0}^{\infty} \int_{k\pi/x}^{(k+1)\pi/x} \sin tx(-f_T'(t))\, dt.$$

The terms of the series alternate in sign, beginning with a positive one, and decrease in magnitude, hence the sum $\geq 0$. [This is the argument indicated for formula (1) of Sec. 6.2.] For $x = 0$, it is trivial that

$$\int_{-\infty}^{\infty} f_T(t)\, dt \geq 0.$$

We have therefore proved that the $p_T$ defined in (4) is positive everywhere, and the proof there shows that $f$ is a ch.f. (cf. Exercise 1 below).

Next we will establish an interesting class of ch.f.'s which are natural extensions of the ch.f.'s corresponding to the normal and the Cauchy distributions.

**Theorem 6.5.4.** For each $\alpha$ in the range (0, 2],

$$f_\alpha(t) = e^{-|t|^\alpha}$$

is a ch.f.

PROOF. For $0 < \alpha \leq 1$, this is a quick consequence of Pólya's theorem above. Other conditions there being obviously satisfied, we need only check that $f_\alpha$ is convex in $[0, \infty)$. This is true because its second derivative is equal to

$$e^{-t^\alpha}\{\alpha^2 t^{2\alpha-2} - \alpha(\alpha - 1)t^{\alpha-2}\} > 0$$

for the range of $\alpha$ in question. No such luck for $1 < \alpha < 2$, and there are several different proofs in this case. Here is the one given by Lévy which works for $0 < \alpha < 2$. Consider the density function

$$p(x) = \begin{cases} \dfrac{\alpha}{2|x|^{\alpha+1}} & \text{if } |x| > 1, \\ 0 & \text{if } |x| \le 1; \end{cases}$$

and compute its ch.f. $f$ as follows, using symmetry:

$$\begin{aligned} 1 - f(t) &= \int_{-\infty}^{\infty} (1 - e^{itx}) p(x)\, dx = \alpha \int_{1}^{\infty} \frac{1 - \cos tx}{x^{\alpha+1}}\, dx \\ &= \alpha |t|^{\alpha} \left\{ \int_{0}^{\infty} \frac{1 - \cos u}{u^{\alpha+1}}\, du - \int_{0}^{t} \frac{1 - \cos u}{u^{\alpha+1}}\, du \right\}, \end{aligned}$$

after the change of variables $tx = u$. Since $1 - \cos u \sim \frac{1}{2}u^2$ near $u = 0$, the first integral in the last member above is finite while the second is asymptotically equivalent to

$$\frac{1}{2} \int_{0}^{t} \frac{u^2}{u^{\alpha+1}}\, du = \frac{1}{2(2-\alpha)} t^{2-\alpha}$$

as $t \downarrow 0$. Therefore we obtain

$$f(t) = 1 - c_{\alpha} |t|^{\alpha} + O(t^2)$$

where $c_{\alpha}$ is a positive constant depending on $\alpha$.

It now follows that

$$f\left(\frac{t}{n^{1/\alpha}}\right)^n = \left\{1 - \frac{c_{\alpha}|t|^{\alpha}}{n} + O\left(\frac{t^2}{n^{2/\alpha}}\right)\right\}^n$$

is also a ch.f. (What is the probabilistic meaning in terms of r.v.'s?) For each $t$, as $n \to \infty$, the limit is equal to $e^{-c_{\alpha}|t|^{\alpha}}$ (the lemma in Sec. 6.4 again!). This, being continuous at $t = 0$, is also a ch.f. by the basic Theorem 6.3.2, and the constant $c_{\alpha}$ may be absorbed by a change of scale. Finally, for $\alpha = 2$, $f_{\alpha}$ is the ch.f. of a normal distribution. This completes the proof of the theorem.

Actually Lévy, who discovered these ch.f.'s around 1923, proved also that there are complex constants $\gamma_{\alpha}$ such that $e^{-\gamma_{\alpha}|t|^{\alpha}}$ is a ch.f., and determined the exact form of these constants (see Gnedenko and Kolmogorov [12]). The corresponding d.f.'s are called *stable distributions*, and those with real positive $\gamma_{\alpha}$ the *symmetric stable* ones. The parameter $\alpha$ is called the *exponent*. These distributions are part of a much larger class called the *infinitely divisible distributions* to be discussed in Chapter 7.

Using the Cauchy ch.f. $e^{-|t|}$ we can settle a question of historical interest. Draw the graph of this function, choose an arbitrary $T > 0$, and draw the tangents at $\pm T$ meeting the abscissa axis at $\pm T'$, where $T' > T$. Now define the function $f_T$ to be $f$ in $[-T, T]$, linear in $[-T', -T]$ and in $[T, T']$, and zero in $(-\infty, -T')$ and $(T', \infty)$. Clearly $f_T$ also satisfies the conditions of Theorem 6.5.3 and so is a ch.f. Furthermore, $f = f_T$ in $[-T, T]$. We have thus established the following theorem and shown indeed how abundant the desired examples are.

**Theorem 6.5.5.** There exist two distinct ch.f.'s that coincide in an interval containing the origin.

That the interval of coincidence can be "arbitrarily large" is, of course, trivial, for if $f_1 = f_2$ in $[-\delta, \delta]$, then $g_1 = g_2$ in $[-n\delta, n\delta]$, where

$$g_1(t) = f_1\left(\frac{t}{n}\right), \quad g_2(t) = f_2\left(\frac{t}{n}\right).$$

**Corollary.** There exist three ch.f.'s, $f_1, f_2, f_3$, such that $f_1 f_3 \equiv f_2 f_3$ but $f_1 \not\equiv f_2$.

To see this, take $f_1$ and $f_2$ as in the theorem, and take $f_3$ to be any ch.f. vanishing outside their interval of coincidence, such as the one described above for a sufficiently small value of $T$. This result shows that in the algebra of ch.f.'s, the cancellation law does not hold.

We end this section by another special but useful way of constructing ch.f.'s.

**Theorem 6.5.6.** If $f$ is a ch.f., then so is $e^{\lambda(f-1)}$ for each $\lambda \geq 0$.

PROOF. For each $\lambda \geq 0$, as soon as the integer $n \geq \lambda$, the function

$$1 - \frac{\lambda}{n} + \frac{\lambda}{n} f = 1 + \frac{\lambda(f-1)}{n}$$

is a ch.f., hence so is its $n$th power; see propositions (iv) and (v) of Sec. 6.1. As $n \to \infty$,

$$\left(1 + \frac{\lambda(f-1)}{n}\right)^n \to e^{\lambda(f-1)}$$

and the limit is clearly continuous. Hence it is a ch.f. by Theorem 6.3.2.

Later we shall see that this class of ch.f.'s is also part of the infinitely divisible family. For $f(t) = e^{it}$, the corresponding

$$e^{\lambda(e^{it}-1)} = \sum_{n=0}^{\infty} \frac{e^{-\lambda}\lambda^n}{n!} e^{itn}$$

is the ch.f. of the Poisson distribution which should be familiar to the reader.

EXERCISES

**1.** If $f$ is continuous in $\mathscr{R}^1$ and satisfies (3) for each $x \in \mathscr{R}^1$ and each $T > 0$, then $f$ is positive definite.

**2.** Show that the following functions are ch.f.'s:

$$\frac{1}{1+|t|}, \quad f(t) = \begin{cases} 1 - |t|^\alpha, & \text{if } |t| \le 1; \\ 0, & \text{if } |t| \ge 1; \end{cases} \quad 0 < \alpha \le 1,$$

$$f(t) = \begin{cases} 1 - |t|, & \text{if } 0 \le |t| \le \frac{1}{2}; \\ \dfrac{1}{4|t|}, & \text{if } |t| \ge \frac{1}{2}. \end{cases}$$

**3.** If $\{X_n\}$ are independent r.v.'s with the same stable distribution of exponent $\alpha$, then $\sum_{k=1}^n X_k/n^{1/\alpha}$ has the same distribution. [This is the origin of the name "stable".]

**4.** If $F$ is a symmetric stable distribution of exponent $\alpha$, $0 < \alpha < 2$, then $\int_{-\infty}^{\infty} |x|^r \, dF(x) < \infty$ for $r < \alpha$ and $= \infty$ for $r \ge \alpha$. [HINT: Use Exercises 7 and 8 of Sec. 6.4.]

***5.** Another proof of Theorem 6.5.3 is as follows. Show that

$$\int_0^\infty t \, df'(t) = 1$$

and define the d.f. $G$ on $\mathscr{R}_+$ by

$$G(u) = \int_{[0,u]} t \, df'(t).$$

Next show that

$$\int_0^\infty \left(1 - \frac{|t|}{u}\right) dG(u) = f(t).$$

Hence if we set

$$f(u, t) = \left(1 - \frac{|t|}{u}\right) \vee 0$$

(see Exercise 2 of Sec. 6.2), then

$$f(t) = \int_{[0,\infty)} f(u, t) \, dG(u).$$

Now apply Exercise 2 of Sec. 6.1.

**6.** Show that there is a ch.f. that has period $2m$, $m$ an integer $\ge 1$, and that is equal to $1 - |t|$ in $[-1, +1]$. [HINT: Compute the Fourier series of such a function to show that the coefficients are positive.]

**7.** Construct a ch.f. that vanishes in $[-b, -a]$ and $[a, b]$, where $0 < a < b$, but nowhere else. [HINT: Let $f_m$ be the ch.f. in Exercise 6 and consider

$$\sum_m p_m f_m, \quad \text{where } p_m \geq 0, \sum_m p_m = 1,$$

and the $p_m$'s are strategically chosen.]

**8.** Suppose $f(t, u)$ is a function on $\mathscr{R}^2$ such that for each $u$, $f(\cdot, u)$ is a ch.f.; and for each $t$, $f(t, \cdot)$ is continuous. Then for any d.f. $G$,

$$\exp\left\{\int_{-\infty}^{\infty} [f(t, u) - 1]\, dG(u)\right\}$$

is a ch.f.

**9.** Show that in Theorem 6.3.2, the hypothesis (a) cannot be relaxed to require convergence of $\{f_n\}$ only in a finite interval $|t| \leq T$.

## 6.6 Multidimensional case; Laplace transforms

We will discuss very briefly the ch.f. of a p.m. in Euclidean space of more than one dimension, which we will take to be two since all extensions are straightforward. The ch.f. of the random vector $(X, Y)$ or of its 2-dimensional p.m. $\mu$ is defined to be the function $f(\cdot, \cdot)$ on $\mathscr{R}^2$:

$$(1) \qquad f(s, t) = f_{(X,Y)}(s, t) = \mathscr{E}(e^{i(sX+tY)}) = \iint_{\mathscr{R}^2} e^{i(sx+ty)} \mu(dx, dy).$$

Propositions (i) to (v) of Sec. 6.1 have their obvious analogues. The inversion formula may be formulated as follows. Call an "interval" (rectangle)

$$\{(x, y): x_1 \leq x \leq x_2, y_1 \leq y \leq y_2\}$$

an *interval of continuity* iff the $\mu$-measure of its boundary (the set of points on its four sides) is zero. For such an interval $I$, we have

$$\mu(I) = \lim_{T\to\infty} \frac{1}{(2\pi)^2} \int_{-T}^{T} \int_{-T}^{T} \frac{e^{-isx_1} - e^{-isx_2}}{is} \frac{e^{-ity_1} - e^{-ity_2}}{it} f(s, t)\, ds\, dt.$$

The proof is entirely similar to that in one dimension. It follows, as there, that $f$ uniquely determines $\mu$. Only the following result is noteworthy.

**Theorem 6.6.1.** Two r.v.'s $X$ and $Y$ are independent if and only if

$$(2) \qquad \forall s, \forall t: \quad f_{(X,Y)}(s, t) = f_X(s) f_Y(t),$$

where $f_X$ and $f_Y$ are the ch.f.'s of $X$ and $Y$, respectively.

The condition (2) is to be contrasted with the following identify in one variable:

$$\forall t: \quad f_{X+Y}(t) = f_X(t) f_Y(t),$$

where $f_{X+Y}$ is the ch.f. of $X + Y$. This holds if $X$ and $Y$ are independent, but the converse is false (see Exercise 14 of Sec. 6.1).

PROOF OF THEOREM 6.6.1. If $X$ and $Y$ are independent, then so are $e^{isX}$ and $e^{itY}$ for every $s$ and $t$, hence

$$\mathscr{E}(e^{i(sX+tY)}) = \mathscr{E}(e^{isX} \cdot e^{itY}) = \mathscr{E}(e^{isX})\mathscr{E}(e^{itY}),$$

which is (2). Conversely, consider the 2-dimensional product measure $\mu_1 \times \mu_2$, where $\mu_1$ and $\mu_2$ are the 1-dimensional p.m.'s of $X$ and $Y$, respectively, and the product is defined in Sec. 3.3. Its ch.f. is given by definition as

$$\iint_{\mathscr{R}^2} e^{i(sx+ty)}(\mu_1 \times \mu_2)(dx, dy) = \iint_{\mathscr{R}_2} e^{isx} \cdot e^{ity} \mu_1(dx)\mu_2(dy)$$

$$= \int_{\mathscr{R}^1} e^{isx} \mu_1(dx) \int_{\mathscr{R}^1} e^{ity} \mu_2(dy) = f_X(s) f_Y(t)$$

(Fubini's theorem!). If (2) is true, then this is the same as $f_{(X,Y)}(s, t)$, so that $\mu_1 \times \mu_2$ has the same ch.f. as $\mu$, the p.m. of $(X, Y)$. Hence, by the uniqueness theorem mentioned above, we have $\mu_1 \times \mu_2 = \mu$. This is equivalent to the independence of $X$ and $Y$.

The multidimensional analogues of the convergence theorem, and some of its applications such as the weak law of large numbers and central limit theorem, are all valid without new difficulties. Even the characterization of Bochner has an easy extension. However, these topics are better pursued with certain objectives in view, such as mathematical statistics, Gaussian processes, and spectral analysis, that are beyond the scope of this book, and so we shall not enter into them.

We shall, however, give a short introduction to an allied notion, that of the Laplace transform, which is sometimes more expedient or appropriate than the ch.f., and is a basic tool in the theory of Markov processes.

Let $X$ be a positive ($\geq 0$) r.v. having the d.f. $F$ so that $F$ has support in $[0, \infty)$, namely $F(0-) = 0$. The *Laplace transform* of $X$ or $F$ is the function $\hat{F}$ on $\mathscr{R}_+ = [0, \infty)$ given by

$$\hat{F}(\lambda) = \mathscr{E}(e^{-\lambda X}) = \int_{[0,\infty)} e^{-\lambda x}\, dF(x). \tag{3}$$

It is obvious (why?) that

$$\hat{F}(0) = \lim_{\lambda \downarrow 0} \hat{F}(\lambda) = 1, \quad \hat{F}(\infty) = \lim_{\lambda \to \infty} \hat{F}(\lambda) = F(0).$$

More generally, we can define the Laplace transform of an s.d.f. or a function $G$ of bounded variation satisfying certain "growth condition" at infinity. In particular, if $F$ is an s.d.f., the Laplace transform of its indefinite integral

$$G(x) = \int_0^x F(u)\,du$$

is finite for $\lambda > 0$ and given by

$$\begin{aligned}
\hat{G}(\lambda) &= \int_{[0,\infty)} e^{-\lambda x} F(x)\,dx = \int_{[0,\infty)} e^{-\lambda x}\,dx \int_{[0,x]} \mu(dy) \\
&= \int_{[0,\infty)} \mu(dy) \int_y^\infty e^{-\lambda x}\,dx = \frac{1}{\lambda}\int_{[0,\infty)} e^{-\lambda y}\mu(dy) = \frac{1}{\lambda}\hat{F}(\lambda),
\end{aligned}$$

where $\mu$ is the s.p.m. of $F$. The calculation above, based on Fubini's theorem, replaces a familiar "integration by parts". However, the reader should beware of the latter operation. For instance, according to the usual definition, as given in Rudin [1] (and several other standard textbooks!), the value of the *Riemann–Stieltjes integral*

$$\int_0^\infty e^{-\lambda x}\,d\delta_0(x)$$

is 0 rather than 1, but

$$\int_0^\infty e^{-\lambda x}\,d\delta_0(x) = \lim_{\substack{\epsilon \downarrow 0 \\ A \uparrow \infty}} \delta_0(x) e^{-\lambda x} \|_\epsilon^A + \int_0^\infty \delta_0(x) \lambda e^{-\lambda x}\,dx$$

is correct only if the left member is taken in the *Lebesgue-Stieltjes* sense, as is always done in this book.

There are obvious analogues of propositions (i) to (v) of Sec. 6.1. However, the inversion formula requiring complex integration will be omitted and the uniqueness theorem, due to Lerch, will be proved by a different method (cf. Exercise 12 of Sec. 6.2).

**Theorem 6.6.2.** Let $\hat{F}_j$ be the Laplace transform of the d.f. $F_j$ with support in $\mathscr{R}_+$, $j = 1, 2$. If $\hat{F}_1 = \hat{F}_2$, then $F_1 = F_2$.

PROOF. We shall apply the Stone–Weierstrass theorem to the algebra generated by the family of functions $\{e^{-\lambda x}, \lambda \geq 0\}$, defined on the *closed positive real line*: $\overline{\mathscr{R}}_+ = [0, \infty]$, namely the *one-point compactification of* $\mathscr{R}_+ = [0, \infty)$. A continuous function of $x$ on $\overline{\mathscr{R}}_+$ is one that is continuous in $\mathscr{R}_+$ and has a finite limit as $x \to \infty$. This family separates points on $\overline{\mathscr{R}}_+$ and vanishes at no point of $\overline{\mathscr{R}}_+$ (at the point $+\infty$, the member $e^{-0x} = 1$ of the family does not vanish!). Hence the set of polynomials in the members of the family, namely the algebra generated by it, is dense in the uniform topology, in the space $C_B(\overline{\mathscr{R}}_+)$ of bounded continuous functions on $\overline{\mathscr{R}}_+$. That is to say, given any $g \in C_B(\overline{\mathscr{R}}_+)$, and $\epsilon > 0$, there exists a polynomial of the form

$$g_\epsilon(x) = \sum_{j=1}^{n} c_j e^{-\lambda_j x},$$

where $c_j$ are real and $\lambda_j \geq 0$, such that

$$\sup_{x \epsilon \overline{\mathscr{R}}+} |g(x) - g_\epsilon(x)| \leq \epsilon.$$

Consequently, we have

$$\int |g(x) - g_\epsilon(x)| \, dF_j(x) \leq \epsilon, \quad j = 1, 2.$$

By hypothesis, we have for each $\lambda \geq 0$:

$$\int e^{-\lambda x} \, dF_1(x) = \int e^{-\lambda x} \, dF_2(x),$$

and consequently,

$$\int g_\epsilon(x) dF_1(x) = \int g_\epsilon(x) dF_2(x).$$

It now follows, as in the proof of Theorem 4.4.1, first that

$$\int g(x) \, dF_1(x) = \int g(x) \, dF_2(x)$$

for each $g \in C_B(\overline{\mathscr{R}}_+)$; second, that this also holds for each $g$ that is the indicator of an interval in $\overline{\mathscr{R}}_+$ (even one of the form $(a, \infty]$); third, that the two p.m.'s induced by $F_1$ and $F_2$ are identical, and finally that $F_1 = F_2$ as asserted.

*Remark.* Although it is not necessary here, we may also extend the domain of definition of a d.f. $F$ to $\overline{\mathscr{R}}_+$; thus $F(\infty) = 1$, but with the new meaning that $F(\infty)$ is actually the value of $F$ at the point $\infty$, rather than a notation for $\lim_{x\to\infty} F(x)$ as previously defined. $F$ is thus continuous at $\infty$. In terms of the p.m. $\mu$, this means we extend its domain to $\overline{\mathscr{R}}_+$ but set $\mu(\{\infty\}) = 0$. On other occasions it may be useful to assign a strictly positive value to $\mu(\{\infty\})$.

Passing to the convergence theorem, we shall prove it in a form closest to Theorem 6.3.2 and refer to Exercise 4 below for improvements.

**Theorem 6.6.3.** Let $\{F_n, 1 \leq n < \infty\}$ be a sequence of s.d.f.'s with supports in $\mathscr{R}_+$ and $\{\hat{F}_n\}$ the corresponding Laplace transforms. Then $F_n \xrightarrow{v} F_\infty$, where $F_\infty$ is a d.f., if and only if:

(a) $\lim_{n\to\infty} \hat{F}_n(\lambda)$ exists for every $\lambda > 0$;
(b) the limit function has the limit 1 as $\lambda \downarrow 0$.

*Remark.* The limit in (a) exists at $\lambda = 0$ and equals 1 if the $F_n$'s are d.f.'s, but even so (b) is not guaranteed, as the example $F_n = \delta_n$ shows.

PROOF. The "only if" part follows at once from Theorem 4.4.1 and the remark that the Laplace transform of an s.d.f. is a continuous function in $\mathscr{R}_+$. Conversely, suppose $\lim \hat{F}_n(\lambda) = G(\lambda)$, $\lambda > 0$; extended $G$ to $\mathscr{R}_+$ by setting $G(0) = 1$ so that $G$ is continuous in $\mathscr{R}_+$ by hypothesis (b). As in the proof of Theorem 6.3.2, consider any vaguely convergent subsequence $F_{n_k}$ with the vague limit $F_\infty$, necessarily an s.d.f. (see Sec. 4.3). Since for each $\lambda > 0$, $e^{-\lambda x} \in C_0$, Theorem 4.4.1 applies to yield $\hat{F}_{n_k}(\lambda) \to \hat{F}_\infty(\lambda)$ for $\lambda > 0$, where $\hat{F}_\infty$ is the Laplace transform of $F_\infty$. Thus $\hat{F}_\infty(\lambda) = G(\lambda)$ for $\lambda > 0$, and consequently for $\lambda \geq 0$ by continuity of $\hat{F}_\infty$ and $G$ at $\lambda = 0$. Hence every vague limit has the same Laplace transform, and therefore is the same $F_\infty$ by Theorem 6.6.2. It follows by Theorem 4.3.4 that $F_n \xrightarrow{v} F_\infty$. Finally, we have $F_\infty(\infty) = \hat{F}_\infty(0) = G(0) = 1$, proving that $F_\infty$ is a d.f.

There is a useful characterization, due to S. Bernstein, of Laplace transforms of measures on $\mathscr{R}_+$. A function is called *completely monotonic* in an interval (finite or infinite, of any kind) iff it has derivatives of all orders there satisfying the condition:

$$(4) \qquad (-1)^n f^{(n)}(\lambda) \geq 0$$

for each $n \geq 0$ and each $\lambda$ in the domain of definition.

**Theorem 6.6.4.** A function $f$ on $(0, \infty)$ is the Laplace transform of a d.f. $F$:

$$(5) \qquad f(\lambda) = \int_{\mathscr{R}+} e^{-\lambda x}\, dF(x),$$

if and only if it is completely monotonic in $(0, \infty)$ with $f(0+) = 1$.

*Remark.* We then extend $f$ to $\mathscr{R}_+$ by setting $f(0) = 1$, and (5) will hold for $\lambda \geq 0$.

PROOF. The "only if" part is immediate, since

$$f^{(n)}(\lambda) = \int_{\mathscr{R}+} (-x)^n e^{-\lambda x}\, dF(x).$$

Turning to the "if" part, let us first prove that $f$ is *quasi-analytic* in $(0, \infty)$, namely it has a convergent Taylor series there. Let $0 < \lambda_0 < \lambda < \mu$, then, by

Taylor's theorem, with the remainder term in the integral form, we have

$$(6)\qquad f(\lambda) = \sum_{j=0}^{k-1} \frac{f^{(j)}(\mu)}{j!}(\lambda - \mu)^j + \frac{(\lambda - \mu)^k}{(k-1)!} \int_0^1 (1-t)^{k-1} f^{(k)}(\mu + (\lambda - \mu)t)\, dt.$$

Because of (4), the last term in (6) is positive and does not exceed

$$\frac{(\lambda - \mu)^k}{(k-1)!} \int_0^1 (1-t)^{k-1} f^{(k)}(\mu + (\lambda_0 - \mu)t)\, dt.$$

For if $k$ is even, then $f^{(k)} \downarrow$ and $(\lambda - \mu)^k \geq 0$, while if $k$ is odd then $f^{(k)} \uparrow$ and $(\lambda - \mu)^k \leq 0$. Now, by (6) with $\lambda$ replaced by $\lambda_0$, the last expression is equal to

$$\left(\frac{\lambda - \mu}{\lambda_0 - \mu}\right)^k \left[ f(\lambda_0) - \sum_{j=0}^{k-1} \frac{f^{(j)}(\mu)}{j!}(\lambda_0 - \mu)^j \right] \leq \left(\frac{\lambda - \mu}{\lambda_0 - \mu}\right)^k f(\lambda_0),$$

where the inequality is trivial, since each term in the sum on the left is positive by (4). Therefore, as $k \to \infty$, the remainder term in (6) tends to zero and the Taylor series for $f(\lambda)$ converges.

Now for each $n \geq 1$, define the discrete s.d.f. $F_n$ by the formula:

$$(7)\qquad F_n(x) = \sum_{j=0}^{[nx]} \frac{n^j}{j!}(-1)^j f^{(j)}(n).$$

This is indeed an s.d.f., since for each $\epsilon > 0$ and $k \geq 1$ we have from (6):

$$1 = f(0+) \geq f(\epsilon) \geq \sum_{j=0}^{k-1} \frac{f^{(j)}(n)}{j!}(\epsilon - n)^j.$$

Letting $\epsilon \downarrow 0$ and then $k \uparrow \infty$, we see that $F_n(\infty) \leq 1$. The Laplace transform of $F_n$ is plainly, for $\lambda > 0$:

$$\int_{\mathscr{R}_+} e^{-\lambda x}\, dF_n(x) = \sum_{j=0}^{\infty} e^{-\lambda(j/n)} \frac{(-n)^j}{j!} f^{(j)}(n)$$
$$= \sum_{j=0}^{\infty} \frac{1}{j!}(n(1 - e^{-\lambda/n}) - n)^j f^{(j)}(n) = f(n(1 - e^{-\lambda/n})),$$

the last equation from the Taylor series. Letting $n \to \infty$, we obtain for the limit of the last term $f(\lambda)$, since $f$ is continuous at each $\lambda$. It follows from

Theorem 6.6.3 that $\{F_n\}$ converges vaguely, say to $F$, and that the Laplace transform of $F$ is $f$. Hence $F(\infty) = f(0) = 1$, and $F$ is a d.f. The theorem is proved.

EXERCISES

**1.** If $X$ and $Y$ are independent r.v.'s with normal d.f.'s of the same variance, then $X + Y$ and $X - Y$ are independent.

$^\star$**2.** Two uncorrelated r.v.'s with a joint normal d.f. of arbitrary parameters are independent. Extend this to any finite number of r.v.'s.

$^\star$**3.** Let $F$ and $G$ be s.d.f.'s. If $\lambda_0 > 0$ and $\hat{F}(\lambda) = \hat{G}(\lambda)$ for all $\lambda \geq \lambda_0$, then $F = G$. More generally, if $\hat{F}(n\lambda_0) = \hat{G}(n\lambda_0)$ for integer $n \geq 1$, then $F = G$. [HINT: In order to apply the Stone–Weierstrass theorem as cited, adjoin the constant 1 to the family $\{e^{-\lambda x}, \lambda \geq \lambda_0\}$; show that a function in $C_0$ can actually be uniformly approximated without a constant.]

$^\star$**4.** Let $\{F_n\}$ be s.d.f.'s. If $\lambda_0 > 0$ and $\lim_{n\to\infty} \hat{F}_n(\lambda)$ exists for all $\lambda \geq \lambda_0$, then $\{F_n\}$ converges vaguely to an s.d.f.

**5.** Use Exercise 3 to prove that for any d.f. whose support is in a finite interval, the moment problem is determinate.

**6.** Let $F$ be an s.d.f. with support in $\mathscr{R}_+$. Define $G_0 = F$,

$$G_n(x) = \int_0^x G_{n-1}(u)\,du$$

for $n \geq 1$. Find $\hat{G}_n(\lambda)$ in terms of $\hat{F}(\lambda)$.

**7.** Let $\hat{f}(\lambda) = \int_0^\infty e^{-\lambda x} f(x)\,dx$ where $f \in L^1(0, \infty)$. Suppose that $f$ has a finite right-hand derivative $f'(0)$ at the origin, then

$$f(0) = \lim_{\lambda\to\infty} \lambda \hat{f}(\lambda),$$

$$f'(0) = \lim_{\lambda\to\infty} \lambda[\lambda \hat{f}(\lambda) - f(0)].$$

$^\star$**8.** In the notation of Exercise 7, prove that for every $\lambda, \mu \in \mathscr{R}_+$:

$$(\mu - \lambda) \int_0^\infty \int_0^\infty e^{-(\lambda s + \mu t)} f(s + t)\,ds\,dt = \hat{f}(\lambda) - \hat{f}(\mu).$$

**9.** Given a function $\sigma$ on $\mathscr{R}_+$ that is finite, continuous, and decreasing to zero at infinity, find a $\sigma$-finite measure $\mu$ on $\mathscr{R}_+$ such that

$$\forall t > 0: \int_{[0,t]} \sigma(t - s)\mu(ds) = 1.$$

[HINT: Assume $\sigma(0) = 1$ and consider the d.f. $1 - \sigma$.]

$^\star$**10.** If $f > 0$ on $(0, \infty)$ and has a derivative $f'$ that is completely monotonic there, then $1/f$ is also completely monotonic.

**11.** If $f$ is completely monotonic in $(0, \infty)$ with $f(0+) = +\infty$, then $f$ is the Laplace transform of an infinite measure $\mu$ on $\mathscr{R}_+$:

$$f(\lambda) = \int_{\mathscr{R}_+} e^{-\lambda x} \mu(dx).$$

[HINT: Show that $F_n(x) \le e^{2x\delta} f(\delta)$ for each $\delta > 0$ and all large $n$, where $F_n$ is defined in (7). Alternatively, apply Theorem 6.6.4 to $f(\lambda + n^{-1})/f(n^{-1})$ for $\lambda \ge 0$ and use Exercise 3.]

**12.** Let $\{g_n, 1 \le n \le \infty\}$ on $\mathscr{R}_+$ satisfy the conditions: (i) for each $n, g_n(\cdot)$ is positive and decreasing; (ii) $g_\infty(x)$ is continuous; (iii) for each $\lambda > 0$,

$$\lim_{n\to\infty} \int_0^\infty e^{-\lambda x} g_n(x)\, dx = \int_0^\infty e^{-\lambda x} g_\infty(x)\, dx.$$

Then

$$\lim_{n\to\infty} g_n(x) = g_\infty(x) \quad \text{for every } x \in \mathscr{R}_+.$$

[HINT: For $\epsilon > 0$ consider the sequence $\int_0^\infty e^{-\epsilon x} g_n(x)\, dx$ and show that

$$\lim_{n\to\infty} \int_a^b e^{-\epsilon x} g_n(x)\, dx = \int_a^b e^{-\epsilon x} g_\infty(x)\, dx, \qquad \overline{\lim_{n\to\infty}}\, g_n(b) \le g_\infty(b-),$$

and so on.]

Formally, the Fourier transform $f$ and the Laplace transform $\hat{F}$ of a p.m. with support in $\mathscr{R}_+$ can be obtained from each other by the substitution $t = i\lambda$ or $\lambda = -it$ in (1) of Sec. 6.1 or (3) of Sec. 6.6. In practice, a certain expression may be derived for one of these transforms that is valid only for the pertinent range, and the question arises as to its validity for the other range. Interesting cases of this will be encountered in Secs. 8.4 and 8.5. The following theorem is generally adequate for the situation.

**Theorem 6.6.5.** The function $h$ of the complex variable $z$ given by

$$h(z) = \int_{\mathscr{R}_+} e^{zx}\, dF(x)$$

is analytic in $\mathbf{R}z < 0$ and continuous in $\mathbf{R}z \le 0$. Suppose that $g$ is another function of $z$ that is analytic in $\mathbf{R}z < 0$ and continuous in $\mathbf{R}z \le 0$ such that

$$\forall t \in \mathscr{R}^1: h(it) = g(it).$$

Then $h(z) \equiv g(z)$ in $\mathbf{R}z \leq 0$; in particular

$$\forall \lambda \in \mathscr{R}_+ \colon h(-\lambda) = g(-\lambda).$$

PROOF. For each integer $m \geq 1$, the function $h_m$ defined by

$$h_m(z) = \int_{[0,m]} e^{zx}\, dF(x) = \sum_{n=0}^{\infty} z^n \int_{[0,m]} \frac{x^n}{n!}\, dF(x)$$

is clearly an entire function of $z$. We have

$$\int_{(m,\infty)} e^{zx}\, dF(x) \leq \int_{(m,\infty)} dF(x)$$

in $\mathbf{R}z \leq 0$; hence the sequence $h_m$ converges uniformly there to $h$, and $h$ is continuous in $\mathbf{R}z \leq 0$. It follows from a basic proposition in the theory of analytic functions that $h$ is analytic in the interior $\mathbf{R}z < 0$. Next, the difference $h - g$ is analytic in $\mathbf{R}z < 0$, continuous is $\mathbf{R}z \leq 0$, and equal to zero on the line $\mathbf{R}z = 0$ by hypothesis. Hence, by Schwarz's reflection principle, it can be analytically continued across the line and over the whole complex plane. The resulting entire functions, being zero on a line, must be identically zero in the plane. In particular $h - g \equiv 0$ in $\mathbf{R}z \leq 0$, proving the theorem.

### Bibliographical Note

For standard references on ch.f.'s, apart from Lévy [7], [11], Cramér [10], Gnedenko and Kolmogorov [12], Loève [14], Rényi [15], Doob [16], we mention:

S. Bochner, *Vorlesungen über Fouriersche Integrale*. Akademische Verlaggesellschaft, Konstanz, 1932.

E. Lukacs, *Characteristic functions*. Charles Griffin, London, 1960.

The proof of Theorem 6.6.4 is taken from

Willy Feller, *Completely monotone functions and sequences*, Duke J. **5** (1939), 661–674.

# 7 Central limit theorem and its ramifications

## 7.1 Liapounov's theorem

The name "central limit theorem" refers to a result that asserts the convergence in dist. of a "normed" sum of r.v.'s, $(S_n - a_n)/b_n$, to the unit normal d.f. $\Phi$. We have already proved a neat, though special, version in Theorem 6.4.4. Here we begin by generalizing the set-up. If we write

$$\frac{S_n - a_n}{b_n} = \left( \sum_{j=1}^{n} \frac{X_j}{b_n} \right) - \frac{a_n}{b_n}, \tag{1}$$

we see that we are really dealing with a double array, as follows. For each $n \geq 1$ let there be $k_n$ r.v.'s $\{X_{nj}, 1 \leq j \leq k_n\}$, where $k_n \to \infty$ as $n \to \infty$:

$$\begin{array}{c} X_{11}, X_{12}, \dots, X_{1k_1}; \\ X_{21}, X_{22}, \dots, X_{2k_2}; \\ \dots\dots\dots\dots\dots\dots \\ X_{n1}, X_{n2}, \dots, X_{nk_n}; \\ \dots\dots\dots\dots\dots\dots \end{array} \tag{2}$$

The r.v.'s with $n$ as first subscript will be referred to as being in the $n$th row. Let $F_{nj}$ be the d.f., $f_{nj}$ the ch.f. of $X_{nj}$; and put

$$S_n = S_{n,k_n} = \sum_{j=1}^{k_n} X_{nj}.$$

The particular case $k_n = n$ for each $n$ yields a triangular array, and if, furthermore, $X_{nj} = X_j$ for every $n$, then it reduces to the initial sections of a single sequence $\{X_j, j \geq 1\}$.

We shall assume from now on that *the r.v.'s in each row in* (2) *are independent*, but those in different rows may be arbitrarily dependent as in the case just mentioned—indeed they may be defined on different probability spaces without any relation to one another. Furthermore, let us introduce the following notation for moments, whenever they are defined, finite or infinite:

$$(3) \qquad \begin{aligned} &\mathscr{E}(X_{nj}) = \alpha_{nj}, && \sigma^2(X_{nj}) = \sigma_{nj}^2, \\ &\mathscr{E}(S_n) = \sum_{j=1}^{k_n} \alpha_{nj} = \alpha_n, && \sigma^2(S_n) = \sum_{j=1}^{k_n} \sigma_{nj}^2 = s_n^2, \\ &\mathscr{E}(|X_{nj}|^3) = \gamma_{nj}, && \Gamma_n = \sum_{j=1}^{k_n} \gamma_{nj}. \end{aligned}$$

In the special case of (1), we have

$$X_{nj} = \frac{X_j}{b_n}, \qquad \sigma^2(X_{nj}) = \frac{\sigma^2(X_j)}{b_n^2}.$$

If we take $b_n = s_n$, then

$$(4) \qquad \sum_{j=1}^{k_n} \sigma^2(X_{nj}) = 1.$$

By considering $X_{nj} - \alpha_{nj}$ instead of $X_{nj}$, we may suppose

$$(5) \qquad \forall n, \forall j: \alpha_{nj} = 0$$

whenever the means exist. The reduction (sometimes called "norming") leading to (4) and (5) is always available if each $X_{nj}$ has a finite second moment, and we shall assume this in the following.

In dealing with the double array (2), it is essential to impose a hypothesis that individual terms in the sum

$$S_n = \sum_{j=1}^{k_n} X_{nj}$$

are "negligible" in comparison with the sum itself. Historically, this arose from the assumption that "small errors" accumulate to cause probabilistically predictable random mass phenomena. We shall see later that such a hypothesis is indeed necessary in a reasonable criterion for the central limit theorem such as Theorem 7.2.1.

In order to clarify the intuitive notion of the negligibility, let us consider the following hierarchy of conditions, each to be satisfied for every $\epsilon > 0$:

(a) $$\forall j: \quad \lim_{n\to\infty} \mathscr{P}\{|X_{nj}| > \epsilon\} = 0;$$

(b) $$\lim_{n\to\infty} \max_{1\le j\le k_n} \mathscr{P}\{|X_{nj}| > \epsilon\} = 0;$$

(c) $$\lim_{n\to\infty} \mathscr{P}\left\{\max_{1\le j\le k_n} |X_{nj}| > \epsilon\right\} = 0;$$

(d) $$\lim_{n\to\infty} \sum_{j=1}^{k_n} \mathscr{P}\{|X_{nj}| > \epsilon\} = 0.$$

It is clear that (d) $\Rightarrow$ (c) $\Rightarrow$ (b) $\Rightarrow$ (a); see Exercise 1 below. It turns out that (b) is the appropriate condition, which will be given a name.

DEFINITION. The double array (2) is said to be *holospoudic** iff (b) holds.

**Theorem 7.1.1.** A necessary and sufficient condition for (2) to be holospoudic is:

(6) $$\forall t \in \mathscr{R}^1: \lim_{n\to\infty} \max_{1\le j\le k_n} |f_{nj}(t) - 1| = 0.$$

PROOF. Assuming that (b) is true, we have

$$\begin{aligned} |f_{nj}(t) - 1| &\le \int |e^{itx} - 1|\, dF_{nj}(x) = \int_{|x|>\epsilon} + \int_{|x|\le\epsilon} \\ &\le \int_{|x|>\epsilon} 2\, dF_{nj}(x) + |t| \int_{|x|\le\epsilon} |x|\, dF_{nj}(x) \\ &\le 2\int_{|x|>\epsilon} dF_{nj}(x) + \epsilon|t|; \end{aligned}$$

and consequently

$$\max_j |f_{nj}(t) - 1| \le 2 \max_j \mathscr{P}\{|X_{nj}| > \epsilon\} + \epsilon|t|.$$

* I am indebted to Professor M. Wigodsky for suggesting this word, the only new term coined in this book.

Letting $n \to \infty$, then $\epsilon \to 0$, we obtain (6). Conversely, we have by the inequality (2) of Sec. 6.3,

$$\int_{|x|>\epsilon} dF_{nj}(x) \leq 2 - \left| \frac{\epsilon}{2} \int_{|t|\leq 2/\epsilon} f_{nj}(t)dt \right| \leq \frac{\epsilon}{2} \int_{|t|\leq 2/\epsilon} |1 - f_{nj}(t)|\, dt;$$

and consequently

$$\max_j \mathscr{P}\{|X_{nj}| > \epsilon\} \leq \frac{\epsilon}{2} \int_{|t|\leq 2/\epsilon} \max_j |1 - f_{nj}(t)|\, dt.$$

Letting $n \to \infty$, the right-hand side tends to 0 by (6) and bounded convergence; hence (b) follows. Note that the basic independence assumption is not needed in this theorem.

We shall first give Liapounov's form of the central limit theorem involving the third moment by a typical use of the ch.f. From this we deduce a special case concerning bounded r.v.'s. From this, we deduce the sufficiency of Lindeberg's condition by direct arguments. Finally we give Feller's proof of the necessity of that condition by a novel application of the ch.f.

In order to bring out the almost mechanical nature of Liapounov's result we state and prove a lemma in calculus.

**Lemma.** Let $\{\theta_{nj}, 1 \leq j \leq k_n, 1 \leq n\}$ be a double array of complex numbers satisfying the following conditions as $n \to \infty$:

- **(i)** $\max_{1\leq j\leq k_n} |\theta_{nj}| \to 0$;
- **(ii)** $\sum_{j=1}^{k_n} |\theta_{nj}| \leq M < \infty$, where $M$ does not depend on $n$;
- **(iii)** $\sum_{j=1}^{k_n} \theta_{nj} \to \theta$, where $\theta$ is a (finite) complex number.

Then we have

$$\prod_{j=1}^{k_n} (1 + \theta_{nj}) \to e^{\theta}. \tag{7}$$

PROOF. By (i), there exists $n_0$ such that if $n \geq n_0$, then $|\theta_{nj}| \leq \frac{1}{2}$ for all $j$, so that $1 + \theta_{nj} \neq 0$. We shall consider only such large values of $n$ below, and we shall denote by log $(1 + \theta_{nj})$ the determination of logarithm with an angle in $(-\pi, \pi]$. Thus

$$\log(1 + \theta_{nj}) = \theta_{nj} + \Lambda|\theta_{nj}|^2, \tag{8}$$

where $\Lambda$ is a complex number depending on various variables but bounded by some *absolute constant* not depending on anything, and is not necessarily the same in each appearance. In the present case we have in fact

$$|\log(1+\theta_{nj}) - \theta_{nj}| = \left|\sum_{m=2}^{\infty} \frac{(-1)^{m-1}}{m}\theta_{nj}^m\right| \leq \sum_{m=2}^{\infty} \frac{|\theta_{nj}|^m}{m}$$
$$\leq \frac{|\theta_{nj}|^2}{2}\sum_{m=2}^{\infty}\left(\frac{1}{2}\right)^{m-2} = |\theta_{nj}|^2 \leq 1,$$

so that the absolute constant mentioned above may be taken to be 1. (The reader will supply such a computation next time.) Hence

$$\sum_{j=1}^{k_n} \log(1+\theta_{nj}) = \sum_{j=1}^{k_n} \theta_{nj} + \Lambda \sum_{j=1}^{k_n} |\theta_{nj}|^2.$$

(This $\Lambda$ is not the same as before, but bounded by the same 1!). It follows from (ii) and (i) that

(9) $$\sum_{j=1}^{k_n} |\theta_{nj}|^2 \leq \max_{1\leq j\leq k_n} |\theta_{nj}| \sum_{j=1}^{k_n} |\theta_{nj}| \leq M \max_{1\leq j\leq k_n} |\theta_{nj}| \to 0;$$

and consequently we have by (iii),

$$\sum_{j=1}^{k_n} \log(1+\theta_{nj}) \to \theta.$$

This is equivalent to (7).

**Theorem 7.1.2.** Assume that (4) and (5) hold for the double array (2) and that $\gamma_{nj}$ is finite for every $n$ and $j$. If

(10) $$\Gamma_n \to 0$$

as $n \to \infty$, then $S_n$ converges in dist. to $\Phi$.

PROOF. For each $n$, the range of $j$ below will be from 1 to $k_n$. It follows from the assumption (10) and Liapounov's inequality that

(11) $$\max_j \sigma_{nj}^3 \leq \max_j \gamma_{nj} \leq \Gamma_n \to 0.$$

By (3′) of Theorem 6.4.2, we have

$$f_{nj}(t) = 1 - \tfrac{1}{2}\sigma_{nj}^2 t^2 + \Lambda_{nj}\gamma_{nj}|t|^3,$$

where $|\Lambda_{nj}| \leq \frac{1}{6}$. We apply the lemma above, for a fixed $t$, to

$$\theta_{nj} = -\tfrac{1}{2}\sigma_{nj}^2 t^2 + \Lambda_{nj}\gamma_{nj}|t|^3.$$

Condition (i) is satisfied since

$$\max_j |\theta_{nj}| \le \frac{t^2}{2} \max_j \sigma_{nj}^2 + \Lambda |t|^3 \max_j \gamma_{nj} \to 0$$

by (11). Condition (ii) is satisfied since

$$\sum_j |\theta_{nj}| \le \frac{t^2}{2} + \Lambda |t|^3 \Gamma_n$$

is bounded by (11); similarly condition (iii) is satisfied since

$$\sum_j \theta_{nj} = -\frac{t^2}{2} + \Lambda |t|^3 \Gamma_n \to -\frac{t^2}{2}.$$

It follows that

$$\prod_{j=1}^{k_n} f_{nj}(t) \to e^{-t^2/2}.$$

This establishes the theorem by the convergence theorem of Sec. 6.3, since the left member is the ch.f. of $S_n$.

**Corollary.** Without supposing that $\mathscr{E}(X_{nj}) = 0$, suppose that for each $n$ and $j$ there is a finite constant $M_{nj}$ such that $|X_{nj}| \le M_{nj}$ a.e., and that

$$\max_{1 \le j \le k_n} M_{nj} \to 0. \tag{12}$$

Then $S_n - \mathscr{E}(S_n)$ converges in dist. to $\Phi$.

This follows at once by the trivial inequality

$$\sum_{j=1}^{k_n} \mathscr{E}(|X_{nj} - \mathscr{E}(X_{nj})|^3) \le 2 \max_{1 \le j \le k_n} M_{nj} \sum_{j=1}^{k_n} \sigma^2(X_{nj})$$
$$= 2 \max_{1 \le j \le k_n} M_{nj}.$$

The usual formulation of Theorem 7.1.2 for a single sequence of independent r.v.'s $\{X_j\}$ with $\mathscr{E}(X_j) = 0$, $\sigma^2(X_j) = \sigma_j^2 < \infty$, $\mathscr{E}(|X_j|^3) = \gamma_j < \infty$,

$$S_n = \sum_{j=1}^n X_j, \quad s_n^2 = \sum_{j=1}^n \sigma_j^2, \quad \Gamma_n = \sum_{j=1}^n \gamma_j, \tag{13}$$

is as follows.

If

$$\frac{\Gamma_n}{s_n^3} \to 0, \tag{14}$$

then $S_n/s_n$ converges in dist. to $\Phi$.

This is obtained by setting $X_{nj} = X_j/s_n$. It should be noticed that the double scheme gets rid of cumbersome fractions in proofs as well as in statements.

We proceed to another proof of Liapounov's theorem for a single sequence by the method of Lindeberg, who actually used it to prove his version of the central limit theorem, see Theorem 7.2.1 below. In recent times this method has been developed into a general tool by Trotter and Feller.

The idea of Lindeberg is to approximate the sum $X_1 + \cdots + X_n$ in (13) successively by replacing one $X$ at a time with a comparable normal (Gaussian) r.v. $Y$, as follows. Let $\{Y_j, j \geq 1\}$ be r.v.'s having the normal distribution $N(0, \sigma_j^2)$; thus $Y_j$ has the same mean and variance as the corresponding $X_j$ above; let all the $X$'s and $Y$'s be totally independent. Now put

$$Z_j = Y_1 + \cdots + Y_{j-1} + X_{j+1} + \cdots + X_n, \quad 1 \leq j \leq n,$$

with the obvious convention that

$$Z_1 = X_2 + \cdots + X_n, \quad Z_n = Y_1 + \cdots + Y_{n-1}.$$

To compare the distribution of $(X_j + Z_j)/s_n$ with that of $(Y_j + Z_j)/s_n$, we use Theorem 6.1.6 by comparing the expectations of test functions. Namely, we estimate the difference below for a suitable class of functions $f$:

$$\begin{aligned} &\mathscr{E}\left\{f\left(\frac{X_1 + \cdots + X_n}{s_n}\right)\right\} - \mathscr{E}\left\{f\left(\frac{Y_1 + \cdots + Y_n}{s_n}\right)\right\} \\ &= \sum_{j=1}^{n} \left[\mathscr{E}\left\{f\left(\frac{X_j + Z_j}{s_n}\right)\right\} - \mathscr{E}\left\{f\left(\frac{Y_j + Z_j}{s_n}\right)\right\}\right]. \end{aligned} \tag{15}$$

This equation follows by telescoping since $Y_j + Z_j = X_{j+1} + Z_{j+1}$. We take $f$ in $C_B^3$, the class of bounded continuous functions with three bounded continuous derivatives. By Taylor's theorem, we have for every $x$ and $y$:

$$\left|f(x+y) - \left[f(x) + f'(x)y + \frac{f''(x)}{2}y^2\right]\right| \leq \frac{M|y|^3}{6}$$

where $M = \sup_{x \in \mathscr{R}^1} |f^{(3)}(x)|$. Hence if $\xi$ and $\eta$ are independent r.v.'s such that $\mathscr{E}\{|\eta|^3\} < \infty$, we have by substitution followed by integration:

$$\begin{aligned} &|\mathscr{E}\{f(\xi + \eta)\} - \mathscr{E}\{f(\xi)\} - \mathscr{E}\{f'(\xi)\}\mathscr{E}\{\eta\} - \frac{1}{2}\mathscr{E}\{f''(\xi)\}\mathscr{E}\{\eta^2\}| \\ &\leq \frac{M}{6}\mathscr{E}\{|\eta|^3\}. \end{aligned} \tag{16}$$

Note that the r.v.'s $f(\xi)$, $f'(\xi)$, and $f''(\xi)$ are bounded hence integrable. If $\zeta$ is another r.v. independent of $\xi$ and having the same mean and variance as $\eta$, and $\mathscr{E}\{|\zeta|^3\} < \infty$, we obtain by replacing $\eta$ with $\zeta$ in (16) and then taking the difference:

$$(17) \qquad |\mathscr{E}\{f(\xi+\eta)\} - \mathscr{E}\{f(\xi+\zeta)\}| \le \frac{M}{6}\mathscr{E}\{|\eta|^3 + |\zeta|^3\}.$$

This key formula is applied to each term on the right side of (15), with $\xi = Z_j/s_n$, $\eta = X_j/s_n$, $\zeta = Y_j/s_n$. The bounds on the right-hand side of (17) then add up to

$$(18) \qquad \frac{M}{6}\sum_{j=1}^{n}\left\{\frac{\gamma_j}{s_n^3} + \frac{c\sigma_j^3}{s_n^3}\right\}$$

where $c = \sqrt{8/\pi}$ since the absolute third moment of $N(0, \sigma^2)$ is equal to $c\sigma_j^3$. By Liapounov's inequality (Sec. 3.2) $\sigma_j^3 \le \gamma_j$, so that the quantity in (18) is $O(\Gamma_n/s_n^3)$. Let us introduce a unit normal r.v. $N$ for convenience of notation, so that $(Y_1 + \cdots + Y_n)/s_n$ may be replaced by $N$ so far as its distribution is concerned. We have thus obtained the following estimate:

$$(19) \qquad \forall f \in C^3: \left|\mathscr{E}\left\{f\left(\frac{S_n}{s_n}\right)\right\} - \mathscr{E}\{f(N)\}\right| \le O\left(\frac{\Gamma_n}{s_n^3}\right).$$

Consequently, under the condition (14), this converges to zero as $n \to \infty$. It follows by the general criterion for vague convergence in Theorem 6.1.6 that $S_n/s_n$ converges in distribution to the unit normal. This is Liapounov's form of the central limit theorem proved above by the method of ch.f.'s. Lindeberg's idea yields a by-product, due to Pinsky, which will be proved under the same assumptions as in Liapounov's theorem above.

**Theorem 7.1.3.** Let $\{x_n\}$ be a sequence of real numbers increasing to $+\infty$ but subject to the growth condition: for some $\epsilon > 0$,

$$(20) \qquad \log\frac{\Gamma_n}{s_n^3} + \frac{x_n^2}{2}(1+\epsilon) \to -\infty$$

as $n \to \infty$. Then for this $\epsilon$, there exists $N$ such that for all $n \ge N$ we have

$$(21) \qquad \exp\left[-\frac{x_n^2}{2}(1+\epsilon)\right] \le \mathscr{P}\{S_n \ge x_n s_n\} \le \exp\left[-\frac{x_n^2}{2}(1-\epsilon)\right].$$

PROOF. This is derived by choosing particular test functions in (19). Let $f \in C^3$ be such that

$$f(x) = 0 \quad \text{for } x \le -\tfrac{1}{2}; \qquad 0 \le f(x) \le 1 \quad \text{for } -\tfrac{1}{2} < x < \tfrac{1}{2};$$

$$f(x) = 1 \quad \text{for } x \ge \tfrac{1}{2};$$

and put for all $x$:

$$f_n(x) = f(x - x_n - \tfrac{1}{2}), \quad g_n(x) = f(x - x_n + \tfrac{1}{2}).$$

Thus we have, denoting by $I_B$ the indicator function of $B \subset \mathscr{R}^1$:

$$I_{[x_n+1,\infty)} \le f_n(x) \le I_{[x_n,\infty)} \le g_n(x) \le I_{[x_n-1,\infty)}.$$

It follows that

$$\mathscr{E}\left\{f_n\left(\frac{S_n}{s_n}\right)\right\} \le \mathscr{P}\{S_n \ge x_n s_n\} \le \mathscr{E}\left\{g_n\left(\frac{S_n}{s_n}\right)\right\} \tag{22}$$

whereas

$$\mathscr{P}\{N \ge x_n + 1\} \le \mathscr{E}\{f_n(N)\} \le \mathscr{E}\{g_n(N)\} \le \mathscr{P}\{N \ge x_n - 1\}. \tag{23}$$

Using (19) for $f = f_n$ and $f = g_n$, and combining the results with (22) and (23), we obtain

$$\begin{aligned} \mathscr{P}\{N \ge x_n + 1\} - O\left(\frac{\Gamma_n}{s_n^3}\right) &\le \mathscr{P}\{S_n \ge x_n s_n\} \\ &\le \mathscr{P}\{N \ge x_n - 1\} + O\left(\frac{\Gamma_n}{s_n^3}\right). \end{aligned} \tag{24}$$

Now an elementary estimate yields for $x \to +\infty$:

$$\mathscr{P}\{N \ge x\} = \frac{1}{\sqrt{2\pi}} \int_x^\infty \exp\left(-\frac{y^2}{2}\right) dy \sim \frac{1}{\sqrt{2\pi} x} \exp\left(-\frac{x^2}{2}\right),$$

(see Exercise 4 of Sec. 7.4), and a quick computation shows further that

$$\mathscr{P}\{N \ge x \pm 1\} = \exp\left[-\frac{x^2}{2}(1 + o(1))\right], \quad x \to +\infty.$$

Thus (24) may be written as

$$\mathscr{P}\{S_n \ge x_n s_n\} = \exp\left[-\frac{x_n^2}{2}(1 + o(1))\right] + O\left(\frac{\Gamma_n}{s_n^3}\right).$$

Suppose $n$ is so large that the $o(1)$ above is strictly less than $\epsilon$ in absolute value; in order to conclude (23) it is sufficient to have

$$\frac{\Gamma_n}{s_n^3} = o\left(\exp\left[-\frac{x_n^2}{2}(1 + \epsilon)\right]\right), \quad n \to \infty.$$

This is the sense of the condition (20), and the theorem is proved.

Recalling that $s_n$ is the standard deviation of $S_n$, we call the probability in (21) that of a "large deviation". Observe that the two estimates of this probability given there have a ratio equal to $e^{\epsilon x_n^2}$ which is large for each $\epsilon$, as $x_n \to +\infty$. Nonetheless, it is small relative to the principal factor $e^{-x_n^2/2}$ on the logarithmic scale, which is just to say that $\epsilon x_n^2$ is small in absolute value compared with $-x_n^2/2$. Thus the estimates in (21) is useful on such a scale, and may be applied in the proof of the law of the iterated logarithm in Sec. 7.5.

EXERCISES

***1.** Prove that for arbitrary r.v.'s $\{X_{nj}\}$ in the array (2), the implications (d) $\Rightarrow$ (c) $\Rightarrow$ (b) $\Rightarrow$ (a) are all strict. On the other hand, if the $X_{nj}$'s are independent in each row, then (d) $\equiv$ (c).

**2.** For any sequence of r.v.'s $\{Y_n\}$, if $Y_n/b_n$ converges in dist. for an increasing sequence of constants $\{b_n\}$, then $Y_n/b'_n$ converges in pr. to 0 if $b_n = o(b'_n)$. In particular, make precise the following statement: "The central limit theorem implies the weak law of large numbers."

**3.** For the double array (2), it is possible that $S_n/b_n$ converges in dist. for a sequence of strictly positive constants $b_n$ tending to a finite limit. Is it still possible if $b_n$ oscillates between finite limits?

**4.** Let $\{X_j\}$ be independent r.v.'s such that $\max_{1\le j\le n} |X_j|/b_n \to 0$ in pr. and $(S_n - a_n)/b_n$ converges to a nondegenerate d.f. Then $b_n \to \infty$, $b_{n+1}/b_n \to 1$, and $(a_{n+1} - a_n)/b_n \to 0$.

***5.** In Theorem 7.1.2 let the d.f. of $S_n$ be $F_n$. Prove that given any $\epsilon > 0$, there exists a $\delta(\epsilon)$ such that $\Gamma_n \le \delta(\epsilon) \Rightarrow L(F_n, \Phi) \le \epsilon$, where $L$ is Levy distance. Strengthen the conclusion to read:

$$\sup_{x \in \mathscr{R}^1} |F_n(x) - \Phi(x)| \le \epsilon.$$

***6.** Prove the assertion made in Exercise 5 of Sec. 5.3 using the methods of this section. [HINT: use Exercise 4 of Sec. 4.3.]

## 7.2 Lindeberg–Feller theorem

We can now state the Lindeberg–Feller theorem for the double array (2) of Sec. 7.1 (with independence in each row).

**Theorem 7.2.1.** Assume $\sigma_{nj}^2 < \infty$ for each $n$ and $j$ and the reduction hypotheses (4) and (5) of Sec. 7.1. In order that as $n \to \infty$ the two conclusions below both hold:

**(i)** $S_n$ converges in dist. to $\Phi$,
**(ii)** the double array (2) of Sec. 7.1 is holospoudic;

it is necessary and sufficient that for each $\eta > 0$, we have

$$\text{(1)} \qquad \sum_{j=1}^{k_n} \int_{|x|>\eta} x^2 \, dF_{nj}(x) \to 0.$$

The condition (1) is called Lindeberg's condition; it is manifestly equivalent to

$$\text{(1')} \qquad \sum_{j=1}^{k_n} \int_{|x|\le\eta} x^2 \, dF_{nj}(x) \to 1.$$

PROOF. *Sufficiency*. By the argument that leads to Chebyshev's inequality, we have

$$\text{(2)} \qquad \mathscr{P}\{|X_{nj}| > \eta\} \le \frac{1}{\eta^2} \int_{|x|>\eta} x^2 \, dF_{nj}(x).$$

Hence (ii) follows from (1); indeed even the stronger form of negligibility (d) in Sec. 7.1 follows. Now for a fixed $\eta$, $0 < \eta < 1$, we truncate $X_{nj}$ as follows:

$$\text{(3)} \qquad X'_{nj} = \begin{cases} X_{nj}, & \text{if } |X_{nj}| \le \eta; \\ 0, & \text{otherwise.} \end{cases}$$

Put $S'_n = \sum_{j=1}^{k_n} X'_{nj}$, $\sigma^2(S'_n) = s'^2_n$. We have, since $\mathscr{E}(X_{nj}) = 0$,

$$\mathscr{E}(X'_{nj}) = \int_{|x|\le\eta} x \, dF_{nj}(x) = -\int_{|x|>\eta} x \, dF_{nj}(x).$$

Hence

$$|\mathscr{E}(X'_{nj})| \le \int_{|x|>\eta} |x| \, dF_{nj}(x) \le \frac{1}{\eta} \int_{|x|>\eta} x^2 \, dF_{nj}(x)$$

and so by (1),

$$|\mathscr{E}(S'_n)| \le \frac{1}{\eta} \sum_{j=1}^{k_n} \int_{|x|>\eta} x^2 \, dF_{nj}(x) \to 0.$$

Next we have by the Cauchy–Schwarz inequality

$$[\mathscr{E}(X'_{nj})]^2 \le \int_{|x|>\eta} x^2 \, dF_{nj}(x) \int_{|x|>\eta} 1 \, dF_{nj}(x),$$

and consequently

$$\sigma^2(X'_{nj}) = \int_{|x|\le\eta} x^2\,dF_{nj}(x) - \mathscr{E}(X'_{nj})^2 \ge \left\{\int_{|x|\le\eta} - \int_{|x|>\eta}\right\} x^2\,dF_{nj}(x).$$

It follows by (1′) that

$$1 = s_n^2 \ge s_n'^2 \ge \sum_{j=1}^{k_n} \left\{\int_{|x|\le\eta} - \int_{|x|>\eta}\right\} x^2\,dF_{nj}(x) \to 1.$$

Thus as $n \to \infty$, we have

$$s'_n \to 1 \quad \text{and} \quad \mathscr{E}(S'_n) \to 0.$$

Since

$$S'_n = \left\{\frac{S'_n - \mathscr{E}(S'_n)}{s'_n} + \frac{\mathscr{E}(S'_n)}{s'_n}\right\} s'_n$$

we conclude (see Theorem 4.4.6, but the situation is even simpler here) that if $[S'_n - \mathscr{E}(S'_n)] \mid /s'_n$ converges in dist., so will $S'_n/s'_n$ to the same d.f.

Now we try to apply the corollary to Theorem 7.1.2 to the double array $\{X'_{nj}\}$. We have $|X'_{nj}| \le \eta$, so that the left member in (12) of Sec. 7.1 corresponding to this array is bounded by $\eta$. But although $\eta$ is at our disposal and may be taken as small as we please, it is fixed with respect to $n$ in the above. Hence we cannot yet make use of the cited corollary. What is needed is the following lemma, which is useful in similar circumstances.

**Lemma 1.** Le t $u(m, n)$ be a function of positive integers $m$ and $n$ such that

$$\forall m: \lim_{n\to\infty} u(m, n) = 0.$$

Then there exists a sequence $\{m_n\}$ increasing to $\infty$ such that

$$\lim_{n\to\infty} u(m_n, n) = 0.$$

PROOF. It is the statement of the lemma and its necessity in our application that requires a certain amount of sophistication; the proof is easy. For each $m$, there is an $n_m$ such that $n \ge n_m \Rightarrow u(m, n) \le 1/m$. We may choose $\{n_m, m \ge 1\}$ inductively so that $n_m$ increases strictly with $m$. Now define $(n_0 = 1)$

$$m_n = m \quad \text{for } n_m \le n < n_{m+1}.$$

Then

$$u(m_n, n) \le \frac{1}{m} \quad \text{for } n_m \le n < n_{m+1},$$

and consequently the lemma is proved.

We apply the lemma to (1) as follows. For each $m \geq 1$, we have

$$\lim_{n\to\infty} m^2 \sum_{j=1}^{k_n} \int_{|x|>1/m} x^2 \, dF_{nj}(x) = 0.$$

It follows that there exists a sequence $\{\eta_n\}$ decreasing to 0 such that

$$\frac{1}{\eta_n^2} \sum_{j=1}^{k_n} \int_{|x|>\eta_n} x^2 \, dF_{nj}(x) \to 0.$$

Now we can go back and modify the definition in (3) by replacing $\eta$ with $\eta_n$. As indicated above, the cited corollary becomes applicable and yields the convergence of $[S'_n - \mathscr{E}(S'_n)]/s'_n$ in dist. to $\Phi$, hence also that of $S'_n/s'_n$ as remarked.

Finally we must go from $S'_n$ to $S_n$. The idea is similar to Theorem 5.2.1 but simpler. Observe that, for the modified $X_{nj}$ in (3) with $\eta$ replaced by $\eta_n$, we have

$$\begin{aligned} \mathscr{P}\{S_n \neq S'_n\} &\leq \mathscr{P}\left\{\bigcup_{j=1}^{k_n} [X_{nj} \neq X'_{nj}]\right\} \leq \sum_{j=1}^{k_n} \mathscr{P}\{|X_{nj}| > \eta_n\} \\ &\leq \sum_{j=1}^{k_n} \frac{1}{\eta_n^2} \int_{|x|>\eta_n} x^2 \, dF_{nj}(x), \end{aligned}$$

the last inequality from (2). As $n \to \infty$, the last term tends to 0 by the above, hence $S_n$ must have the same limit distribution as $S'_n$ (why?) and the sufficiency of Lindeberg's condition is proved. Although this method of proof is somewhat longer than a straightforward approach by means of ch.f.'s (Exercise 4 below), it involves several good ideas that can be used on other occasions. Indeed the sufficiency part of the most general form of the central limit theorem (see below) can be proved in the same way.

*Necessity*. Here we must resort entirely to manipulations with ch.f.'s. By the convergence theorem of Sec. 6.3 and Theorem 7.1.1, the conditions (i) and (ii) are equivalent to:

$$\forall t\colon \lim_{n\to\infty} \prod_{j=1}^{k_n} f_{nj}(t) = e^{-t^2/2}; \tag{4}$$

$$\forall t\colon \lim_{n\to\infty} \max_{1\leq j\leq k_n} |f_{nj}(t) - 1| = 0. \tag{5}$$

By Theorem 6.3.1, the convergence in (4) is uniform in $|t| \leq T$ for each finite $T$ : similarly for (5) by Theorem 7.1.1. Hence for each $T$ there exists $n_0(T)$

such that if $n \geq n_0(T)$, then

$$\max_{|t|\leq T} \max_{1\leq j\leq k_n} |f_{nj}(t) - 1| \leq \tfrac{1}{2}.$$

We shall consider only such values of $n$ below. We may take the distinguished logarithms (see Theorems 7.6.2 and 7.6.3 below) to conclude that

$$\lim_{n\to\infty} \sum_{j=1}^{k_n} \log f_{nj}(t) = -\frac{t^2}{2}. \tag{6}$$

By (8) and (9) of Sec. 7.1, we have

$$\log f_{nj}(t) = f_{nj}(t) - 1 + \Lambda |f_{nj}(t) - 1|^2; \tag{7}$$

$$\sum_{j=1}^{k_n} |f_{nj}(t) - 1|^2 \leq \max_{1\leq j\leq k_n} |f_{nj}(t) - 1| \sum_{j=1}^{k_n} |f_{nj}(t) - 1|. \tag{8}$$

Now the last-written sum is, with some $\theta$, $|\theta| \leq 1$:

$$\begin{aligned} \sum_j \left| \int_{-\infty}^{\infty} (e^{itx} - 1)\, dF_{nj}(x) \right| &= \sum_j \left| \int_{-\infty}^{\infty} \left( itx + \theta \frac{t^2 x^2}{2} \right) dF_{nj}(x) \right| \\ &\leq \frac{t^2}{2} \sum_j \int_{-\infty}^{\infty} x^2\, dF_{nj}(x) = \frac{t^2}{2}. \end{aligned} \tag{9}$$

Hence it follows from (5) and (9) that the left member of (8) tends to 0 as $n \to \infty$. From this, (7), and (6) we obtain

$$\lim_{n\to\infty} \sum_j \{f_{nj}(t) - 1\} = -\frac{t^2}{2}.$$

Taking real parts, we have

$$\lim_{n\to\infty} \sum_j \int_{-\infty}^{\infty} (1 - \cos tx)\, dF_{nj}(x) = \frac{t^2}{2}.$$

Hence for each $\eta > 0$, if we split the integral into two parts and transpose one of them, we obtain

$$\begin{aligned} &\overline{\lim_{n\to\infty}} \left| \frac{t^2}{2} - \sum_j \int_{|x|\leq\eta} (1 - \cos tx)\, dF_{nj}(x) \right| \\ &= \overline{\lim_{n\to\infty}} \left| \sum_j \int_{|x|>\eta} (1 - \cos tx)\, dF_{nj}(x) \right| \end{aligned}$$

$$\leq \overline{\lim_{n\to\infty}} \sum_j \int_{|x|>\eta} 2\, dF_{nj}(x)$$

$$\leq \overline{\lim_{n\to\infty}}\, 2 \sum_j \frac{{\sigma_{nj}}^2}{\eta^2} = \frac{2}{\eta^2},$$

the last inequality above by Chebyshev's inequality. Since $0 \leq 1 - \cos\theta \leq \theta^2/2$ for every real $\theta$, this implies

$$\frac{2}{\eta^2} \geq \overline{\lim_{n\to\infty}} \left\{ \frac{t^2}{2} - \sum_j \frac{t^2}{2} \int_{|x|\leq n} x^2\, dF_{nj}(x) \right\} \geq 0,$$

the quantity in braces being clearly positive. Thus

$$\frac{4}{t^2\eta^2} \geq \overline{\lim_{n\to\infty}} \left\{ 1 - \sum_{j=1}^{k_n} \int_{|x|\leq\eta} x^2\, dF_{nj}(x) \right\};$$

$t$ being arbitrarily large while $\eta$ is fixed; this implies Lindeberg's condition (1$'$). Theorem 7.2.1 is completely proved.

Lindeberg's theorem contains both the identically distributed case (Theorem 6.4.4) and Liapounov's theorem. Let us derive the latter, which assets that, under (4) and (5) of Sec. 7.1, the condition below for any one value of $\delta > 0$ is a sufficient condition for $S_n$ to converge in dist. to $\Phi$:

$$\sum_{j=1}^{k_n} \int_{-\infty}^{\infty} |x|^{2+\delta}\, dF_{nj}(x) \to 0. \tag{10}$$

For $\delta = 1$ this condition is just (10) of Sec. 7.1. In the general case the assertion follows at once from the following inequalities:

$$\sum_j \int_{|x|>\eta} x^2\, dF_{nj}(x) \leq \sum_j \int_{|x|>\eta} \frac{|x|^{2+\delta}}{\eta^\delta}\, dF_{nj}(x)$$

$$\leq \frac{1}{\eta^\delta} \sum_j \int_{-\infty}^{\infty} |x|^{2+\delta}\, dF_{nj}(x),$$

showing that (10) implies (1).

The essence of Theorem 7.2.1 is the assumption of the finiteness of the second moments together with the "classical" norming factor $s_n$, which is the standard deviation of the sum $S_n$; see Exercise 10 below for an interesting possibility. In the most general case where "nothing is assumed," we have the following criterion, due to Feller and in a somewhat different form to Lévy.

**Theorem 7.2.2.** For the double array (2) of Sec. 7.1 (with independence in each row), in order that there exists a sequence of constants $\{a_n\}$ such that (i) $\sum_{j=1}^{k_n} X_{nj} - a_n$ converges in dist. to $\Phi$, and (ii) the array is holospoudic, it is necessary and sufficient that the following two conditions hold for every $\eta > 0$:

(a) $\sum_{j=1}^{k_n} \int_{|x|>\eta} dF_{nj}(x) \to 0$;

(b) $\sum_{j=1}^{k_n} \{\int_{|x|\le\eta} x^2\, dF_{nj}(x) - (\int_{|x|\le\eta} x\, dF_{nj}(x))^2\} \to 1$.

We refer to the monograph by Gnedenko and Kolmogorov [12] for this result and its variants, as well as the following criterion for a single sequence of independent, identically distributed r.v.'s due to Lévy.

**Theorem 7.2.3.** Let $\{X_j, j \ge 1\}$ be independent r.v.'s having the common d.f. $F$; and $S_n = \sum_{j=1}^{n} X_j$. In order that there exist constants $a_n$ and $b_n > 0$ (necessarily $b_n \to +\infty$) such that $(S_n - a_n)/b_n$ converges in dist. to $\Phi$, it is necessary and sufficient that we have, as $y \to +\infty$:

(11)
$$y^2 \int_{|x|>y} dF(x) = o\left(\int_{|x|\le y} x^2\, dF(x)\right).$$

The central limit theorem, applied to concrete cases, leads to asymptotic formulas. The best-known one is perhaps the following, for $0 < p < 1$, $p + q = 1$, and $x_1 < x_2$, as $n \to \infty$:

(12)
$$\sum_{x_1\sqrt{npq} \le k - np \le x_2\sqrt{npq}} \binom{n}{k} p^k q^{n-k} \sim \Phi(x_2) - \Phi(x_1) = \frac{1}{\sqrt{2\pi}} \int_{x_1}^{x_2} e^{-y^2/2}\, dy.$$

This formula, due to DeMoivre, can be derived from Stirling's formula for factorials in a rather messy way. But it is just an application of Theorem 6.4.4 (or 7.1.2), where each $X_j$ has the Bernoullian d.f. $p\delta_1 + q\delta_0$.

More interesting and less expected applications to combinatorial analysis will be illustrated by the following example, which incidentally demonstrates the logical necessity of a double array even in simple situations.

Consider all $n!$ distinct permutations $(a_1, a_2, \dots, a_n)$ of the $n$ integers $(1, 2, \dots, n)$. The sample space $\Omega = \Omega_n$ consists of these $n!$ points, and $\mathscr{P}$ assigns probability $1/n!$ to each of the points. For each $j$, $1 \le j \le n$, and each $\omega = (a_1, a_2, \dots, a_n)$ let $X_{nj}$ be the number of "inversions" caused by $j$ in $\omega$; namely $X_{nj}(\omega) = m$ if and only if $j$ precedes exactly $m$ of the integers $1, \dots, j-1$ in the permutation $\omega$. The basic structure of the sequence $\{X_{nj}, 1 \le j \le n\}$ is contained in the lemma below.

**Lemma 2.** For each $n$, the r.v.'s $\{X_{nj}, 1 \le j \le n\}$ are independent with the following distributions:

$$\mathscr{P}\{X_{nj} = m\} = \frac{1}{j} \quad \text{for } 0 \le m \le j-1.$$

The lemma is a striking example that stochastic independence need not be an obvious phenomenon even in the simplest problems and may require verification as well as discovery. It is often disposed of perfunctorily but a formal proof is lengthier than one might think. Observe first that the values of $X_{n1}, \dots, X_{nj}$ are determined as soon as the positions of the integers $\{1, \dots, j\}$ are known, irrespective of those of the rest. Given $j$ arbitrary positions among $n$ ordered slots, the number of $\omega$'s in which $\{1, \dots, j\}$ occupy these positions in some order is $j!(n-j)!$. Among these the number of $\omega$'s in which $j$ occupies the $(j-m)$th place, where $0 \le m \le j-1$, (in order from left to right) of the given positions is $(j-1)!(n-j)!$. This position being fixed for $j$, the integers $\{1, \dots, j-1\}$ may occupy the remaining given positions in $(j-1)!$ distinct ways, each corresponding uniquely to a possible value of the random vector

$$\{X_{n1}, \dots, X_{n,j-1}\}.$$

That this correspondence is $1-1$ is easily seen if we note that the total number of possible values of this vector is precisely $1 \cdot 2 \cdots (j-1) = (j-1)!$. It follows that for any such value $(c_1, \dots, c_{j-1})$ the number of $\omega$'s in which, first, $\{1, \dots, j\}$ occupy the given positions and, second, $X_{n1}(\omega) = c_1, \dots, X_{n,j-1}(\omega) = c_{j-1}, X_{nj}(\omega) = m$, is equal to $(n-j)!$. Hence the number of $\omega$'s satisfying the second condition alone is equal to $\binom{n}{j}(n-j)! = n!/j!$. Summing over $m$ from 0 to $j-1$, we obtain the number of $\omega$'s in which $X_{n1}(\omega) = c_1, \dots, X_{n,j-1}(\omega) = c_{j-1}$ to be $jn!/j! = n!/(j-1)!$. Therefore we have

$$\frac{\mathscr{P}\{X_{n1} = c_1, \dots, X_{n,j-1} = c_{j-1}, X_{nj} = m\}}{\mathscr{P}\{X_{n1} = c_1, \dots, X_{n,j-1} = c_{j-1}\}} = \frac{\dfrac{n!}{j!}}{\dfrac{n!}{(j-1)!}} = \frac{1}{j}.$$

This, as we know, establishes the lemma. The reader is urged to see for himself whether he can shorten this argument while remaining scrupulous.

The rest is simple algebra. We find

$$\mathscr{E}(X_{nj}) = \frac{j-1}{2}, \qquad \sigma_{nj}^2 = \frac{j^2-1}{12}, \qquad \mathscr{E}(S_n) \sim \frac{n^2}{4}, \qquad s_n^2 \sim \frac{n^3}{36}.$$

For each $\eta > 0$, and sufficiently large $n$, we have

$$|X_{nj}| \le j-1 \le n-1 < \eta s_n.$$

Hence Lindeberg's condition is satisfied for the double array

$$\left\{\frac{X_{nj}}{s_n}; 1 \le j \le n, 1 \le n\right\}$$

(in fact the Corollary to Theorem 7.1.2 is also applicable) and we obtain the central limit theorem:

$$\lim_{n\to\infty} \mathscr{P}\left\{\frac{S_n - \dfrac{n^2}{4}}{\dfrac{n^{3/2}}{6}} \le x\right\} = \Phi(x).$$

Here for each permutation $\omega$, $S_n(\omega) = \sum_{j=1}^n X_{nj}(\omega)$ is the total number of inversions in $\omega$; and the result above asserts that among the $n!$ permutations on $\{1, \dots, n\}$, the number of those in which there are $\le n^2/4 + x(n^{3/2})/6$ inversions has a proportion $\Phi(x)$, as $n \to \infty$. In particular, for example, the number of those with $\le n^2/4$ inversions has an asymptotic proportion of $\frac{1}{2}$.

## EXERCISES

**1.** Restate Theorem 7.1.2 in terms of normed sums of a single sequence.

**2.** Prove that Lindeberg's condition (1) implies that

$$\max_{1 \le j \le k_n} \sigma_{nj} \to 0.$$

***3.** Prove that in Theorem 7.2.1, (i) does not imply (1). [HINT: Consider r.v.'s with normal distributions.]

***4.** Prove the sufficiency part of Theorem 7.2.1 without using Theorem 7.1.2, but by elaborating the proof of the latter. [HINT: Use the expansion

$$e^{itx} = 1 + itx + \theta\frac{(tx)^2}{2} \quad \text{for } |x| > \eta$$

and

$$e^{itx} = 1 + itx - \frac{(tx)^2}{2} + \theta'\frac{|tx|^3}{6} \quad \text{for } |x| \le \eta.$$

As a matter of fact, Lindeberg's original proof does not even use ch.f.'s; see Feller [13, vol. 2].]

**5.** Derive Theorem 6.4.4 from Theorem 7.2.1.

**6.** Prove that if $\delta < \delta'$, then the condition (10) implies the similar one when $\delta$ is replaced by $\delta'$.

**★7.** Find an example where Lindeberg's condition is satisfied but Liapounov's is not for any $\delta > 0$.

In Exercises 8 to 10 below $\{X_j, j \geq 1\}$ is a sequence of independent r.v.'s.

**8.** For each $j$ let $X_j$ have the uniform distribution in $[-j, j]$. Show that Lindeberg's condition is satisfied and state the resulting central limit theorem.

**9.** Let $X_j$ be defined as follows for some $\alpha > 1$:

$$X_j = \begin{cases} \pm j^{\alpha}, & \text{with probability } \dfrac{1}{6j^{2(\alpha-1)}} \text{ each;} \\[2ex] 0, & \text{with probability } 1 - \dfrac{1}{3j^{2(\alpha-1)}}. \end{cases}$$

Prove that Lindeberg's condition is satisfied if and only if $\alpha < 3/2$.

**★10.** It is important to realize that the failure of Lindeberg's condition means only the failure of either (i) or (ii) in Theorem 7.2.1 *with the specified constants* $s_n$. A central limit theorem may well hold with a different sequence of constants. Let

$$X_j = \begin{cases} \pm j^2, & \text{with probability } \dfrac{1}{12j^2} \text{ each;} \\[2ex] \pm j, & \text{with probability } \dfrac{1}{12} \text{ each;} \\[2ex] 0, & \text{with probability } 1 - \dfrac{1}{6} - \dfrac{1}{6j^2}. \end{cases}$$

Prove that Lindeberg's condition is not satisfied. Nonetheless if we take $b_n^2 = n^3/18$, then $S_n/b_n$ converges in dist. to $\Phi$. The point is that abnormally large values may not count! [HINT: Truncate out the abnormal value.]

**11.** Prove that $\int_{-\infty}^{\infty} x^2\, dF(x) < \infty$ implies the condition (11), but not vice versa.

**★12.** The following combinatorial problem is similar to that of the number of inversions. Let $\Omega$ and $\mathscr{P}$ be as in the example in the text. It is standard knowledge that each permutation

$$\begin{pmatrix} 1 & 2 & \cdots & n \\ a_1 & a_2 & \cdots & a_n \end{pmatrix}$$

can be uniquely decomposed into the product of *cycles*, as follows. Consider the permutation as a mapping $\pi$ from the set $(1, \ldots, n)$ onto itself such that $\pi(j) = a_j$. Beginning with 1 and applying the mapping successively, $1 \to \pi(1) \to \pi^2(1) \to \cdots$, until the first $k$ such that $\pi^k(1) = 1$. Thus

$$(1, \pi(1), \pi^2(1), \ldots, \pi^{k-1}(1))$$

is the first cycle of the decomposition. Next, begin with the least integer, say $b$, not in the first cycle and apply $\pi$ to it successively; and so on. We say $1 \to \pi(1)$ is the first step, $\dots$, $\pi^{k-1}(1) \to 1$ the $k$th step, $b \to \pi(b)$ the $(k+1)$st step of the decomposition, and so on. Now define $X_{nj}(\omega)$ to be equal to 1 if in the decomposition of $\omega$, a cycle is completed at the $j$th step; otherwise to be 0. Prove that for each $n$, $\{X_{nj}, 1 \le j \le n\}$ is a set of independent r.v.'s with the following distributions:

$$\mathscr{P}\{X_{nj} = 1\} = \frac{1}{n - j + 1},$$

$$\mathscr{P}\{X_{nj} = 0\} = 1 - \frac{1}{n - j + 1}.$$

Deduce the central limit theorem for the number of cycles of a permutation.

## 7.3 Ramifications of the central limit theorem

As an illustration of a general method of extending the central limit theorem to certain classes of dependent r.v.'s, we prove the following result. Further elaboration along the same lines is possible, but the basic idea, attributed to S. Bernstein, consists always in separating into blocks and neglecting small ones.

Let $\{X_n, n \ge 1\}$ be a sequence of r.v.'s; let $\mathscr{F}_n$ be the Borel field generated by $\{X_k, 1 \le k \le n\}$, and $\mathscr{F}'_n$ that by $\{X_k, n < k < \infty\}$. The sequence is called *m-dependent* iff there exists an integer $m \ge 0$ such that for every $n$ the fields $\mathscr{F}_n$ and $\mathscr{F}'_{n+m}$ are independent. When $m = 0$, this reduces to independence.

**Theorem 7.3.1.** Suppose that $\{X_n\}$ is a sequence of $m$-dependent, uniformly bounded r.v.'s such that

$$\frac{\sigma(S_n)}{n^{1/3}} \to +\infty$$

as $n \to \infty$. Then $[S_n - \mathscr{E}(S_n)]/\sigma(S_n)$ converges in dist. to $\Phi$.

PROOF. Let the uniform bound be $M$. Without loss of generality we may suppose that $\mathscr{E}(X_n) = 0$ for each $n$. For an integer $k \ge 1$ let $n_j = [jn/k]$, $0 \le j \le k$, and put for large values of $n$:

$$Y_j = X_{n_j+1} + X_{n_j+2} + \cdots + X_{n_{j+1}-m};$$

$$Z_j = X_{n_{j+1}-m+1} + X_{n_{j+1}-m+2} + \cdots + X_{n_{j+1}}.$$

We have then

$$S_n = \sum_{j=0}^{k-1} Y_j + \sum_{j=0}^{k-1} Z_j = S'_n + S''_n, \text{ say.}$$

It follows from the hypothesis of $m$-dependence and Theorem 3.3.2 that the $Y_j$'s are independent; so are the $Z_j$'s, provided $n_{j+1} - m + 1 - n_j > m$, which is the case if $n/k$ is large enough. Although $S'_n$ and $S''_n$ are not independent of each other, we shall show that the latter is comparatively negligible so that $S_n$ behaves like $S'_n$. Observe that each term $X_r$ in $S''_n$ is independent of every term $X_s$ in $S'_n$ except at most $m$ terms, and that $\mathscr{E}(X_r X_s) = 0$ when they are independent, while $|\mathscr{E}(X_r X_s)| \leq M^2$ otherwise. Since there are $km$ terms in $S''_n$, it follows that

$$|\mathscr{E}(S'_n S''_n)| \leq km \cdot m \cdot M^2 = k(mM)^2.$$

We have also

$$\mathscr{E}({S''}^2_n) = \sum_{j=0}^{k-1} \mathscr{E}(Z_j^2) \leq k(mM)^2.$$

From these inequalities and the identity

$$\mathscr{E}(S_n^2) = \mathscr{E}({S'}^2_n) + 2\mathscr{E}(S'_n S''_n) + \mathscr{E}({S''}^2_n)$$

we obtain

$$|\mathscr{E}(S_n^2) - \mathscr{E}({S'}^2_n)| \leq 3km^2M^2.$$

Now we choose $k = k_n = [n^{2/3}]$ and write $s_n^2 = \mathscr{E}(S_n^2) = \sigma^2(S_n)$, ${s'}^2_n = \mathscr{E}({S'}^2_n) = \sigma^2(S'_n)$. Then we have, as $n \to \infty$.

$$\frac{s'_n}{s_n} \to 1, \tag{1}$$

and

$$\mathscr{E}\left\{\left(\frac{S''_n}{s_n}\right)^2\right\} \to 0. \tag{2}$$

Hence, first, $S''_n/s_n \to 0$ in pr. (Theorem 4.1.4) and, second, since

$$\frac{S_n}{s_n} = \frac{s'_n}{s_n}\frac{S'_n}{s'_n} + \frac{S''_n}{s_n},$$

$S_n/s_n$ will converge in dist. to $\Phi$ if $S'_n/s'_n$ does.

Since $k_n$ is a function of $n$, in the notation above $Y_j$ should be replaced by $Y_{nj}$ to form the double array $\{Y_{nj}, 0 \leq j \leq k_n - 1, 1 \leq n\}$, which retains independence in each row. We have, since each $Y_{nj}$ is the sum of no more than $[n/k_n] + 1$ of the $X_n$'s,

$$|Y_{nj}| \leq \left(\frac{n}{k_n} + 1\right) M = O(n^{1/3}) = o(s_n) = o(s'_n),$$

the last relation from (1) and the one preceding it from a hypothesis of the theorem. Thus for each $\eta > 0$, we have for all sufficiently large $n$:

$$\int_{|x|>\eta s'_n} x^2 \, dF_{nj}(x) = 0, \quad 0 \le j \le k_n - 1,$$

where $F_{nj}$ is the d.f. of $Y_{nj}$. Hence Lindeberg's condition is satisfied for the double array $\{Y_{nj}/s'_n\}$, and we conclude that $S'_n/s'_n$ converges in dist. to $\Phi$. This establishes the theorem as remarked above.

The next extension of the central limit theorem is to the case of a random number of terms (cf. the second part of Sec. 5.5). That is, we shall deal with the r.v. $S_{\nu_n}$ whose value at $\omega$ is given by $S_{\nu_n(\omega)}(\omega)$, where

$$S_n(\omega) = \sum_{j=1}^{n} X_j(\omega)$$

as before, and $\{\nu_n(\omega), n \ge 1\}$ is a sequence of r.v.'s. The simplest case, but not a very useful one, is when all the r.v.'s in the "double family" $\{X_n, \nu_n, n \ge 1\}$ are independent. The result below is more interesting and is distinguished by the simple nature of its hypothesis. The proof relies essentially on Kolmogorov's inequality given in Theorem 5.3.1.

**Theorem 7.3.2.** Let $\{X_j, j \ge 1\}$ be a sequence of independent, identically distributed r.v.'s with mean 0 and variance 1. Let $\{\nu_n, n \ge 1\}$ be a sequence of r.v.'s taking only strictly positive integer values such that

$$\frac{\nu_n}{n} \to c \qquad \text{in pr.,} \tag{3}$$

where $c$ is a constant: $0 < c < \infty$. Then $S_{\nu_n}/\sqrt{\nu_n}$ converges in dist. to $\Phi$.

PROOF. We know from Theorem 6.4.4 that $S_n/\sqrt{n}$ converges in dist. to $\Phi$, so that our conclusion means that we can substitute $\nu_n$ for $n$ there. The remarkable thing is that no kind of independence is assumed, but only the limit property in (3). First of all, we observe that in the result of Theorem 6.4.4 we may substitute $[cn]$ (= integer part of $cn$) for $n$ to conclude the convergence in dist. of $S_{[cn]}/\sqrt{[cn]}$ to $\Phi$ (why?). Next we write

$$\frac{S_{\nu_n}}{\sqrt{\nu_n}} = \left( \frac{S_{[cn]}}{\sqrt{[cn]}} + \frac{S_{\nu_n} - S_{[cn]}}{\sqrt{[cn]}} \right) \sqrt{\frac{[cn]}{\nu_n}}.$$

The second factor on the right converges to 1 in pr., by (3). Hence a simple argument used before (Theorem 4.4.6) shows that the theorem will be proved if we show that

$$\frac{S_{\nu_n} - S_{[cn]}}{\sqrt{[cn]}} \to 0 \qquad \text{in pr.} \tag{4}$$

Let $\epsilon$ be given, $0 < \epsilon < 1$; put

$$a_n = [(1 - \epsilon^3)[cn]], \quad b_n = [(1 + \epsilon^3)[cn]] - 1.$$

By (3), there exists $n_0(\epsilon)$ such that if $n \geq n_0(\epsilon)$, then the set

$$\Lambda = \{\omega: a_n \leq \nu_n(\omega) \leq b_n\}$$

has probability $\geq 1 - \epsilon$. If $\omega$ is in this set, then $S_{\nu_n(\omega)}(\omega)$ is one of the sums $S_j$ with $a_n \leq j \leq b_n$. For $[cn] < j \leq b_n$, we have

$$S_j - S_{[cn]} = X_{[cn]+1} + X_{[cn]+2} + \cdots + X_j;$$

hence by Kolmogorov's inequality

$$\mathscr{P}\left\{\max_{[cn] \leq j \leq b_n} |S_j - S_{[cn]}| > \epsilon\sqrt{cn}\right\} \leq \frac{\sigma^2(S_{b_n} - S_{[cn]})}{\epsilon^2 cn} \leq \frac{\epsilon^3[cn]}{\epsilon^2 cn} \leq \epsilon.$$

A similar inequality holds for $a_n \leq j < [cn]$; combining the two, we obtain

$$\mathscr{P}\{\max_{a_n \leq j \leq b_n} |S_j - S_{[cn]}| > \epsilon\sqrt{cn}\} \leq 2\epsilon.$$

Now we have, if $n \geq n_0(\epsilon)$:

$$\begin{aligned}
&\mathscr{P}\left\{\left|\frac{S_{\nu_n} - S_{[cn]}}{\sqrt{[cn]}}\right| > \epsilon\right\} \\
&\quad = \sum_{j=1}^{\infty} \mathscr{P}\left\{\nu_n = j; \left|\frac{S_{\nu_n} - S_{[cn]}}{\sqrt{[cn]}}\right| > \epsilon\right\} \\
&\quad \leq \sum_{a_n \leq j \leq b_n} \mathscr{P}\left\{\nu_n = j; \max_{a_n \leq j \leq b_n} |S_j - S_{[cn]}| > \epsilon\sqrt{[cn]}\right\} + \sum_{j \notin [a_n, b_n]} \mathscr{P}\{\nu_n = j\} \\
&\quad \leq \mathscr{P}\left\{\max_{a_n \leq j \leq b_n} |S_j - S_{[cn]}| > \epsilon\sqrt{[cn]}\right\} + \mathscr{P}\{\nu_n \notin [a_n, b_n]\} \\
&\quad \leq 2\epsilon + 1 - \mathscr{P}\{\Lambda\} \leq 3\epsilon.
\end{aligned}$$

Since $\epsilon$ is arbitrary, this proves (4) and consequently the theorem.

As a third, somewhat deeper application of the central limit theorem, we give an instance of another limiting distribution inherently tied up with the normal. In this application the role of the central limit theorem, indeed in high dimensions, is that of an underlying source giving rise to multifarious manifestations. The topic really belongs to the theory of Brownian motion process, to which one must turn for a true understanding.

Let $\{X_j, j \geq 1\}$ be a sequence of independent and identically distributed r.v.'s with mean 0 and variance 1; and

$$S_n = \sum_{j=1}^{n} X_j.$$

It will become evident that these assumptions may be considerably weakened, and the basic hypothesis is that the central limit theorem should be applicable. Consider now the infinite sequence of successive sums $\{S_n, n \geq 1\}$. This is the kind of stochastic process to which an entire chapter will be devoted later (Chapter 8). The classical limit theorems we have so far discussed deal with the individual terms of the sequence $\{S_n\}$ itself, but there are various other sequences derived from it that are no less interesting and useful. To give a few examples:

$$\max_{1 \leq m \leq n} S_m, \qquad \min_{1 \leq m \leq n} S_m, \qquad \max_{1 \leq m \leq n} |S_m|, \qquad \max_{1 \leq m \leq n} \frac{|S_m|}{\sqrt{m}},$$

$$\sum_{m=1}^{n} \delta_a(S_m), \qquad \sum_{m=1}^{n} \gamma(S_m, S_{m+1});$$

where $\gamma(a, b) = 1$ if $ab < 0$ and 0 otherwise. Thus the last two examples represent, respectively, the "number of sums $\geq a$" and the "number of changes of sign". Now the central idea, originating with Erdös and Kac, is that the asymptotic behavior of these functionals of $S_n$ should be the same regardless of the special properties of $\{X_j\}$, so long as the central limit theorem applies to it (at least when certain regularity conditions are satisfied, such as the finiteness of a higher moment). Thus, in order to obtain the asymptotic distribution of one of these functionals one may calculate it in a very particular case where the calculations are feasible. We shall illustrate this method, which has been called an "invariance principle", by carrying it out in the case of max $S_m$; for other cases see Exercises 6 and 7 below.

Let us therefore put, for a given $x$:

$$P_n(x) = \mathscr{P}\left\{\max_{1 \leq m \leq n} S_m \leq x\sqrt{n}\right\}.$$

For an integer $k \geq 1$ let $n_j = [jn/k]$, $0 \leq j \leq k$, and define

$$R_{nk}(x) = \mathscr{P}\left\{\max_{1 \leq j \leq k} S_{n_j} \leq x\sqrt{n}\right\}.$$

Let also

$$E_j = \{\omega: S_m(\omega) \leq x\sqrt{n}, 1 \leq m < j; S_j(\omega) > x\sqrt{n}\};$$

and for each $j$, define $\ell(j)$ by

$$n_{\ell(j)-1} < j \le n_{\ell(j)}.$$

Now we write, for $0 < \epsilon < x$:

$$\mathscr{P}\left\{\max_{1\le m\le n} S_m > x\sqrt{n}\right\} = \sum_{j=1}^{n} \mathscr{P}\{E_j; |S_{n\ell(j)} - S_j| \le \epsilon\sqrt{n}\}$$

$$+ \sum_{j=1}^{n} \mathscr{P}\{E_j; |S_{n\ell(j)} - S_j| > \epsilon\sqrt{n}\} = \sum_1 + \sum_2,$$

say. Since $E_j$ is independent of $\{|S_{n\ell(j)} - S_j| > \epsilon\sqrt{n}\}$ and $\sigma^2(S_{n\ell(j)} - S_j) \le n/k$, we have by Chebyshev's inequality:

$$\sum_2 \le \sum_{j=1}^{n} \mathscr{P}(E_j)\frac{\frac{n}{k}}{\epsilon^2 n} \le \frac{1}{\epsilon^2 k}.$$

On the other hand, since $S_j > x\sqrt{n}$ and $|S_{n\ell(j)} - S_j| \le \epsilon\sqrt{n}$ imply $S_{n\ell(j)} > (x-\epsilon)\sqrt{n}$, we have

$$\sum_1 \le \mathscr{P}\left\{\max_{1\le \ell\le k} S_{n\ell} > (x-\epsilon)\sqrt{n}\right\} = 1 - R_{nk}(x-\epsilon).$$

It follows that

$$P_n(x) = 1 - \sum_1 - \sum_2 \ge R_{nk}(x-\epsilon) - \frac{1}{\epsilon^2 k}. \tag{5}$$

Since it is trivial that $P_n(x) \le R_{nk}(x)$, we obtain from (5) the following inequalities:

$$P_n(x) \le R_{nk}(x) \le P_n(x+\epsilon) + \frac{1}{\epsilon^2 k}. \tag{6}$$

We shall show that for fixed $x$ and $k$, $\lim_{n\to\infty} R_{nk}(x)$ exists. Since

$$R_{nk}(x) = \mathscr{P}\{S_{n_1} \le x\sqrt{n}, S_{n_2} \le x\sqrt{n}, \ldots, S_{n_k} \le x\sqrt{n}\},$$

it is sufficient to show that the sequence of $k$-dimensional random vectors

$$\left(\sqrt{\frac{k}{n}}S_{n_1}, \sqrt{\frac{k}{n}}S_{n_2}, \ldots, \sqrt{\frac{k}{n}}S_{n_k}\right)$$

converges in distribution as $n \to \infty$. Now its ch.f. $f(t_1, \dots, t_k)$ is given by

$$\begin{aligned}\mathscr{E}\{\exp(i\sqrt{k/n}(t_1 S_{n_1} + \cdots + t_k S_{n_k}))\} = \mathscr{E}\{&\exp(i\sqrt{k/n}[(t_1 + \cdots + t_k)S_{n_1} \\ &+ (t_2 + \cdots + t_k)(S_{n_2} - S_{n_1}) \\ &+ \cdots + t_k(S_{n_k} - S_{n_{k-1}})])\},\end{aligned}$$

which converges to

$$\exp\left[-\tfrac{1}{2}(t_1 + \cdots + t_k)^2\right] \exp\left[-\tfrac{1}{2}(t_2 + \cdots + t_k)^2\right] \cdots \exp\left(-\tfrac{1}{2}t_k^2\right), \tag{7}$$

since the ch.f.'s of

$$\sqrt{\frac{k}{n}}S_{n_1}, \sqrt{\frac{k}{n}}(S_{n_2} - S_{n_1}), \dots, \sqrt{\frac{k}{n}}(S_{n_k} - S_{n_{k-1}})$$

all converge to $e^{-t^2/2}$ by the central limit theorem (Theorem 6.4.4), $n_{j+1} - n_j$ being asymptotically equal to $n/k$ for each $j$. It is well known that the ch.f. given in (7) is that of the $k$-dimensional normal distribution, but for our purpose it is sufficient to know that the convergence theorem for ch.f.'s holds in any dimension, and so $R_{nk}$ converges vaguely to $R_{\infty k}$, where $R_{\infty k}$ is some fixed $k$-dimensional distribution.

Now suppose for a special sequence $\{\tilde{X}_j\}$ satisfying the same conditions as $\{X_j\}$, the corresponding $\tilde{P}_n$ can be shown to converge ("pointwise" in fact, but "vaguely" if need be):

$$\forall x: \lim_{n \to \infty} \tilde{P}_n(x) = G(x). \tag{8}$$

Then, applying (6) with $P_n$ replaced by $\tilde{P}_n$ and letting $n \to \infty$, we obtain, since $R_{\infty k}$ is a fixed distribution:

$$G(x) \le R_{\infty k}(x) \le G(x + \epsilon) + \frac{1}{\epsilon^2 k}.$$

Substituting back into the original (6) and taking upper and lower limits, we have

$$\begin{aligned}G(x - \epsilon) - \frac{1}{\epsilon^2 k} \le R_{\infty k}(x - \epsilon) - \frac{1}{\epsilon^2 k} &\le \overline{\lim_n} P_n(x) \\ &\le R_{\infty k}(x) \le G(x + \epsilon) + \frac{1}{\epsilon^2 k}.\end{aligned}$$

Letting $k \to \infty$, we conclude that $P_n$ converges vaguely to $G$, since $\epsilon$ is arbitrary.

It remains to prove (8) for a special choice of $\{X_j\}$ and determine $G$. This can be done most expeditiously by taking the common d.f. of the $X_j$'s to be the symmetric Bernoullian $\frac{1}{2}(\delta_1 + \delta_{-1})$. In this case we can indeed compute the more specific probability

$$\mathscr{P}\left\{\max_{1\le m\le n} S_m < x; S_n = y\right\}, \tag{9}$$

where $x$ and $y$ are two integers such that $x > 0, x > y$. If we observe that in our particular case $\max_{1\le m\le n} S_m \ge x$ if and only if $S_j = x$ for some $j$, $1 \le j \le n$, the probability in (9) is seen to be equal to

$$\begin{aligned}
&\mathscr{P}\{S_n = y\} - \mathscr{P}\left\{\max_{1\le m\le n} S_m \ge x; S_n = y\right\}\\
&\quad= \mathscr{P}\{S_n = y\} - \sum_{j=1}^{n} \mathscr{P}\{S_m < x, 1 \le m < j; S_j = x; S_n = y\}\\
&\quad= \mathscr{P}\{S_n = y\} - \sum_{j=1}^{n} \mathscr{P}\{S_m < x, 1 \le m < j; S_j = x; S_n - S_j = y - x\}\\
&\quad= \mathscr{P}\{S_n = y\} - \sum_{j=1}^{n} \mathscr{P}\{S_m < x, 1 \le m < j; S_j = x\}\mathscr{P}\{S_n - S_j = y - x\},
\end{aligned}$$

where the last step is by independence. Now, the r.v.

$$S_n - S_j = \sum_{m=j+1}^{n} X_m$$

being symmetric, we have $\mathscr{P}\{S_n - S_j = y - x\} = \mathscr{P}\{S_n - S_j = x - y\}$. Substituting this and reversing the steps, we obtain

$$\begin{aligned}
&\mathscr{P}\{S_n = y\} - \sum_{j=1}^{n} \mathscr{P}\{S_m < x, 1 \le m < j; S_j = x\}\mathscr{P}\{S_n - S_j = x - y\}\\
&\quad= \mathscr{P}\{S_n = y\} - \sum_{j=1}^{n} \mathscr{P}\{S_m < x, 1 \le m < j; S_j = x; S_n - S_j = x - y\}\\
&\quad= \mathscr{P}\{S_n = y\} - \sum_{j=1}^{n} \mathscr{P}\{S_m < x, 1 \le m < j; S_j = x; S_n = 2x - y\}\\
&\quad= \mathscr{P}\{S_n = y\} - \mathscr{P}\left\{\max_{1\le m\le n} S_m \ge x; S_n = 2x - y\right\}.
\end{aligned}$$

Since $2x - y > x$, $S_n = 2x - y$ implies $\max_{1\le m\le n} S_m \ge x$, hence the last line reduces to

$$\mathscr{P}\{S_n = y\} - \mathscr{P}\{S_n = 2x - y\}, \tag{10}$$

and we have proved that the value of the probability in (9) is given by (10). The trick used above, which consists in changing the sign of every $X_j$ after the first time when $S_n$ reaches the value $x$, or, geometrically speaking, reflecting the path $\{(j, S_j), j \geq 1\}$ about the line $S_j = x$ upon first reaching it, is called the "reflection principle". It is often attributed to Desiré André in its combinatorial formulation of the so-called "ballot problem" (see Exercise 5 below).

The value of (10) is, of course, well known in the Bernoullian case, and summing over $y$ we obtain, if $n$ is even:

$$\begin{aligned}
\mathscr{P}\left\{\max_{1 \leq m \leq n} S_m < x\right\} &= \sum_{y<x} \frac{1}{2^n}\left\{\binom{n}{\frac{n-y}{2}} - \binom{n}{\frac{n-2x+y}{2}}\right\} \\
&= \sum_{y<x} \frac{1}{2^n}\left\{\binom{n}{\frac{n-y}{2}} - \binom{n}{\frac{n+2x-y}{2}}\right\} \\
&= \frac{1}{2^n} \sum_{\frac{n-x}{2} < j \leq \frac{n+x}{2}} \binom{n}{j} \\
&= \frac{1}{2^n} \sum_{|j-\frac{n}{2}|<\frac{x}{2}} \binom{n}{j} + \frac{1}{2^n}\binom{n}{\frac{n+x}{2}},
\end{aligned}$$

where $\binom{n}{j} = 0$ if $|j| > n$ or if $j$ is not an integer. Replacing $x$ by $x\sqrt{n}$ (or $[x\sqrt{n}]$ if one is pedantic) in the last expression, and using the central limit theorem for the Bernoullian case in the form of (12) of Sec. 7.2 with $p = q = \frac{1}{2}$, we see that the preceding probability tends to the limit

$$\frac{1}{\sqrt{2\pi}} \int_{-x}^{x} e^{-y^2/2}\, dy = \sqrt{\frac{2}{\pi}} \int_0^x e^{-y^2/2}\, dy$$

as $n \to \infty$. It should be obvious, even without a similar calculation, that the same limit obtains for odd values of $n$. Finally, since this limit as a function of $x$ is a d.f. with support in $(0, \infty)$, the corresponding limit for $x \leq 0$ must be 0. We state the final result below.

**Theorem 7.3.3.** Let $\{X_j, j \geq 0\}$ be independent and identically distributed r.v.'s with mean 0 and variance 1, then $(\max_{1 \leq m \leq n} S_m)/\sqrt{n}$ converges in dist. to the "positive normal d.f." $G$, where

$$\forall x\colon G(x) = (2\Phi(x) - 1) \vee 0.$$

EXERCISES

**1.** Let $\{X_j, j \geq 1\}$ be a sequence of independent r.v.'s, and $f$ a Borel measurable function of $m$ variables. Then if $\xi_k = f(X_{k+1}, \dots, X_{k+m})$, the sequence $\{\xi_k, k \geq 1\}$ is $(m-1)$-dependent.

$^\star$**2.** Let $\{X_j, j \geq 1\}$ be a sequence of independent r.v.'s having the Bernoullian d.f. $p\delta_1 + (1-p)\delta_0$, $0 < p < 1$. An *r-run* of successes in the sample sequence $\{X_j(\omega), j \geq 1\}$ is defined to be a sequence of $r$ consecutive "ones" preceded and followed by "zeros". Let $N_n$ be the number of $r$-runs in the first $n$ terms of the sample sequence. Prove a central limit theorem for $N_n$.

**3.** Let $\{X_j, \nu_j, j \geq 1\}$ be independent r.v.'s such that the $\nu_j$'s are integer-valued, $\nu_j \to \infty$ a.e., and the central limit theorem applies to $(S_n - a_n)/b_n$, where $S_n = \sum_{j=1}^n X_j$, $a_n$, $b_n$ are real constants, $b_n \to \infty$. Then it also applies to $(S_{\nu_n} - a_{\nu_n})/b_{\nu_n}$.

$^\star$**4.** Give an example of a sequence of independent and identically distributed r.v.'s $\{X_n\}$ with mean 0 and variance 1 and a sequence of positive integer-valued r.v.'s $\nu_n$ tending to $\infty$ a.e. such that $S_{\nu_n}/s_{\nu_n}$ does not converge in distribution. [HINT: The easiest way is to use Theorem 8.3.3 below.]

$^\star$**5.** There are $a$ ballots marked A and $b$ ballots marked B. Suppose that these $a+b$ ballots are counted in random order. What is the probability that the number of ballots for A always leads in the counting?

**6.** If $\{X_n\}$ are independent, identically distributed symmetric r.v.'s, then for every $x \geq 0$,

$$\mathscr{P}\{|S_n| > x\} \geq \tfrac{1}{2}\mathscr{P}\{\max_{1\leq k\leq n} |X_k| > x\} \geq \tfrac{1}{2}[1 - e^{-n\mathscr{P}\{|X_1|>x\}}].$$

**7.** Deduce from Exercise 6 that for a symmetric stable r.v. $X$ with exponent $\alpha$, $0 < \alpha < 2$ (see Sec. 6.5), there exists a constant $c > 0$ such that $\mathscr{P}\{|X| > n^{1/\alpha}\} \geq c/n$. [This is due to Feller; use Exercise 3 of Sec. 6.5.]

**8.** Under the same hypothesis as in Theorem 7.3.3, prove that $\max_{1\leq m\leq n} |S_m|/\sqrt{n}$ converges in dist. to the d.f. $H$, where $H(x) = 0$ for $x \leq 0$ and

$$H(x) = \frac{4}{\pi}\sum_{k=0}^{\infty} \frac{(-1)^k}{2k+1} \exp\left[-\frac{(2k+1)^2\pi^2}{8x^2}\right] \quad \text{for } x > 0.$$

[HINT: There is no difficulty in proving that the limiting distribution is the same for any sequence satisfying the hypothesis. To find it for the symmetric Bernoullian case, show that for $0 < z < x$ we have

$$\mathscr{P}\left\{-z < \min_{1\leq m\leq n} S_m \leq \max_{1\leq m\leq n} S_m < x - z; S_n = y - z\right\}$$

$$= \frac{1}{2^n} \sum_{k=-\infty}^{\infty} \left\{ \binom{n}{\frac{n+2kx+y-z}{2}} - \binom{n}{\frac{n+2kx-y-z}{2}} \right\}.$$

This can be done by starting the sample path at $z$ and reflecting at both barriers 0 and $x$ (Kelvin's method of images). Next, show that

$$\lim_{n\to\infty} \mathscr{P}\{-z\sqrt{n} < S_m < (x-z)\sqrt{n} \quad \text{for } 1 \le m \le n\}$$

$$= \frac{1}{\sqrt{2\pi}} \sum_{k=-\infty}^{\infty} \left\{ \int_{2kx-z}^{(2k+1)x-z} - \int_{(2k-1)x-z}^{2kx-z} \right\} e^{y^2/2}\, dy.$$

Finally, use the Fourier series for the function $h$ of period $2x$:

$$h(y) = \begin{cases} -1, & \text{if } -x-z < y < -z; \\ +1, & \text{if } -z < y < x-z; \end{cases}$$

to convert the above limit to

$$\frac{4}{\pi} \sum_{k=0}^{\infty} \frac{1}{2k+1} \sin \frac{(2k+1)\pi z}{x} \exp\left[-\frac{(2k+1)^2\pi^2}{2x^2}\right].$$

This gives the asymptotic joint distribution of

$$\min_{1\le m\le n} S_m \quad \text{and} \quad \max_{1\le m\le n} S_m,$$

of which that of $\max_{1\le m\le n} |S_m|$ is a particular case.

**9.** Let $\{X_j \ge 1\}$ be independent r.v.'s with the symmetric Bernoullian distribution. Let $N_n(\omega)$ be the number of zeros in the first $n$ terms of the sample sequence $\{S_j(\omega),\ j \ge 1\}$. Prove that $N_n/\sqrt{n}$ converges in dist. to the same $G$ as in Theorem 7.3.3. [HINT: Use the method of moments. Show that for each integer $r \ge 1$:

$$\mathscr{E}(N_n^r) \sim r! \sum_{0<j_1<\cdots<j_r\le n/2} p_{2j_1} p_{2(j_2-j_1)} \cdots p_{2(j_r-j_{r-1})}$$

where

$$p_{2j} = \mathscr{P}\{S_{2j} = 0\} = \binom{2j}{j} \frac{1}{2^{2j}} \sim \frac{1}{\sqrt{\pi j}}.$$

as $j \to \infty$. To evaluate the multiple sum, say $\sum(r)$, use induction on $r$ as follows. If

$$\sum(r) \sim c_r \left(\frac{n}{2}\right)^{r/2}$$

as $n \to \infty$, then

$$\sum(r+1) \sim \frac{c_r}{\sqrt{\pi}} \int_0^1 z^{-1/2}(1-z)^{r/2} dz \left(\frac{n}{2}\right)^{(r+1)/2}.$$

Thus

$$c_{r+1} = c_r \frac{\Gamma\left(\frac{r}{2}+1\right)}{\Gamma\left(\frac{r+1}{2}+1\right)}.$$

Finally

$$\mathscr{E}\left\{\left(\frac{N_n}{\sqrt{n}}\right)^r\right\} \to \frac{\Gamma(r+1)}{2^{r/2}\Gamma\left(\frac{r}{2}+1\right)} = \frac{2^{r/2}\Gamma\left(\frac{r+1}{2}\right)}{\Gamma\left(\frac{1}{2}\right)} = \int_0^\infty x^r dG(x).$$

This result remains valid if the common d.f. $F$ of $X_j$ is of the integer lattice type with mean 0 and variance 1. If $F$ is not of the lattice type, no $S_n$ need ever be zero—but the "next nearest thing", to wit the number of changes of sign of $S_n$, is asymptotically distributed as $G$, at least under the additional assumption of a finite third absolute moment.]

## 7.4 Error estimation

Questions of convergence lead inevitably to the question of the "speed" of convergence—in other words, to an investigation of the difference between the approximating expression and its limit. Specifically, if a sequence of d.f.'s $F_n$ converge to the unit normal d.f. $\Phi$, as in the central limit theorem, what can one say about the "remainder term" $F_n(x) - \Phi(x)$? An adequate estimate of this term is necessary in many mathematical applications, as well as for numerical computation. Under Liapounov's condition there is a neat "order bound" due to Berry and Esseen, who improved upon Liapounov's older result, as follows.

**Theorem 7.4.1.** Under the hypotheses of Theorem 7.1.2, there is a universal constant $A_0$ such that

(1) $$\sup_x |F_n(x) - \Phi(x)| \le A_0 \Gamma_n$$

where $F_n$ is the d.f. of $S_n$.

In the case of a single sequence of independent and identically distributed r.v.'s $\{X_j, j \geq 1\}$ with mean 0, variance $\sigma^2$, and third absolute moment $\gamma < \infty$, the right side of (1) reduces to

$$A_0 \frac{n\gamma}{(n\sigma^2)^{3/2}} = \frac{A_0 \gamma}{\sigma^3} \frac{1}{n^{1/2}}.$$

H. Cramér and P. L. Hsu have shown that under somewhat stronger conditions, one may even obtain an asymptotic expansion of the form:

$$F_n(x) = \Phi(x) + \frac{H_1(x)}{n^{1/2}} + \frac{H_2(x)}{n} + \frac{H_3(x)}{n^{3/2}} + \cdots,$$

where the $H$'s are explicit functions involving the Hermite polynomials. We shall not go into this, as the basic method of obtaining such an expansion is similar to the proof of the preceding theorem, although considerable technical complications arise. For this and other variants of the problem see Cramér [10], Gnedenko and Kolmogorov [12], and Hsu's paper cited at the end of this chapter.

We shall give the proof of Theorem 7.4.1 in a series of lemmas, in which the machinery of operating with ch.f.'s is further exploited and found to be efficient.

**Lemma 1.** Let $F$ be a d.f., $G$ a real-valued function satisfying the conditions below:

**(i)** $\lim_{x \to -\infty} G(x) = 0$, $\lim_{x \to +\infty} G(x) = 1$;
**(ii)** $G$ has a derivative that is bounded everywhere: $\sup_x |G'(x)| \leq M$.

Set

$$\Delta = \frac{1}{2M} \sup_x |F(x) - G(x)|. \tag{2}$$

Then there exists a real number $a$ such that we have for every $T > 0$:

$$\begin{aligned} &2MT\Delta \left\{ 3 \int_0^{T\Delta} \frac{1 - \cos x}{x^2} \, dx - \pi \right\} \\ &\leq \left| \int_{-\infty}^{\infty} \frac{1 - \cos Tx}{x^2} \{F(x + a) - G(x + a)\} \, dx \right|. \end{aligned} \tag{3}$$

PROOF. Clearly the $\Delta$ in (2) is finite, since $G$ is everywhere bounded by (i) and (ii). We may suppose that the left member of (3) is strictly positive, for otherwise there is nothing to prove; hence $\Delta > 0$. Since $F - G$ vanishes at $\pm\infty$ by (i), there exists a sequence of numbers $\{x_n\}$ converging to a finite limit

$b$ such that $F(x_n) - G(x_n)$ converges to $2M\Delta$ or $-2M\Delta$. Hence either $F(b) - G(b) = 2M\Delta$ or $F(b-) - G(b) = -2M\Delta$. The two cases being similar, we shall treat the second one. Put $a = b - \Delta$; then if $|x| < \Delta$, we have by (ii) and the mean value theorem of differential calculus:

$$G(x + a) \geq G(b) + (x - \Delta)M$$

and consequently

$$F(x + a) - G(x + a) \leq F(b-) - [G(b) + (x - \Delta)M] = -M(x + \Delta).$$

It follows that

$$\int_{-\Delta}^{\Delta} \frac{1 - \cos Tx}{x^2}\{F(x + a) - G(x + a)\}\,dx \leq -M \int_{-\Delta}^{\Delta} \frac{1 - \cos Tx}{x^2}(x + \Delta)\,dx$$

$$= -2M\Delta \int_0^{\Delta} \frac{1 - \cos Tx}{x^2}\,dx;$$

$$\left|\left\{\int_{-\infty}^{-\Delta} + \int_{\Delta}^{\infty}\right\} \frac{1 - \cos Tx}{x^2}\{F(x + a) - G(x + a)\}\,dx\right|$$

$$\leq 2M\Delta \left(\int_{-\infty}^{-\Delta} + \int_{\Delta}^{\infty}\right) \frac{1 - \cos Tx}{x^2}\,dx = 4M\Delta \int_{\Delta}^{\infty} \frac{1 - \cos Tx}{x^2}\,dx.$$

Adding these inequalities, we obtain

$$\int_{-\infty}^{\infty} \frac{1 - \cos Tx}{x^2}\{F(x + a) - G(x + a)\}\,dx \leq 2M\Delta \left\{-\int_0^{\Delta} + 2\int_{\Delta}^{\infty}\right\}$$

$$\frac{1 - \cos Tx}{x^2}\,dx = 2M\Delta \left\{-3\int_0^{\Delta} + 2\int_0^{\infty}\right\} \frac{1 - \cos Tx}{x^2}\,dx.$$

This reduces to (3), since

$$\int_0^{\infty} \frac{1 - \cos Tx}{x^2}\,dx = \frac{\pi T}{2}$$

by (3) of Sec. 6.2, provided that $T$ is so large that the left member of (3) is positive; otherwise (3) is trivial.

**Lemma 2.** In addition to the assumptions of Lemma 1, we assume that

**(iii)** $G$ is of bounded variation in $(-\infty, \infty)$;
**(iv)** $\int_{-\infty}^{\infty} |F(x) - G(x)|\,dx < \infty$.

Let

$$f(t) = \int_{-\infty}^{\infty} e^{itx}\,dF(x), \quad g(t) = \int_{-\infty}^{\infty} e^{itx}\,dG(x).$$

Then we have

$$\Delta \leq \frac{1}{\pi M} \int_0^T \frac{|f(t) - g(t)|}{t} \, dt + \frac{12}{\pi T}. \tag{4}$$

PROOF. That the integral on the right side of (4) is finite will soon be apparent, and as a Lebesgue integral the value of the integrand at $t = 0$ may be overlooked. We have, by a partial integration, which is permissible on account of condition (iii):

$$f(t) - g(t) = -it \int_{-\infty}^{\infty} \{F(x) - G(x)\} e^{itx} \, dx; \tag{5}$$

and consequently

$$\frac{f(t) - g(t)}{-it} e^{-ita} = \int_{-\infty}^{\infty} \{F(x+a) - G(x+a)\} e^{itx} \, dx.$$

In particular, the left member above is bounded for all $t \neq 0$ by condition (iv). Multiplying by $T - |t|$ and integrating, we obtain

$$\begin{aligned} \int_{-T}^{T} \frac{f(t) - g(t)}{-it} e^{-ita} (T - |t|) \, dt \\ = \int_{-T}^{T} \int_{-\infty}^{\infty} \{F(x+a) - G(x+a)\} e^{itx} (T - |t|) \, dx \, dt. \end{aligned} \tag{6}$$

We may invert the repeated integral by Fubini's theorem and condition (iv), and obtain (cf. Exercise 2 of Sec. 6.2):

$$\left| \int_{-\infty}^{\infty} \{F(x+a) - G(x+a)\} \frac{1 - \cos Tx}{x^2} \, dx \right| \leq T \int_0^T \frac{|f(t) - g(t)|}{t} dt.$$

In conjunction with (3), this yields

$$2M\Delta \left\{ 3 \int_0^{T\Delta} \frac{1 - \cos x}{x^2} \, dx - \pi \right\} \leq \int_0^T \frac{|f(t) - g(t)|}{t} dt. \tag{7}$$

The quantity in braces in (7) is not less than

$$3 \int_0^{\infty} \frac{1 - \cos x}{x^2} \, dx - 3 \int_{T\Delta}^{\infty} \frac{2}{x^2} \, dx - \pi = \frac{\pi}{2} - \frac{6}{T\Delta}.$$

Using this in (7), we obtain (4).

Lemma 2 bounds the maximum difference between two d.f.'s satisfying certain regularity conditions by means of a certain average difference between their ch.f.'s (It is stated for a function of bounded variation, since this is needed for the asymptotic expansion mentioned above). This should be compared with the general discussion around Theorem 6.3.4. We shall now apply the lemma to the specific case of $F_n$ and $\Phi$ in Theorem 7.4.1. Let the ch.f. of $F_n$ be $f_n$, so that

$$f_n(t) = \prod_{j=1}^{k_n} f_{nj}(t).$$

**Lemma 3.** For $|t| < 1/(2\Gamma_n^{1/3})$, we have

$$|f_n(t) - e^{-t^2/2}| \leq \Gamma_n |t|^3 e^{-t^2/2}. \tag{8}$$

PROOF. We shall denote by $\theta$ below a "generic" complex number with $|\theta| \leq 1$, otherwise unspecified and not necessarily the same at each appearance. By Taylor's expansion:

$$f_{nj}(t) = 1 - \frac{\sigma_{nj}^2}{2} t^2 + \theta \frac{\gamma_{nj}}{6} t^3.$$

For the range of $t$ given in the lemma, we have by Liapounov's inequality:

$$|\sigma_{nj} t| \leq |\gamma_{nj}{}^{1/3} t| \leq |\Gamma_n{}^{1/3} t| < \tfrac{1}{2}, \tag{9}$$

so that

$$\left| -\frac{\sigma_{nj}^2}{2} t^2 + \frac{\theta \gamma_{nj} t^3}{6} \right| < \frac{1}{8} + \frac{1}{48} < \frac{1}{4}.$$

Using (8) of Sec. 7.1 with $\Lambda = \theta/2$, we may write

$$\log f_{nj}(t) = -\frac{\sigma_{nj}^2}{2} t^2 + \frac{\theta \gamma_{nj}}{6} t^3 + \frac{\theta}{2} \left( -\frac{\sigma_{nj}^2 t^2}{2} + \frac{\theta \gamma_{nj} t^3}{6} \right)^2.$$

The absolute value of the last term above is less than

$$\frac{\sigma_{nj}^4 t^4}{4} + \frac{\gamma_{nj}^2 t^6}{36} \leq \left( \frac{\sigma_{nj}|t|}{4} + \frac{\gamma_{nj}|t|^3}{36} \right) \gamma_{nj}|t|^3 \leq \left( \frac{1}{4 \cdot 2} + \frac{1}{36 \cdot 8} \right) \gamma_{nj}|t|^3$$

by (9); hence

$$\log f_{nj}(t) = -\frac{\sigma_{nj}^2}{2} t^2 + \theta \left( \frac{1}{6} + \frac{1}{8} + \frac{1}{288} \right) \gamma_{nj}|t|^3 = -\frac{\sigma_{nj}^2}{2} t^2 + \frac{\theta}{2} \gamma_{nj} t^3.$$

Summing over $j$, we have

$$\log f_n(t) = -\frac{t^2}{2} + \frac{\theta}{2} \Gamma_n t^3,$$

or explicitly:

$$\left|\log f_n(t) + \frac{t^2}{2}\right| \le \frac{1}{2}\Gamma_n|t|^3.$$

Since $|e^u - 1| \le |u|e^{|u|}$ for all $u$, it follows that

$$|f_n(t)e^{t^2/2} - 1| \le \frac{\Gamma_n|t|^3}{2}\exp\left[\frac{\Gamma_n|t|^3}{2}\right].$$

Since $\Gamma_n|t|^3/2 \le 1/16$ and $e^{1/16} \le 2$, this implies (8).

**Lemma 4.** For $|t| < 1/(4\Gamma_n)$, we have

$$|f_n(t)| \le e^{-t^2/3}. \tag{10}$$

PROOF. We symmetrize (see end of Sec. 6.2) to facilitate the estimation. We have

$$|f_{nj}(t)|^2 = \int_{-\infty}^{\infty}\int_{-\infty}^{\infty} \cos t(x-y)\,dF_{nj}(x)\,dF_{nj}(y),$$

since $|f_{nj}|^2$ is real. Using the elementary inequalities

$$\left|\cos u - 1 + \frac{u^2}{2}\right| \le \frac{|u|^3}{6},$$

$$|x-y|^3 \le 4(|x|^3 + |y|^3);$$

we see that the double integral above does not exceed

$$\int_{-\infty}^{\infty}\int_{-\infty}^{\infty}\left\{1 - \frac{t^2}{2}(x^2 - 2xy + y^2) + \frac{2}{3}|t|^3(|x|^3 + |y|^3)\right\} dF_{nj}(x)\,dF_{nj}(y)$$

$$= 1 - \sigma_{nj}^2 t^2 + \frac{4}{3}\gamma_{nj}|t|^3 \le \exp\left(-\sigma_{nj}^2 t^2 + \frac{4}{3}\gamma_{nj}|t|^3\right).$$

Multiplying over $j$, we obtain

$$|f_n(t)|^2 \le \exp\left(-t^2 + \frac{4}{3}\Gamma_n|t|^3\right) \le e^{-(2/3)t^2}$$

for the range of $t$ specified in the lemma, proving (10).

Note that Lemma 4 is weaker than Lemma 3 but valid in a wider range. We now combine them.

**Lemma 5.** For $|t| < 1/(4\Gamma_n)$, we have

$$|f_n(t) - e^{-t^2/2}| \le 16\Gamma_n|t|^3 e^{-t^2/3}. \tag{11}$$

PROOF. If $|t| < 1/(2\Gamma_n{}^{1/3})$, this is implied by (8). If $1/(2\Gamma_n{}^{1/3}) \le |t| < 1/(4\Gamma_n)$, then $1 \le 8\Gamma_n|t|^3$, and so by (10):

$$|f_n(t) - e^{-t^2/2}| \le |f_n(t)| + e^{-t^2/2} \le 2e^{-t^2/3} \le 16\Gamma_n|t|^3e^{-t^2/3}.$$

PROOF OF THEOREM 7.4.1. Apply Lemma 2 with $F = F_n$ and $G = \Phi$. The $M$ in condition (ii) of Lemma 1 may be taken to be $\frac{1}{2}\pi$, since both $F_n$ and $\Phi$ have mean 0 and variance 1, it follows from Chebyshev's inequality that

$$F(x) \vee G(x) \le \frac{1}{x^2}, \qquad \text{if } x < 0,$$
$$(1 - F(x)) \vee (1 - G(x)) \le \frac{1}{x^2}, \qquad \text{if } x > 0;$$

and consequently

$$\forall x: |F(x) - G(x)| \le \frac{1}{x^2}.$$

Thus condition (iv) of Lemma 2 is satisfied. In (4) we take $T = 1/(4\Gamma_n)$; we have then from (4) and (11):

$$\begin{aligned}
\sup_x |F_n(x) - \Phi(x)| &\le \frac{2}{\pi}\int_0^{1/(4\Gamma_n)} \frac{|f_n(t) - e^{-t^2/2}|}{t}dt + \frac{96}{\sqrt{2\pi^3}}\Gamma_n \\
&\le \frac{32\Gamma_n}{\pi}\int_0^{1/(4\Gamma_n)} t^2e^{-t^2/3}dt + \frac{96}{\sqrt{2\pi^3}}\Gamma_n \\
&\le \Gamma_n\left\{\frac{32}{\pi}\int_0^{\infty} t^2e^{-t^2/3}dt + \frac{96}{\sqrt{2\pi^3}}\right\}.
\end{aligned}$$

This establishes (1) with a numerical value for $A_0$ (which may be somewhat improved).

Although Theorem 7.4.1 gives the best possible uniform estimate of the remainder $F_n(x) - \Phi(x)$, namely one that does not depend on $x$, it becomes less useful if $x = x_n$ increases with $n$ even at a moderate speed. For instance, we have

$$1 - F_n(x_n) = \frac{1}{\sqrt{2\pi}}\int_{x_n}^{\infty} e^{-y^2/2}dy + O(\Gamma_n),$$

where the first "principal" term on the right is asymptotically equal to

$$\frac{1}{\sqrt{2\pi}x_n}e^{-x_n^2/2}.$$

Hence already when $x_n = \sqrt{2\log(1/\Gamma_n)}$ this will be $o(\Gamma_n)$ for $\Gamma_n \to 0$ and absorbed by the remainder. For such "large deviations", what is of interest is an asymptotic evaluation of

$$\frac{1 - F_n(x_n)}{1 - \Phi(x_n)}$$

as $x_n \to \infty$ more rapidly than indicated above. This requires a different type of approximation by means of "bilateral Laplace transforms", which will not be discussed here.

EXERCISES

**1.** If $F$ and $G$ are d.f.'s with finite first moments, then

$$\int_{-\infty}^{\infty} |F(x) - G(x)| \, dx < \infty.$$

[HINT: Use Exercise 18 of Sec. 3.2.]

**2.** If $f$ and $g$ are ch.f.'s such that $f(t) = g(t)$ for $|t| \leq T$, then

$$\int_{-\infty}^{\infty} |F(x) - G(x)| \, dx \leq \frac{\pi}{T}.$$

This is due to Esseen (*Acta Math*, **77**(1944)).

$^\star$**3.** There exists a universal constant $A_1 > 0$ such that for any sequence of independent, identically distributed integer-valued r.v.'s $\{X_j\}$ with mean 0 and variance 1, we have

$$\sup_x |F_n(x) - \Phi(x)| \geq \frac{A_1}{n^{1/2}},$$

where $F_n$ is the d.f. of $(\sum_{j=1}^n X_j)/\sqrt{n}$. [HINT: Use Exercise 24 of Sec. 6.4.]

**4.** Prove that for every $x > 0$:

$$\frac{x}{1 + x^2} e^{-x^2/2} \leq \int_x^{\infty} e^{-y^2/2} \, dy \leq \frac{1}{x} e^{-x^2/2}.$$

## 7.5 Law of the iterated logarithm

*The law of the iterated logarithm* is a crowning achievement in classical probability theory. It had its origin in attempts to perfect Borel's theorem on normal numbers (Theorem 5.1.3). In its simplest but basic form, this asserts: if $N_n(\omega)$ denotes the number of occurrences of the digit 1 in the first $n$ places of the binary (dyadic) expansion of the real number $\omega$ in [0, 1], then $N_n(\omega) \sim n/2$ for almost every $\omega$ in Borel measure. What can one say about the deviation $N_n(\omega) - n/2$? The order bounds $O(n^{(1/2)+\epsilon})$, $\epsilon > 0$; $O((n \log n)^{1/2})$ (cf.

Theorem 5.4.1); and $O((n \log \log n)^{1/2})$ were obtained successively by Hausdorff (1913), Hardy and Littlewood (1914), and Khintchine (1922); but in 1924 Khintchine gave the definitive answer:

$$\overline{\lim_{n\to\infty}} \frac{N_n(\omega) - \dfrac{n}{2}}{\sqrt{\dfrac{1}{2} n \log \log n}} = 1$$

for almost every $\omega$. This sharp result with such a fine order of infinity as "log log" earned its celebrated name. No less celebrated is the following extension given by Kolmogorov (1929). Let $\{X_n, n \geq 1\}$ be a sequence of independent r.v.'s, $S_n = \sum_{j=1}^{n} X_j$; suppose that $\mathscr{E}(X_n) = 0$ for each $n$ and

$$(1) \qquad \sup_{\omega} |X_n(\omega)| = o\left(\frac{s_n}{\sqrt{\log \log s_n}}\right),$$

where $s_n^2 = \sigma^2(S_n)$, then we have for almost every $\omega$:

$$(2) \qquad \overline{\lim_{n\to\infty}} \frac{S_n(\omega)}{\sqrt{2s_n^2 \log \log s_n}} = 1.$$

The condition (1) was shown by Marcinkiewicz and Zygmund to be of the best possible kind, but an interesting complement was added by Hartman and Wintner that (2) also holds if the $X_n$'s are identically distributed with a finite second moment. Finally, further sharpening of (2) was given by Kolmogorov and by Erdös in the Bernoullian case, and in the general case under exact conditions by Feller; the "last word" being as follows: for any increasing sequence $\varphi_n$, we have

$$\mathscr{P}\{S_n(\omega) > s_n \varphi_n \text{ i.o.}\} = \begin{cases} 0 \\ 1 \end{cases}$$

according as the series

$$\sum_{n=1}^{\infty} \frac{\varphi_n}{n} e^{-\varphi_n{}^2/2} \begin{cases} < \\ = \end{cases} \infty.$$

We shall prove the result (2) under a condition different from (1) and apparently overlapping it. This makes it possible to avoid an intricate estimate concerning "large deviations" in the central limit theorem and to replace it by an immediate consequence of Theorem 7.4.1.* It will become evident that the

* An alternative which bypasses Sec. 7.4 is to use Theorem 7.1.3; the details are left as an exercise.

proof of such a "strong limit theorem" (bounds with probability one) as the law of the iterated logarithm depends essentially on the corresponding "weak limit theorem" (convergence of distributions) with a sufficiently good estimate of the remainder term.

The vital link just mentioned will be given below as Lemma 1. In the rest of this section "$A$" will denote a generic strictly positive constant, not necessarily the same at each appearance, and $\Lambda$ will denote a constant such that $|\Lambda| \leq A$. We shall also use the notation in the preceding statement of Kolmogorov's theorem, and

$$\gamma_n = \mathscr{E}(|X_n|^3), \quad \Gamma_n = \sum_{j=1}^{n} \gamma_j$$

as in Sec. 7.4, but for a single sequence of r.v.'s. Let us set also

$$\varphi(\lambda, x) = \sqrt{2\lambda x^2 \log\log x}, \quad \lambda > 0, x > 0.$$

**Lemma 1.** Suppose that for some $\epsilon$, $0 < \epsilon < 1$, we have

$$\frac{\Gamma_n}{s_n^3} \leq \frac{A}{(\log s_n)^{1+\epsilon}}. \tag{3}$$

Then for each $\delta$, $0 < \delta < \epsilon$, we have

$$\mathscr{P}\{S_n > \varphi(1+\delta, s_n)\} \leq \frac{A}{(\log s_n)^{1+\delta}}; \tag{4}$$

$$\mathscr{P}\{S_n > \varphi(1-\delta, s_n)\} \geq \frac{A}{(\log s_n)^{1-(\delta/2)}}. \tag{5}$$

PROOF. By Theorem 7.4.1, we have for each $x$:

$$\mathscr{P}\{S_n > xs_n\} = \frac{1}{\sqrt{2\pi}} \int_x^\infty e^{-y^2/2}\, dy + \Lambda \frac{\Gamma_n}{s_n^3}. \tag{6}$$

We have as $x \to \infty$:

$$\int_x^\infty e^{-y^2/2}\, dy \sim \frac{e^{-x^2/2}}{x}. \tag{7}$$

(See Exercise 4 of Sec. 7.4). Substituting $x = \sqrt{2(1 \pm \delta)\log\log s_n}$, the first term on the right side of (6) is, by (7), asymptotically equal to

$$\frac{1}{\sqrt{4\pi(1 \pm \delta)\log\log s_n}} \frac{1}{(\log s_n)^{1\pm\delta}}.$$

This dominates the second (remainder) term on the right side of (6), by (3) since $0 < \delta < \epsilon$. Hence (4) and (5) follow as rather weak consequences.

To establish (2), let us write for each fixed $\delta, 0 < \delta < \epsilon$:

$$E_n{}^+ = \{\omega: S_n(\omega) > \varphi(1+\delta, s_n)\},$$

$$E_n{}^- = \{\omega: S_n(\omega) > \varphi(1-\delta, s_n)\},$$

and proceed by steps.

1°. We prove first that

$$\mathscr{P}\{E_n{}^+ \text{ i.o.}\} = 0 \tag{8}$$

in the notation introduced in Sec. 4.2, by using the convergence part of the Borel–Cantelli lemma there. But it is evident from (4) that the series $\sum_n \mathscr{P}(E_n{}^+)$ is far from convergent, since $s_n$ is expected to be of the order of $\sqrt{n}$. The main trick is to apply the lemma to a crucial subsequence $\{n_k\}$ (see Theorem 5.1.2 for a crude form of the trick) chosen to have two properties: first, $\sum_k \mathscr{P}(E_{n_k}{}^+)$ converges, and second, "$E_n{}^+$ i.o." already implies "$E_{n_k}{}^+$ i.o." *nearly*, namely if the given $\delta$ is slightly decreased. This modified implication is a consequence of a simple but essential probabilistic argument spelled out in Lemma 2 below.

Given $c > 1$, let $n_k$ be the largest value of $n$ satisfying $s_n \le c^k$, so that

$$s_{n_k} \le c^k < s_{n_k+1}.$$

Since $(\max_{1\le j\le n} \sigma_j)/s_n \to 0$ (why?), we have $s_{n_k+1}/s_{n_k} \to 1$, and so

$$s_{n_k} \sim c^k \tag{9}$$

as $k \to \infty$. Now for each $k$, consider the range of $j$ below:

$$n_k \le j < n_{k+1} \tag{10}$$

and put

$$F_j = \{\omega: |S_{n_{k+1}} - S_j| < s_{n_{k+1}}\}. \tag{11}$$

By Chebyshev's inequality, we have

$$\mathscr{P}(F_j) \ge 1 - \frac{s^2_{n_{k+1}} - s^2_{n_k}}{s^2_{n_{k+1}}} \to \frac{1}{c^2};$$

hence $\mathscr{P}(F_j) \ge A > 0$ for all sufficiently large $k$.

**Lemma 2.** Let $\{E_j\}$ and $\{F_j\}$, $1 \le j \le n < \infty$, be two sequences of events. Suppose that for each $j$, the event $F_j$ is independent of $E_1^c \cdots E_{j-1}^c E_j$, and that there exists a constant $A > 0$ such that $\mathscr{P}(F_j) \geq A$ for every $j$. Then we have

$$(12) \qquad \mathscr{P}\left(\bigcup_{j=1}^{n} E_j F_j\right) \geq A\mathscr{P}\left(\bigcup_{j=1}^{n} E_j\right).$$

PROOF. The left member in (12) is equal to

$$\mathscr{P}\left(\bigcup_{j=1}^{n}[(E_1F_1)^c \cdots (E_{j-1}F_{j-1})^c(E_jF_j)]\right)$$

$$\geq \mathscr{P}\left(\bigcup_{j=1}^{n}[E_1^c \cdots E_{j-1}^c E_j F_j]\right) = \sum_{j=1}^{n} \mathscr{P}(E_1^c \cdots E_{j-1}^c E_j)\mathscr{P}(F_j)$$

$$\geq \sum_{j=1}^{n} \mathscr{P}(E_1^c \cdots E_{j-1}^c E_j) \cdot A,$$

which is equal to the right member in (12).

Applying Lemma 2 to $E_j{}^+$ and the $F_j$ in (11), we obtain

$$(13) \qquad \mathscr{P}\left\{\bigcup_{j=n_k}^{n_{k+1}-1} E_j{}^+ F_j\right\} \geq A\mathscr{P}\left\{\bigcup_{j=n_k}^{n_{k+1}-1} E_j{}^+\right\}.$$

It is clear that the event $E_j{}^+ \cap F_j$ implies

$$S_{n_{k+1}} > S_j - s_{n_{k+1}} > \varphi(1+\delta, s_j) - s_{n_{k+1}},$$

which is, by (9) and (10), asymptotically greater than

$$\frac{1}{c}\varphi\left(1 + \frac{3}{4}\delta, s_{n_{k+1}}\right).$$

Choose $c$ so close to 1 that $(1 + 3/4\delta)/c^2 > 1 + (\delta/2)$ and put

$$G_k = \left\{\omega: S_{n_{k+1}} > \varphi\left(1 + \frac{\delta}{2}, s_{n_{k+1}}\right)\right\};$$

note that $G_k$ is just $E_{n_{k+1}}{}^+$ with $\delta$ replaced by $\delta/2$. The above implication may be written as

$$E_j{}^+ F_j \subset G_k$$

for sufficiently large $k$ and all $j$ in the range given in (10); hence we have

$$(14) \qquad \bigcup_{j=n_k}^{n_{k+1}-1} E_j^+ F_j \subset G_k.$$

It follows from (4) that

$$\sum_k \mathscr{P}(G_k) \le \sum_k \frac{A}{(\log s_{n_k})^{1+(\delta/2)}} \le A \sum_k \frac{1}{(k \log c)^{1+(\delta/2)}} < \infty.$$

In conjunction with (13) and (14) we conclude that

$$\sum_k \mathscr{P}\left\{\bigcup_{j=n_k}^{n_{k+1}-1} E_j^{+}\right\} < \infty;$$

and consequently by the Borel–Cantelli lemma that

$$\mathscr{P}\left\{\bigcup_{j=n_k}^{n_{k+1}-1} E_j^{+}\text{i.o.}\right\} = 0.$$

This is equivalent (why?) to the desired result (8).

2°. Next we prove that with the same subsequence $\{n_k\}$ but an arbitrary $c$, if we put $t_k^2 = s_{n_{k+1}}^2 - s_{n_k}^2$, and

$$D_k = \left\{\omega\colon S_{n_{k+1}}(\omega) - S_{n_k}(\omega) > \varphi\left(1 - \frac{\delta}{2}, t_k\right)\right\},$$

then we have

$$\mathscr{P}(D_k \text{ i.o.}) = 1. \tag{15}$$

Since the differences $S_{n_{k+1}} - S_{n_k}$, $k \ge 1$, are independent r.v.'s, the divergence part of the Borel–Cantelli lemma is applicable to them. To estimate $\mathscr{P}(D_k)$, we may apply Lemma 1 to the sequence $\{X_{n_k+j}, j \ge 1\}$. We have

$$t_k^2 \sim \left(1 - \frac{1}{c^2}\right) s_{n_{k+1}}^2 \sim \left(1 - \frac{1}{c^2}\right) c^{2(k+1)},$$

and consequently by (3):

$$\frac{\Gamma_{n_{k+1}} - \Gamma_{n_k}}{t_k^3} \le \frac{A\Gamma_{n_{k+1}}}{S_{n_{k+1}}^3} \le \frac{A}{(\log t_k)^{1+\epsilon}}.$$

Hence by (5),

$$\mathscr{P}(D_k) \ge \frac{A}{(\log t_k)^{1-(\delta/4)}} \ge \frac{A}{k^{1-(\delta/4)}}$$

and so $\sum_k \mathscr{P}(D_k) = \infty$ and (15) follows.

3°. We shall use (8) and (15) together to prove

$$\mathscr{P}(E_n^{-} \text{ i.o.}) = 1. \tag{16}$$

This requires just a rough estimate entailing a suitable choice of $c$. By (8) applied to $\{-X_n\}$ and (15), for almost every $\omega$ the following two assertions are true:

**(i)** $S_{n_{k+1}}(\omega) - S_{n_k}(\omega) > \varphi(1-(\delta/2), t_k)$ for infinitely many $k$;
**(ii)** $S_{n_k}(\omega) \geq -\varphi(2, s_{n_k})$ for all sufficiently large $k$.

For such an $\omega$, we have then

$$(17) \qquad S_{n_{k+1}}(\omega) > \varphi\left(1-\frac{\delta}{2}, t_k\right) - \varphi(2, s_{n_k}) \qquad \text{for infinitely many } k.$$

Using (9) and $\log\log t_k^2 \sim \log\log s_{n_{k+1}}^2$, we see that the expression in the right side of (17) is asymptotically greater than

$$\left[\sqrt{\left(1-\frac{\delta}{2}\right)\left(1-\frac{1}{c^2}\right)} - \frac{\sqrt{2}}{c}\right]\varphi(1, s_{n_{k+1}}) > \varphi(1-\delta, s_{n_{k+1}}),$$

provided that $c$ is chosen sufficiently large. Doing this, we have therefore proved that

$$(18) \qquad \mathscr{P}(E_{n_{k+1}}{}^{-} \text{ i.o.}) = 1,$$

which certainly implies (16).

4°. The truth of (8) and (16), for each fixed $\delta$, $0 < \delta < \in$, means exactly the conclusion (2), by an argument that should by now be familiar to the reader.

**Theorem 7.5.1.** Under the condition (3), the lim sup and lim inf, as $n \to \infty$, of $S_n/\sqrt{2s_n^2 \log\log s_n}$ are respectively $+1$ and $-1$, with probability one.

The assertion about lim inf follows, of course, from (2) if we apply it to $\{-X_j, j \geq 1\}$. Recall that (3) is more than sufficient to ensure the validity of the central limit theorem, namely that $S_n/s_n$ converges in dist. to $\Phi$. Thus the law of the iterated logarithm complements the central limit theorem by circumscribing the extraordinary fluctuations of the sequence $\{S_n, n \geq 1\}$. An immediate consequence is that for almost every $\omega$, the sample sequence $S_n(\omega)$ changes sign infinitely often. For much more precise results in this direction see Chapter 8.

In view of the discussion preceding Theorem 7.3.3, one may wonder about the almost everywhere bounds for

$$\max_{1\leq m\leq n} S_m, \qquad \max_{1\leq m\leq n} |S_m|, \qquad \text{and so on.}$$

It is interesting to observe that as far as the $\limsup_n$ is concerned, these two functionals behave exactly like $S_n$ itself (Exercise 2 below). However, the question of $\liminf_n$ is quite different. In the case of $\max_{1\le m\le n}|S_m|$, another law of the (inverted) iterated logarithm holds as follows. For almost every $\omega$, we have

$$\underline{\lim}_{n\to\infty} \frac{\max_{1\le m\le n}|S_m(\omega)|}{\sqrt{\dfrac{\pi^2 {s_n}^2}{8\log\log s_n}}} = 1;$$

under a condition analogous to but stronger than (3). Finally, one may wonder about an asymptotic lower bound for $|S_n|$. It is rather trivial to see that this is always $o(s_n)$ when the central limit theorem is applicable; but actually it is even $o(s_n^{-1})$ in some general cases. Indeed in the integer lattice case, under the conditions of Exercise 9 of 7.3, we have "$S_n = 0$ i.o. a.e." This kind of phenomenon belongs really to the recurrence properties of the sequence $\{S_n\}$, to be discussed in Chapter 8.

EXERCISES

**1.** Show that condition (3) is fulfilled if the $X_j$'s have a common d.f. with a finite third moment.

$^\star$**2.** Prove that whenever (2) holds, then the analogous relations with $S_n$ replaced by $\max_{1\le m\le n} S_m$ or $\max_{1\le m\le n}|S_m|$ also hold.

$^\star$**3.** Let $\{X_j, j\ge 1\}$ be a sequence of independent, identically distributed r.v.'s with mean 0 and variance 1, and $S_n = \sum_{j=1}^n X_j$. Then

$$\mathscr{P}\left\{\underline{\lim}_{n\to\infty} \frac{|S_n(\omega)|}{\sqrt{n}} = 0\right\} = 1.$$

[HINT: Consider $S_{n_{k+1}} - S_{n_k}$ with $n_k \sim k^k$. A quick proof follows from Theorem 8.3.3 below.]

**4.** Prove that in Exercise 9 of Sec. 7.3 we have $\mathscr{P}\{S_n = 0 \text{ i.o.}\} = 1$.

$^\star$**5.** The law of the iterated logarithm may be used to supply certain counterexamples. For instance, if the $X_n$'s are independent and $X_n = \pm n^{1/2}/\log\log n$ with probability $\frac{1}{2}$ each, then $S_n/n \to 0$ a.e., but Kolmogorov's sufficient condition (see case (i) after Theorem 5.4.1) $\sum_n \mathscr{E}(X_n^2)/n^2 < \infty$ fails.

**6.** Prove that $\mathscr{P}\{|S_n| > \varphi(1-\delta, s_n)\text{i.o.}\} = 1$, without use of (8), as follows. Let

$$e_k = \{\omega: |S_{n_k}(\omega)| < \varphi(1-\delta, s_{n_k})\};$$

$$f_k = \left\{\omega: S_{n_{k+1}}(\omega) - S_{n_k}(\omega) > \varphi\left(1-\frac{\delta}{2}, s_{n_{k+1}}\right)\right\}.$$

Show that for sufficiently large $k$ the event $e_k \cap f_k$ implies the complement of $e_{k+1}$; hence deduce

$$\mathscr{P}\left(\bigcap_{j=j_0}^{k+1} e_j\right) \le \mathscr{P}(e_{j_0}) \prod_{j=j_0}^{k} [1 - \mathscr{P}(f_j)]$$

and show that the product $\to 0$ as $k \to \infty$.

## 7.6 Infinite divisibility

The weak law of large numbers and the central limit theorem are concerned, respectively, with the convergence in dist. of sums of independent r.v.'s to a degenerate and a normal d.f. It may seem strange that we should be so much occupied with these two apparently unrelated distributions. Let us point out, however, that in terms of ch.f.'s these two may be denoted, respectively, by $e^{ait}$ and $e^{ait-b^2t^2}$ — exponentials of polynomials of the first and second degree in $(it)$. This explains the considerable similarity between the two cases, as evidenced particularly in Theorems 6.4.3 and 6.4.4.

Now the question arises: what other limiting d.f.'s are there when small independent r.v.'s are added? Specifically, consider the double array (2) in Sec. 7.1, in which independence in each row and holospoudicity are assumed. Suppose that for some sequence of constants $a_n$,

$$S_n - a_n = \sum_{j=1}^{k_n} X_{nj} - a_n$$

converges in dist. to $F$. What is the class of such $F$'s, and when does such a convergence take place? For a single sequence of independent r.v.'s $\{X_j, j \ge 1\}$, similar questions may be posed for the "normed sums" $(S_n - a_n)/b_n$.
These questions have been answered completely by the work of Lévy, Khintchine, Kolmogorov, and others; for a comprehensive treatment we refer to the book by Gnedenko and Kolmogorov [12]. Here we must content ourselves with a modest introduction to this important and beautiful subject.

We begin by recalling other cases of the above-mentioned limiting distributions, conveniently displayed by their ch.f.'s:

$$e^{\lambda(e^{it}-1)}, \qquad \lambda > 0; \qquad e^{-c|t|^\alpha}, \qquad 0 < \alpha < 2, c > 0.$$

The former is the Poisson distribution; the latter is called the *symmetric stable distribution of exponent* $\alpha$ (see the discussion in Sec. 6.5), including the Cauchy distribution for $\alpha = 1$. We may include the normal distribution among the latter for $\alpha = 2$. All these are exponentials and have the further property that their "$n$th roots":

$$e^{ait/n}, \qquad e^{(1/n)(ait - b^2t^2)}, \qquad e^{(\lambda/n)(e^{it}-1)}, \qquad e^{-(c/n)|t|^\alpha},$$

are also ch.f.'s. It is remarkable that this simple property already characterizes the class of distributions we are looking for (although we shall prove only part of this fact here).

DEFINITION OF INFINITE DIVISIBILITY. A ch.f. $f$ is called *infinitely divisible* iff for each integer $n \geq 1$, there exists a ch.f. $f_n$ such that

$$f = (f_n)^n. \tag{1}$$

In terms of d.f.'s, this becomes in obvious notation:

$$F = F_n^{n*} = \underbrace{F_n * F_n * \cdots * F_n}_{(n \text{ factors})}.$$

In terms of r.v.'s this means, for each $n \geq 1$, in a suitable probability space (why "suitable"?): there exist r.v.'s $X$ and $X_{nj}$, $1 \leq j \leq n$, the latter being independent among themselves, such that $X$ has ch.f. $f$, $X_{nj}$ has ch.f. $f_n$, and

$$X = \sum_{j=1}^{n} X_{nj}. \tag{2}$$

$X$ is thus "divisible" into $n$ independent and identically distributed parts, for each $n$. That such an $X$, and only such a one, can be the "limit" of sums of small independent terms as described above seems at least plausible.

A vital, though not characteristic, property of an infinitely divisible ch.f. will be given first.

**Theorem 7.6.1.** An infinitely divisible ch.f. never vanishes (for real $t$).

PROOF. We shall see presently that a complex-valued ch.f. is trouble-some when its "$n$th root" is to be extracted. So let us avoid this by going to the real, using the Corollary to Theorem 6.1.4. Let $f$ and $f_n$ be as in (1) and write

$$g = |f|^2, \qquad g_n = |f_n|^2.$$

For each $t \in \mathscr{R}^1$, $g(t)$ being real and positive, though conceivably vanishing, its *real positive nth root* is uniquely defined; let us denote it by $[g(t)]^{1/n}$. Since by (1) we have

$$g(t) = [g_n(t)]^n,$$

and $g_n(t) \geq 0$, it follows that

$$\forall t: g_n(t) = [g(t)]^{1/n}. \tag{3}$$

But $0 \le g(t) \le 1$, hence $\lim_{n\to\infty}[g(t)]^{1/n}$ is 0 or 1 according as $g(t) = 0$ or $g(t) \neq 0$. Thus $\lim_{n\to\infty} g_n(t)$ exists for every $t$, and the limit function, say $h(t)$, can take at most the two possible values 0 and 1. Furthermore, since $g$ is continuous at $t = 0$ with $g(0) = 1$, there exists a $t_0 > 0$ such that $g(t) \neq 0$ for $|t| \le t_0$. It follows that $h(t) = 1$ for $|t| \le t_0$. Therefore the sequence of ch.f.'s $g_n$ converges to the function $h$, which has just been shown to be continuous at the origin. By the convergence theorem of Sec. 6.3, $h$ must be a ch.f. and so continuous everywhere. Hence $h$ is identically equal to 1, and so by the remark after (3) we have

$$\forall t\colon |f(t)|^2 = g(t) \neq 0.$$

The theorem is proved.

Theorem 7.6.1 immediately shows that the uniform distribution on $[-1, 1]$ is not infinitely divisible, since its ch.f. is $\sin t/t$, which vanishes for some $t$, although in a literal sense it has infinitely many divisors! (See Exercise 8 of Sec. 6.3.) On the other hand, the ch.f.

$$\frac{2 + \cos t}{3}$$

never vanishes, but for it (1) fails even when $n = 2$; when $n \ge 3$, the failure of (1) for this ch.f. is an immediate consequence of Exercise 7 of Sec. 6.1, if we notice that the corresponding p.m. consists of exactly 3 atoms.

Now that we have proved Theorem 7.6.1, it seems natural to go back to an arbitrary infinitely divisible ch.f. and establish the generalization of (3):

$$f_n(t) = [f(t)]^{1/n}$$

for some "determination" of the multiple-valued $n$th root on the right side. This can be done by a simple process of "continuous continuation" of a complex-valued function of a real variable. Although merely an elementary exercise in "complex variables", it has been treated in the existing literature in a cavalier fashion and then misused or abused. For this reason the main propositions will be spelled out in meticulous detail here.

**Theorem 7.6.2.** Let a complex-valued function $f$ of the real variable $t$ be given. Suppose that $f(0) = 1$ and that for some $T > 0$, $f$ is continuous in $[-T, T]$ and does not vanish in the interval. Then there exists a unique (single-valued) function $\lambda$ of $t$ in $[-T, T]$ with $\lambda(0) = 0$ that is continuous there and satisfies

$$f(t) = e^{\lambda(t)}, \qquad -T \le t \le T. \tag{4}$$

The corresponding statement when $[-T, T]$ is replaced by $(-\infty, \infty)$ is also true.

PROOF. Consider the range of $f(t), t \in [-T, T]$; this is a closed set of points in the complex plane. Since it does not contain the origin, we have

$$\inf_{-T \le t \le T} |f(t) - 0| = \rho_T > 0.$$

Next, since $f$ is uniformly continuous in $[-T, T]$, there exists a $\delta_T$, $0 < \delta_T < \rho_T$, such that if $t$ and $t'$ both belong to $[-T, T]$ and $|t - t'| \le \delta_T$, then $|f(t) - f(t')| \le \rho_T/2 \le \frac{1}{2}$. Now divide $[-T, T]$ into equal parts of length less than $\delta_T$, say:

$$-T = t_{-\ell} < \cdots < t_{-1} < t_0 = 0 < t_1 < \cdots < t_\ell = T.$$

For $t_{-1} \le t \le t_1$, we define $\lambda$ as follows:

$$\lambda(t) = \sum_{j=1}^{\infty} \frac{(-1)^j}{j} \{f(t) - 1\}^j. \tag{5}$$

This is a continuous function of $t$ in $[t_{-1}, t_1]$, representing that determination of $\log f(t)$ which equals 0 for $t = 0$. Suppose that $\lambda$ has already been defined in $[t_{-k}, t_k]$; then we define $\lambda$ in $[t_k, t_{k+1}]$ as follows:

$$\lambda(t) = \lambda(t_k) + \sum_{j=1}^{\infty} \frac{(-1)^j}{j} \left( \frac{f(t) - f(t_k)}{f(t_k)} \right)^j; \tag{6}$$

similarly in $[t_{-k-1}, t_{-k}]$ by replacing $t_k$ with $t_{-k}$ everywhere on the right side above. Since we have, for $t_k \le t \le t_{k+1}$,

$$\left| \frac{f(t) - f(t_k)}{f(t_k)} \right| \le \frac{\frac{\rho_T}{2}}{\rho_T} = \frac{1}{2},$$

the power series in (6) converges uniformly in $[t_k, t_{k+1}]$ and represents a continuous function there equal to that determination of the logarithm of the function $f(t)/f(t_k) - 1$ which is 0 for $t = t_k$. Specifically, for the "schlicht neighborhood" $|z - 1| \le \frac{1}{2}$, let

$$L(z) = \sum_{j=1}^{\infty} \frac{(-1)^j}{j} (z - 1)^j \tag{7}$$

be the unique determination of $\log z$ vanishing at $z = 1$. Then (5) and (6) become, respectively:

$$\lambda(t) = L(f(t)), \qquad t_{-1} \le t \le t_1;$$

$$\lambda(t) = \lambda(t_k) + L\left( \frac{f(t)}{f(t_k)} \right), \qquad t_k \le t \le t_{k+1};$$

with a similar expression for $t_{-k-1} \le t \le t_{-k}$. Thus (4) is satisfied in $[t_{-1}, t_1]$, and if it is satisfied for $t = t_k$, then it is satisfied in $[t_k, t_{k+1}]$, since

$$e^{\lambda(t)} = e^{\lambda(t_k)+L(f(t)/f(t_k))} = f(t_k)\frac{f(t)}{f(t_k)} = f(t).$$

Thus (4) is satisfied in $[-T, T]$ by induction, and the theorem is proved for such an interval. To prove it for $(-\infty, \infty)$ let us observe that, having defined $\lambda$ in $[-n, n]$, we can *extend* it to $[-n-1, n+1]$ with the previous method, by dividing $[n, n+1]$, for example, into small equal parts whose length must be chosen dependent on $n$ at each stage (why?). The continuity of $\lambda$ is clear from the construction.

To prove the uniqueness of $\lambda$, suppose that $\lambda'$ has the same properties as $\lambda$. Since both satisfy equation (4), it follows that for each $t$, there exists an integer $m(t)$ such that

$$\lambda(t) - \lambda'(t) = 2\pi i\, m(t).$$

The left side being continuous in $t$, $m(\cdot)$ must be a constant (why?), which must be equal to $m(0) = 0$. Thus $\lambda(t) = \lambda'(t)$.

*Remark.* It may not be amiss to point out that $\lambda(t)$ is a single-valued function of $t$ but not of $f(t)$; see Exercise 7 below.

**Theorem 7.6.3.** For a fixed $T$, let each ${}_kf$, $k \ge 1$, as well as $f$ satisfy the conditions for $f$ in Theorem 7.6.2, and denote the corresponding $\lambda$ by ${}_k\lambda$. Suppose that ${}_kf$ converges uniformly to $f$ in $[-T, T]$, then ${}_k\lambda$ converges uniformly to $\lambda$ in $[-T, T]$.

PROOF. Let $L$ be as in (7), then there exists a $\delta$, $0 < \delta < \frac{1}{2}$, such that

$$|L(z)| \le 1, \qquad \text{if } |z - 1| \le \delta.$$

By the hypothesis of uniformity, there exists $k_1(T)$ such that if $k \ge k_1(T)$, then we have

$$\text{(8)} \qquad \sup_{|t| \le T} \left| \frac{{}_kf(t)}{f(t)} - 1 \right| \le \delta,$$

and consequently

$$\text{(9)} \qquad \sup_{|t| \le T} \left| L\left(\frac{{}_kf(t)}{f(t)}\right) \right| \le 1.$$

Since for each $t$, the exponentials of ${}_k\lambda(t) - \lambda(t)$ and $L({}_kf(t)/f(t))$ are equal, there exists an integer-valued function ${}_km(t)$, $|t| \le T$, such that

$$\text{(10)} \qquad L\left(\frac{{}_kf(t)}{f(t)}\right) = {}_k\lambda(t) - \lambda(t) + 2\pi i\, {}_km(t), \quad |t| \le T.$$

Since $L$ is continuous in $|z-1| \le \delta$, it follows that ${}_km(\cdot)$ is continuous in $|t| \le T$. Since it is integer-valued and equals 0 at $t = 0$, it is identically zero. Thus (10) reduces to

$$(11) \qquad {}_k\lambda(t) - \lambda(t) = L\left(\frac{{}_kf(t)}{f(t)}\right), \qquad |t| \le T.$$

The function $L$ being continuous at $z = 1$, the uniform convergence of ${}_kf/f$ to 1 in $[-T, T]$ implies that of ${}_k\lambda - \lambda$ to 0, as asserted by the theorem.

Thanks to Theorem 7.6.1, Theorem 7.6.2 is applicable to each infinitely divisible ch.f. $f$ in $(-\infty, \infty)$. Henceforth we shall call the corresponding $\lambda$ the *distinguished logarithm*, and $e^{\lambda(t)/n}$ the *distinguished nth root* of $f$. We can now extract the correct $n$th root in (1) above.

**Theorem 7.6.4.** For each $n$, the $f_n$ in (1) is just the distinguished $n$th root of $f$.

PROOF. It follows from Theorem 7.6.1 and (1) that the ch.f. $f_n$ never vanishes in $(-\infty, \infty)$, hence its distinguished logarithm $\lambda_n$ is defined. Taking multiple-valued logarithms in (1), we obtain as in (10):

$$\forall t: \lambda(t) - n\lambda_n(t) = 2\pi i m_n(t),$$

where $m_n(\cdot)$ takes only integer values. We conclude as before that $m_n(\cdot) \equiv 0$, and consequently

$$(12) \qquad f_n(t) = e^{\lambda_n(t)} = e^{\lambda(t)/n}$$

as asserted.

**Corollary.** If $f$ is a positive infinitely divisible ch.f., then for every $t$ the $f_n(t)$ in (1) is just the real positive $n$th root of $f(t)$.

PROOF. Elementary analysis shows that the real-valued logarithm of a real number $x$ in $(0, \infty)$ is a continuous function of $x$. It follows that this is a continuous solution of equation (4) in $(-\infty, \infty)$. The uniqueness assertion in Theorem 7.6.2 then identifies it with the distinguished logarithm of $f$, and the corollary follows, since the real positive $n$th root is the exponential of $1/n$ times the real logarithm.

As an immediate consequence of (12), we have

$$\forall t: \lim_{n\to\infty} f_n(t) = 1.$$

Thus by Theorem 7.1.1 the double array $\{X_{nj}, 1 \le j \le n, 1 \le n\}$ giving rise to (2) is holospoudic. We have therefore proved that each infinitely divisible distribution can be obtained as the limiting distribution of $S_n = \sum_{j=1}^{n} X_{nj}$ in such an array.

It is trivial that the product of two infinitely divisible ch.f.'s is again such a one, for we have in obvious notation:

$$_1f \cdot {_2f} = ({_1f_n})^n \cdot ({_2f_n})^n = ({_1f_n} \cdot {_2f_n})^n.$$

The next proposition lies deeper.

**Theorem 7.6.5.** Let $\{_kf, k \geq 1\}$ be a sequence of infinitely divisible ch.f.'s converging everywhere to the ch.f. $f$. Then $f$ is infinitely divisible.

PROOF. The difficulty is to prove first that $f$ never vanishes. Consider, as in the proof of Theorem 7.6.1: $g = |f|^2, {_kg} = |{_kf}|^2$. For each $n > 1$, let $x^{1/n}$ denote the real positive $n$th root of a real positive $x$. Then we have, by the hypothesis of convergence and the continuity of $x^{1/n}$ as a function of $x$,

$$\forall t\colon [{_kg}(t)]^{1/n} \to [g(t)]^{1/n}. \tag{13}$$

By the Corollary to Theorem 7.6.4, the left member in (13) is a ch.f. The right member is continuous everywhere. It follows from the convergence theorem for ch.f.'s that $[g(\cdot)]^{1/n}$ is a ch.f. Since $g$ is its $n$th power, and this is true for each $n \geq 1$, we have proved that $g$ is infinitely divisible and so never vanishes. Hence $f$ never vanishes and has a distinguished logarithm $\lambda$ defined everywhere. Let that of $_kf$ be $_k\lambda$. Since the convergence of $_kf$ to $f$ is necessarily uniform in each finite interval (see Sec. 6.3), it follows from Theorem 7.6.3 that $_k\lambda \to \lambda$ everywhere, and consequently

$$\exp({_k\lambda}(t)/n) \to \exp(\lambda(t)/n) \tag{14}$$

for every $t$. The left member in (14) is a ch.f. by Theorem 7.6.4, and the right member is continuous by the definition of $\lambda$. Hence it follows as before that $e^{\lambda(t)/n}$ is a ch.f. and $f$ is infinitely divisible.

The following alternative proof is interesting. There exists a $\delta > 0$ such that $f$ does not vanish for $|t| \leq \delta$, hence $\lambda$ is defined in this interval. For each $n$, (14) holds uniformly in this interval by Theorem 7.6.3. By Exercise 6 of Sec. 6.3, this is sufficient to ensure the existence of a *subsequence* from $\{\exp({_k\lambda}(t)/n), k \geq 1\}$ converging everywhere to *some* ch.f. $\varphi_n$. The $n$th power of this subsequence then converges to $(\varphi_n)^n$, but being a subsequence of $\{_kf\}$ it also converges to $f$. Hence $f = (\varphi_n)^n$, and we conclude again that $f$ is infinitely divisible.

Using the preceding theorem, we can construct a wide class of ch.f.'s that are infinitely divisible. For each $a$ and real $u$, the function

$$\mathfrak{P}(t; a, u) = e^{a(e^{itu}-1)} \tag{15}$$

is an infinitely divisible ch.f., since it is obtained from the Poisson ch.f. with parameter $a$ by substituting $ut$ for $t$. We shall call such a ch.f. a *generalized Poisson ch.f.* A finite product of these:

$$\prod_{j=1}^{k} \mathfrak{P}(t; a_j, u_j) = \exp\left[\sum_{j=1}^{k} a_j(e^{itu_j} - 1)\right] \tag{16}$$

is then also infinitely divisible. Now if $G$ is any bounded increasing function, the integral $\int_{-\infty}^{\infty}(e^{itu} - 1)dG(u)$ may be approximated by sums of the kind appearing as exponent in the right member of (16), for all $t$ in $\mathscr{R}^1$ and indeed uniformly so in every finite interval (why?). It follows that for each such $G$, the function

$$f(t) = \exp\left[\int_{-\infty}^{\infty}(e^{itu} - 1)\,dG(u)\right] \tag{17}$$

is an infinitely divisible ch.f. Now it turns out that although this falls somewhat short of being the most general form of an infinitely divisible ch.f., we have nevertheless the following qualititive result, which is a complete generalization of (16).

**Theorem 7.6.6.** For each infinitely divisible ch.f. $f$, there exists a double array of pairs of real constants $(a_{nj}, u_{nj})$, $1 \le j \le k_n$, $1 \le n$, where $a_j > 0$, such that

$$f(t) = \lim_{n\to\infty} \prod_{j=1}^{k_n} \mathfrak{P}(t; a_{nj}, u_{nj}). \tag{18}$$

The converse is also true. Thus the class of infinitely divisible d.f.'s coincides with the closure, with respect to vague convergence, of convolutions of a finite number of generalized Poisson d.f.'s.

PROOF. Let $f$ and $f_n$ be as in (1) and let $\lambda$ be the distinguished logarithm of $f$, $F_n$ the d.f. corresponding to $f_n$. We have for each $t$, as $n \to \infty$:

$$n[f_n(t) - 1] = n[e^{\lambda(t)/n} - 1] \to \lambda(t)$$

and consequently

$$e^{n[f_n(t)-1]} \to e^{\lambda(t)} = f(t). \tag{19}$$

Actually the first member in (19) is a ch.f. by Theorem 6.5.6, so that the convergence is uniform in each finite interval, but this fact alone is neither necessary nor sufficient for what follows. We have

$$n[f_n(t) - 1] = \int_{-\infty}^{\infty}(e^{itu} - 1)n\,dF_n(u).$$

For each $n$, $nF_n$ is a bounded increasing function, hence there exists

$$\{a_{nj}, u_{nj}; 1 \leq j \leq k_n\}$$

where $-\infty < u_{n1} < u_{n2} < \cdots < u_{nk_n} < \infty$ and $a_{nj} = n[F_n(u_{n,j}) - F_n(u_{n,j-1})]$, such that

$$\text{(20)} \qquad \sup_{|t|\leq n} \left| \sum_{j=1}^{k_n} (e^{itu_{nj}} - 1)a_{nj} - \int_{-\infty}^{\infty} (e^{itu} - 1)n\, dF_n(u) \right| \leq \frac{1}{n}.$$

(Which theorem in Chapter 6 implies this?) Taking exponentials and using the elementary inequality $|e^z - e^{z'}| \leq |e^z|(e^{|z-z'|} - 1)$, we conclude that as $n \to \infty$,

$$\text{(21)} \qquad \sup_{|t|\leq n} \left| e^{n[f_n(t)-1]} - \prod_{j=1}^{k_n} \mathfrak{P}(t; a_{nj}, u_{nj}) \right| = O\left(\frac{1}{n}\right).$$

This and (19) imply (18). The converse is proved at once by Theorem 7.6.5.

We are now in a position to state the fundamental theorem on infinitely divisible ch.f.'s, due to P. Lévy and Khintchine.

**Theorem 7.6.7.** Every infinitely divisible ch.f. $f$ has the following canonical representation:

$$f(t) = \exp\left[ait + \int_{-\infty}^{\infty} \left(e^{itu} - 1 - \frac{itu}{1+u^2}\right) \frac{1+u^2}{u^2}\, dG(u)\right]$$

where $a$ is a real constant, $G$ is a bounded increasing function in $(-\infty, \infty)$, and the integrand is defined by continuity to be $-t^2/2$ at $u = 0$. Furthermore, the class of infinitely divisible ch.f.'s coincides with the class of limiting ch.f.'s of $\sum_{j=1}^{k_n} X_{nj} - a_n$ in a holospoudic double array

$$\{X_{nj}, 1 \leq j \leq k_n, 1 \leq n\},$$

where $k_n \to \infty$ and for each $n$, the r.v.'s $\{X_{nj}, 1 \leq j \leq k_n\}$ are independent.

Note that we have proved above that every infinitely divisible ch.f. is in the class of limiting ch.f.'s described here, although we did not establish the canonical representation. Note also that if the hypothesis of "holospoudicity" is omitted, then every ch.f. is such a limit, trivially (why?). For a complete proof of the theorem, various special cases, and further developments, see the book by Gnedenko and Kolmogorov [12].

Let us end this section with an interesting example. Put $s = \sigma + it$, $\sigma > 1$ and $t$ real; consider the *Riemann zeta function*:

$$\zeta(s) = \sum_{n=1}^{\infty} \frac{1}{n^s} = \prod_p \left(1 - \frac{1}{p^s}\right)^{-1},$$

where $p$ ranges over all prime numbers. Fix $\sigma > 1$ and define

$$f(t) = \frac{\zeta(\sigma + it)}{\zeta(\sigma)}.$$

We assert that $f$ is an infinitely divisible ch.f. For each $p$ and every real $t$, the complex number $1 - p^{-\sigma - it}$ lies within the circle $\{z : |z - 1| < \frac{1}{2}\}$. Let $\log z$ denote that determination of the logarithm with an angle in $(-\pi, \pi]$. By looking at the angles, we see that

$$\begin{aligned}
\log \frac{1 - p^{-\sigma}}{1 - p^{-\sigma - it}} &= \log(1 - p^{-\sigma}) - \log(1 - p^{-\sigma - it}) \\
&= \sum_{m=1}^{\infty} \frac{1}{m p^{m\sigma}} (e^{-(m \log p)it} - 1) \\
&= \sum_{m=1}^{\infty} \log \mathfrak{P}(t; m^{-1} p^{-m\sigma}, -m \log p).
\end{aligned}$$

Since

$$f(t) = \lim_{n \to \infty} \prod_{p \leq n} \mathfrak{P}(t; m^{-1} p^{-m\sigma}, -m \log p),$$

it follows that $f$ is an infinitely divisible ch.f.

So far as known, this famous relationship between two "big names" has produced no important issue.

## EXERCISES

**1.** Is the convex combination of infinitely divisible ch.f.'s also infinitely divisible?

**2.** If $f$ is an infinitely divisible ch.f. and $\lambda$ its distinguished logarithm, $r > 0$, then the $r$th power of $f$ is defined to be $e^{r\lambda(t)}$. Prove that for each $r > 0$ it is an infinitely divisible ch.f.

$^\star$**3.** Let $f$ be a ch.f. such that there exists a sequence of positive integers $n_k$ going to infinity and a sequence of ch.f.'s $\varphi_k$ satisfying $f = (\varphi_k)^{n_k}$; then $f$ is infinitely divisible.

**4.** Give another proof that the right member of (17) is an infinitely divisible ch.f. by using Theorem 6.5.6.

**5.** Show that $f(t) = (1 - b)/(1 - be^{it})$, $0 < b < 1$, is an infinitely divisible ch.f. [HINT: Use canonical form.]

**6.** Show that the d.f. with density $\beta^\alpha \Gamma(\alpha)^{-1} x^{\alpha-1} e^{-\beta x}$, $\alpha > 0$, $\beta > 0$, in $(0, \infty)$, and 0 otherwise, is infinitely divisible.

**7.** Carry out the proof of Theorem 7.6.2 specifically for the "trivial" but instructive case $f(t) = e^{ait}$, where $a$ is a fixed real number.

***8.** Give an example to show that in Theorem 7.6.3, if the uniformity of convergence of ${}_k f$ to $f$ is omitted, then ${}_k\lambda$ need not converge to $\lambda$. [HINT: ${}_k f(t) = \exp\{2\pi i(-1)^k kt(1 + kt)^{-1}\}$.]

**9.** Let $f(t) = 1 - t$, $f_k(t) = 1 - t + (-1)^k itk^{-1}$, $0 \le t \le 2$, $k \ge 1$. Then $f_k$ never vanishes and converges uniformly to $f$ in $[0, 2]$. Let $\sqrt{f_k}$ denote the distinguished square root of $f_k$ in $[0, 2]$. Show that $\sqrt{f_k}$ does not converge in any neighborhood of $t = 1$. Why is Theorem 7.6.3 not applicable? [This example is supplied by E. Reich].

***10.** Some writers have given the proof of Theorem 7.6.6 by apparently considering each fixed $t$ and using an analogue of (21) without the "$\sup_{|t|\le n}$" there. Criticize this "quick proof". [HINT: Show that the two relations

$$\forall t \text{ and } \forall m: \qquad \lim_{m\to\infty} u_{mn}(t) = u_{.n}(t),$$

$$\forall t: \qquad \lim_{n\to\infty} u_{.n}(t) = u(t),$$

do not imply the existence of a sequence $\{m_n\}$ such that

$$\forall t: \lim_{n\to\infty} u_{m_n n}(t) = u(t).$$

Indeed, they do not even imply the existence of two subsequences $\{m_\nu\}$ and $\{n_\nu\}$ such that

$$\forall t: \lim_{\nu\to\infty} u_{m_\nu n_\nu}(t) = u(t).$$

Thus the extension of Lemma 1 in Sec. 7.2 is false.]

The three "counterexamples" in Exercises 7 to 9 go to show that the cavalierism alluded to above is not to be shrugged off easily.

**11.** Strengthening Theorem 6.5.5, show that two infinitely divisible ch.f.'s may coincide in a neighborhood of 0 without being identical.

***12.** Reconsider Exercise 17 of Sec. 6.4 and try to apply Theorem 7.6.3. [HINT: The latter is not immediately applicable, owing to the lack of uniform convergence. However, show first that if $e^{c_n it}$ converges for $t \in A$, where $m(A) > 0$, then it converges for all $t$. This follows from a result due to Steinhaus, asserting that the difference set $A - A$ contains a neighborhood of 0 (see, e.g., Halmos [4, p. 68]), and from the equation $e^{c_n it} e^{c_n it'} = e^{c_n i(t+t')}$. Let $\{b_n\}$, $\{b'_n\}$ be any two subsequences of $c_n$, then $e^{(b_n - b'_n)it} \to 1$ for all $t$. Since 1 is a ch.f., the convergence is uniform in every finite interval by the convergence theorem for ch.f.'s. Alternatively, if

$$\varphi(t) = \lim_{n\to\infty} e^{c_n it}$$

then $\varphi$ satisfies Cauchy's functional equation and must be of the form $e^{cit}$, which is a ch.f. These approaches are fancier than the simple one indicated in the hint for the said exercise, but they are interesting. There is no known quick proof by "taking logarithms", as some authors have done.]

### Bibliographical Note

The most comprehensive treatment of the material in Secs. 7.1, 7.2, and 7.6 is by Gnedenko and Kolmogorov [12]. In this as well as nearly all other existing books on the subject, the handling of logarithms must be strengthened by the discussion in Sec. 7.6.

For an extensive development of Lindeberg's method (the operator approach) to infinitely divisible laws, see Feller [13, vol. 2].

Theorem 7.3.2 together with its proof as given here is implicit in

W. Doeblin, *Sur deux problèmes de M. Kolmogoroff concernant les chaines dénombrables*, Bull. Soc. Math. France **66** (1938), 210–220.

It was rediscovered by F. J. Anscombe. For the extension to the case where the constant $c$ in (3) of Sec. 7.3 is replaced by an arbitrary r.v., see H. Wittenberg, *Limiting distributions of random sums of independent random variables*, Z. Wahrscheinlichkeitstheorie **1** (1964), 7–18.

Theorem 7.3.3 is contained in

P. Erdös and M. Kac, *On certain limit theorems of the theory of probability*, Bull. Am. Math. Soc. **52** (1946) 292–302.

The proof given for Theorem 7.4.1 is based on

P. L. Hsu, *The approximate distributions of the mean and variance of a sample of independent variables*, Ann. Math. Statistics **16** (1945), 1–29.

This paper contains historical references as well as further extensions.

For the law of the iterated logarithm, see the classic

A. N. Kolmogorov, *Über das Gesetz des iterierten Logarithmus*, Math. Annalen **101** (1929), 126–136.

For a survey of "classical limit theorems" up to 1945, see

W. Feller, *The fundamental limit theorems in probability*, Bull. Am. Math. Soc. **51** (1945), 800–832.

Kai Lai Chung, *On the maximum partial sums of sequences of independent random variables*. Trans. Am. Math. Soc. **64** (1948), 205–233.

Infinitely divisible laws are treated in Chapter 7 of Lévy [11] as the analytical counterpart of his full theory of additive processes, or processes with independent increments (later supplemented by J. L. Doob and K. Ito).

New developments in limit theorems arose from connections with the Brownian motion process in the works by Skorohod and Strassen. For an exposition see references [16] and [22] of General Bibliography.

# 8 Random walk

## 8.1 Zero-or-one laws

In this chapter we adopt the notation $N$ for the set of strictly positive integers, and $N^0$ for the set of positive integers; used as an index set, each is endowed with the natural ordering and interpreted as a discrete time parameter. Similarly, for each $n \in N$, $N_n$ denotes the ordered set of integers from 1 to $n$ (both inclusive); $N_n^0$ that of integers from 0 to $n$ (both inclusive); and $N'_n$ that of integers beginning with $n+1$.

On the probability triple $(\Omega, \mathscr{F}, \mathscr{P})$, a sequence $\{X_n, n \in N\}$ where each $X_n$ is an r.v. (defined on $\Omega$ and finite a.e.), will be called a (*discrete parameter*) *stochastic process*. Various Borel fields connected with such a process will now be introduced. For any sub-B.F. $\mathscr{G}$ of $\mathscr{F}$, we shall write

$$(1) \qquad X \in \mathscr{G}$$

and use the expression "$X$ belongs to $\mathscr{G}$" or "$\mathscr{G}$ contains $X$" to mean that $X^{-1}(\mathscr{B}) \subset \mathscr{G}$ (see Sec. 3.1 for notation): in the customary language $X$ is said to be "measurable with respect to $\mathscr{G}$". For each $n \in N$, we define two B.F.'s as follows:

$\mathscr{F}_n$ = the augmented B.F. generated by the family of r.v.'s $\{X_k, k \in N_n\}$; that is, $\mathscr{F}_n$ is the smallest B.F. $\mathscr{G}$ containing all $X_k$ in the family and all null sets;

$\mathscr{F}'_n$ = the augmented B.F. generated by the family of r.v.'s $\{X_k, k \in N'_n\}$.

Recall that the union $\bigcup_{n=1}^{\infty} \mathscr{F}_n$ is a field but not necessarily a B.F. The smallest B.F. containing it, or equivalently containing every $\mathscr{F}_n$, $n \in N$, is denoted by

$$\mathscr{F}_\infty = \bigvee_{n=1}^{\infty} \mathscr{F}_n;$$

it is the B.F. generated by the stochastic process $\{X_n, n \in N\}$. On the other hand, the intersection $\bigcap_{n=1}^{\infty} \mathscr{F}'_n$ is a B.F. denoted also by $\bigwedge_{n=1}^{\infty} \mathscr{F}'_n$. It will be called the *remote field* of the stochastic process and a member of it a *remote event*.

Since $\mathscr{F}_\infty \subset \mathscr{F}$, $\mathscr{P}$ is defined on $\mathscr{F}_\infty$. For the study of the process $\{X_n, n \in N\}$ alone, it is sufficient to consider the reduced triple $(\Omega, \mathscr{F}_\infty, \mathscr{P}\mid_{\mathscr{F}_\infty})$. The following approximation theorem is fundamental.

**Theorem 8.1.1.** Given $\epsilon > 0$ and $\Lambda \in \mathscr{F}_\infty$, there exists $\Lambda_\epsilon \in \bigcup_{n=1}^{\infty} \mathscr{F}_n$ such that

$$(2) \qquad \mathscr{P}(\Lambda \Delta \Lambda_\epsilon) \leq \epsilon.$$

PROOF. Let $\mathscr{G}$ be the collection of sets $\Lambda$ for which the assertion of the theorem is true. Suppose $\Lambda_k \in \mathscr{G}$ for each $k \in N$ and $\Lambda_k \uparrow \Lambda$ or $\Lambda_k \downarrow \Lambda$. Then $\Lambda$ also belongs to $\mathscr{G}$, as we can easily see by first taking $k$ large and then applying the asserted property to $\Lambda_k$. Thus $\mathscr{G}$ is a monotone class. Since it is trivial that $\mathscr{G}$ contains the field $\bigcup_{n=1}^{\infty} \mathscr{F}_n$ that generates $\mathscr{F}_\infty$, $\mathscr{G}$ must contain $\mathscr{F}_\infty$ by the Corollary to Theorem 2.1.2, proving the theorem.

Without using Theorem 2.1.2, one can verify that $\mathscr{G}$ is closed with respect to complementation (trivial), finite union (by Exercise 1 of Sec. 2.1), and countable union (as increasing limit of finite unions). Hence $\mathscr{G}$ is a B.F. that must contain $\mathscr{F}_\infty$.

It will be convenient to use a particular type of sample space $\Omega$. In the notation of Sec. 3.4, let

$$\Omega = \bigtimes_{n=1}^{\infty} \Omega_n,$$

where each $\Omega_n$ is a "copy" of the real line $\mathscr{R}^1$. Thus $\Omega$ is just the space of all infinite sequences of real numbers. A point $\omega$ will be written as $\{\omega_n, n \in N\}$, and $\omega_n$ as a function of $\omega$ will be called the $n$th *coordinate* (*function*) of $\omega$.

Each $\Omega_n$ is endowed with the Euclidean B.F. $\mathscr{B}^1$, and the product Borel field $\mathscr{F}(=\mathscr{F}_\infty$ in the notation above) on $\Omega$ is defined to be the B.F. generated by the finite-product sets of the form

$$\text{(3)} \qquad \bigcap_{j=1}^{k} \{\omega: \omega_{n_j} \in B_{n_j}\}$$

where $(n_1, \dots, n_k)$ is an arbitrary finite subset of $N$ and where each $B_{n_j} \in \mathscr{B}^1$. In contrast to the case discussed in Sec. 3.3, however, no restriction is made on the p.m. $\mathscr{P}$ on $\mathscr{F}$. We shall not enter here into the matter of a general construction of $\mathscr{P}$. The Kolmogorov extension theorem (Theorem 3.3.6) asserts that on the concrete space-field $(\Omega, \mathscr{F})$ just specified, there exists a p.m. $\mathscr{P}$ whose projection on each finite-dimensional subspace may be arbitrarily preassigned, subject only to consistency (where one such subspace contains the other). Theorem 3.3.4 is a particular case of this theorem.

In this chapter we shall use the concrete probability space above, of the so-called "function space type", to simplify the exposition, but all the results below remain valid without this specification. This follows from a measure-preserving homomorphism between an abstract probability space and one of the function-space type; see Doob [17, chap. 2].

The chief advantage of this concrete representation of $\Omega$ is that it enables us to define an important mapping on the space.

DEFINITION OF THE SHIFT. The *shift* $\tau$ is a mapping of $\Omega$ such that

$$\tau: \omega = \{\omega_n, n \in N\} \to \tau\omega = \{\omega_{n+1}, n \in N\};$$

in other words, the image of a point has as its $n$th coordinate the $(n+1)$st coordinate of the original point.

Clearly $\tau$ is an $\infty$-to-1 mapping and it is from $\Omega$ *onto* $\Omega$. Its iterates are defined as usual by composition: $\tau^0 =$ identity, $\tau^k = \tau \circ \tau^{k-1}$ for $k \geq 1$. It induces a direct set mapping $\tau$ and an inverse set mapping $\tau^{-1}$ according to the usual definitions. Thus

$$\tau^{-1}\Lambda = \{\omega: \tau\omega \in \Lambda\}$$

and $\tau^{-n}$ is the $n$th iterate of $\tau^{-1}$. If $\Lambda$ is the set in (3), then

$$\text{(4)} \qquad \tau^{-1}\Lambda = \bigcap_{j=1}^{k} \{\omega: \omega_{n_j+1} \in B_{n_j}\}.$$

It follows from this that $\tau^{-1}$ maps $\mathscr{F}$ into $\mathscr{F}$; more precisely,

$$\forall \Lambda \in \mathscr{F}: \tau^{-n}\Lambda \in \mathscr{F}'_n, \quad n \in N,$$

where $\mathscr{F}_n'$ is the Borel field generated by $\{\omega_k, k > n\}$. This is obvious by (4) if $\Lambda$ is of the form above, and since the class of $\Lambda$ for which the assertion holds is a B.F., the result is true in general.

DEFINITION. A set in $\mathscr{F}$ is called *invariant* (*under the shift*) iff $\Lambda = \tau^{-1}\Lambda$. An r.v. $Y$ on $\Omega$ is invariant iff $Y(\omega) = Y(\tau\omega)$ for every $\omega \in \Omega$.

Observe the following general relation, valid for each point mapping $\tau$ and the associated inverse set mapping $\tau^{-1}$, each function $Y$ on $\Omega$ and each subset $A$ of $\mathscr{R}^1$:

$$\tau^{-1}\{\omega: Y(\omega) \in A\} = \{\omega: Y(\tau\omega) \in A\}. \tag{5}$$

This follows from $\tau^{-1} \circ Y^{-1} = (Y \circ \tau)^{-1}$.

We shall need another kind of mapping of $\Omega$. A *permutation on* $N_n$ is a 1-to-1 mapping of $N_n$ to itself, denoted as usual by

$$\begin{pmatrix} 1, & 2, \dots, & n \\ \sigma 1, & \sigma 2, \dots, & \sigma n \end{pmatrix}.$$

The collection of such mappings forms a group with respect to composition. A *finite permutation* on $N$ is by definition a permutation on a certain "initial segment" $N_n$ of $N$. Given such a permutation as shown above, we define $\sigma\omega$ to be the point in $\Omega$ whose coordinates are obtained from those of $\omega$ by the corresponding permutation, namely

$$(\sigma\omega)_j = \begin{cases} \omega_{\sigma j}, & \text{if } j \in N_n; \\ \omega_j, & \text{if } j \in N_n'. \end{cases}$$

As usual, $\sigma$ induces a direct set mapping $\sigma$ and an inverse set mapping $\sigma^{-1}$, the latter being also the direct set mapping induced by the "group inverse" $\sigma^{-1}$ of $\sigma$. In analogy with the preceding definition we have the following.

DEFINITION. A set in $\mathscr{F}$ is called permutable iff $\Lambda = \sigma\Lambda$ for every finite permutation $\sigma$ on $N$. A function $Y$ on $\Omega$ is permutable iff $Y(\omega) = Y(\sigma\omega)$ for every finite permutation $\sigma$ and every $\omega \in \Omega$.

It is fairly obvious that an invariant set is remote and a remote set is permutable; also that each of the collections: all invariant events, all remote events, all permutable events, forms a sub-B.F. of $\mathscr{F}$. If each of these B.F.'s is augmented (see Exercise 20 of Sec. 2.2), the resulting augmented B.F.'s will be called "almost invariant", "almost remote", and "almost permutable", respectively. Finally, the collection of all sets in $\mathscr{F}$ of probability either 0 or

1 clearly forms a B.F., which may be called the "all-or-nothing" field. This B.F. will also be referred to as "almost trivial".

Now that we have defined all these concepts for the general stochastic process, it must be admitted that they are not very useful without further specifications of the process. We proceed at once to the particular case below.

DEFINITION. A sequence of independent r.v.'s will be called an *independent process*; it is called a *stationary independent process* iff the r.v.'s have a common distribution.

Aspects of this type of process have been our main object of study, usually under additional assumptions on the distributions. Having christened it, we shall henceforth focus our attention on "the evolution of the process as a whole"—whatever this phrase may mean. For this type of process, the specific probability triple described above has been constructed in Sec. 3.3. Indeed, $\mathscr{F} = \mathscr{F}_\infty$, and the sequence of independent r.v.'s is just that of the successive coordinate functions $\{\omega_n, n \in N\}$, which, however, will also be interchangeably denoted by $\{X_n, n \in N\}$. If $\varphi$ is any Borel measurable function, then $\{\varphi(X_n), n \in N\}$ is another such process.

The following result is called Kolmogorov's "zero-or-one law".

**Theorem 8.1.2.** For an independent process, each remote event has probability zero or one.

PROOF. Let $\Lambda \in \bigcap_{n=1}^{\infty} \mathscr{F}'_n$ and suppose that $\mathscr{P}(\Lambda) > 0$; we are going to prove that $\mathscr{P}(\Lambda) = 1$. Since $\mathscr{F}_n$ and $\mathscr{F}'_n$ are independent fields (see Exercise 5 of Sec. 3.3), $\Lambda$ is independent of every set in $\mathscr{F}_n$ for each $n \in N$; namely, if $\mathrm{M} \in \bigcup_{n=1}^{\infty} \mathscr{F}_n$, then

(6)
$$\mathscr{P}(\Lambda \cap \mathrm{M}) = \mathscr{P}(\Lambda)\mathscr{P}(\mathrm{M}).$$

If we set

$$\mathscr{P}_\Lambda(\mathrm{M}) = \frac{\mathscr{P}(\Lambda \cap \mathrm{M})}{\mathscr{P}(\Lambda)}$$

for $\mathrm{M} \in \mathscr{F}$, then $\mathscr{P}_\Lambda(\cdot)$ is clearly a p.m. (the conditional probability relative to $\Lambda$; see Chapter 9). By (6) it coincides with $\mathscr{P}$ on $\bigcup_{n=1}^{\infty} \mathscr{F}_n$ and consequently also on $\mathscr{F}$ by Theorem 2.2.3. Hence we may take M to be $\Lambda$ in (6) to conclude that $\mathscr{P}(\Lambda) = \mathscr{P}(\Lambda)^2$ or $\mathscr{P}(\Lambda) = 1$.

The usefulness of the notions of shift and permutation in a stationary independent process is based on the next result, which says that both $\tau^{-1}$ and $\sigma$ are "measure-preserving".

**Theorem 8.1.3.** For a stationary independent process, if $\Lambda \in \mathscr{F}$ and $\sigma$ is any finite permutation, we have

$$\mathscr{P}(\tau^{-1}\Lambda) = \mathscr{P}(\Lambda); \tag{7}$$

$$\mathscr{P}(\sigma\Lambda) = \mathscr{P}(\Lambda). \tag{8}$$

PROOF. Define a set function $\tilde{\mathscr{P}}$ on $\mathscr{F}$ as follows:

$$\tilde{\mathscr{P}}(\Lambda) = \mathscr{P}(\tau^{-1}\Lambda).$$

Since $\tau^{-1}$ maps disjoint sets into disjoint sets, it is clear that $\tilde{\mathscr{P}}$ is a p.m. For a finite-product set $\Lambda$, such as the one in (3), it follows from (4) that

$$\tilde{\mathscr{P}}(\Lambda) = \prod_{j=1}^{k} \mu(B_{n_j}) = \mathscr{P}(\Lambda).$$

Hence $\mathscr{P}$ and $\tilde{\mathscr{P}}$ coincide also on each set that is the union of a finite number of such disjoint sets, and so on the B.F. $\mathscr{F}$ generated by them, according to Theorem 2.2.3. This proves (7); (8) is proved similarly.

The following companion to Theorem 8.1.2, due to Hewitt and Savage, is very useful.

**Theorem 8.1.4.** For a stationary independent process, each permutable event has probability zero or one.

PROOF. Let $\Lambda$ be a permutable event. Given $\epsilon > 0$, we may choose $\epsilon_k > 0$ so that

$$\sum_{k=1}^{\infty} \epsilon_k \le \epsilon.$$

By Theorem 8.1.1, there exists $\Lambda_k \in \mathscr{F}_{n_k}$ such that $\mathscr{P}(\Lambda \,\Delta\, \Lambda_k) \le \epsilon_k$, and we may suppose that $n_k \uparrow \infty$. Let

$$\sigma = \begin{pmatrix} 1, \ldots, n_k, n_k + 1, \ldots, 2n_k \\ n_k + 1, \ldots, 2n_k, \quad 1, \ldots, n_k \end{pmatrix},$$

and $\mathrm{M}_k = \sigma\Lambda_k$. Then clearly $\mathrm{M}_k \in \mathscr{F}'_{n_k}$. It follows from (8) that

$$\mathscr{P}(\Lambda \,\Delta\, \mathrm{M}_k) \le \epsilon_k.$$

For any sequence of sets $\{E_k\}$ in $\mathscr{F}$, we have

$$\mathscr{P}(\limsup_k E_k) = \mathscr{P}\left(\bigcap_{m=1}^{\infty} \bigcup_{k=m}^{\infty} E_k\right) \le \sum_{k=1}^{\infty} \mathscr{P}(E_k).$$

Applying this to $E_k = \Lambda \triangle \mathrm{M}_k$, and observing the simple identity

$$\limsup(\Lambda \triangle \mathrm{M}_k) = (\Lambda \backslash \liminf \mathrm{M}_k) \cup (\limsup \mathrm{M}_k \backslash \Lambda),$$

we deduce that

$$\mathscr{P}(\Lambda \triangle \limsup \mathrm{M}_k) \leq \mathscr{P}(\limsup(\Lambda \triangle \mathrm{M}_k)) \leq \epsilon.$$

Since $\bigcup_{k=m}^{\infty} \mathrm{M}_k \in \mathscr{F}'_{n_m}$, the set $\limsup \mathrm{M}_k$ belongs to $\bigcap_{k=m}^{\infty} \mathscr{F}'_{n_k}$, which is seen to coincide with the remote field. Thus $\limsup \mathrm{M}_k$ has probability zero or one by Theorem 8.1.2, and the same must be true of $\Lambda$, since $\epsilon$ is arbitrary in the inequality above.

Here is a more transparent proof of the theorem based on the metric on the measure space $(\Omega, \mathscr{F}, \mathscr{P})$ given in Exercise 8 of Sec. 3.2. Since $\Lambda_k$ and $\mathrm{M}_k$ are independent, we have

$$\mathscr{P}(\Lambda_k \cap \mathrm{M}_k) = \mathscr{P}(\Lambda_k)\mathscr{P}(\mathrm{M}_k).$$

Now $\Lambda_k \to \Lambda$ and $\mathrm{M}_k \to \Lambda$ in the metric just mentioned, hence also

$$\Lambda_k \cap \mathrm{M}_k \to \Lambda \cap \Lambda$$

in the same sense. Since convergence of events in this metric implies convergence of their probabilities, it follows that $\mathscr{P}(\Lambda \cap \Lambda) = \mathscr{P}(\Lambda)\mathscr{P}(\Lambda)$, and the theorem is proved.

**Corollary.** For a stationary independent process, the B.F.'s of almost permutable or almost remote or almost invariant sets all coincide with the all-or-nothing field.

## EXERCISES

$\Omega$ and $\mathscr{F}$ are the infinite product space and field specified above.

**1.** Find an example of a remote field that is not the trivial one; to make it interesting, insist that the r.v.'s are not identical.

**2.** An r.v. belongs to the all-or-nothing field if and only if it is constant a.e.

**3.** If $\Lambda$ is invariant then $\Lambda = \tau\Lambda$; the converse is false.

**4.** An r.v. is invariant [permutable] if and only if it belongs to the invariant [permutable] field.

**5.** The set of convergence of an arbitrary sequence of r.v.'s $\{Y_n, n \in N\}$ or of the sequence of their partial sums $\sum_{j=1}^{n} Y_j$ are both permutable. Their limits are permutable r.v.'s with domain the set of convergence.

**★6.** If $a_n > 0$, $\lim_{n\to\infty} a_n$ exists $> 0$ finite or infinite, and $\lim_{n\to\infty}(a_{n+1}/a_n) = 1$, then the set of convergence of $\{{a_n}^{-1}\sum_{j=1}^{n} Y_j\}$ is invariant. If $a_n \to +\infty$, the upper and lower limits of this sequence are invariant r.v.'s.

**★7.** The set $\{Y_{2n} \in A \text{ i.o.}\}$, where $A \in \mathscr{B}^1$, is remote but not necessarily invariant; the set $\{\sum_{j=1}^{n} Y_j \; A \text{ i.o.}\}$ is permutable but not necessarily remote. Find some other essentially different examples of these two kinds.

**8.** Find trivial examples of independent processes where the three numbers $\mathscr{P}(\tau^{-1}\Lambda)$, $\mathscr{P}(\Lambda)$, $\mathscr{P}(\tau\Lambda)$ take the values 1, 0, 1; or 0, $\frac{1}{2}$, 1.

**9.** Prove that an invariant event is remote and a remote event is permutable.

**★10.** Consider the *bi-infinite product space* of all bi-infinite sequences of real numbers $\{\omega_n, n \in \overline{N}\}$, where $\overline{N}$ is the set of all integers in its natural (algebraic) ordering. Define the shift as in the text with $\overline{N}$ replacing $N$, and show that it is 1-to-1 on this space. Prove the analogue of (7).

**11.** Show that the conclusion of Theorem 8.1.4 holds true for a sequence of independent r.v.'s, not necessarily stationary, but satisfying the following condition: for every $j$ there exists a $k > j$ such that $X_k$ has the same distribution as $X_j$. [This remark is due to Susan Horn.]

**12.** Let $\{X_n, n \geq 1\}$ be independent r.v.'s with $\mathscr{P}\{X_n = 4^{-n}\} = \mathscr{P}\{X_n = -4^{-n}\} = \frac{1}{2}$. Then the remote field of $\{S_n, n \geq 1\}$, where $S_n = \sum_{j=1}^{n} X_j$, is not trivial.

## 8.2 Basic notions

From now on we consider only a stationary independent process $\{X_n, n \in N\}$ on the concrete probability triple specified in the preceding section. The common distribution of $X_n$ will be denoted by $\mu$ (p.m.) or $F$ (d.f.); when only this is involved, we shall write $X$ for a representative $X_n$, thus $\mathscr{E}(X)$ for $\mathscr{E}(X_n)$.

Our interest in such a process derives mainly from the fact that it underlies another process of richer content. This is obtained by forming the successive partial sums as follows:

$$S_n = \sum_{j=1}^{n} X_j, \quad n \in N. \tag{1}$$

An initial r.v. $S_0 \equiv 0$ is adjoined whenever this serves notational convenience, as in $X_n = S_n - S_{n-1}$ for $n \in N$. The sequence $\{S_n, n \in N\}$ is then a very familiar object in this book, but now we wish to find a proper name for it. An officially correct one would be "stochastic process with stationary independent differences"; the name "homogeneous additive process" can also be used. We have, however, decided to call it a "random walk (process)", although the use of this term is frequently restricted to the case when $\mu$ is of the integer lattice type or even more narrowly a Bernoullian distribution.

DEFINITION OF RANDOM WALK. A *random walk* is the process $\{S_n, n \in N\}$ defined in (1) where $\{X_n, n \in N\}$ is a stationary independent process. By convention we set also $S_0 \equiv 0$.

A similar definition applies in a Euclidean space of any dimension, but we shall be concerned only with $\mathscr{R}^1$ except in some exercises later.

Let us observe that even for an independent process $\{X_n, n \in N\}$, its remote field is in general different from the remote field of $\{S_n, n \in N\}$, where $S_n = \sum_{j=1}^{n} X_j$. They are almost the same, being both almost trivial, for a stationary independent process by virtue of Theorem 8.1.4, since the remote field of the random walk is clearly contained in the permutable field of the corresponding stationary independent process.

We add that, while the notion of remoteness applies to any process, "(shift)-invariant" and "permutable" will be used here only for the underlying "coordinate process" $\{\omega_n, n \in N\}$ or $\{X_n, n \in N\}$.

The following relation will be much used below, for $m < n$:

$$S_{n-m}(\tau^m \omega) = \sum_{j=1}^{n-m} X_j(\tau^m \omega) = \sum_{j=1}^{n-m} X_{j+m}(\omega) = S_n(\omega) - S_m(\omega).$$

It follows from Theorem 8.1.3 that $S_{n-m}$ and $S_n - S_m$ have the same distribution. This is obvious directly, since it is just $\mu^{(n-m)*}$.

As an application of the results of Sec. 8.1 to a random walk, we state the following consequence of Theorem 8.1.4.

**Theorem 8.2.1.** Let $B_n \in \mathscr{B}^1$ for each $n \in N$. Then

$$\mathscr{P}\{S_n \in B_n \text{ i.o.}\}$$

is equal to zero or one.

PROOF. If $\sigma$ is a permutation on $N_m$, then $S_n(\sigma\omega) = S_n(\omega)$ for $n \geq m$, hence the set

$$\Lambda_m = \bigcup_{n=m}^{\infty} \{S_n \in B_n\}$$

is unchanged under $\sigma^{-1}$ or $\sigma$. Since $\Lambda_m$ decreases as $m$ increases, it follows that

$$\bigcap_{m=1}^{\infty} \Lambda_m$$

is permutable, and the theorem is proved.

Even for a fixed $B = B_n$ the result is significant, since it is by no means evident that the set $\{S_n > 0 \text{ i.o.}\}$, for instance, is even vaguely invariant or remote with respect to $\{X_n, n \in N\}$ (cf. Exercise 7 of Sec. 8.1). Yet the preceding theorem implies that it is in fact almost invariant. This is the strength of the notion of permutability as against invariance or remoteness.

For any serious study of the random walk process, it is imperative to introduce the concept of an "optional r.v." This notion has already been used more than once in the book (where?) but has not yet been named. Since the basic inferences are very simple and are supposed to be intuitively obvious, it has been the custom in the literature until recently not to make the formal introduction at all. However, the reader will profit by meeting these fundamental ideas for the theory of stochastic processes at the earliest possible time. They will be needed in the next chapter, too.

DEFINITION OF OPTIONAL r.v. An r.v. $\alpha$ is called *optional* relative to the arbitrary stochastic process $\{Z_n, n \in N\}$ iff it takes strictly positive integer values or $+\infty$ and satisfies the following condition:

$$\forall n \in N \cup \{\infty\}: \{\omega: \alpha(\omega) = n\} \in \mathscr{F}_n, \tag{2}$$

where $\mathscr{F}_n$ is the B.F. generated by $\{Z_k, k \in N_n\}$.

Similarly if the process is indexed by $N^0$ (as in Chapter 9), then the range of $\alpha$ will be $N^0$. Thus if the index $n$ is regarded as the time parameter, then $\alpha$ effects a choice of time (an "option") for each sample point $\omega$. One may think of this choice as a time to "stop", whence the popular alias "stopping time", but this is usually rather a momentary pause after which the process proceeds again: time marches on!

Associated with each optional r.v. $\alpha$ there are two or three important objects. First, the *pre-$\alpha$ field* $\mathscr{F}_\alpha$ is the collection of all sets in $\mathscr{F}_\infty$ of the form

$$\bigcup_{1 \le n \le \infty} [\{\alpha = n\} \cap \Lambda_n], \tag{3}$$

where $\Lambda_n \in \mathscr{F}_n$ for each $n \in N \cup \{\infty\}$. This collection is easily verified to be a B.F. (how?). If $\Lambda \in \mathscr{F}_\alpha$, then we have clearly $\Lambda \cap \{\alpha = n\} \in \mathscr{F}_n$, for every $n$. This property also characterizes the members of $\mathscr{F}_\alpha$ (see Exercise 1 below). Next, the *post-$\alpha$ process* is the process | $\{Z_{\alpha+n}, n \in N\}$ defined on the trace of the original probability triple on the set $\{\alpha < \infty\}$, where

$$\forall n \in N: Z_{\alpha+n}(\omega) = Z_{\alpha(\omega)+n}(\omega). \tag{4}$$

Each $Z_{\alpha+n}$ is seen to be an r.v. with domain $\{\alpha < \infty\}$; indeed it is finite a.e. there provided the original $Z_n$'s are finite a.e. It is easy to see that $Z_\alpha \in \mathscr{F}_\alpha$. The *post-$\alpha$ field* $\mathscr{F}'_\alpha$ is the B.F. generated by the post-$\alpha$ process: it is a sub-B.F. of $\{\alpha < \infty\} \cap \mathscr{F}_\infty$.

Instead of requiring $\alpha$ to be defined on all $\Omega$ but possibly taking the value $\infty$, we may suppose it to be defined on a set $\Delta$ in $\mathscr{F}_\infty$. Note that a strictly positive integer $n$ is an optional r.v. The concepts of pre-$\alpha$ and post-$\alpha$ fields reduce in this case to the previous $\mathscr{F}_n$ and $\mathscr{F}'_n$.

A vital example of optional r.v. is that of the *first entrance time into a given Borel set* $A$:

$$\alpha_A(\omega) = \begin{cases} \min\{n \in N: Z_n(\omega) \in A\} & \text{on } \bigcup_{n=1}^{\infty} \{\omega: Z_n(\omega) \in A\}; \\ +\infty & \text{elsewhere.} \end{cases} \tag{5}$$

To see that this is optional, we need only observe that for each $n \in N$:

$$\{\omega: \alpha_A(\omega) = n\} = \{\omega: Z_j(\omega) \in A^c, 1 \leq j \leq n-1; Z_n(\omega) \in A\}$$

which clearly belongs to $\mathscr{F}_n$; similarly for $n = \infty$.

Concepts connected with optionality have everyday counterparts, implicit in phrases such as "within thirty days of the accident (should it occur)". Historically, they arose from "gambling systems", in which the gambler chooses opportune times to enter his bets according to previous observations, experiments, or whatnot. In this interpretation, $\alpha + 1$ is the time chosen to gamble and is determined by events strictly prior to it. Note that, along with $\alpha$, $\alpha + 1$ is also an optional r.v., but the converse is false.

So far the notions are valid for an arbitrary process on an arbitrary triple. We now return to a stationary independent process on the specified triple and extend the notion of "shift" to an "*$\alpha$-shift*" as follows: $\tau^\alpha$ is a mapping on $\{\alpha < \infty\}$ such that

$$\tau^\alpha \omega = \tau^n \omega \qquad \text{on } \{\omega: \alpha(\omega) = n\}. \tag{6}$$

Thus the post-$\alpha$ process is just the process $\{X_n(\tau^\alpha \omega), n \in N\}$. Recalling that $X_n$ is a mapping on $\Omega$, we may also write

$$X_{\alpha+n}(\omega) = X_n(\tau^\alpha \omega) = (X_n \circ \tau^\alpha)(\omega) \tag{7}$$

and regard $X_n \circ \tau^\alpha$, $n \in N$, as the r.v.'s of the new process. The inverse set mapping $(\tau^\alpha)^{-1}$, to be written more simply as $\tau^{-\alpha}$, is defined as usual:

$$\tau^{-\alpha}\Lambda = \{\omega: \tau^\alpha \omega \in \Lambda\}.$$

Let us now prove the fundamental theorem about "stopping" a stationary independent process.

**Theorem 8.2.2.** For a stationary independent process and an almost everywhere finite optional r.v. $\alpha$ relative to it, the pre-$\alpha$ and post-$\alpha$ fields are independent. Furthermore the post-$\alpha$ process is a stationary independent process with the same common distribution as the original one.

PROOF. Both assertions are summarized in the formula below. For any $\Lambda \in \mathscr{F}_\alpha$, $k \in N$, $B_j \in \mathscr{B}^1$, $1 \le j \le k$, we have

$$\mathscr{P}\{\Lambda; X_{\alpha+j} \in B_j, 1 \le j \le k\} = \mathscr{P}\{\Lambda\} \prod_{j=1}^{k} \mu(B_j). \tag{8}$$

To prove (8), we observe that it follows from the definition of $\alpha$ and $\mathscr{F}_\alpha$ that

$$\Lambda \cap \{\alpha = n\} = \Lambda_n \cap \{\alpha = n\} \in \mathscr{F}_n, \tag{9}$$

where $\Lambda_n \in \mathscr{F}_n$ for each $n \in N$. Consequently we have

$$\begin{aligned}\mathscr{P}\{\Lambda; \alpha = n; X_{\alpha+j} \in B_j, 1 \le j \le k\} &= \mathscr{P}\{\Lambda_n; \alpha = n; X_{n+j} \in B_j, 1 \le j \le k\} \\ &= \mathscr{P}\{\Lambda; \alpha = n\} \mathscr{P}\{X_{n+j} \in B_j, 1 \le j \le k\} \\ &= \mathscr{P}\{\Lambda; \alpha = n\} \prod_{j=1}^{k} \mu(B_j),\end{aligned}$$

where the second equation is a consequence of (9) and the independence of $\mathscr{F}_n$ and $\mathscr{F}_n'$. Summing over $n \in N$, we obtain (8).

An immediate corollary is the extension of (7) of Sec. 8.1 to an $\alpha$-shift.

**Corollary.** For each $\Lambda \in \mathscr{F}_\infty$ we have

$$\mathscr{P}(\tau^{-\alpha}\Lambda) = \mathscr{P}(\Lambda). \tag{10}$$

Just as we iterate the shift $\tau$, we can iterate $\tau^\alpha$. Put $\alpha^1 = \alpha$, and define $\alpha^k$ inductively by

$$\alpha^{k+1}(\omega) = \alpha^k(\tau^\alpha \omega), \quad k \in N.$$

Each $\alpha^k$ is finite a.e. if $\alpha$ is. Next, define $\beta_0 = 0$, and

$$\beta_k = \sum_{j=1}^{k} \alpha^j, \quad k \in N.$$

We are now in a position to state the following result, which will be needed later.

**Theorem 8.2.3.** Let $\alpha$ be an a.e. finite optional r.v. relative to a stationary independent process. Then the random vectors $\{V_k, k \in N\}$, where

$$V_k(\omega) = (\alpha^k(\omega), X_{\beta_{k-1}+1}(\omega), \ldots, X_{\beta_k}(\omega)),$$

are independent and identically distributed.

PROOF. The independence follows from Theorem 8.2.2 by our showing that $V_1, \dots, V_{k-1}$ belong to the pre-$\beta_{k-1}$ field, while $V_k$ belongs to the post-$\beta_{k-1}$ field. The details are left to the reader; cf. Exercise 6 below.

To prove that $V_k$ and $V_{k+1}$ have the same distribution, we may suppose that $k = 1$. Then for each $n \in N$, and each $n$-dimensional Borel set $A$, we have

$$\begin{aligned}\{\omega: \alpha^2(\omega) = n; (X_{\alpha^1+1}(\omega), \dots, X_{\alpha^1+\alpha^2}(\omega)) \in A\}\\ = \{\omega: \alpha^1(\tau^\alpha\omega) = n; (X_1(\tau^\alpha\omega), \dots, X_{\alpha^1}(\tau^\alpha\omega) \in A\},\end{aligned}$$

since

$$\begin{aligned}X_{\alpha^1}(\tau^\alpha\omega) &= X_{\alpha^1(\tau^\alpha\omega)}(\tau^\alpha\omega) = X_{\alpha^2(\omega)}(\tau^\alpha\omega)\\ &= X_{\alpha^1(\omega)+\alpha^2(\omega)}(\omega) = X_{\alpha^1+\alpha^2}(\omega)\end{aligned}$$

by the quirk of notation that denotes by $X_\alpha(\cdot)$ the function whose value at $\omega$ is given by $X_{\alpha(\omega)}(\omega)$ and by (7) with $n = \alpha^2(\omega)$. By (5) of Sec. 8.1, the preceding set is the $\tau^{-\alpha}$-image (inverse image under $\tau^\alpha$) of the set

$$\{\omega: \alpha^1(\omega) = n; (X_1(\omega), \dots, X_{\alpha^1}(\omega)) \in A\},$$

and so by (10) has the same probability as the latter. This proves our assertion.

**Corollary.** The r.v.'s $\{Y_k, k \in N\}$, where

$$Y_k(\omega) = \sum_{n=\beta_{k-1}+1}^{\beta_k} \varphi(X_n(\omega))$$

and $\varphi$ is a Borel measurable function, are independent and identically distributed.

For $\varphi \equiv 1$, $Y_k$ reduces to $\alpha^k$. For $\varphi(x) \equiv x$, $Y_k = S_{\beta_k} - S_{\beta_{k-1}}$. The reader is advised to get a clear picture of the quantities $\alpha^k$, $\beta_k$, and $Y_k$ before proceeding further, perhaps by considering a special case such as (5).

We shall now apply these considerations to obtain results on the "global behavior" of the random walk. These will be broad qualitative statements distinguished by their generality without any additional assumptions.

The optional r.v. to be considered is the first entrance time into the strictly positive half of the real line, namely $A = (0, \infty)$ in (5) above. Similar results hold for $[0, \infty)$; and then by taking the negative of each $X_n$, we may deduce the corresponding result for $(-\infty, 0]$ or $(-\infty, 0)$. Results obtained in this way will be labeled as "dual" below. Thus, omitting $A$ from the notation:

$$(11)\qquad \alpha(\omega) = \begin{cases} \min\{n \in N: S_n > 0\} & \text{on } \bigcup_{n=1}^{\infty}\{\omega: S_n(\omega) > 0\};\\ +\infty & \text{elsewhere;}\end{cases}$$

and

$$\forall n \in N: \{\alpha = n\} = \{S_j \leq 0 \text{ for } 1 \leq j \leq n-1; S_n > 0\}.$$

Define also the r.v. $M_n$ as follows:

$$\forall n \in N^0: \quad M_n(\omega) = \max_{0 \leq j \leq n} S_j(\omega). \tag{12}$$

The inclusion of $S_0$ above in the maximum may seem artificial, but it does not affect the next theorem and will be essential in later developments in the next section. Since each $X_n$ is assumed to be finite a.e., so is each $S_n$ and $M_n$. Since $M_n$ increases with $n$, it tends to a limit, finite or positive infinite, to be denoted by

$$M(\omega) = \lim_{n \to \infty} M_n(\omega) = \sup_{0 \leq j < \infty} S_j(\omega). \tag{13}$$

**Theorem 8.2.4.** The statements (a), (b), and (c) below are equivalent; the statements (a′), (b′), and (c′) are equivalent.

(a) $\mathscr{P}\{\alpha < +\infty\} = 1$; (a′) $\mathscr{P}\{\alpha < +\infty\} < 1$;

(b) $\mathscr{P}\{\overline{\lim_{n \to \infty}} S_n = +\infty\} = 1$; (b′) $\mathscr{P}\{\overline{\lim_{n \to \infty}} S_n = +\infty\} = 0$;

(c) $\mathscr{P}\{M = +\infty\} = 1$; (c′) $\mathscr{P}\{M = +\infty\} = 0$.

PROOF. If (a) is true, we may suppose $\alpha < \infty$ everywhere. Consider the r.v. $S_\alpha$: it is strictly positive by definition and so $0 < \mathscr{E}(S_\alpha) \leq +\infty$. By the Corollary to Theorem 8.2.3, $\{S_{\beta_{k+1}} - S_{\beta_k}, k \geq 1\}$ is a sequence of independent and identically distributed r.v.'s. Hence the strong law of large numbers (Theorem 5.4.2 supplemented by Exercise 1 of Sec. 5.4) asserts that, if $\alpha^0 \equiv 0$ and $S_\alpha^0 \equiv 0$:

$$\frac{S_{\beta_n}}{n} = \frac{1}{n} \sum_{k=0}^{n-1} (S_{\beta_{k+1}} - S_{\beta_k}) \to \mathscr{E}(S_\alpha) > 0 \quad \text{a.e.}$$

This implies (b). Since $\overline{\lim}_{n \to \infty} S_n \leq M$, (b) implies (c). It is trivial that (c) implies (a). We have thus proved the equivalence of (a), (b), and (c). If (a′) is true, then (a) is false, hence (b) is false. But the set

$$\{\overline{\lim_{n \to \infty}} S_n = +\infty\}$$

is clearly permutable (it is even invariant, but this requires a little more reflection), hence (b′) is true by Theorem 8.1.4. Now any numerical sequence with finite upper limit is bounded above, hence (b′) implies (c′). Finally, if (c′) is true then (c) is false, hence (a) is false, and (a′) is true. Thus (a′), (b′), and (c′) are also equivalent.

**Theorem 8.2.5.** For the general random walk, there are four mutually exclusive possibilities, each taking place a.e.:

**(i)** $\forall n \in N: S_n = 0$;

**(ii)** $S_n \to -\infty$;

**(iii)** $S_n \to +\infty$;

**(iv)** $-\infty = \underline{\lim}_{n\to\infty} S_n < \overline{\lim}_{n\to\infty} S_n = +\infty$.

PROOF. If $X = 0$ a.e., then (i) happens. Excluding this, let $\varphi_1 = \overline{\lim}_n S_n$. Then $\varphi_1$ is a permutable r.v., hence a constant $c$, possibly $\pm\infty$, a.e. by Theorem 8.1.4. Since

$$\overline{\lim_n} S_n = X_1 + \overline{\lim_n}(S_n - X_1),$$

we have $\varphi_1 = X_1 + \varphi_2$, where $\varphi_2(\omega) = \varphi_1(\tau\omega) = c$ a.e. Since $X_1 \not\equiv 0$, it follows that $c = +\infty$ or $-\infty$. This means that

$$\text{either} \quad \overline{\lim_n} S_n = +\infty \text{ or } \lim_n S_n = -\infty.$$

By symmetry we have also

$$\text{either} \quad \underline{\lim_n} S_n = -\infty \text{ or } \lim_n S_n = +\infty.$$

These double alternatives yield the new possibilities (ii), (iii), or (iv), other combinations being impossible.

This last possibility will be elaborated upon in the next section.

## EXERCISES

In Exercises 1–6, the stochastic process is arbitrary.

$^\star$**1.** $\alpha$ is optional if and only if $\forall n \in N: \{\alpha \le n\} \in \mathscr{F}_n$.

$^\star$**2.** For each optional $\alpha$ we have $\alpha \in \mathscr{F}_\alpha$ and $X_\alpha \in \mathscr{F}_\alpha$. If $\alpha$ and $\beta$ are both optional and $\alpha \le \beta$, then $\mathscr{F}_\alpha \subset \mathscr{F}_\beta$.

**3.** If $\alpha_1$ and $\alpha_2$ are both optional, then so is $\alpha_1 \wedge \alpha_2, \alpha_1 \vee \alpha_2, \alpha_1 + \alpha_2$. If $\alpha$ is optional and $\Delta \in \mathscr{F}_\alpha$, then $\alpha_\Delta$ defined below is also optional:

$$\alpha_\Delta = \begin{cases} \alpha & \text{on } \Delta \\ +\infty & \text{on } \Omega\backslash\Delta. \end{cases}$$

$^\star$**4.** If $\alpha$ is optional and $\beta$ is optional relative to the post-$\alpha$ process, then $\alpha + \beta$ is optional (relative to the original process).

**5.** $\forall k \in N: \alpha^1 + \cdots + \alpha^k$ is optional. [For the $\alpha$ in (11), this has been called the $k$th *ladder* variable.]

*6. Prove the following relations:

$$\alpha^k = \alpha \circ \tau^{\beta_{k-1}}; \quad \tau^{\beta_{k-1}} \circ \tau^\alpha = \tau^{\beta_k}; \quad X_{\beta_{k-1}+j} \circ \tau^\alpha = X_{\beta_k+j}.$$

**7.** If $\alpha$ and $\beta$ are any two optional r.v.'s, then

$$X_{\beta+j}(\tau^\alpha \omega) = X_{\alpha(\omega)+\beta(\tau^\alpha\omega)+j}(\omega);$$

$$(\tau^\beta \circ \tau^\alpha)(\omega) = \tau^{\beta(\tau^\alpha\omega)+\alpha(\omega)}(\omega) \neq \tau^{\beta+\alpha}(\omega) \text{ in general.}$$

*8. Find an example of two optional r.v.'s $\alpha$ and $\beta$ such that $\alpha \leq \beta$ but $\mathscr{F}'_\alpha \not\supset \mathscr{F}'_\beta$. However, if $\gamma$ is optional relative to the post-$\alpha$ process and $\beta = \alpha + \gamma$, then indeed $\mathscr{F}'_\alpha \supset \mathscr{F}'_\beta$. As a particular case, $\mathscr{F}'_{\beta_k}$ is decreasing (while $\mathscr{F}_{\beta_k}$ is increasing) as $k$ increases.

**9.** Find an example of two optional r.v.'s $\alpha$ and $\beta$ such that $\alpha < \beta$ but $\beta - \alpha$ is not optional.

**10.** Generalize Theorem 8.2.2 to the case where the domain of definition and finiteness of $\alpha$ is $\Delta$ with $0 < \mathscr{P}(\Delta) < 1$. [This leads to a useful extension of the notion of independence. For a given $\Delta$ in $\mathscr{F}$ with $\mathscr{P}(\Delta) > 0$, two events $\Lambda$ and M, where $\text{M} \subset \Delta$, are said to be *independent relative* to $\Delta$ iff $\mathscr{P}\{\Lambda \cap \Delta \cap \text{M}\} = \mathscr{P}\{\Lambda \cap \Delta\}\mathscr{P}_\Delta\{\text{M}\}$.]

*11. Let $\{X_n, n \in N\}$ be a stationary independent process and $\{\alpha_k, k \in N\}$ a sequence of strictly increasing finite optional r.v.'s. Then $\{X_{\alpha_k+1}, k \in N\}$ is a stationary independent process with the same common distribution as the original process. [This is the gambling-system theorem first given by Doob in 1936.]

**12.** Prove the Corollary to Theorem 8.2.2.

**13.** State and prove the analogue of Theorem 8.2.4 with $\alpha$ replaced by $\alpha_{[0,\infty)}$. [The inclusion of 0 in the set of entrance causes a small difference.]

**14.** In an independent process where all $X_n$ have a common bound, $\mathscr{E}\{\alpha\} < \infty$ implies $\mathscr{E}\{S_\alpha\} < \infty$ for each optional $\alpha$ [cf. Theorem 5.5.3].

## 8.3 Recurrence

A basic question about the random walk is the *range* of the whole process: $\bigcup_{n=1}^\infty S_n(\omega)$ for a.e. $\omega$; or, "where does it ever go?" Theorem 8.2.5 tells us that, ignoring the trivial case where it stays put at 0, it either goes off to $-\infty$ or $+\infty$, or fluctuates between them. But how does it fluctuate? Exercise 9 below will show that the random walk can take large leaps from one end to the other without stopping in any middle range more than a finite number of times. On the other hand, it may revisit every neighborhood of every point an infinite number of times. The latter circumstance calls for a definition.

DEFINITION. The number $x \in \mathscr{R}^1$ is called a *recurrent value* of the random walk $\{S_n, n \in N\}$, iff for every $\epsilon > 0$ we have

$$(1) \qquad \mathscr{P}\{|S_n - x| < \epsilon \text{ i.o.}\} = 1.$$

The set of all recurrent values will be denoted by $\Re$.

Taking a sequence of $\epsilon$ decreasing to zero, we see that (1) implies the apparently stronger statement that the random walk is in each neighborhood of $x$ i.o. a.e.

Let us also call the number $x$ a *possible value* of the random walk iff for every $\epsilon > 0$, there exists $n \in N$ such that $\mathscr{P}\{|S_n - x| < \epsilon\} > 0$. Clearly a recurrent value is a possible value of the random walk (see Exercise 2 below).

**Theorem 8.3.1.** The set $\Re$ is either empty or a closed additive group of real numbers. In the latter case it reduces to the singleton $\{0\}$ if and only if $X \equiv 0$ a.e.; otherwise $\Re$ is either the whole $\mathscr{R}^1$ or the infinite cyclic group generated by a nonzero number $c$, namely $\{\pm nc: n \in N^0\}$.

PROOF. Suppose $\Re \neq \phi$ throughout the proof. To prove that $\Re$ is a group, let us show that if $x$ is a possible value and $y \in \Re$, then $y - x \in \Re$. Suppose not; then there is a strictly positive probability that from a certain value of $n$ on, $S_n$ will not be in a certain neighborhood of $y - x$. Let us put for $z \in \mathscr{R}^1$:

$$(2) \qquad p_{\epsilon,m}(z) = \mathscr{P}\{|S_n - z| \geq \epsilon \text{ for all } n \geq m\};$$

so that $p_{2\epsilon,m}(y - x) > 0$ for some $\epsilon > 0$ and $m \in N$. Since $x$ is a possible value, for the same $\epsilon$ we have a $k$ such that $\mathscr{P}\{|S_k - x| < \epsilon\} > 0$. Now the two independent events $|S_k - x| < \epsilon$ and $|S_n - S_k - (y - x)| \geq 2\epsilon$ together imply that $|S_n - y| > \epsilon$; hence

$$(3) \qquad p_{\epsilon,k+m}(y) = \mathscr{P}\{|S_n - y| \geq \epsilon \text{ for all } n \geq k + m\}$$
$$\geq \mathscr{P}\{|S_k - x| < \epsilon\}\mathscr{P}\{|S_n - S_k - (y - x)| \geq 2\epsilon \text{ for all } n \geq k + m\}.$$

The last-written probability is equal to $p_{2\epsilon,m}(y - x)$, since $S_n - S_k$ has the same distribution as $S_{n-k}$. It follows that the first term in (3) is strictly positive, contradicting the assumption that $y \in \Re$. We have thus proved that $\Re$ is an additive subgroup of $\mathscr{R}^1$. It is trivial that $\Re$ as a subset of $\mathscr{R}^1$ is closed in the Euclidean topology. A well-known proposition (proof?) asserts that the only closed additive subgroups of $\mathscr{R}^1$ are those mentioned in the second sentence of the theorem. Unless $X \equiv 0$ a.e., it has at least one possible value $x \neq 0$, and the argument above shows that $-x = 0 - x \in \Re$ and consequently also $x = 0 - (-x) \in \Re$. Suppose $\Re$ is not empty, then $0 \in \Re$. Hence $\Re$ is not a singleton. The theorem is completely proved.

It is clear from the preceding theorem that the key to recurrence is the value 0, for which we have the criterion below.

**Theorem 8.3.2.** If for some $\epsilon > 0$ we have

$$\sum_n \mathscr{P}\{|S_n| < \epsilon\} < \infty, \tag{4}$$

then

$$\mathscr{P}\{|S_n| < \epsilon \text{ i.o.}\} = 0 \tag{5}$$

(for the same $\epsilon$) so that $0 \notin \Re$. If for every $\epsilon > 0$ we have

$$\sum_n \mathscr{P}\{|S_n| < \epsilon\} = \infty, \tag{6}$$

then

$$\mathscr{P}\{|S_n| < \epsilon \text{ i.o.}\} = 1 \tag{7}$$

for every $\epsilon > 0$ and so $0 \in \Re$.

*Remark.* Actually if (4) or (6) holds for any $\epsilon > 0$, then it holds for every $\epsilon > 0$; this fact follows from Lemma 1 below but is not needed here.

PROOF. The first assertion follows at once from the convergence part of the Borel–Cantelli lemma (Theorem 4.2.1). To prove the second part consider

$$F = \liminf_n \{|S_n| \geq \epsilon\};$$

namely $F$ is the event that $|S_n| < \epsilon$ for only a finite number of values of $n$. For each $\omega$ on $F$, there is an $m(\omega)$ such that $|S_n(\omega)| \geq \epsilon$ for all $n \geq m(\omega)$; it follows that if we consider "*the last time* that $|S_n| < \epsilon$", we have

$$\mathscr{P}(F) = \sum_{m=0}^{\infty} \mathscr{P}\{|S_m| < \epsilon; |S_n| \geq \epsilon \text{ for all } n \geq m+1\}.$$

Since the two independent events $|S_m| < \epsilon$ and $|S_n - S_m| \geq 2\epsilon$ together imply that $|S_n| \geq \epsilon$, we have

$$\begin{aligned} 1 \geq \mathscr{P}(F) &\geq \sum_{m=1}^{\infty} \mathscr{P}\{|S_m| < \epsilon\}\mathscr{P}\{|S_n - S_m| \geq 2\epsilon \text{ for all } n \geq m+1\} \\ &= \sum_{m=1}^{\infty} \mathscr{P}\{|S_m| < \epsilon\} p_{2\epsilon,1}(0) \end{aligned}$$

by the previous notation (2), since $S_n - S_m$ has the same distribution as $S_{n-m}$. Consequently (6) cannot be true unless $p_{2\epsilon,1}(0) = 0$. We proceed to extend this to show that $p_{2\epsilon,k} = 0$ for every $k \in N$. To this aim we fix $k$ and consider the event

$$A_m = \{|S_m| < \epsilon, |S_n| \geq \epsilon \text{ for all } n \geq m + k\};$$

then $A_m$ and $A_{m'}$ are disjoint whenever $m' \geq m + k$ and consequently (why?)

$$k \geq \sum_{m=1}^{\infty} \mathscr{P}(A_m).$$

The argument above for the case $k = 1$ can now be repeated to yield

$$k \geq \sum_{m=1}^{\infty} \mathscr{P}\{|S_m| < \epsilon\} p_{2\epsilon,k}(0),$$

and so $p_{2\epsilon,k}(0) = 0$ for every $\epsilon > 0$. Thus

$$\mathscr{P}(F) = \lim_{k \to \infty} p_{\epsilon,k}(0) = 0,$$

which is equivalent to (7).

A simple sufficient condition for $0 \in \mathfrak{R}$, or equivalently for $\mathfrak{R} \neq \phi$, will now be given.

**Theorem 8.3.3.** If the weak law of large numbers holds for the random walk $\{S_n, n \in N\}$ in the form that $S_n/n \to 0$ in pr., then $\mathfrak{R} \neq \phi$.

PROOF. We need two lemmas, the first of which is also useful elsewhere.

**Lemma 1.** For any $\epsilon > 0$ and $m \in N$ we have

$$\sum_{n=0}^{\infty} \mathscr{P}\{|S_n| < m\epsilon\} \leq 2m \sum_{n=0}^{\infty} \mathscr{P}\{|S_n| < \epsilon\}. \tag{8}$$

PROOF OF LEMMA 1. It is sufficient to prove that if the right member of (8) is finite, then so is the left member and (8) is true. Put

$$I = (-\epsilon, \epsilon), \quad J = [j\epsilon, (j+1)\epsilon),$$

for a fixed $j \in N$; and denote by $\varphi_I, \varphi_J$ the respective indicator functions. Denote also by $\alpha$ the first entrance time into $J$, as defined in (5) of Sec. 8.2 with $Z_n$ replaced by $S_n$ and $A$ by $J$. We have

$$\mathscr{E}\left\{\sum_{n=1}^{\infty} \varphi_J(S_n)\right\} = \sum_{k=1}^{\infty} \int_{\{\alpha=k\}} \sum_{n=1}^{\infty} \varphi_J(S_n) d\mathscr{P}. \tag{9}$$

The typical integral on the right side is by definition of $\alpha$ equal to

$$\int_{\{\alpha=k\}} \left\{1 + \sum_{n=k+1}^{\infty} \varphi_J(S_n)\right\} d\mathscr{P} \leq \int_{\{\alpha=k\}} \left\{1 + \sum_{n=k+1}^{\infty} \varphi_I(S_n - S_k)\right\} d\mathscr{P},$$

since $\{\alpha = k\} \subset \{S_k \in J\}$ and $\{S_k \in J\} \cap \{S_n \in J\} \subset \{S_n - S_k \in I\}$. Now $\{\alpha = k\}$ and $S_n - S_k$ are independent, hence the last-written integral is equal to

$$\mathscr{P}(\alpha = k)\left\{1 + \int_{\Omega} \sum_{n=k+1}^{\infty} \varphi_I(S_n - S_k)\, d\mathscr{P}\right\} = \mathscr{P}(\alpha = k)\mathscr{E}\left\{\sum_{n=0}^{\infty} \varphi_I(S_n)\right\},$$

since $\varphi_I(S_0) = 1$. Summing over $k$ and observing that $\varphi_J(0) = 1$ only if $j = 0$, in which case $J \subset I$ and the inequality below is trivial, we obtain for each $j$:

$$\mathscr{E}\left\{\sum_{n=0}^{\infty} \varphi_J(S_n)\right\} \leq \mathscr{E}\left\{\sum_{n=0}^{\infty} \varphi_I(S_n)\right\}.$$

Now if we write $J_j$ for $J$ and sum over $j$ from $-m$ to $m - 1$, the inequality (8) ensues in disguised form.

This lemma is often demonstrated by a geometrical type of argument. We have written out the preceding proof in some detail as an example of the maxim: whatever can be shown by drawing pictures can also be set down in symbols!

**Lemma 2.** Let the positive numbers $\{u_n(m)\}$, where $n \in N$ and $m$ is a real number $\geq 1$, satisfy the following conditions:

- **(i)** $\forall_n: u_n(m)$ is increasing in $m$ and tends to 1 as $m \to \infty$;
- **(ii)** $\exists c > 0: \sum_{n=0}^{\infty} u_n(m) \leq cm \sum_{n=0}^{\infty} u_n(1)$ for all $m \geq 1$
- **(iii)** $\forall \delta > 0: \lim_{n\to\infty} u_n(\delta_n) = 1$.

Then we have

$$\sum_{n=0}^{\infty} u_n(1) = \infty. \tag{10}$$

*Remark.* If (ii) is true for all integer $m \geq 1$, then it is true for all real $m \geq 1$, with $c$ doubled.

PROOF OF LEMMA 2. Suppose not; then for every $A > 0$:

$$\infty > \sum_{n=0}^{\infty} u_n(1) \geq \frac{1}{cm} \sum_{n=0}^{\infty} u_n(m) \geq \frac{1}{cm} \sum_{n=0}^{[Am]} u_n(m)$$

$$\geq \frac{1}{cm} \sum_{n=0}^{[Am]} u_n \left(\frac{n}{A}\right).$$

Letting $m \to \infty$ and applying (iii) with $\delta = A^{-1}$, we obtain

$$\sum_{n=0}^{\infty} u_n(1) \geq \frac{A}{c};$$

since $A$ is arbitrary, this is a contradiction, which proves the lemma

To return to the proof of Theorem 8.3.3, we apply Lemma 2 with

$$u_n(m) = \mathscr{P}\{|S_n| < m\}.$$

Then condition (i) is obviously satisfied and condition (ii) with $c = 2$ follows from Lemma 1. The hypothesis that $S_n/n \to 0$ in pr. may be written as

$$u_n(\delta n) = \mathscr{P}\left\{\left|\frac{S_n}{n}\right| < \delta\right\} \to 1$$

for every $\delta > 0$ as $n \to \infty$, hence condition (iii) is also satisfied. Thus Lemma 2 yields

$$\sum_{n=0}^{\infty} \mathscr{P}\{|S_n| < 1\} = +\infty.$$

Applying this to the "magnified" random walk with each $X_n$ replaced by $X_n/\epsilon$, which does not disturb the hypothesis of the theorem, we obtain (6) for every $\epsilon > 0$, and so the theorem is proved.

In practice, the following criterion is more expedient (Chung and Fuchs, 1951).

**Theorem 8.3.4.** Suppose that at least one of $\mathscr{E}(X^+)$ and $\mathscr{E}(X^-)$ is finite. The $\mathfrak{R} \neq \phi$ if and only if $\mathscr{E}(X) = 0$; otherwise case (ii) or (iii) of Theorem 8.2.5 happens according as $\mathscr{E}(X) < 0$ or $> 0$.

PROOF. If $-\infty \leq \mathscr{E}(X) < 0$ or $0 < \mathscr{E}(X) \leq +\infty$, then by the strong law of large numbers (as amended by Exercise 1 of Sec. 5.4), we have

$$\frac{S_n}{n} \to \mathscr{E}(X) \text{ a.e.},$$

so that either (ii) or (iii) happens as asserted. If $\mathscr{E}(X) = 0$, then the same law or its weaker form Theorem 5.2.2 applies; hence Theorem 8.3.3 yields the conclusion.

DEFINITION OF RECURRENT RANDOM WALK. A random walk will be called *recurrent* iff $\mathfrak{R} \neq \varnothing$; it is *degenerate* iff $\mathfrak{R} = \{0\}$; and it is *of the lattice type* iff $\mathfrak{R}$ is generated by a nonzero element.

The most exact kind of recurrence happens when the $X_n$'s have a common distribution which is concentrated on the integers, such that every integer is a possible value (for some $S_n$), and which has mean zero. In this case for each integer $c$ we have $\mathscr{P}\{S_n = c \text{ i.o.}\} = 1$. For the symmetrical Bernoullian random walk this was first proved by Pólya in 1921.

We shall give another proof of the recurrence part of Theorem 8.3.4, namely that $\mathscr{E}(X) = 0$ is a sufficient condition for the random walk to be recurrent as just defined. This method is applicable in cases not covered by the two preceding theorems (see Exercises 6–9 below), and the analytic machinery of ch.f.'s which it employs opens the way to the considerations in the next section.

The starting point is the integrated form of the inversion formula in Exercise 3 of Sec. 6.2. Talking $x = 0$, $u = \epsilon$, and $F$ to be the d.f. of $S_n$, we have

$$(11) \qquad \mathscr{P}(|S_n| < \epsilon) \geq \frac{1}{\epsilon}\int_0^\epsilon \mathscr{P}(|S_n| < u)\,du = \frac{1}{\pi\epsilon}\int_{-\infty}^{\infty} \frac{1 - \cos\epsilon t}{t^2} f(t)^n\,dt.$$

Thus the series in (6) may be bounded below by summing the last expression in (11). The latter does not sum well as it stands, and it is natural to resort to a summability method. The Abelian method suits it well and leads to, for $0 < r < 1$:

$$(12) \qquad \sum_{n=0}^{\infty} r^n \mathscr{P}(|S_n| < \epsilon) \geq \frac{1}{\pi\epsilon}\int_{-\infty}^{\infty} \frac{1 - \cos\epsilon t}{t^2} \mathbf{R}\frac{1}{1 - rf(t)}\,dt,$$

where $\mathbf{R}$ and $\mathbf{I}$ later denote the real and imaginary parts or a complex quantity. Since

$$(13) \qquad \mathbf{R}\frac{1}{1 - rf(t)} \geq \frac{1 - r}{|1 - rf(t)|^2} > 0$$

and $(1 - \cos\epsilon t)/t^2 \geq C\epsilon^2$ for $|\epsilon t| < 1$ and some constant $C$, it follows that for $\eta < 1\backslash\epsilon$ the right member of (12) is not less than

$$(14) \qquad \frac{C\epsilon}{\pi}\int_{-\eta}^{\eta} \mathbf{R}\frac{1}{1 - rf(t)}\,dt.$$

Now the existence of $\mathscr{E}(|X|)$ implies by Theorem 6.4.2 that $1 - f(t) = o(t)$ as $t \to 0$. Hence for any given $\delta > 0$ we may choose the $\eta$ above so that

$$|1 - rf(t)|^2 \leq (1 - r + r[1 - \mathbf{R}f(t)])^2 + (r\mathbf{I}f(t))^2$$

$$\leq 2(1-r)^2 + 2(r\delta t)^2 + (r\delta t)^2 = 2(1-r)^2 + 3r^2\delta^2 t^2.$$

The integral in (14) is then not less than

$$\int_{-\eta}^{\eta} \frac{(1-r)\,dt}{2(1-r)^2 + 3r^2\delta^2 t^2}\,dt \geq \frac{1}{3}\int_{-\eta(1-r)^{-1}}^{\eta(1-r)^{-1}} \frac{ds}{1+\delta^2 s^2}.$$

As $r \uparrow 1$, the right member above tends to $\pi\backslash 3\delta$; since $\delta$ is arbitrary, we have proved that the right member of (12) tends to $+\infty$ as $r \uparrow 1$. Since the series in (6) dominates that on the left member of (12) for every $r$, it follows that (6) is true and so $0 \in \Re$ by Theorem 8.3.2.

## EXERCISES

$f$ is the ch.f. of $\mu$.

**1.** Generalize Theorems 8.3.1 and 8.3.2 to $\mathscr{R}^d$. (For $d \geq 3$ the generalization is illusory; see Exercise 12 below.)

***2.** If a random walk in $\mathscr{R}^d$ is recurrent, then every possible value is a recurrent value.

**3.** Prove the Remark after Theorem 8.3.2.

***4.** Assume that $\mathscr{P}\{X_1 = 0\} < 1$. Prove that $x$ is a recurrent value of the random walk if and only if

$$\sum_{n=1}^{\infty} \mathscr{P}\{|S_n - x| < \epsilon\} = \infty \quad \text{for every } \epsilon > 0.$$

**5.** For a recurrent random walk that is neither degenerate nor of the lattice type, the countable set of points $\{S_n(\omega), n \in N\}$ is everywhere dense in $\mathscr{R}^1$ for a.e.$\omega$. Hence prove the following result in Diophantine approximation: if $\gamma$ is irrational, then given any real $x$ and $\epsilon > 0$ there exist integers $m$ and $n$ such that $|m\gamma + n - x| < \epsilon$.

***6.** If there exists a $\delta > 0$ such that (the integral below being real-valued)

$$\overline{\lim_{r\uparrow 1}} \int_{-\delta}^{\delta} \frac{dt}{1 - rf(t)} = \infty,$$

then the random walk is recurrent.

***7.** If there exists a $\delta > 0$ such that

$$\sup_{0<r<1} \int_{-\delta}^{\delta} \frac{dt}{1 - rf(t)} < \infty,$$

then the random walk is not recurrent. [HINT: Use Exercise 3 of Sec. 6.2 to show that there exists a constant $C(\epsilon)$ such that

$$\mathscr{P}(|S_n| < \epsilon) \le C(\epsilon) \int_{\mathscr{R}^1} \frac{1 - \cos \epsilon^{-1} x}{x^2} \mu_n(dx) \le \frac{C(\epsilon)}{2} \int_0^{1/\epsilon} du \int_{-u}^{u} f(t)^n \, dt,$$

where $\mu_n$ is the distribution of $S_n$, and use (13). [Exercises 6 and 7 give, respectively, a necessary and a sufficient condition for recurrence. If $\mu$ is of the integer lattice type with span one, then

$$\int_{-\pi}^{\pi} \frac{1}{1 - f(t)} \, dt = \infty$$

is such a condition according to Kesten and Spitzer.]

★**8.** Prove that the random walk with $f(t) = e^{-|t|}$ (Cauchy distribution) is recurrent.

★**9.** Prove that the random walk with $f(t) = e^{-|t|^\alpha}$, $0 < \alpha < 1$, (Stable law) is not recurrent, but (iv) of Theorem 8.2.5 holds.

**10.** Generalize Exercises 6 and 7 above to $\mathscr{R}^d$.

★**11.** Prove that in $\mathscr{R}^2$ if the common distribution of the random vector $(X, Y)$ has mean zero and finite second moment, namely:

$$\mathscr{E}(X) = 0, \quad \mathscr{E}(Y) = 0, \quad 0 < \mathscr{E}(X^2 + Y^2) < \infty,$$

then the random walk is recurrent. This implies by Exercise 5 above that almost every Brownian motion path is everywhere dense in $\mathscr{R}^2$. [HINT: Use the generalization of Exercise 6 and show that

$$\mathbf{R} \frac{1}{1 - f(t_1, t_2)} \ge \frac{c}{t_1^2 + t_2^2}$$

for sufficiently small $|t_1| + |t_2|$. One can also make a direct estimate:

$$\mathscr{P}(|S_n| < \epsilon) \ge \frac{c'}{n}.]$$

★**12.** Prove that no truly 3-dimensional random walk, namely one whose common distribution does not have its support in a plane, is recurrent. [HINT: There exists $A > 0$ such that

$$\iiint_{-A}^{A} \left( \sum_{i=1}^{3} t_i x_i \right)^2 \mu(dx)$$

is a strictly positive quadratic form $Q$ in $(t_1, t_2, t_3)$. If

$$\sum_{i=1}^{3} |t_i| < A^{-1},$$

then

$$\mathbf{R}\{1 - f(t_1, t_2, t_3)\} \geq CQ(t_1, t_2, t_3).]$$

**13.** Generalize Lemma 1 in the proof of Theorem 8.3.3 to $\mathscr{R}^d$. For $d = 2$ the constant $2m$ in the right member of (8) is to be replaced by $4m^2$, and "$S_n < \epsilon$" means $S_n$ is in the open square with center at the origin and side length $2\epsilon$.

**14.** Extend Lemma 2 in the proof of Theorem 8.3.3 as follows. Keep condition (i) but replace (ii) and (iii) by

(ii′) $\sum_{n=0}^{x} u_n(m) \leq cm^2 \sum_{n=0}^{x} u_n(1)$;

There exists $d > 0$ such that for every $b > 1$ and $m \geq m(b)$:

(iii′) $u_n(m) \geq \dfrac{dm^2}{n}$ for $m^2 \leq n \leq (bm)^2$

Then (10) is true.*

**15.** Generalize Theorem 8.3.3 to $\mathscr{R}^2$ as follows. If the central limit theorem applies in the form that $S_n \mid \sqrt{n}$ converges in dist. to the unit normal, then the random walk is recurrent. [HINT: Use Exercises 13 and 14 and Exercise 4 of § 4.3. This is sharper than Exercise 11. No proof of Exercise 12 using a similar method is known.]

**16.** Suppose $\mathscr{E}(X) = 0$, $0 < \mathscr{E}(X^2) < \infty$, and $\mu$ is of the integer lattice type, then

$$\mathscr{P}\{S_{n^2} = 0 \text{ i.o.}\} = 1.$$

**17.** The basic argument in the proof of Theorem 8.3.2 was the "last time in $(-\epsilon, \epsilon)$". A harder but instructive argument using the "first time" may be given as follows.

$$f_n^{m} = \mathscr{P}\{|S_j| \geq \epsilon \text{ for } m \leq j \leq n-1; |S_n| < \epsilon\}. \quad g_n(\epsilon) = \mathscr{P}\{|S_n| < \epsilon\}.$$

Show that for $1 \leq m < M$:

$$\sum_{n=m}^{M} g_n(\epsilon) \leq \sum_{n=m}^{M} f_n^{(m)} \sum_{n=0}^{M} g_n(2\epsilon).$$

* This form of condition (iii′) is due to Hsu Pei; see also Chung and Lindvall, Proc. Amer. Math. Soc. Vol. 78 (1980), p. 285.

It follows by a form of Lemma 1 in this section that

$$\lim_{m\to x}\sum_{n=m}^{x} f_n^{(m)} \geq \frac{1}{4};$$

now use Theorem 8.1.4.

**18.** For an arbitrary random walk, if $\mathscr{P}\{\forall n \in N : S_n > 0\} > 0$, then

$$\sum_n \mathscr{P}\{S_n \leq 0, S_{n+1} > 0\} < \infty.$$

Hence if in addition $\mathscr{P}\{\forall n \in N : S_n \leq 0\} > 0$, then

$$\sum_n |\mathscr{P}\{S_n > 0\} - \mathscr{P}\{S_{n+1} > 0\}| < \infty;$$

and consequently

$$\sum_n \frac{(-1)^n}{n}\mathscr{P}\{S_n > 0\} < \infty.$$

[HINT: For the first series, consider the last time that $S_n \leq 0$; for the third series, apply Du Bois–Reymond's test. Cf. Theorem 8.4.4 below; this exercise will be completed in Exercise 15 of Sec. 8.5.]

## 8.4 Fine structure

In this section we embark on a probing in depth of the r.v. $\alpha$ defined in (11) of Sec. 8.2 and some related r.v.'s.

The r.v. $\alpha$ being optional, the key to its introduction is to break up the time sequence into a pre-$\alpha$ and a post-$\alpha$ era, as already anticipated in the terminology employed with a general optional r.v. We do this with a sort of characteristic functional of the process which has already made its appearance in the last section:

$$\mathscr{E}\left\{\sum_{n=0}^{\infty} r^n e^{itS_n}\right\} = \sum_{n=0}^{\infty} r^n f(t)^n = \frac{1}{1 - rf(t)}, \tag{1}$$

where $0 < r < 1$, $t$ is real, and $f$ is the ch.f. of $X$. Applying the principle just enunciated, we break this up into two parts:

$$\mathscr{E}\left\{\sum_{n=0}^{\alpha-1} r^n e^{itS_n}\right\} + \mathscr{E}\left\{\sum_{n=\alpha}^{\infty} r^n e^{itS_n}\right\}$$

with the understanding that on the set $\{\alpha = \infty\}$, the first sum above is $\sum_{n=0}^{\infty}$ while the second is empty and hence equal to zero. Now the second part may be written as

$$(2) \qquad \mathscr{E}\left\{\sum_{n=0}^{\infty} r^{\alpha+n} e^{itS_{\alpha+n}}\right\} = \mathscr{E}\left\{r^{\alpha} e^{itS_{\alpha}} \sum_{n=0}^{\infty} r^{n} e^{it(S_{\alpha+n}-S_{\alpha})}\right\}.$$

It follows from (7) of Sec. 8.2 that

$$S_{\alpha+n} - S_{\alpha} = S_{n} \circ \tau^{\alpha}$$

has the same distribution as $S_n$, and by Theorem 8.2.2 that for each $n$ it is independent of $S_\alpha$. [Note that the same fact has been used more than once before, but for a constant $\alpha$.] Hence the right member of (2) is equal to

$$\mathscr{E}\{r^{\alpha} e^{itS_{\alpha}}\}\mathscr{E}\left\{\sum_{n=0}^{\infty} r^{n} e^{itS_{n}}\right\} = \mathscr{E}\{r^{\alpha} e^{itS_{\alpha}}\}\frac{1}{1 - rf(t)},$$

where $r^{\alpha} e^{itS_{\alpha}}$ is taken to be 0 for $\alpha = \infty$. Substituting this into (1), we obtain

$$(3) \qquad \frac{1}{1 - rf(t)}[1 - \mathscr{E}\{r^{\alpha} e^{itS_{\alpha}}\}] = \mathscr{E}\left\{\sum_{n=0}^{\alpha-1} r^{n} e^{itS_{n}}\right\}.$$

We have

$$(4) \qquad 1 - \mathscr{E}\{r^{\alpha} e^{itS_{\alpha}}\} = 1 - \sum_{n=1}^{\infty} r^{n} \int_{\{\alpha=n\}} e^{itS_{n}} d\mathscr{P};$$

and

$$(5) \qquad \mathscr{E}\left\{\sum_{n=0}^{\alpha-1} r^{n} e^{itS_{n}}\right\} = \sum_{k=1}^{\infty} \int_{\{\alpha=k\}} \sum_{n=0}^{k-1} r^{n} e^{itS_{n}}\, d\mathscr{P}$$

$$= 1 + \sum_{n=1}^{\infty} r^{n} \int_{\{\alpha>n\}} e^{itS_{n}}\, d\mathscr{P}$$

by an interchange of summation. Let us record the two power series appearing in (4) and (5) as

$$P(r,t) = 1 - \mathscr{E}\{r^{\alpha} e^{itS_{\alpha}}\} = \sum_{n=0}^{\infty} r^{n} p_{n}(t);$$

$$Q(r,t) = \mathscr{E}\left\{\sum_{n=0}^{\alpha-1} r^{n} e^{itS_{n}}\right\} = \sum_{n=0}^{\infty} r^{n} q_{n}(t),$$

where

$$p_0(t) \equiv 1,\ p_n(t) = -\int_{\{\alpha=n\}} e^{itS_n}\, d\mathscr{P} = -\int_{\mathscr{R}^1} e^{itx} U_n(dx);$$

$$q_0(t) \equiv 1,\ q_n(t) = \int_{\{\alpha>n\}} e^{itS_n}\, d\mathscr{P} = \int_{\mathscr{R}^1} e^{itx} V_n(dx).$$

Now $U_n(\cdot) = \mathscr{P}\{\alpha = n; S_n \in \cdot\}$ is a finite measure with support in $(0, \infty)$, while $V_n(\cdot) = \mathscr{P}\{\alpha > n; S_n \in \cdot\}$ is a finite measure with support in $(-\infty, 0]$. Thus each $p_n$ is the Fourier transform of a finite measure in $(0, \infty)$, while each $q_n$ is the Fourier transform of a finite measure in $(-\infty, 0]$. Returning to (3), we may now write it as

(6) $$\frac{1}{1 - rf(t)} P(r, t) = Q(r, t).$$

The next step is to observe the familiar Taylor series:

$$\frac{1}{1-x} = \exp\left\{\sum_{n=1}^{\infty} \frac{x^n}{n}\right\}, \qquad |x| < 1.$$

Thus we have

(7) $$\begin{aligned} \frac{1}{1 - rf(t)} &= \exp\left\{\sum_{n=1}^{\infty} \frac{r^n}{n} f(t)^n\right\} \\ &= \exp\left\{\sum_{n=1}^{\infty} \frac{r^n}{n}\left[\int_{\{S_n>0\}} e^{itS_n}\, d\mathscr{P} + \int_{\{S_n \le 0\}} e^{itS_n}\, d\mathscr{P}\right]\right\} \\ &= f_+(r, t)^{-1} f_-(r, t), \end{aligned}$$

where

$$f_+(r, t) = \exp\left\{-\sum_{n=1}^{\infty} \frac{r^n}{n} \int_{(0,\infty)} e^{itx} \mu_n(dx)\right\},$$

$$f_-(r, t) = \exp\left\{+\sum_{n=1}^{\infty} \frac{r^n}{n} \int_{(-\infty,0]} e^{itx} \mu_n(dx)\right\},$$

and $\mu_n(\cdot) = \mathscr{P}\{S_n \in \cdot\}$ is the distribution of $S_n$. Since the convolution of two measures both with support in $(0, \infty)$ has support in $(0, \infty)$, and the convolution of two measures both with support in $(-\infty, 0]$ has support in $(-\infty, 0]$, it follows by expansion of the exponential functions above and rearrangements of the resulting double series that

$$f_+(r,t) = 1 + \sum_{n=1}^{\infty} r^n \varphi_n(t), \quad f_-(r,t) = 1 + \sum_{n=1}^{\infty} r^n \psi_n(t),$$

where each $\varphi_n$ is the Fourier transform of a measure in $(0, \infty)$, while each $\psi_n$ is the Fourier transform of a measure in $(-\infty, 0]$. Substituting (7) into (6) and multiplying through by $f_+(r,t)$, we obtain

$$P(r,t)f_-(r,t) = Q(r,t)f_+(r,t). \tag{8}$$

The next theorem below supplies the basic analytic technique for this development, known as the Wiener–Hopf technique.

**Theorem 8.4.1.** Let

$$P(r,t) = \sum_{n=0}^{\infty} r^n p_n(t), \quad Q(r,t) = \sum_{n=0}^{\infty} r^n q_n(t),$$

$$P^*(r,t) = \sum_{n=0}^{\infty} r^n p_n^*(t), \quad Q^*(r,t) = \sum_{n=0}^{\infty} r^n q_n^*(t),$$

where $p_0(t) \equiv q_0(t) \equiv p_0^*(t) \equiv q_0^*(t) \equiv 1$; and for $n \geq 1$, $p_n$ and $p_n^*$ as functions of $t$ are Fourier transforms of measures with support in $(0, \infty)$; $q_n$ and $q_n^*$ as functions of $t$ are Fourier transforms of measures in $(-\infty, 0]$. Suppose that for some $r_0 > 0$ the four power series converge for $r$ in $(0, r_0)$ and all real $t$, and the identity

$$P(r,t)Q^*(r,t) \equiv P^*(r,t)Q(r,t) \tag{9}$$

holds there. Then

$$P \equiv P^*, Q \equiv Q^*.$$

The theorem is also true if $(0, \infty)$ and $(-\infty, 0]$ are replaced by $[0, \infty)$ and $(-\infty, 0)$, respectively.

PROOF. It follows from (9) and the identity theorem for power series that for every $n \geq 0$:

$$\sum_{k=0}^{n} p_k(t)q_{n-k}^*(t) = \sum_{k=0}^{n} p_k^*(t)q_{n-k}(t). \tag{10}$$

Then for $n = 1$ equation (10) reduces to the following:

$$p_1(t) - p_1^*(t) = q_1(t) - q_1^*(t).$$

By hypothesis, the left member above is the Fourier transform of a finite signed measure $\nu_1$ with support in $(0, \infty)$, while the right member is the Fourier transform of a finite signed measure with support in $(-\infty, 0]$. It follows from the uniqueness theorem for such transforms (Exercise 13 of Sec. 6.2) that we must have $\nu_1 \equiv \nu_2$, and so both must be identically zero since they have disjoint supports. Thus $p_1 \equiv p_1^*$ and $q_1 \equiv q_1^*$. To proceed by induction on $n$, suppose that we have proved that $p_j \equiv p_j^*$ and $q_j \equiv q_j^*$ for $0 \leq j \leq n-1$. Then it follows from (10) that

$$p_n(t) + q_n^*(t) \equiv p_n^*(t) + q_n(t).$$

Exactly the same argument as before yields that $p_n \equiv p_n^*$ and $q_n \equiv q_n^*$. Hence the induction is complete and the first assertion of the theorem is proved; the second is proved in the same way.

Applying the preceding theorem to (8), we obtain the next theorem in the case $A = (0, \infty)$; the rest is proved in exactly the same way.

**Theorem 8.4.2.** If $\alpha = \alpha_A$ is the first entrance time into $A$, where $A$ is one of the four sets: $(0, \infty)$, $[0, \infty)$, $(-\infty, 0)$, $(-\infty, 0]$, then we have

$$1 - \mathscr{E}\{r^\alpha e^{itS_\alpha}\} = \exp\left\{-\sum_{n=1}^{\infty} \frac{r^n}{n} \int_{\{S_n \in A\}} e^{itS_n} \, d\mathscr{P}\right\}; \tag{11}$$

$$\mathscr{E}\left\{\sum_{n=0}^{\alpha-1} r^n e^{itS_n}\right\} = \exp\left\{+\sum_{n=1}^{\infty} \frac{r^n}{n} \int_{\{S_n \in A^c\}} e^{itS_n} \, d\mathscr{P}\right\}. \tag{12}$$

From this result we shall deduce certain analytic expressions involving the r.v. $\alpha$. Before we do that, let us list a number of classical Abelian and Tauberian theorems below for ready reference.

**(A)** If $c_n \geq 0$ and $\sum_{n=0}^{\infty} c_n r^n$ converges for $0 \leq r < 1$, then

$$\lim_{r \uparrow 1} \sum_{n=0}^{\infty} c_n r^n = \sum_{n=0}^{\infty} c_n \tag{$*$}$$

finite or infinite.

**(B)** If $c_n$ are complex numbers and $\sum_{n=0}^{\infty} c_n r^n$ converges for $0 \leq r \leq 1$, then (*) is true.

**(C)** If $c_n$ are complex numbers such that $c_n = o(n^{-1})$ [or just $O(n^{-1})$] as $n \to \infty$, and the limit in the left member of $(*)$ exists and is finite, then $(*)$ is true.

**(D)** If $c_n^{(i)} \geq 0$, $\sum_{n=0}^{\infty} c_n^{(i)} r^n$ converges for $0 \leq r < 1$ and diverges for $r = 1$, $i = 1, 2$; and

$$\sum_{k=0}^{n} c_k{}^{(1)} \sim K \sum_{k=0}^{n} c_n^{(2)}$$

[or more particularly $c_n^{(1)} \sim K c_n^{(2)}$] as $n \to \infty$, where $0 \leq K \leq +\infty$, then

$$\sum_{n=0}^{\infty} c_n^{(1)} r^n \sim K \sum_{n=0}^{\infty} c_n^{(2)} r^n$$

as $r \uparrow 1$.

**(E)** If $c_n \geq 0$ and

$$\sum_{n=0}^{\infty} c_n r^n \sim \frac{1}{1-r}$$

as $r \uparrow 1$, then

$$\sum_{k=0}^{n-1} c_k \sim n.$$

Observe that (C) is a partial converse of (B), and (E) is a partial converse of (D). There is also an analogue of (D), which is actually a consequence of (B): if $c_n$ are complex numbers converging to a finite limit $c$, then as $r \uparrow 1$,

$$\sum_{n=0}^{\infty} c_n r^n \sim \frac{c}{1-r}.$$

Proposition (A) is trivial. Proposition (B) is Abel's theorem, and proposition (C) is Tauber's theorem in the "little $o$" version and Littlewood's theorem in the "big $O$" version; only the former will be needed below. Proposition (D) is an Abelian theorem, and proposition (E) a Tauberian theorem, the latter being sometimes referred to as that of Hardy-Littlewood-Karamata. All four can be found in the admirable book by Titchmarsh, *The theory of functions* (2nd ed., Oxford University Press, Inc., New York, 1939, pp. 9–10, 224 ff.).

**Theorem 8.4.3.** The generating function of $\alpha$ in Theorem 8.4.2 is given by

$$\text{(13)} \qquad \begin{aligned} \mathscr{E}\{r^\alpha\} &= 1 - \exp\left\{-\sum_{n=1}^{\infty} \frac{r^n}{n} \mathscr{P}[S_n \in A]\right\} \\ &= 1 - (1-r)\exp\left\{\sum_{n=1}^{\infty} \frac{r^n}{n} \mathscr{P}[S_n \in A^c]\right\}. \end{aligned}$$

We have

$$\text{(14)} \qquad \mathscr{P}\{\alpha < \infty\} = 1 \text{ if and only if } \sum_{n=1}^{\infty} \frac{1}{n} \mathscr{P}[S_n \in A] = \infty;$$

in which case

$$\text{(15)} \qquad \mathscr{E}\{\alpha\} = \exp\left\{\sum_{n=1}^{\infty} \frac{1}{n} \mathscr{P}[S_n \in A^c]\right\}.$$

PROOF. Setting $t = 0$ in (11), we obtain the first equation in (13), from which the second follows at once through

$$\frac{1}{1-r} = \exp\left\{\sum_{n=1}^{\infty} \frac{r^n}{n}\right\} = \exp\left\{\sum_{n=1}^{\infty} \frac{r^n}{n} \mathscr{P}[S_n \in A] + \sum_{n=1}^{\infty} \frac{r^n}{n} \mathscr{P}[S_n \in A^c]\right\}.$$

Since

$$\lim_{r \uparrow 1} \mathscr{E}\{r^\alpha\} = \lim_{r \uparrow 1} \sum_{n=1}^{\infty} \mathscr{P}\{\alpha = n\} r^n = \sum_{n=1}^{\infty} \mathscr{P}\{\alpha = n\} = \mathscr{P}\{\alpha < \infty\}$$

by proposition (A), the middle term in (13) tends to a finite limit, hence also the power series in $r$ there (why?). By proposition (A), the said limit may be obtained by setting $r = 1$ in the series. This establishes (14). Finally, setting $t = 0$ in (12), we obtain

$$\text{(16)} \qquad \mathscr{E}\left\{\sum_{n=0}^{\alpha-1} r^n\right\} = \exp\left\{+\sum_{n=1}^{\infty} \frac{r^n}{n} \mathscr{P}[S_n \in A^c]\right\}.$$

Rewriting the left member in (16) as in (5) and letting $r \uparrow 1$, we obtain

$$\text{(17)} \qquad \lim_{r \uparrow 1} \sum_{n=0}^{\infty} r^n \mathscr{P}[\alpha > n] = \sum_{n=0}^{\infty} \mathscr{P}[\alpha > n] = \mathscr{E}\{\alpha\} \leq \infty$$

by proposition (A). The right member of (16) tends to the right member of (15) by the same token, proving (15).

When is $\mathscr{E}\{\alpha\}$ in (15) finite? This is answered by the next theorem.*

**Theorem 8.4.4.** Suppose that $X \not\equiv 0$ and at least one of $\mathscr{E}(X^+)$ and $\mathscr{E}(X^-)$ is finite; then

$$\text{(18)} \qquad \mathscr{E}(X) > 0 \Rightarrow \mathscr{E}\{\alpha_{(0,\infty)}\} < \infty;$$

$$\text{(19)} \qquad \mathscr{E}(X) \leq 0 \Rightarrow \mathscr{E}\{\alpha_{[0,\infty)}\} = \infty.$$

* It can be shown that $S_n \to +\infty$ a.e. if and only if $\mathscr{E}\{\alpha_{(0,\infty)}\} < \infty$; see A. J. Lemoine, *Annals of Probability* **2**(1974).

PROOF. If $\mathscr{E}(X) > 0$, then $\mathscr{P}\{S_n \to +\infty\} = 1$ by the strong law of large numbers. Hence $\mathscr{P}\{\lim_{n\to\infty} S_n = -\infty\} = 0$, and this implies by the dual of Theorem 8.2.4 that $\mathscr{P}\{\alpha_{(-\infty,0)} < \infty\} < 1$. Let us sharpen this slightly to $\mathscr{P}\{\alpha_{(-\infty,0]} < \infty\} < 1$. To see this, write $\alpha' = \alpha_{(-\infty,0]}$ and consider $S_{\beta'_n}$ as in the proof of Theorem 8.2.4. Clearly $S_{\beta'_n} \le 0$, and so if $\mathscr{P}\{\alpha' < \infty\} = 1$, one would have $\mathscr{P}\{S_n \le 0 \text{ i.o.}\} = 1$, which is impossible. Now apply (14) to $\alpha_{(-\infty,0]}$ and (15) to $\alpha_{(0,\infty)}$ to infer

$$\mathscr{E}\{\alpha_{(0,\infty)}\} = \exp\left\{\sum_{n=1}^{\infty} \frac{1}{n}\mathscr{P}[S_n \le 0]\right\} < \infty,$$

proving (18). Next if $\mathscr{E}(X) = 0$, then $\mathscr{P}\{\alpha_{(-\infty,0)} < \infty\} = 1$ by Theorem 8.3.4. Hence, applying (14) to $\alpha_{(-\infty,0)}$ and (15) to $\alpha_{[0,\infty)}$, we infer

$$\mathscr{E}\{\alpha_{[0,\infty)}\} = \exp\left\{\sum_{n=1}^{\infty} \frac{1}{n}\mathscr{P}[S_n < 0]\right\} = \infty.$$

Finally, if $\mathscr{E}(X) < 0$, then $\mathscr{P}\{\alpha_{[0,\infty)} = \infty\} > 0$ by an argument dual to that given above for $\alpha_{(-\infty,0]}$, and so $\mathscr{E}\{\alpha_{[0,\infty)}\} = \infty$ trivially.

Incidentally we have shown that the two r.v.'s $\alpha_{(0,\infty)}$ and $\alpha_{[0,\infty)}$ have both finite or both infinite expectations. Comparing this remark with (15), we derive an analytic by-product as follows.

**Corollary.** We have

$$\sum_{n=1}^{\infty} \frac{1}{n}\mathscr{P}[S_n = 0] < \infty. \tag{20}$$

This can also be shown, purely analytically, by means of Exercise 25 of Sec. 6.4.

The astonishing part of Theorem 8.4.4 is the case when the random walk is recurrent, which is the case if $\mathscr{E}(X) = 0$ by Theorem 8.3.4. Then the set $[0, \infty)$, which is more than half of the whole range, is revisited an infinite number of times. Nevertheless (19) says that the expected time for even one visit is infinite! This phenomenon becomes more paradoxical if one reflects that the same is true for the other half $(-\infty, 0]$, and yet in a single step one of the two halves will certainly be visited. Thus we have:

$$\alpha_{(-\infty,0]} \wedge \alpha_{[0,\infty)} = 1, \qquad \mathscr{E}\{\alpha_{(-\infty,0]}\} = \mathscr{E}\{\alpha_{[0,\infty)}\} = \infty.$$

Another curious by-product concerning the strong law of large numbers is obtained by combining Theorem 8.4.3 with Theorem 8.2.4.

**Theorem 8.4.5.** $S_n/n \to m$ a.e. for a finite constant $m$ if and only if for every $\epsilon > 0$ we have

$$\sum_{n=1}^{\infty} \frac{1}{n} \mathscr{P}\left\{\left|\frac{S_n}{n} - m\right| > \epsilon\right\} < \infty. \tag{21}$$

PROOF. Without loss of generality we may suppose $m = 0$. We know from Theorem 5.4.2 that $S_n/n \to 0$ a.e. if and only if $\mathscr{E}(|X|) < \infty$ and $\mathscr{E}(X) = 0$. If this is so, consider the stationary independent process $\{X'_n, n \in N\}$, where $X'_n = X_n - \epsilon, \epsilon > 0$; and let $S'_n$ and $\alpha' = \alpha'_{(0,\infty)}$ be the corresponding r.v.'s for this modified process. Since $\mathscr{E}(X') = -\epsilon$, it follows from the strong law of large numbers that $S'_n \to -\infty$ a.e., and consequently by Theorem 8.2.4 we have $\mathscr{P}\{\alpha' < \infty\} < 1$. Hence we have by (14) applied to $\alpha'$:

$$\sum_{n=1}^{\infty} \frac{1}{n} \mathscr{P}[S_n - n\epsilon > 0] < \infty. \tag{22}$$

By considering $X_n + \epsilon$ instead of $X_n - \epsilon$, we obtain a similar result with "$S_n - n\epsilon > 0$" in (22) replaced by "$S_n + n\epsilon < 0$". Combining the two, we obtain (21) when $m = 0$.

Conversely, if (21) is true with $m = 0$, then the argument above yields $\mathscr{P}\{\alpha' < \infty\} < 1$, and so by Theorem 8.2.4, $\mathscr{P}\{\overline{\lim}_{n\to\infty} S'_n = +\infty\} = 0$. *A fortiori* we have

$$\forall \epsilon > 0 : \mathscr{P}\{S'_n > n\epsilon \text{ i.o.}\} = \mathscr{P}\{S_n > 2n\epsilon \text{ i.o.}\} = 0.$$

Similarly we obtain $\forall \epsilon > 0 : \mathscr{P}\{S_n < -2n\epsilon \text{ i.o.}\} = 0$, and the last two relations together mean exactly $S_n/n \to 0$ a.e. (cf. Theorem 4.2.2).

Having investigated the stopping time $\alpha$, we proceed to investigate the *stopping place* $S_\alpha$, where $\alpha < \infty$. The crucial case will be handled first.

**Theorem 8.4.6.** If $\mathscr{E}(X) = 0$ and $0 < \mathscr{E}(X^2) = \sigma^2 < \infty$, then

$$\mathscr{E}\{S_\alpha\} = \frac{\sigma}{\sqrt{2}} \exp\left\{\sum_{n=1}^{\infty} \frac{1}{n}\left[\frac{1}{2} - \mathscr{P}(S_n \in A)\right]\right\} < \infty. \tag{23}$$

PROOF. Observe that $\mathscr{E}(X) = 0$ implies each of the four r.v.'s $\alpha$ is finite a.e. by Theorem 8.3.4. We now switch from Fourier transform to Laplace transform in (11), and suppose for the sake of definiteness that $A = (0, \infty)$. It is readily verified that Theorem 6.6.5 is applicable, which yields

$$1 - \mathscr{E}\{r^\alpha e^{-\lambda S_\alpha}\} = \exp\left\{-\sum_{n=1}^{\infty} \frac{r^n}{n} \int_{\{S_n > 0\}} e^{-\lambda S_n}\, d\mathscr{P}\right\} \tag{24}$$

for $0 \le r < 1$, $0 \le \lambda < \infty$. Letting $r \uparrow 1$ in (24), we obtain an expression for the Laplace transform of $S_\alpha$, but we must go further by differentiating (24) with respect to $\lambda$ to obtain

$$\text{(25)}\quad \mathscr{E}\{r^\alpha e^{-\lambda S_\alpha} S_\alpha\}$$

$$= \sum_{n=1}^{\infty} \frac{r^n}{n} \int_{\{S_n > 0\}} S_n e^{-\lambda S_n}\, d\mathscr{P} \cdot \exp\left\{ -\sum_{n=1}^{\infty} \frac{r^n}{n} \int_{\{S_n > 0\}} e^{-\lambda S_n}\, d\mathscr{P} \right\},$$

the justification for termwise differentiation being easy, since $\mathscr{E}\{|S_n|\} \le n\mathscr{E}\{|X|\}$. If we now set $\lambda = 0$ in (25), the result is

$$\text{(26)}\qquad \mathscr{E}\{r^\alpha S_\alpha\} = \sum_{n=1}^{\infty} \frac{r^n}{n} \mathscr{E}\{S_n^+\} \exp\left\{ -\sum_{n=1}^{\infty} \frac{r^n}{n} \mathscr{P}[S_n > 0] \right\}.$$

By Exercise 2 of Sec. 6.4, we have as $n \to \infty$,

$$\mathscr{E}\left\{ \frac{S_n^+}{n} \right\} \sim \frac{\sigma}{\sqrt{2\pi n}},$$

so that the coefficients in the first power series in (26) are asymptotically equal to $\sigma/\sqrt{2}$ times those of

$$(1-r)^{-1/2} = \sum_{n=0}^{\infty} \frac{1}{2^{2n}} \binom{2n}{n} r^n,$$

since

$$\frac{1}{2^{2n}} \binom{2n}{n} \sim \frac{1}{\sqrt{\pi n}}.$$

It follows from proposition (D) above that

$$\sum_{n=1}^{\infty} \frac{r^n}{n} \mathscr{E}\{S_n^+\} \sim \frac{\sigma}{\sqrt{2}} (1-r)^{-1/2} = \frac{\sigma}{\sqrt{2}} \exp\left\{ +\sum_{n=1}^{\infty} \frac{r^n}{2n} \right\}.$$

Substituting into (26), and observing that as $r \uparrow 1$, the left member of (26) tends to $\mathscr{E}\{S_\alpha\} \le \infty$ by the monotone convergence theorem, we obtain

$$\text{(27)}\qquad \mathscr{E}\{S_\alpha\} = \frac{\sigma}{\sqrt{2}} \lim_{r \uparrow 1} \exp\left\{ \sum_{n=1}^{\infty} \frac{r^n}{n} \left[ \frac{1}{2} - \mathscr{P}(S_n > 0) \right] \right\}.$$

It remains to prove that the limit above is finite, for then the limit of the power series is also finite (why? it is precisely here that the Laplace transform saves the day for us), and since the coefficients are $o(1/n)$ by the central limit theorem, and certainly $O(1/n)$ in any event, proposition (C) above will identify it as the right member of (23) with $A = (0, \infty)$.

Now by analogy with (27), replacing $(0, \infty)$ by $(-\infty, 0]$ and writing $\alpha_{(-\infty,0]}$ as $\beta$, we have

$$\mathscr{E}\{S_\beta\} = \frac{\sigma}{\sqrt{2}} \lim_{r \uparrow 1} \exp\left\{\sum_{n=1}^{\infty} \frac{r^n}{n}\left[\frac{1}{2} - \mathscr{P}(S_n \le 0)\right]\right\}. \tag{28}$$

Clearly the product of the two exponentials in (27) and (28) is just $\exp 0 = 1$, hence if the limit in (27) were $+\infty$, that in (28) would have to be 0. But since $\mathscr{E}(X) = 0$ and $\mathscr{E}(X^2) > 0$, we have $\mathscr{P}(X < 0) > 0$, which implies at once $\mathscr{P}(S_\beta < 0) > 0$ and consequently $\mathscr{E}\{S_\beta\} < 0$. This contradiction proves that the limits in (27) and (28) must both be finite, and the theorem is proved.

**Theorem 8.4.7.** Suppose that $X \not\equiv 0$ and at least one of $\mathscr{E}(X^+)$ and $\mathscr{E}(X^-)$ is finite; and let $\alpha = \alpha_{(0,\infty)}$, $\beta = \alpha_{(-\infty,0]}$.

**(i)** If $\mathscr{E}(X) > 0$ but may be $+\infty$, then $\mathscr{E}(S_\alpha) = \mathscr{E}(\alpha)\mathscr{E}(X)$.

**(ii)** If $\mathscr{E}(X) = 0$, then $\mathscr{E}(S_\alpha)$ and $\mathscr{E}(S_\beta)$ are both finite if and only if $\mathscr{E}(X^2) < \infty$.

PROOF. The assertion (i) is a consequence of (18) and Wald's equation (Theorem 5.5.3 and Exercise 8 of Sec. 5.5). The "if" part of assertion (ii) has been proved in the preceding theorem; indeed we have even "evaluated" $\mathscr{E}(S_\alpha)$. To prove the "only if" part, we apply (11) to both $\alpha$ and $\beta$ in the preceding proof and multiply the results together to obtain the remarkable equation:

$$[1 - \mathscr{E}\{r^\alpha e^{itS_\alpha}\}][1 - \mathscr{E}\{r^\beta e^{itS_\beta}\}] = \exp\left\{-\sum_{n=1}^{\infty} \frac{r^n}{n} f(t)^n\right\} = 1 - rf(t). \tag{29}$$

Setting $r = 1$, we obtain for $t \neq 0$:

$$\frac{1 - f(t)}{t^2} = \frac{1 - \mathscr{E}\{e^{itS_\alpha}\}}{-it} \frac{1 - \mathscr{E}\{e^{itS_\beta}\}}{+it}.$$

Letting $t \downarrow 0$, the right member above tends to $\mathscr{E}\{S_\alpha\}\mathscr{E}\{-S_\beta\}$ by Theorem 6.4.2. Hence the left member has a real and finite limit. This implies $\mathscr{E}(X^2) < \infty$ by Exercise 1 of Sec. 6.4.

## 8.5 Continuation

Our next study concerns the r.v.'s $M_n$ and $M$ defined in (12) and (13) of Sec. 8.2. It is convenient to introduce a new r.v. $L_n$, which is *the first time* (necessarily belonging to $N_n^0$) *that* $M_n$ *is attained*:

(1) $$\forall n \in N^0: \quad L_n(\omega) = \min\{k \in N_n^0 : S_k(\omega) = M_n(\omega)\};$$

note that $L_0 = 0$. We shall also use the abbreviation

(2) $$\alpha = \alpha_{(0,\infty)}, \quad \beta = \alpha_{(-\infty,0]}$$

as in Theorems 8.4.6 and 8.4.7.

For each $n$ consider the special permutation below:

(3) $$\rho_n = \begin{pmatrix} 1, & 2, & \dots, & n \\ n, & n-1, & \dots, & 1 \end{pmatrix},$$

which amounts to a "reversal" of the ordered set of indices in $N_n$. Recalling that $\beta \circ \rho_n$ is the r.v. whose value at $\omega$ is $\beta(\rho_n \omega)$, and $S_k \circ \rho_n = S_n - S_{n-k}$ for $1 \le k \le n$, we have

$$\{\beta \circ \rho_n > n\} = \bigcap_{k=1}^{n} \{S_k \circ \rho_n > 0\}$$

$$= \bigcap_{k=1}^{n} \{S_n > S_{n-k}) = \bigcap_{k=0}^{n-1} \{S_n > S_k\} = \{L_n = n\}.$$

It follows from (8) of Sec. 8.1 that

(4) $$\int_{\{\beta > n\}} e^{itS_n} \, d\mathscr{P} = \int_{\{\beta \circ \rho_n > n\}} e^{it(S_n \circ \rho_n)} \, d\mathscr{P} = \int_{\{L_n = n\}} e^{itS_n} \, d\mathscr{P}.$$

Applying (5) and (12) of Sec. 8.4 to $\beta$ and substituting (4), we obtain

(5) $$\sum_{n=0}^{\infty} r^n \int_{\{L_n = n\}} e^{itS_n} \, d\mathscr{P} = \exp\left\{+\sum_{n=1}^{\infty} \frac{r^n}{n} \int_{\{S_n > 0\}} e^{itS_n} \, d\mathscr{P}\right\};$$

applying (5) and (12) of Sec. 8.4 to $\alpha$ and substituting the obvious relation $\{\alpha > n\} = \{L_n = 0\}$, we obtain

(6) $$\sum_{n=0}^{\infty} r^n \int_{\{L_n = 0\}} e^{itS_n} \, d\mathscr{P} = \exp\left\{+\sum_{n=1}^{\infty} \frac{r^n}{n} \int_{\{S_n \le 0\}} e^{itS_n} \, d\mathscr{P}\right\}.$$

We are ready for the main results for $M_n$ and $M$, known as Spitzer's identity:

**Theorem 8.5.1.** We have for $0 < r < 1$:

(7) $$\sum_{n=0}^{\infty} r^n \mathscr{E}\{e^{itM_n}\} = \exp\left\{\sum_{n=1}^{\infty} \frac{r^n}{n} \mathscr{E}(e^{itS_n^+})\right\}.$$

$M$ is finite a.e. if and only if

$$\sum_{n=1}^{\infty} \frac{1}{n} \mathscr{P}\{S_n > 0\} < \infty, \tag{8}$$

in which case we have

$$\mathscr{E}\{e^{itM}\} = \exp\left\{\sum_{n=1}^{\infty} \frac{1}{n}[\mathscr{E}(e^{itS_n^+}) - 1]\right\}. \tag{9}$$

PROOF. Observe the basic equation that follows at once from the meaning of $L_n$:

$$\{L_n = k\} = \{L_k = k\} \cap \{L_{n-k} \circ \tau^k = 0\}, \tag{10}$$

where $\tau^k$ is the $k$th iterate of the shift. Since the two events on the right side of (10) are independent, we obtain for each $n \in N^0$, and real $t$ and $u$:

$$\begin{aligned} \mathscr{E}\{e^{itM_n} e^{iu(S_n - M_n)}\} &= \sum_{k=0}^{n} \int_{\{L_n = k\}} e^{itS_k} e^{iu(S_n - S_k)}\, d\mathscr{P} \\ &= \sum_{k=0}^{n} \int_{\{L_k = k\}} e^{itS_k}\, d\mathscr{P} \int_{\{L_{n-k} \circ \tau^k = 0\}} e^{iu(S_{n-k} \circ \tau^k)}\, d\mathscr{P} \\ &= \sum_{k=0}^{n} \int_{\{L_k = k\}} e^{itS_k}\, d\mathscr{P} \int_{\{L_{n-k} = 0\}} e^{iuS_{n-k}}\, d\mathscr{P}. \end{aligned} \tag{11}$$

It follows that

$$\begin{aligned} &\sum_{n=0}^{\infty} r^n \mathscr{E}\{e^{itM_n} e^{iu(S_n - M_n)}\} \\ &\qquad = \sum_{n=0}^{\infty} r^n \int_{\{L_n = n\}} e^{itS_n}\, d\mathscr{P} \cdot \sum_{n=0}^{\infty} r^n \int_{\{L_n = 0\}} e^{iuS_n}\, d\mathscr{P}. \end{aligned} \tag{12}$$

Setting $u = 0$ in (12) and using (5) as it stands and (6) with $t = 0$, we obtain

$$\sum_{n=0}^{\infty} r^n \mathscr{E}\{e^{itM_n}\} = \exp\left\{\sum_{n=1}^{\infty} \frac{r^n}{n} \left[\int_{\{S_n > 0\}} e^{itS_n}\, dP + \int_{\{S_n \le 0\}} 1 d\mathscr{P}\right]\right\},$$

which reduces to (7).

Next, by Theorem 8.2.4, $M < \infty$ a.e. if and only if $\mathscr{P}\{\alpha < \infty\} < 1$ or equivalently by (14) of Sec. 8.4 if and only if (8) holds. In this case the convergence theorem for ch.f.'s asserts that

$$\mathscr{E}\{e^{itM}\} = \lim_{n \to \infty} \mathscr{E}\{e^{itM_n}\},$$

and consequently

$$\begin{aligned}
\mathscr{E}\{e^{itM}\} &= \lim_{r\uparrow 1}(1-r)\sum_{n=0}^{\infty} r^n \mathscr{E}\{e^{itM_n}\} \\
&= \lim_{r\uparrow 1} \exp\left\{-\sum_{n=1}^{\infty}\frac{r^n}{n}\right\}\exp\left\{\sum_{n=1}^{\infty}\frac{r^n}{n}\mathscr{E}(e^{itS_n^+})\right\} \\
&= \lim_{r\uparrow 1} \exp\left\{\sum_{n=1}^{\infty}\frac{r^n}{n}[\mathscr{E}(e^{itS_n^+})-1]\right\},
\end{aligned}$$

where the first equation is by proposition (B) in Sec. 8.4. Since

$$\sum_{n=1}^{\infty}\frac{1}{n}|\mathscr{E}(e^{itS_n^+})-1| \le \sum_{n=1}^{\infty}\frac{2}{n}\mathscr{P}[S_n>0] < \infty$$

by (8), the last-written limit above is equal to the right member of (9), by proposition (B). Theorem 8.5.1. is completely proved.

By switching to Laplace transforms in (7) as done in the proof of Theorem 8.4.6 and using proposition (E) in Sec. 8.4, it is possible to give an "analytic" derivation of the second assertion of Theorem 8.5.1 without recourse to the "probabilistic" result in Theorem 8.2.4. This kind of mathematical gambit should appeal to the curious as well as the obstinate; see Exercise 9 below. Another interesting exercise is to establish the following result:

$$\mathscr{E}(M_n) = \sum_{k=1}^{n}\frac{1}{k}\mathscr{E}(S_k^+) \tag{13}$$

by differentiating (7) with respect to $t$. But a neat little formula such as (13) deserves a simpler proof, so here it is.

*Proof of* (13). Writing

$$M_n = \bigvee_{j=0}^{n} S_j = \left(\bigvee_{j=1}^{n} S_j\right)^+$$

and dropping "$d\mathscr{P}$" in the integrals below:

$$\begin{aligned}
\mathscr{E}(M_n) &= \int_{\{M_n>0\}} \bigvee_{k=1}^{n} S_k \\
&= \int_{\{M_n>0;S_n>0\}}\left[X_1 + 0\bigvee_{k=2}^{n}(S_k - X_1)\right] + \int_{\{M_n>0;S_n\le 0\}}\bigvee_{k=1}^{n} S_k \\
&= \int_{\{S_n>0\}} X_1 + \int_{\{S_n>0\}}\left[0\bigvee_{k=2}^{n}(S_k-X_1)\right] + \int_{\{M_{n-1}>0;S_n\le 0\}}\bigvee_{k=1}^{n-1} S_k.
\end{aligned}$$

Call the last three integrals $\int_1$, $\int_2$, and $\int_3$. We have on grounds of symmetry

$$\int_1 = \frac{1}{n}\int_{\{S_n>0\}} S_n.$$

Apply the cyclical permutation

$$\begin{pmatrix} 1, 2, \dots, n \\ 2, 3, \dots, 1 \end{pmatrix}$$

to $\int_2$ to obtain

$$\int_2 = \int_{\{S_n>0\}} M_{n-1}.$$

Obviously, we have

$$\int_3 = \int_{\{M_{n-1}>0;S_n\le 0\}} M_{n-1} = \int_{\{S_n\le 0\}} M_{n-1}.$$

Gathering these, we obtain

$$\begin{aligned}\mathscr{E}(M_n) &= \frac{1}{n}\int_{\{S_n>0\}} S_n + \int_{\{S_n>0\}} M_{n-1} + \int_{\{S_n\le 0\}} M_{n-1} \\ &= \frac{1}{n}\mathscr{E}(S_n^+) + \mathscr{E}(M_{n-1}),\end{aligned}$$

and (13) follows by recursion.

Another interesting quantity in the development was the "number of strictly positive terms" in the random walk. We shall treat this by combinatorial methods as an antidote to the analytic skulduggery above. Let us define

$$\nu_n(\omega) = \text{the number of } k \text{ in } N_n \text{ such that } S_k(\omega) > 0;$$

$$\nu_n'(\omega) = \text{the number of } k \text{ in } N_n \text{ such that } S_k(\omega) \le 0.$$

For easy reference let us repeat two previous definitions together with two new ones below, for $n \in N^0$:

$$M_n(\omega) = \max_{0\le j\le n} S_j(\omega); \quad L_n(\omega) = \min\{j \in N_n^0 : S_j(\omega) = M_n(\omega)\};$$

$$M_n'(\omega) = \min_{0\le j\le n} S_j(\omega); \quad L_n'(\omega) = \max\{j \in N_n^0 : S_j(\omega) = M_n'(\omega)\}.$$

Since the reversal given in (3) is a 1-to-1 measure preserving mapping that leaves $S_n$ unchanged, it is clear that for each $\Lambda \in \mathscr{F}_\infty$, the measures on $\mathscr{B}^1$ below are equal:

$$\mathscr{P}\{\Lambda; S_n \in \cdot\} = \mathscr{P}\{\rho_n\Lambda; S_n \in \cdot\}. \tag{14}$$

**Lemma.** For each $n \in N^0$, the two random vectors

$$(L_n, S_n) \quad \text{and} \quad (n - L'_n, S_n)$$

have the same distribution.

PROOF. This will follow from (14) if we show that

$$\forall k \in N_n^0: \quad \rho_n^{-1}\{L_n = k\} = \{L'_n = n - k\}. \tag{15}$$

Now for $\rho_n\omega$ the first $n+1$ partial sums $S_j(\rho_n\omega)$, $j \in N_n^0$, are

$$0, \omega_n, \omega_n + \omega_{n-1}, \ldots, \omega_n + \cdots + \omega_{n-j+1}, \ldots, \omega_n + \cdots + \omega_1,$$

which is the same as

$$S_n - S_n, S_n - S_{n-1}, S_n - S_{n-2}, \ldots, S_n - S_{n-j}, \ldots, S_n - S_0,$$

from which (15) follows by inspection.

**Theorem 8.5.2.** For each $n \in N^0$, the random vectors

$$(L_n, S_n) \quad \text{and} \quad (\nu_n, S_n) \tag{16}$$

have the same distribution; and the random vectors

$$(L'_n, S_n) \quad \text{and} \quad (\nu'_n, S_n) \tag{16$'$}$$

have the same distribution.

PROOF. For $n = 0$ there is nothing to prove; for $n = 1$, the assertion about (16) is trivially true since $\{L_1 = 0\} = \{S_1 \le 0\} = \{\nu_1 = 0\}$; similarly for (16′). We shall prove the general case by simultaneous induction, supposing that both assertions have been proved when $n$ is replaced by $n-1$. For each $k \in N_{n-1}^0$ and $y \in \mathscr{R}^1$, let us put

$$G(y) = \mathscr{P}\{L_{n-1} = k; S_{n-1} \le y\}, \quad H(y) = \mathscr{P}\{\nu_{n-1} = k; S_{n-1} \le y\}.$$

Then the induction hypothesis implies that $G \equiv H$. Since $X_n$ is independent of $\mathscr{F}_{n-1}$ and so of the vector $(L_{n-1}, S_{n-1})$, we have for each $x \in \mathscr{R}^1$:

$$\begin{aligned} \mathscr{P}\{L_{n-1} = k; S_n \le x\} &= \int_{-\infty}^{\infty} F(x-y)\,dG(y) \\ &= \int_{-\infty}^{\infty} F(x-y)\,dH(y) \\ &= \mathscr{P}\{\nu_{n-1} = k; S_n \le x\}, \end{aligned} \tag{17}$$

where $F$ is the common d.f. of each $X_n$. Now observe that on the set $\{S_n \leq 0\}$ we have $L_n = L_{n-1}$ by definition, hence if $x \leq 0$:

$$(18) \qquad \{\omega : L_n(\omega) = k; S_n(\omega) \leq x\} = \{\omega : L_{n-1}(\omega) = k; S_n(\omega) \leq x\}.$$

On the other hand, on the set $\{S_n \leq 0\}$ we have $\nu_{n-1} = \nu_n$, so that if $x \leq 0$,

$$(19) \qquad \{\omega : \nu_n(\omega) = k; S_n(\omega) \leq x\} = \{\omega : \nu_{n-1}(\omega) = k; S_n(\omega) \leq x\}.$$

Combining (17) with (19), we obtain

$$(20) \qquad \forall k \in N_n^0, x \leq 0 : \mathscr{P}\{L_n = k; S_n \leq x\} = \mathscr{P}\{\nu_n = k; S_n \leq x\}.$$

Next if $k \in N_n^0$, and $x \geq 0$, then by similar arguments:

$$\{\omega : L'_n(\omega) = n - k; S_n(\omega) > x\} = \{\omega : L'_{n-1}(\omega) = n - k; S_n(\omega) > x\},$$

$$\{\omega : \nu'_n(\omega) = n - k; S_n(\omega) > x\} = \{\omega : \nu'_{n-1}(\omega) = n - k; S_n(\omega) > x\}.$$

Using (16′) when $n$ is replaced by $n - 1$, we obtain the analogue of (20):

$$(21) \qquad \forall k \in N_n^0, x \geq 0 : \mathscr{P}\{L'_n = n - k; S_n > x\} = \mathscr{P}\{\nu'_n = n - k; S_n > x\}.$$

The left members in (21) and (22) below are equal by the lemma above, while the right members are trivially equal since $\nu_n + \nu'_n = n$:

$$(22) \qquad \forall k \in N_n^0, x \geq 0 : \mathscr{P}\{L_n = k; S_n > x\} = \mathscr{P}\{\nu_n = k; S_n > x\}.$$

Combining (20) and (22) for $x = 0$, we have

$$(23) \qquad \forall k \in N_n^0 : \mathscr{P}\{L_n = k\} = \mathscr{P}\{\nu_n = k\};$$

subtracting (22) from (23), we obtain the equation in (20) for $x \geq 0$; hence it is true for every $x$, proving the assertion about (16). Similarly for (16′), and the induction is complete.

As an immediate consequence of Theorem 8.5.2, the obvious relation (10) is translated into the by-no-means obvious relation (24) below:

**Theorem 8.5.3.** We have for $k \in N_n^0$:

$$(24) \qquad \mathscr{P}\{\nu_n = k\} = \mathscr{P}\{\nu_k = k\}\mathscr{P}\{\nu_{n-k} = 0\}.$$

If the common distribution of each $X_n$ is symmetric with no atom at zero, then

$$(25) \qquad \forall k \in N_n^0 : \mathscr{P}\{\nu_n = k\} = (-1)^n \binom{-\frac{1}{2}}{k}\binom{-\frac{1}{2}}{n-k}.$$

PROOF. Let us denote the number on right side of (25), which is equal to

$$\frac{1}{2^{2n}}\binom{2k}{k}\binom{2n-2k}{n-k},$$

by $a_n(k)$. Then for each $n \in N$, $\{a_n(k), k \in N_n^0\}$ is a well-known probability distribution. For $n = 1$ we have

$$\mathscr{P}\{\nu_1 = 0\} = \mathscr{P}\{\nu_1 = 1\} = \tfrac{1}{2} = a_n(0) = a_n(n),$$

so that (25) holds trivially. Suppose now that it holds when $n$ is replaced by $n-1$; then for $k \in N_{n-1}$ we have by (24):

$$\mathscr{P}\{\nu_n = k\} = (-1)^k\binom{-\frac{1}{2}}{k}(-1)^{n-k}\binom{-\frac{1}{2}}{n-k} = a_n(k).$$

It follows that

$$\mathscr{P}\{\nu_n = 0\} + \mathscr{P}\{\nu_n = n\} = 1 - \sum_{k=1}^{n-1}\mathscr{P}\{\nu_n = k\}$$

$$= 1 - \sum_{k=1}^{n-1} a_n(k) = a_n(0) + a_n(n).$$

Under the hypotheses of the theorem, it is clear by considering the dual random walk that the two terms in the first member above are equal; since the two terms in the last member are obviously equal, they are all equal and the theorem is proved.

Stirling's formula and elementary calculus now lead to the famous "arcsin law", first discovered by Paul Lévy (1939) for Brownian motion.

**Theorem 8.5.4.** If the common distribution of a stationary independent process is symmetric, then we have

$$\forall x \in [0,1]: \lim_{n\to\infty}\mathscr{P}\left\{\frac{\nu_n}{n} \le x\right\} = \frac{2}{\pi}\arcsin\sqrt{x} = \frac{1}{\pi}\int_0^x \frac{du}{\sqrt{u(1-u)}}.$$

This limit theorem also holds for an independent, not necessarily stationary process, in which each $X_n$ has mean 0 and variance 1 and such that the classical central limit theorem is applicable. This can be proved by the same method (invariance principle) as Theorem 7.3.3.

EXERCISES

**1.** Derive (3) of Sec. 8.4 by considering

$$\mathscr{E}\{r^{\alpha}e^{itS_{\alpha}}\} = \sum_{n=1}^{\infty} r^n \left[\int_{\{\alpha>n-1\}} e^{itS_n}\,d\mathscr{P} - \int_{\{\alpha>n\}} e^{itS_n}\,d\mathscr{P}\right].$$

***2.** Under the conditions of Theorem 8.4.6, show that

$$\mathscr{E}\{S_{\alpha}\} = -\lim_{n\to\infty}\int_{\{\alpha>n\}} S_n\,d\mathscr{P}.$$

**3.** Find an expression for the Laplace transform of $S_{\alpha}$. Is the corresponding formula for the Fourier transform valid?

***4.** Prove (13) of Sec. 8.5 by differentiating (7) there; justify the steps.

**5.** Prove that

$$\sum_{n=0}^{\infty} r^n \mathscr{P}[M_n = 0] = \exp\left\{\sum_{n=1}^{\infty}\frac{r^n}{n}\mathscr{P}[S_n \le 0]\right\}$$

and deduce that

$$\mathscr{P}[M = 0] = \exp\left\{-\sum_{n=1}^{\infty}\frac{1}{n}\mathscr{P}[S_n > 0]\right\}.$$

**6.** If $M < \infty$ a.e., then it has an infinitely divisible distribution.

**7.** Prove that

$$\sum_{n=0}^{\infty}\mathscr{P}\{L_n = 0; S_n = 0\} = \exp\left\{\sum_{n=1}^{\infty}\frac{1}{n}\mathscr{P}[S_n = 0]\right\}.$$

[HINT: One way to deduce this is to switch to Laplace transforms in (6) of Sec. 8.5 and let $\lambda \to \infty$.]

**8.** Prove that the left member of the equation in Exercise 7 is equal to

$$\left[1 - \sum_{n=1}^{\infty}\mathscr{P}\{\alpha' = n; S_n = 0\}\right]^{-1},$$

where $\alpha' = \alpha_{[0,\infty)}$; hence prove its convergence.

***9.** Prove that

$$\sum_{n=1}^{\infty}\frac{1}{n}\mathscr{P}[S_n > 0] < \infty$$

implies $\mathscr{P}(M < \infty) = 1$ as follows. From the Laplace transform version of (7) of Sec. 8.5, show that for $\lambda > 0$,

$$\lim_{r \uparrow 1}(1 - r)\sum_{n=0}^{\infty} r^n \mathscr{E}\{e^{-\lambda M_n}\}$$

exists and is finite, say $= \chi(\lambda)$. Now use proposition (E) in Sec. 8.4 and apply the convergence theorem for Laplace transforms (Theorem 6.6.3).

*__10.__ Define a sequence of r.v.'s $\{Y_n, n \in N^0\}$ as follows:

$$Y_0 = 0, \quad Y_{n+1} = (Y_n + X_{n+1})^+, \quad n \in N^0,$$

where $\{X_n, n \in N\}$ is a stationary independent sequence. Prove that for each $n$, $Y_n$ and $M_n$ have the same distribution. [This approach is useful in queuing theory.]

**11.** If $-\infty \leq \mathscr{E}(X) < 0$, then

$$\mathscr{E}(X) = \mathscr{E}(-V^-), \quad \text{where } V = \sup_{1 \leq j < \infty} S_j.$$

[HINT: If $V_n = \max_{1 \leq j \leq n} S_j$, then

$$\mathscr{E}(e^{itV_n^+}) + \mathscr{E}(e^{itV_n^-}) - 1 = \mathscr{E}(e_n^{itV}) = \mathscr{E}(e^{itV_{n-1}^+})f(t);$$

let $n \to \infty$, then $t \downarrow 0$. For the case $\mathscr{E}(X) = -\infty$, truncate. This result is due to S. Port.]

*__12.__ If $\mathscr{P}\{\alpha_{(0,\infty)} < \infty\} < 1$, then $v_n \to v$, $L_n \to L$, both limits being finite a.e. and having the generating functions:

$$\mathscr{E}\{r^\nu\} = \mathscr{E}\{r^L\} = \exp\left\{\sum_{n=1}^{\infty} \frac{r^n - 1}{n}\mathscr{P}[S_n > 0]\right\}.$$

[HINT: Consider $\lim_{m \to \infty} \sum_{n=0}^{\infty} \mathscr{P}\{v_m = n\}r^n$ and use (24) of Sec. 8.5.]

**13.** If $\mathscr{E}(X) = 0$, $\mathscr{E}(X^2) = \sigma^2$, $0 < \sigma^2 < \infty$, then as $n \to \infty$ we have

$$\mathscr{P}[\nu_n = 0] \sim \frac{e^c}{\sqrt{\pi n}}, \quad \mathscr{P}[\nu_n = n] \sim \frac{e^{-c}}{\sqrt{\pi n}},$$

where

$$c = \sum_{n=1}^{\infty} \frac{1}{n}\left\{\frac{1}{2} - \mathscr{P}[S_n > 0]\right\}.$$

[HINT: Consider

$$\lim_{r \uparrow 1}(1 - r)^{1/2}\sum_{n=0}^{\infty} r^n \mathscr{P}[\nu_n = 0]$$

as in the proof of Theorem 8.4.6, and use the following lemma: if $p_n$ is a decreasing sequence of positive numbers such that

$$\sum_{k=1}^{n} p_k \sim 2n^{1/2},$$

then $p_n \sim n^{-1/2}$.]

**★14.** Prove Theorem 8.5.4.

**15.** For an arbitrary random walk, we have

$$\sum_{n} \frac{(-1)^n}{n} \mathscr{P}\{S_n > 0\} < \infty.$$

[HINT: Half of the result is given in Exercise 18 of Sec. 8.3. For the remaining case, apply proposition (C) in the $O$-form to equation (5) of Sec. 8.5 with $L_n$ replaced by $\nu_n$ and $t = 0$. This result is due to D. L. Hanson and M. Katz.]

### Bibliographical Note

For Sec. 8.3, see

K. L. Chung and W. H. J. Fuchs, *On the distribution of values of sums of random variables*, Mem. Am. Math. Soc. **6** (1951), 1–12.

K. L. Chung and D. Ornstein, *On the recurrence of sums of random variables*, Bull. Am. Math. Soc. **68** (1962) 30–32.

Theorems 8.4.1 and 8.4.2 are due to G. Baxter; see

G. Baxter, *An analytical approach to finite fluctuation problems in probability*, J. d'Analyse Math. **9** (1961), 31–70.

Historically, recurrence problems for Bernoullian random walks were first discussed by Pólya in 1921.

The rest of Sec. 8.4, as well as Theorem 8.5.1, is due largely to F. L. Spitzer; see

F. L. Spitzer, *A combinatorial lemma and its application to probability theory*, Trans. Am. Math. Soc. **82** (1956), 323–339.

F. L. Spitzer, *A Tauberian theorem and its probability interpretation*, Trans. Am. Math. Soc. **94** (1960), 150–169.

See also his book below—which, however, treats only the lattice case:

F. L. Spitzer, *Principles of random walk*. D. Van Nostrand Co., Inc., Princeton, N.J., 1964.

The latter part of Sec. 8.5 is based on

W. Feller, *On combinatorial methods in fluctuation theory*, Surveys in Probability and Statistics (the Harald Cramér volume), 75–91. Almquist & Wiksell, Stockholm, 1959.

Theorem 8.5.3 is due to E. S. Andersen. Feller [13] also contains much material on random walks.

# 9 Conditioning. Markov property. Martingale

## 9.1 Basic properties of conditional expectation

If $\Lambda$ is any set in $\mathscr{F}$ with $\mathscr{P}(\Lambda)>0$, we define $\mathscr{P}_\Lambda(\cdot)$ on $\mathscr{F}$ as follows:

(1) $$\mathscr{P}_\Lambda(E) = \frac{\mathscr{P}(\Lambda \cap E)}{\mathscr{P}(\Lambda)}$$

Clearly $\mathscr{P}_\Lambda$ is a p.m. on $\mathscr{F}$; it is called the "conditional probability relative to $\Lambda$". The integral with respect to this p.m. is called the "conditional expectation relative to $\Lambda$":

(2) $$\mathscr{E}_\Lambda(Y) = \int_\Omega Y(\omega)\mathscr{P}_\Lambda(d\omega) = \frac{1}{\mathscr{P}(\Lambda)}\int_\Lambda Y(\omega)\mathscr{P}(d\omega).$$

If $\mathscr{P}(\Lambda) = 0$, we decree that $\mathscr{P}_\Lambda(E) = 0$ for every $E \in \mathscr{F}$. This convention is expedient as in (3) and (4) below.

Let now $\{\Lambda_n, n \geq 1\}$ be a countable measurable partition of $\Omega$, namely:

$$\Omega = \bigcup_{n=1}^{\infty} \Lambda_n, \qquad \Lambda_n \in \mathscr{F}, \qquad \Lambda_m \cap \Lambda_n = \varnothing, \qquad \text{if } m \neq n.$$

Then we have

$$\mathscr{P}(E) = \sum_{n=1}^{\infty} \mathscr{P}(\Lambda_n \cap E) = \sum_{n=1}^{\infty} \mathscr{P}(\Lambda_n)\mathscr{P}_{\Lambda_n}(E); \tag{3}$$

$$\mathscr{E}(Y) = \sum_{n=1}^{\infty} \int_{\Lambda_n} Y(\omega)\mathscr{P}(d\omega) = \sum_{n=1}^{\infty} \mathscr{P}(\Lambda_n)\mathscr{E}_{\Lambda_n}(Y), \tag{4}$$

provided that $\mathscr{E}(Y)$ is defined. We have already used such decompositions before, for example in the proof of Kolmogorov's inequality (Theorem 5.3.1):

$$\int_{\Lambda} S_n^2 \, d\mathscr{P} = \sum_{k=1}^{n} \mathscr{P}(\Lambda_k)\mathscr{E}_{\Lambda_k}(S_n^2).$$

Another example is in the proof of Wald's equation (Theorem 5.5.3), where

$$\mathscr{E}(S_N) = \sum_{k=1}^{\infty} \mathscr{P}(N = k)\mathscr{E}_{\{N=k\}}(S_N).$$

Thus the notion of conditioning gives rise to a decomposition when a given event or r.v. is considered on various parts of the sample space, on each of which some particular information may be obtained.

Let us, however, reflect for a few moments on the even more elementary example below. From a pack of 52 playing cards one card is drawn and seen to be a spade. What is the probability that a second card drawn from the remaining deck will also be a spade? Since there are 51 cards left, among which are 12 spades, it is clear that the required probability is 12/51. But is this the conditional probability $\mathscr{P}_{\Lambda}(E)$ defined above, where $\Lambda$ = "first card is a spade" and $E$ = "second card is a spade"? According to the definition,

$$\frac{\mathscr{P}(\Lambda \cap E)}{\mathscr{P}(\Lambda)} = \frac{\dfrac{13.12}{52.51}}{\dfrac{13}{52}} = \frac{12}{51},$$

where the denominator and numerator have been separately evaluated by elementary combinatorial formulas. Hence the answer to the question above is indeed "yes"; but this verification would be futile if we did not have another way to evaluate $\mathscr{P}_{\Lambda}(E)$ as first indicated. Indeed, conditional probability is often used to evaluate a joint probability by turning the formula (1) around as follows:

$$\mathscr{P}(\Lambda \cap E) = \mathscr{P}(\Lambda)\mathscr{P}_{\Lambda}(E) = \frac{13}{52} \cdot \frac{12}{51}.$$

In general, it is used to reduce the calculation of a probability or expectation to a modified one which can be handled more easily.

Let $\mathscr{G}$ be the Borel field generated by a countable partition $\{\Lambda_n\}$, for example that by a discrete r.v. $X$, where $\Lambda_n = \{X = a_n\}$. Given an integrable r.v. $Y$, we define the *function* $\mathscr{E}_{\mathscr{G}}(Y)$ on $\Omega$ by:

$$\mathscr{E}_{\mathscr{G}}(Y) = \sum_n \mathscr{E}_{\Lambda_n}(Y) 1_{\Lambda_n}(\cdot). \tag{5}$$

Thus $\mathscr{E}_{\mathscr{G}}(Y)$ is a discrete r.v. that assumes that value $\mathscr{E}_{\Lambda_n}(Y)$ on the set $\Lambda_n$, for each $n$. Now we can rewrite (4) as follows:

$$\mathscr{E}(Y) = \sum_n \int_{\Lambda_n} \mathscr{E}_{\mathscr{G}}(Y)\, d\mathscr{P} = \int_{\Omega} \mathscr{E}_{\mathscr{G}}(Y)\, d\mathscr{P}.$$

Furthermore, for any $\Lambda \in \mathscr{G}$, $\Lambda$ is the union of a subcollection of the $\Lambda_n$'s (see Exercise 9 of Sec. 2.1), and the same manipulation yields

$$\forall \Lambda \in \mathscr{G}: \int_{\Lambda} Y\, d\mathscr{P} = \int_{\Lambda} \mathscr{E}_{\mathscr{G}}(Y)\, d\mathscr{P}. \tag{6}$$

In particular, this shows that $\mathscr{E}_{\mathscr{G}}(Y)$ is integrable. Formula (6) equates two integrals over the same set with an essential difference: while the integrand $Y$ on the left belongs to $\mathscr{F}$, the integrand $\mathscr{E}_{\mathscr{G}}(Y)$ on the right belongs to the subfield $\mathscr{G}$. [The fact that $\mathscr{E}_{\mathscr{G}}(Y)$ is discrete is incidental to the nature of $\mathscr{G}$.] It holds for every $\Lambda$ in the subfield $\mathscr{G}$, but not necessarily for a set in $\mathscr{F}\backslash\mathscr{G}$. Now suppose that there are two functions $\varphi_1$ and $\varphi_2$, both belonging to $\mathscr{G}$, such that

$$\forall \Lambda \in \mathscr{G}: \int_{\Lambda} Y\, d\mathscr{P} = \int_{\Lambda} \varphi_i\, d\mathscr{P}, \quad i = 1, 2.$$

Let $\Lambda = \{\omega : \varphi_1(\omega) > \varphi_2(\omega)\}$, then $\Lambda \in \mathscr{G}$ and so

$$\int_{\Lambda} (\varphi_1 - \varphi_2)\, d\mathscr{P} = 0.$$

Hence $\mathscr{P}(\Lambda) = 0$; interchanging $\varphi_1$ and $\varphi_2$ above we conclude that $\varphi_1 = \varphi_2$ a.e. We have therefore proved that the $\mathscr{E}_{\mathscr{G}}(Y)$ in (6) is unique up to an equivalence. Let us agree to use $\mathscr{E}_{\mathscr{G}}(Y)$ or $\mathscr{E}(Y \mid \mathscr{G})$ to denote the corresponding equivalence class, and let us call any particular member of the class a "version" of the conditional expectation.

The results above are valid for an arbitrary Borel subfield $\mathscr{G}$ and will be stated in the theorem below.

**Theorem 9.1.1.** If $\mathscr{E}(|Y|) < \infty$ and $\mathscr{G}$ is a Borel subfield of $\mathscr{F}$, then there exists a unique equivalence class of integrable r.v.'s $\mathscr{E}(Y \mid \mathscr{G})$ belonging to $\mathscr{G}$ such that (6) holds.

PROOF. Consider the set function $\nu$ on $\mathscr{G}$:

$$\forall \Lambda \in \mathscr{G}: \nu(\Lambda) = \int_\Lambda Y \, d\mathscr{P}.$$

It is finite-valued and countably additive, hence a "signed measure" on $\mathscr{G}$. If $\mathscr{P}(\Lambda) = 0$. then $\nu(\Lambda) = 0$; hence it is absolutely continuous with respect to $\mathscr{P}: \nu \ll \mathscr{P}$. The theorem then follows from the Radon–Nikodym theorem (see, e.g., Royden [5] or Halmos [4]), the resulting "derivative" $d\nu/d\mathscr{P}$ being what we have denoted by $\mathscr{E}(Y \mid \mathscr{G})$.

Having established the existence and uniqueness, we may repeat the definition as follows.

DEFINITION OF CONDITIONAL EXPECTATION. Given an integrable r.v. $Y$ and a Borel subfield $\mathscr{G}$, the *conditional expectation* $\mathscr{E}(Y \mid \mathscr{G})$ of $Y$ *relative to* $\mathscr{G}$ is any one of the equivalence class of r.v.'s on $\Omega$ satisfying the two properties:

(a) it belongs to $\mathscr{G}$;
(b) it has the same integral as $Y$ over any set in $\mathscr{G}$.

We shall refer to (b), or equivalently formula (6) above, as the "defining relation" of the conditional expectation. In practice as well as in theory, the identification of conditional expectations or relations between them is established by verifying the two properties listed above. When $Y = 1_\Delta$, where $\Delta \in \mathscr{F}$, we write

$$\mathscr{P}(\Delta \mid \mathscr{G}) = \mathscr{E}(1_\Delta \mid \mathscr{G})$$

and call it the "conditional probability of $\Delta$ relative to $\mathscr{G}$". Specifically, $\mathscr{P}(\Delta \mid \mathscr{G})$ is any one of the equivalence class of r.v.'s belonging to $\mathscr{G}$ and satisfying the condition

$$\forall \Lambda \in \mathscr{G}: \mathscr{P}(\Delta \cap \Lambda) = \int_\Lambda \mathscr{P}(\Delta \mid \mathscr{G}) \, d\mathscr{P}. \tag{7}$$

It follows from the definition that for an integrable r.v. $Y$ and a Borel subfield $\mathscr{G}$, we have

$$\int_\Lambda [Y - \mathscr{E}(Y \mid \mathscr{G})] \, d\mathscr{P} = 0,$$

for every $\Lambda \in \mathscr{G}$, and consequently also

$$\mathscr{E}\{[Y - \mathscr{E}(Y \mid \mathscr{G})]Z\} = 0$$

for every bounded $Z \in \mathscr{G}$ (why?). This implies the decomposition:

$$Y = Y' + Y'' \quad \text{where} \quad Y' = \mathscr{E}(Y \mid \mathscr{G}) \text{ and } Y'' \perp \mathscr{G},$$

where "$Y'' \perp \mathscr{G}$" means that $\mathscr{E}(Y''Z) = 0$ for every bounded $Z \in \mathscr{G}$. In the language of Banach space, $Y'$ is the "projection" of $Y$ on $\mathscr{G}$ and $Y''$ its "orthogonal complement".

For the Borel field $\mathscr{F}\{X\}$ generated by the r.v. $X$, we write also $\mathscr{E}(Y \mid X)$ for $\mathscr{E}(Y \mid \mathscr{F}\{X\})$; similarly for $\mathscr{E}(Y \mid X_1, \dots, X_n)$. The next theorem clarifies certain useful connections.

**Theorem 9.1.2.** One version of the conditional expectation $\mathscr{E}(Y \mid X)$ is given by $\varphi(X)$, where $\varphi$ is a Borel measurable function on $\mathscr{R}^1$. Furthermore, if we define the signed measure $\lambda$ on $\mathscr{B}^1$ by

$$\forall B \in \mathscr{B}^1 : \lambda(B) = \int_{X^{-1}(B)} Y \, d\mathscr{P},$$

and the p.m. of $X$ by $\mu$, then $\varphi$ is one version of the Radon–Nikodym derivative $d\lambda/d\mu$.

PROOF. The first assertion of the theorem is a particular case of the following lemma.

**Lemma.** If $Z \in \mathscr{F}\{X\}$, then $Z = \varphi(X)$ for some extended-valued Borel measurable function $\varphi$.

PROOF OF THE LEMMA. It is sufficient to prove this for a bounded positive $Z$ (why?). Then there exists a sequence of simple functions $Z_m$ which increases to $Z$ everywhere, and each $Z_m$ is of the form

$$\sum_{j=1}^{\ell} c_j 1_{\Lambda_j}$$

where $\Lambda_j \in \mathscr{F}\{X\}$. Hence $\Lambda_j = X^{-1}(B_j)$ for some $B_j \in \mathscr{B}^1$ (see Exercise 11 of Sec. 3.1). Thus if we take

$$\varphi_m = \sum_{j=1}^{\ell} c_j 1_{B_j},$$

we have $Z_m = \varphi_m(X)$. Since $\varphi_m(X) \to Z$, it follows that $\varphi_m$ converges on the range of $X$. But this range need not be Borel or even Lebesgue measurable (Exercise 6 of Sec. 3.1). To overcome this nuisance, we put,

$$\forall x \in \mathscr{R}^1 : \varphi(x) = \overline{\lim_{m \to \infty}} \varphi_m(x).$$

Then $Z = \overline{\lim}_m \varphi_m(X) = \varphi(X)$, and $\varphi$ is Borel measurable, proving the lemma.

To prove the second assertion: given any $B \in \mathscr{B}^1$, let $\Lambda = X^{-1}(B)$, then by Theorem 3.2.2 we have

$$\int_\Lambda \mathscr{E}(Y \mid X) \, d\mathscr{P} = \int_\Omega 1_B(X)\varphi(X) \, d\mathscr{P} = \int_{\mathscr{R}^1} 1_B(x)\varphi(x) \, d\mu = \int_B \varphi(x) \, d\mu.$$

Hence by (6),

$$\lambda(B) = \int_{\Lambda} Y \, d\mathscr{P} = \int_{B} \varphi(x) \, d\mu.$$

This being true for every $B$ in $\mathscr{B}^1$, it follows that $\varphi$ is a version of the derivative $d\lambda/d\mu$. Theorem 9.1.2 is proved.

As a consequence of the theorem, the function $\mathscr{E}(Y \mid X)$ of $\omega$ is constant a.e. on each set on which $X(\omega)$ is constant. By an abuse of notation, the $\varphi(x)$ above is sometimes written as $\mathscr{E}(Y \mid X = x)$. We may then write, for example, for each real $c$:

$$\int_{\{X \leq c\}} Y \, d\mathscr{P} = \int_{(-\infty, c]} \mathscr{E}(Y \mid X = x) \, d\mathscr{P}\{X \leq x\}.$$

Generalization to a finite number of $X$'s is straightforward. Thus one version of $\mathscr{E}(Y \mid X_1, \ldots, X_n)$ is $\varphi(X_1, \ldots, X_n)$, where $\varphi$ is an $n$-dimensional Borel measurable function, and by $\mathscr{E}(Y \mid X_1 = x_1, \ldots, X_n = x_n)$ is meant $\varphi(x_1, \ldots, x_n)$.

It is worthwhile to point out the extreme cases of $\mathscr{E}(Y \mid \mathscr{G})$:

$$\mathscr{E}(Y \mid \mathscr{J}) = \mathscr{E}(Y), \qquad \mathscr{E}(Y \mid \mathscr{F}) = Y; \qquad \text{a.e.}$$

where $\mathscr{J}$ is the trivial field $\{\varnothing, \Omega\}$. If $\mathscr{G}$ is the field generated by one set $\Lambda : \{\varnothing, \Lambda, \Lambda^c, \Omega\}$, then $\mathscr{E}(Y \mid \mathscr{G})$ is equal to $\mathscr{E}(Y \mid \Lambda)$ on $\Lambda$ and $\mathscr{E}(Y \mid \Lambda^c)$ on $\Lambda^c$. All these equations, as hereafter, are between equivalent classes of r.v.'s.

We shall suppose the pair $(\mathscr{F}, \mathscr{P})$ to be complete and each Borel subfield $\mathscr{G}$ of $\mathscr{F}$ to be augmented (see Exercise 20 of Sec. 2.2). But even if $\mathscr{G}$ is not augmented and $\overline{\mathscr{G}}$ is its augmentation, it follows from the definition that $\mathscr{E}(Y \mid \mathscr{G}) = \mathscr{E}(Y \mid \overline{\mathscr{G}})$, since an r.v. belonging to $\overline{\mathscr{G}}$ is equal to one belonging to $\mathscr{G}$ almost everywhere (why ?). Finally, if $\mathscr{G}_0$ is a field generating $\mathscr{G}$, or just a collection of sets whose finite disjoint unions form such a field, then the validity of (6) for each $\Lambda$ in $\mathscr{G}_0$ is sufficient for (6) as it stands. This follows easily from Theorem 2.2.3.

The next result is basic.

**Theorem 9.1.3.** Let $Y$ and $YZ$ be integrable r.v.'s and $Z \in \mathscr{G}$; then we have

$$\mathscr{E}(YZ \mid \mathscr{G}) = Z\mathscr{E}(Y \mid \mathscr{G}) \quad \text{a.e.} \tag{8}$$

[Here "a.e." is necessary, since we have not stipulated to regard Z as an equivalence class of r.v.'s, although conditional expectations are so regarded by definition. Nevertheless we shall sometimes omit such obvious "a.e.'s" from now on.]

PROOF. As usual we may suppose $Y \geq 0, Z \geq 0$ (see property (ii) below). The proof consists in observing that the right member of (8) belongs to $\mathscr{G}$ and satisfies the defining relation for the left member, namely:

$$\forall \Lambda \in \mathscr{G}: \int_\Lambda Z\mathscr{E}(Y \mid \mathscr{G})\, d\mathscr{P} = \int_\Lambda ZY\, d\mathscr{P}. \tag{9}$$

For (9) is true if $Z = 1_\Delta$, where $\Delta \in \mathscr{G}$, hence it is true if $Z$ is a simple r.v. belonging to $\mathscr{G}$ and consequently also for each $Z$ in $\mathscr{G}$ by monotone convergence, whether the limits are finite or positive infinite. Note that the integrability of $Z\mathscr{E}(Y \mid \mathscr{G})$ is part of the assertion of the theorem.

Recall that when $\mathscr{G}$ is generated by a partition $\{\Lambda_n\}$, we have exhibited a specific version (5) of $\mathscr{E}(Y \mid \mathscr{G})$. Now consider the corresponding $\mathscr{P}(\mathrm{M} \mid \mathscr{G})$ as a function of the pair $(\mathrm{M}, \omega)$:

$$\mathscr{P}(\mathrm{M} \mid \mathscr{G})(\omega) = \sum_n \mathscr{P}(\mathrm{M} \mid \Lambda_n) 1_{\Lambda_n}(\omega).$$

For each fixed M, as a function of $\omega$ this is a specific version of $\mathscr{P}(\mathrm{M} \mid \mathscr{G})$. For each fixed $\omega_0$, as a function of M this is a p.m. on $\mathscr{F}$ given by $\mathscr{P}\{\cdot \mid \Lambda_m\}$ for $\omega_0 \in \Lambda_m$. Let us denote for a moment the family of p.m.'s arising in this manner by $C(\omega_0, \cdot)$. We have then for each integrable r.v. $Y$ and each $\omega_0$ in $\Omega$:

$$\mathscr{E}(Y \mid \mathscr{G})(\omega_0) = \sum_n \mathscr{E}(Y \mid \Lambda_n) 1_{\Lambda_n}(\omega_0) = \int_\Omega Y C(\omega_0, d\omega). \tag{10}$$

Thus the specific version of $\mathscr{E}(Y \mid \mathscr{G})$ may be evaluated, at each $\omega_0 \in \Omega$, by integrating $Y$ with respect to the p.m. $C(\omega_0, \cdot)$. In this case the conditional expectation $\mathscr{E}(\cdot \mid \mathscr{G})$ as a functional on integrable r.v.'s is an integral in the literal sense. But in general such a representation is impossible (see Doob [16, Sec. 1.9]) and we must fall back on the defining relations to deduce its properties, usually from the unconditional analogues. Below are some of the simplest examples, in which $X$ and $X_n$ are integrable r.v.'s.

**(i)** If $X \in \mathscr{G}$, then $\mathscr{E}(X \mid \mathscr{G}) = X$ a.e.; this is true in particular if $X$ is a constant a.e.

**(ii)** $\mathscr{E}(X_1 + X_2 \mid \mathscr{G}) = \mathscr{E}(X_1 \mid \mathscr{G}) + \mathscr{E}(X_2 \mid \mathscr{G})$.

**(iii)** If $X_1 \leq X_2$, then $\mathscr{E}(X_1 \mid \mathscr{G}) \leq \mathscr{E}(X_2 \mid \mathscr{G})$.

**(iv)** $|\mathscr{E}(X \mid \mathscr{G})| \leq \mathscr{E}(|X| \mid \mathscr{G})$.

**(v)** If $X_n \uparrow X$, then $\mathscr{E}(X_n \mid \mathscr{G}) \uparrow \mathscr{E}(X \mid \mathscr{G})$.

**(vi)** If $X_n \downarrow X$, then $\mathscr{E}(X_n \mid \mathscr{G}) \downarrow \mathscr{E}(X \mid \mathscr{G})$.

**(vii)** If $|X_n| \leq Y$ where $\mathscr{E}(Y) < \infty$ and $X_n \to X$, then $\mathscr{E}(X_n \mid \mathscr{G}) \to \mathscr{E}(X \mid \mathscr{G})$.

To illustrate, (iii) is proved by observing that for each $\Lambda \in \mathscr{G}$:

$$\int_\Lambda \mathscr{E}(X_1 \mid \mathscr{G})\, d\mathscr{P} = \int_\Lambda X_1\, d\mathscr{P} \leq \int_\Lambda X_2\, d\mathscr{P} = \int_\Lambda \mathscr{E}(X_2 \mid \mathscr{G})\, d\mathscr{P}.$$

Hence if $\Lambda = \{\mathscr{E}(X_1 \mid \mathscr{G}) > \mathscr{E}(X_2 \mid \mathscr{G})\}$, we have $\mathscr{P}(\Lambda) = 0$. The inequality (iv) may be proved by (ii), (iii), and the equation $X = X^+ - X^-$. To prove (v), let the limit of $\mathscr{E}(X_n \mid \mathscr{G})$ be $Z$, which exists a.e. by (iii). Then for each $\Lambda \in \mathscr{G}$, we have by the monotone convergence theorem:

$$\int_\Lambda Z\, d\mathscr{P} = \lim_n \int_\Lambda \mathscr{E}(X_n \mid \mathscr{G})\, d\mathscr{P} = \lim_n \int_\Lambda X_n\, d\mathscr{P} = \int_\Lambda X\, d\mathscr{P}.$$

Thus $Z$ satisfies the defining relation for $\mathscr{E}(X \mid \mathscr{G})$, and it belongs to $\mathscr{G}$ with the $\mathscr{E}(X_n \mid \mathscr{G})$'s, hence $Z = \mathscr{E}(X \mid \mathscr{G})$.

To appreciate the caution that is necessary in handling conditional expectations, let us consider the Cauchy–Schwarz inequality:

$$\mathscr{E}(|XY| \mid \mathscr{G})^2 \leq \mathscr{E}(X^2 \mid \mathscr{G})\mathscr{E}(Y^2 \mid \mathscr{G}).$$

If we try to extend one of the usual proofs based on the positiveness of the quadratic form in $\lambda : \mathscr{E}((X + \lambda Y)^2 \mid \mathscr{G})$, the question arises that for each $\lambda$ the quantity is defined only up to a null set $N_\lambda$, and the union of these over all $\lambda$ cannot be ignored without comment. The reader is advised to think this difficulty through to its logical end, and then get out of it by restricting the $\lambda$'s to the rationals. Here is another way out: start from the following trivial inequality:

$$\frac{|X||Y|}{\alpha\beta} \leq \frac{X^2}{2\alpha^2} + \frac{Y^2}{2\beta^2},$$

where $\alpha = \mathscr{E}(X^2 \mid \mathscr{G})^{1/2}$, $\beta = \mathscr{E}(Y^2 \mid \mathscr{G})^{1/2}$, and $\alpha\beta>0$; apply the operation $\mathscr{E}\{- \mid \mathscr{G}\}$ using (ii) and (iii) above to obtain

$$\mathscr{E}\left\{\frac{|XY|}{\alpha\beta}\middle| \mathscr{G}\right\} \leq \frac{1}{2}\mathscr{E}\left\{\frac{X^2}{\alpha^2}\middle| \mathscr{G}\right\} + \frac{1}{2}\mathscr{E}\left\{\frac{Y^2}{\beta^2}\middle| \mathscr{G}\right\}.$$

Now use Theorem 9.1.3 to infer that this can be reduced to

$$\frac{1}{\alpha\beta}\mathscr{E}\{|XY| \mid \mathscr{G}\} \leq \frac{1}{2}\frac{\alpha^2}{\alpha^2} + \frac{1}{2}\frac{\beta^2}{\beta^2} = 1,$$

the desired inequality.

The following theorem is a generalization of Jensen's inequality in Sec. 3.2.

**Theorem 9.1.4.** If $\varphi$ is a convex function on $\mathscr{R}^1$ and $X$ and $\varphi(X)$ are integrable r.v.'s, then for each $\mathscr{G}$:

$$\varphi(\mathscr{E}(X \mid \mathscr{G})) \leq \mathscr{E}(\varphi(X) \mid \mathscr{G}). \tag{11}$$

PROOF. If $X$ is a simple r.v. taking the values $\{y_j\}$ on the sets $\{\Lambda_j\}$, $1 \leq j \leq n$, which forms a partition of $\Omega$, we have

$$\mathscr{E}(X \mid \mathscr{G}) = \sum_{j=1}^{n} y_j \mathscr{P}(\Lambda_j \mid \mathscr{G}),$$

$$\mathscr{E}(\varphi(X) \mid \mathscr{G}) = \sum_{j=1}^{n} \varphi(y_j) \mathscr{P}(\Lambda_j \mid \mathscr{G}),$$

where $\sum_{j=1}^{n} \mathscr{P}(\Lambda_j \mid \mathscr{G}) = 1$ a.e. Hence (11) is true in this case by the property of convexity. In general let $\{X_m\}$ be a sequence of simple r.v.'s converging to $X$ a.e. and satisfying $|X_m| \leq |X|$ for all $m$ (see Exercise 7 of Sec. 3.2). If we let $m \to \infty$ below:

$$\varphi(\mathscr{E}(X_m \mid \mathscr{G})) \leq \mathscr{E}(\varphi(X_m) \mid \mathscr{G}), \tag{12}$$

the left-hand member converges to the left-hand member of (11) by the continuity of $\varphi$, but we need dominated convergence on the right-hand side. To get this we first consider $\varphi_n$ which is obtained from $\varphi$ by replacing the graph of $\varphi$ outside $(-n, n)$ with tangential lines. Thus for each $n$ there is a constant $C_n$ such that

$$\forall x \in \mathscr{R}^1: |\varphi_n(x)| \leq C_n(|x| + 1).$$

Consequently, we have

$$|\varphi_n(X_m)| \leq C_n(|X_m| + 1) \leq C_n(|X| + 1)$$

and the last term is integrable by hypothesis. It now follows from property (vii) of conditional expectations that

$$\lim_{m \to \infty} \mathscr{E}(\varphi_n(X_m) \mid \mathscr{G}) = \mathscr{E}(\varphi_n(X) \mid \mathscr{G}).$$

This establishes (11) when $\varphi$ is replaced by $\varphi_n$. Letting $n \to \infty$ we have $\varphi_n \uparrow \varphi$ and $\varphi_n(X)$ is integrable; hence (11) follows for a general convex $\varphi$, by monotone convergence (v).

Here is an alternative proof, slightly more elegant and more delicate. We have for any $x$ and $y$:

$$\varphi(x) - \varphi(y) \geq \varphi'(y)(x - y)$$

where $\varphi'$ is the right-hand derivative of $\varphi$. Hence

$$\varphi(X) - \varphi(\mathscr{E}(X \mid \mathscr{G})) \geq \varphi'(\mathscr{E}(X \mid \mathscr{G}))[X - \mathscr{E}(X \mid \mathscr{G})].$$

The right member may not be integrable; but let $\Lambda = \{\omega : |\mathscr{E}(X \mid \mathscr{G})| \leq A\}$ for $A>0$. Replace $X$ by $X1_\Lambda$ in the above, take expectations of both sides, and let $A \uparrow \infty$. Observe that

$$\mathscr{E}\{\varphi(X1_\Lambda) \mid \mathscr{G}\} = \mathscr{E}\{\varphi(X)1_\Lambda + \varphi(0)1_{\Lambda^c} \mid \mathscr{G}\} = \mathscr{E}\{\varphi(X) \mid \mathscr{G}\}1_\Lambda + \varphi(0)1_{\Lambda^c}.$$

We now come to the most important property of conditional expectation relating to changing fields. Note that when $\Lambda = \Omega$, the defining relation (6) may be written as

$$\mathscr{E}_{\mathscr{J}}(\mathscr{E}_{\mathscr{G}}(Y)) = \mathscr{E}_{\mathscr{J}}(Y) = \mathscr{E}_{\mathscr{G}}(\mathscr{E}_{\mathscr{J}}(Y)).$$

This has an immediate generalization.

**Theorem 9.1.5.** If $Y$ is integrable and $\mathscr{F}_1 \subset \mathscr{F}_2$, then

$$\mathscr{E}_{\mathscr{F}_1}(Y) = \mathscr{E}_{\mathscr{F}_2}(Y) \qquad \text{if and only if } \mathscr{E}_{\mathscr{F}_2}(Y) \in \mathscr{F}_1; \tag{13}$$

and

$$\mathscr{E}_{\mathscr{F}_1}(\mathscr{E}_{\mathscr{F}_2}(Y)) = \mathscr{E}_{\mathscr{F}_1}(Y) = \mathscr{E}_{\mathscr{F}_2}(\mathscr{E}_{\mathscr{F}_1}(Y)). \tag{14}$$

PROOF. Since $Y$ satisfies trivially the defining relation for $\mathscr{E}(Y \mid \mathscr{F}_1)$, it will be equal to the latter if and only if $Y \in \mathscr{F}_1$. Now if we replace our basic $\mathscr{F}$ by $\mathscr{F}_2$ and $Y$ by $\mathscr{E}_{\mathscr{F}_2}(Y)$, the assertion (13) ensues. Next, since

$$\mathscr{E}_{\mathscr{F}_1}(Y) \in \mathscr{F}_1 \subset \mathscr{F}_2,$$

the second equation in (14) follows from the same observation. It remains to prove the first equation in (14). Let $\Lambda \in \mathscr{F}_1$, then $\Lambda \in \mathscr{F}_2$; applying the defining relation twice, we obtain

$$\int_\Lambda \mathscr{E}_{\mathscr{F}_1}(\mathscr{E}_{\mathscr{F}_2}(Y))\, d\mathscr{P} = \int_\Lambda \mathscr{E}_{\mathscr{F}_2}(Y)\, d\mathscr{P} = \int_\Lambda Y\, d\mathscr{P}.$$

Hence $\mathscr{E}_{\mathscr{F}_1}(\mathscr{E}_{\mathscr{F}_2}(Y))$ satisfies the defining relation for $\mathscr{E}_{\mathscr{F}_1}(Y)$; since it belongs to $\mathscr{F}_1$, it is equal to the latter.

As a particular case, we note, for example,

$$\mathscr{E}\{\mathscr{E}(Y \mid X_1, X_2) \mid X_1\} = \mathscr{E}(Y \mid X_1) = \mathscr{E}\{\mathscr{E}(Y \mid X_1) \mid X_1, X_2\}. \tag{15}$$

To understand the meaning of this formula, we may think of $X_1$ and $X_2$ as discrete, each producing a countable partition. The superimposition of both partitions yields sets of the form $\{\Lambda_j \cap \text{M}_k\}$. The "inner" expectation on the left of (15) is the result of replacing $Y$ by its "average" over each $\Lambda_j \cap \text{M}_k$. Now if we replace this average r.v. again by its average over each $\Lambda_j$, then the result is the same as if we had simply replaced $Y$ by its average over each $\Lambda_j$. The second equation has a similar interpretation.

Another kind of simple situation is afforded by the probability triple $(\mathscr{U}^n, \mathscr{B}^n, m^n)$ discussed in Example 2 of Sec. 3.3. Let $x_1, \dots, x_n$ be the coordinate r.v.'s ●●● $= f(x_1, \dots, x_n)$, where $f$ is (Borel) measurable and integrable. It is easy to see that for $1 \leq k \leq n-1$,

$$\mathscr{E}(y \mid x_1, \dots, x_k) = \int_0^1 \cdots \int_0^1 f(x_1, \dots, x_n)\, dx_{k+1} \cdots dx_n,$$

while for $k = n$, the left side is just $y$ (a.e.). Thus, taking conditional expectation with respect to certain coordinate r.v.'s here amounts to *integrating out the other r.v.'s*. The first equation in (15) in this case merely asserts the possibility of iterated integration, while the second reduces to a banality, which we leave to the reader to write out.

## EXERCISES

**1.** Prove Lemma 2 in Sec. 7.2 by using conditional probabilities.

**2.** Let $\{\Lambda_n\}$ be a countable measurable partition of $\Omega$, and $E \in \mathscr{F}$ with $\mathscr{P}(E)>0$; then we have for each $m$:

$$\mathscr{P}_E(\Lambda_m) = \frac{\mathscr{P}(\Lambda_m)\mathscr{P}_{\Lambda_m}(E)}{\sum_n \mathscr{P}(\Lambda_n)\mathscr{P}_{\Lambda_n}(E)}.$$

[This is Bayes' rule.]

***3.** If $X$ is an integrable r.v., $Y$ a bounded r.v., and $\mathscr{G}$ a Borel subfield, then we have

$$\mathscr{E}\{\mathscr{E}(X \mid \mathscr{G})Y\} = \mathscr{E}\{X\mathscr{E}(Y \mid \mathscr{G})\}.$$

**4.** Prove Fatou's lemma and Lebesgue's dominated convergence theorem for conditional expectations.

***5.** Give an example where $\mathscr{E}(\mathscr{E}(Y \mid X_1) \mid X_2) \neq \mathscr{E}(\mathscr{E}(Y \mid X_2) \mid X_1)$. [HINT: It is sufficient to give an example where $\mathscr{E}(X \mid Y) \neq \mathscr{E}\{\mathscr{E}(X \mid Y) \mid X\}$; consider an $\Omega$ with three points.]

***6.** Prove that $\sigma^2(\mathscr{E}_{\mathscr{G}}(Y)) \leq \sigma^2(Y)$, where $\sigma^2$ is the variance.

**7.** If the random vector has the probability density function $p(\cdot, \cdot)$ and $X$ is integrable, then one version of $\mathscr{E}(X \mid X + Y = z)$ is given by

$$\int xp(x, z-x)\, dx \Big/ \int p(x, z-x)\, dx.$$

★**8.** In the case above, there exists an integrable function $\varphi(\cdot, \cdot)$ with the property that for each $B \in \mathscr{B}^1$,

$$\int_B \varphi(x, y)\, dy$$

is a version of $\mathscr{P}\{Y \in B \mid X = x\}$. [This is called a "conditional density function in the wide sense" and

$$\Phi(x, \eta) = \int_{-\infty}^{\eta} \varphi(x, y)\, dy$$

is the corresponding conditional distribution in the wide sense:

$$\Phi(x, \eta) = \mathscr{P}\{Y \leq \eta \mid X = x\}.$$

The term "wide sense" refers to the fact that these are functions on $\mathscr{R}^1$ rather than on $\Omega$; see Doob [1, Sec. I.9].]

**9.** Let the $p(\cdot, \cdot)$ above be the 2-dimensional normal density:

$$\frac{1}{2\pi\sigma_1\sigma_2\sqrt{1-\rho^2}} \exp\left\{-\frac{1}{2(1-\rho^2)}\left(\frac{x^2}{\sigma_1^2} - \frac{2\rho xy}{\sigma_1\sigma_2} + \frac{y^2}{\sigma_2^2}\right)\right\},$$

where $\sigma_1>0$, $\sigma_2>0$, $0 < \rho < 1$. Find the $\varphi$ mentioned in Exercise 8 and

$$\int_{-\infty}^{\infty} y\varphi(x, y)\, dy.$$

The latter should be a version of $\mathscr{E}(Y \mid X = x)$; verify it.

**10.** Let $\mathscr{G}$ be a B.F., $X$ and $Y$ two r.v.'s such that

$$\mathscr{E}(Y^2 \mid \mathscr{G}) = X^2, \qquad \mathscr{E}(Y \mid \mathscr{G}) = X.$$

Then $Y = X$ a.e.

**11.** As in Exercise 10 but suppose now for any $f \in C_K$;

$$\mathscr{E}\{X^2 \mid f(X)\} = \mathscr{E}\{Y^2 \mid f(X)\}; \qquad \mathscr{E}\{X \mid f(X)\} = \mathscr{E}\{Y \mid f(X)\}.$$

Then $Y = X$ a.e. [HINT: By a monotone class theorem the equations hold for $f = 1_B$, $B \in \mathscr{B}^1$; now apply Exercise 10 with $\mathscr{G} = \mathscr{F}\{X\}$.]

**12.** Recall that $X_n$ in $L^1$ converges weakly in $L^1$ to $X$ iff $\mathscr{E}(X_n Y) \to \mathscr{E}(XY)$ for every bounded r.v. $Y$. Prove that this implies $\mathscr{E}(X_n \mid \mathscr{G})$ converges weakly in $L^1$ to $\mathscr{E}(X \mid \mathscr{G})$ for any Borel subfield $\mathscr{G}$ of $\mathscr{F}$.

★**13.** Let $S$ be an r.v. such that $\mathscr{P}\{S > t\} = e^{-t}$, $t>0$. Compute $\mathscr{E}\{S \mid S\Lambda t\}$ and $\mathscr{E}\{S \mid S \vee t\}$ for each $t > 0$.

## 9.2 Conditional independence; Markov property

In this section we shall first apply the concept of conditioning to independent r.v.'s and to random walks, the two basic types of stochastic processes that have been extensively studied in this book; then we shall generalize to a Markov process. Another important generalization, the martingale theory, will be taken up in the next three sections.

All the B.F.'s (Borel fields) below will be subfields of $\mathscr{F}$. The B.F.'s $\{\mathscr{F}_\alpha, \alpha \in A\}$, where $A$ is an arbitrary index set, are said to be *conditionally independent relative to the B.F.* $\mathscr{G}$, iff for any finite collection of sets $\Lambda_1, \ldots, \Lambda_n$ such that $\Lambda_j \in \mathscr{F}_{\alpha_j}$ and the $\alpha_j$'s are distinct indices from $A$, we have

$$\mathscr{P}\left(\bigcap_{j=1}^{n} \Lambda_j \mid \mathscr{G}\right) = \prod_{j=1}^{n} \mathscr{P}(\Lambda_j \mid \mathscr{G}).$$

When $\mathscr{G}$ is the trivial B.F., this reduces to unconditional independence.

**Theorem 9.2.1.** For each $\alpha \in A$ let $\mathscr{F}^{(\alpha)}$ denote the smallest B.F. containing all $\mathscr{F}_\beta$, $\beta \in A - \{\alpha\}$. Then the $\mathscr{F}_\alpha$'s are conditionally independent relative to $\mathscr{G}$ if and only if for each $\alpha$ and $\Lambda_\alpha \in \mathscr{F}_\alpha$ we have

$$\mathscr{P}(\Lambda_\alpha \mid \mathscr{F}^{(\alpha)} \vee \mathscr{G}) = \mathscr{P}(\Lambda_\alpha \mid \mathscr{G}),$$

where $\mathscr{F}^{(\alpha)} \vee \mathscr{G}$ denotes the smallest B.F. containing $\mathscr{F}^{(\alpha)}$ and $\mathscr{G}$.

PROOF. It is sufficient to prove this for two B.F.'s $\mathscr{F}_1$ and $\mathscr{F}_2$, since the general result follows by induction (how?). Suppose then that for each $\Lambda \in \mathscr{F}_1$ we have

$$(1) \qquad \mathscr{P}(\Lambda \mid \mathscr{F}_2 \vee \mathscr{G}) = \mathscr{P}(\Lambda \mid \mathscr{G}).$$

Let $M \in \mathscr{F}_2$, then

$$\begin{aligned}\mathscr{P}(\Lambda M \mid \mathscr{G}) &= \mathscr{E}\{\mathscr{P}(\Lambda M \mid \mathscr{F}_2 \vee \mathscr{G}) \mid \mathscr{G}\} = \mathscr{E}\{\mathscr{P}(\Lambda \mid \mathscr{F}_2 \vee \mathscr{G})1_M \mid \mathscr{G}\} \\ &= \mathscr{E}\{\mathscr{P}(\Lambda \mid \mathscr{G})1_M \mid \mathscr{G}\} = \mathscr{P}(\Lambda \mid \mathscr{G})\mathscr{P}(M \mid \mathscr{G}),\end{aligned}$$

where the first equation follows from Theorem 9.1.5, the second and fourth from Theorem 9.1.3, and the third from (1). Thus $\mathscr{F}_1$ and $\mathscr{F}_2$ are conditionally independent relative to $\mathscr{G}$. Conversely, suppose the latter assertion is true, then

$$\begin{aligned}\mathscr{E}\{\mathscr{P}(\Lambda \mid \mathscr{G})1_M \mid \mathscr{G}\} &= \mathscr{P}(\Lambda \mid \mathscr{G})\mathscr{P}(M \mid \mathscr{G}) \\ &= \mathscr{P}(\Lambda M \mid \mathscr{G}) = \mathscr{E}\{\mathscr{P}(\Lambda \mid \mathscr{F}_2 \vee \mathscr{G})1_M \mid \mathscr{G}\},\end{aligned}$$

where the first equation follows from Theorem 9.1.3, the second by hypothesis, and the third as shown above. Hence for every $\Delta \in \mathscr{G}$, we have

$$\int_{M\Delta} \mathscr{P}(\Lambda \mid \mathscr{G})\, d\mathscr{P} = \int_{M\Delta} \mathscr{P}(\Lambda \mid \mathscr{F}_2 \vee \mathscr{G})\, d\mathscr{P} = \mathscr{P}(\Lambda M \Delta)$$

It follows from Theorem 2.1.2 (take $\mathscr{F}_0$ to be finite disjoint unions of sets like $M\Delta$) or more quickly from Exercise 10 of Sec. 2.1 that this remains true if $M\Delta$ is replaced by any set in $\mathscr{F}_2 \vee \mathscr{G}$. The resulting equation implies (1), and the theorem is proved.

When $\mathscr{G}$ is trivial and each $\mathscr{F}_\alpha$ is generated by a single r.v., we have the following corollary.

**Corollary.** Let $\{X_\alpha, \alpha \in A\}$ be an arbitrary set of r.v.'s. For each $\alpha$ let $\mathscr{F}^{(\alpha)}$ denote the Borel field generated by all the r.v.'s in the set except $X_\alpha$. Then the $X_\alpha$'s are independent if and only if: for each $\alpha$ and each $B \in \mathscr{B}^1$, we have

$$\mathscr{P}\{X_\alpha \in B \mid \mathscr{F}^{(\alpha)}\} = \mathscr{P}\{X_\alpha \in B\} \qquad \text{a.e.}$$

An equivalent form of the corollary is as follows: for each integrable r.v. $Y$ belonging to the Borel field generated by $X_\alpha$, we have

$$\mathscr{E}\{Y \mid \mathscr{F}^{(\alpha)}\} = \mathscr{E}\{Y\}. \tag{2}$$

This is left as an exercise.

Roughly speaking, independence among r.v.'s is equivalent to the lack of effect by conditioning relative to one another. However, such intuitive statements must be treated with caution, as shown by the following example which will be needed later.

If $\mathscr{F}_1$, $\mathscr{F}_2$, and $\mathscr{F}_3$ are three Borel fields such that $\mathscr{F}_1 \vee \mathscr{F}_2$ is independent of $\mathscr{F}_3$, then for each integrable $X \in \mathscr{F}_1$, we have

$$\mathscr{E}\{X \mid \mathscr{F}_2 \vee \mathscr{F}_3\} = \mathscr{E}\{X \mid \mathscr{F}_2\}. \tag{3}$$

Instead of a direct verification, which is left to the reader, it is interesting to deduce this from Theorem 9.2.1 by proving the following proposition.

If $\mathscr{F}_1 \vee \mathscr{F}_2$ is independent of $\mathscr{F}_3$, then $\mathscr{F}_1$ and $\mathscr{F}_3$ are conditionally independent relative to $\mathscr{F}_2$.

To see this, let $\Lambda_1 \in \mathscr{F}_1$, $\Lambda_3 \in \mathscr{F}_3$. Since

$$\mathscr{P}(\Lambda_1\Lambda_2\Lambda_3) = \mathscr{P}(\Lambda_1\Lambda_2)\mathscr{P}(\Lambda_3) = \int_{\Lambda_2} \mathscr{P}(\Lambda_1 \mid \mathscr{F}_2)\mathscr{P}(\Lambda_3)\, d\mathscr{P}$$

for every $\Lambda_2 \in \mathscr{F}_2$, we have

$$\mathscr{P}(\Lambda_1\Lambda_3 \mid \mathscr{F}_2) = \mathscr{P}(\Lambda_1 \mid \mathscr{F}_2)\mathscr{P}(\Lambda_3) = \mathscr{P}(\Lambda_1 \mid \mathscr{F}_2)\mathscr{P}(\Lambda_3 \mid \mathscr{F}_2),$$

which proves the proposition.

Next we ask: if $X_1$ and $X_2$ are independent, what is the effect of conditioning $X_1 + X_2$ by $X_1$?

**Theorem 9.2.2.** Let $X_1$ and $X_2$ be independent r.v.'s with p.m.'s $\mu_1$ and $\mu_2$; then for each $B \in \mathscr{B}^1$:

$$\mathscr{P}\{X_1 + X_2 \in B \mid X_1\} = \mu_2(B - X_1) \quad \text{a.e.} \tag{4}$$

More generally, if $\{X_n, n \geq 1\}$ is a sequence of independent r.v.'s with p.m.'s $\{\mu_n, n \geq 1\}$, and $S_n = \sum_{j=1}^{n} X_j$, then for each $B \in \mathscr{B}^1$:

$$\mathscr{P}\{S_n \in B \mid S_1, \dots, S_{n-1}\} = \mu_n(B - S_{n-1}) = \mathscr{P}\{S_n \in B \mid S_{n-1}\} \quad \text{a.e.} \tag{5}$$

PROOF. To prove (4), since its right member belongs to $\mathscr{F}\{X_1\}$, it is sufficient to verify that it satisfies the defining relation for its left member. Let $\Lambda \in \mathscr{F}\{X_1\}$, then $\Lambda = X_1^{-1}(A)$ for some $A \in \mathscr{B}^1$. It follows from Theorem 3.2.2 that

$$\int_\Lambda \mu_2(B - X_1)\, d\mathscr{P} = \int_A \mu_2(B - x_1)\mu_1(dx_1).$$

Writing $\mu = \mu_1 \times \mu_2$ and applying Fubini's theorem to the right side above, then using Theorem 3.2.3, we obtain

$$\int_A \mu_1(dx_1) \int_{x_1 + x_2 \in B} \mu_2(dx_2) = \iint\limits_{\substack{x_1 \in A \\ x_1 + x_2 \in B}} \mu(dx_1, dx_2)$$

$$= \int\limits_{\substack{X_1 \in A \\ X_1 + X_2 \in B}} d\mathscr{P} = \mathscr{P}\{X_1 \in A; X_1 + X_2 \in B\}.$$

This establishes (4).

To prove (5), we begin by observing that the second equation has just been proved. Next we observe that since $\{X_1, \dots, X_n\}$ and $\{S_1, \dots, S_n\}$ obviously generate the same Borel field, the left member of (5) is just

$$\mathscr{P}\{S_n = B \mid X_1, \dots, X_{n-1}\}.$$

Now it is trivial that as a function of $(X_1, \dots, X_{n-1})$, $S_n$ "depends on them only through their sum $S_{n-1}$". It thus appears obvious that the first term in (5) should depend only on $S_{n-1}$, namely belong to $\mathscr{F}\{S_{n-1}\}$ (rather than the larger $\mathscr{F}\{S_1, \dots, S_{n-1}\}$). Hence the equality of the first and third terms in (5) should be a consequence of the assertion (13) in Theorem 9.1.5. This argument, however, is not rigorous, and requires the following formal substantiation.

Let $\mu^{(n)} = \mu_1 \times \cdots \times \mu_n = \mu^{(n-1)} \times \mu_n$ and

$$\Lambda = \bigcap_{j=1}^{n-1} S_j^{-1}(B_j),$$

where $B_n = B$. Sets of the form of $\Lambda$ generate the Borel field $\mathscr{F}\{S_1, \dots, S_{n-1}\}$. It is therefore sufficient to verify that for each such $\Lambda$, we have

$$\int_\Lambda \mu_n(B_n - S_{n-1})\, d\mathscr{P} = \mathscr{P}\{\Lambda; S_n \in B_n\}.$$

If we proceed as before and write $s_n = \sum_{j=1}^n x_j$, the left side above is equal to

$$\int \cdots \int_{s_j \in B_j, 1 \le j \le n-1} \mu_n(B_n - s_{n-1})\mu^{(n-1)}(dx_1, \dots, dx_{n-1})$$

$$= \int \cdots \int_{s_j \in B_j, 1 \le j \le n} \mu^{(n)}(dx_1, \dots, dx_n) = \mathscr{P}\left\{\bigcap_{j=1}^n [S_j \in B_j]\right\},$$

as was to be shown. In the first equation above, we have made use of the fact that the set of $(x_1, \dots, x_n)$ in $\mathscr{R}^n$ for which $x_n \in B_n - s_{n-1}$ is exactly the set for which $s_n \in B_n$, which is the formal counterpart of the heuristic argument above. The theorem is proved.

The fact of the equality of the two extreme terms in (5) is a fundamental property of the sequence $\{S_n\}$. We have indeed made frequent use of this in the study of sums of independent r.v.'s, particularly for a random walk (Chapter 8), although no explicit mention has been made of it. It would be instructive for the reader to review the material there and locate some instances of its application. As an example, we prove the following proposition, where the intuitive picture of conditioning is particularly clear.

**Theorem 9.2.3.** Let $\{X_n, n \ge 1\}$ be an independent (but not necessarily stationary) process such that for $A>0$ there exists $\delta>0$ satisfying

$$\inf_n \mathscr{P}\{X_n \ge A\} > \delta.$$

Then we have

$$\forall n \ge 1: \mathscr{P}\{S_j \in (0, A] \text{ for } 1 \le j \le n\} \le (1-\delta)^n.$$

Furthermore, given any finite interval $I$, there exists an $\epsilon>0$ such that

$$\mathscr{P}\{S_j \in I, \text{ for } 1 \le j \le n\} \le (1-\epsilon)^n.$$

PROOF. We write $\Lambda_n$ for the event that $S_j \in (0, A]$ for $1 \le j \le n$; then

$$\mathscr{P}\{\Lambda_n\} = \mathscr{P}\{\Lambda_{n-1}; 0 < S_n \le A\}.$$

By the definition of conditional probability and (5), the last-written probability is equal to

$$\int_{\Lambda_{n-1}} \mathscr{P}\{0 < S_n \le A \mid S_1, \dots, S_{n-1}\}\, d\mathscr{P}$$

$$= \int_{\Lambda_{n-1}} [F_n(A - S_{n-1}) - F_n(0 - S_{n-1})]\, d\mathscr{P},$$

where $F_n$ is the d.f. of $\mu_n$. [A quirk of notation forbids us to write the integrand on the right as $\mathscr{P}\{-S_{n-1} < X_n \le A - S_{n-1}\}$!] Now for each $\omega_0$ in $\Lambda_{n-1}$, $S_{n-1}(\omega_0) > 0$, hence

$$F_n(A - S_{n-1}(\omega_0)) \le \mathscr{P}\{X_n < A\} \le 1 - \delta$$

by hypothesis. It follows that

$$\mathscr{P}\{\Lambda_n\} \le \int_{\Lambda_{n-1}} (1 - \delta)\, d\mathscr{P} = (1 - \delta)\mathscr{P}\{\Lambda_{n-1}\},$$

and the first assertion of the theorem follows by iteration. The second is proved similarly by observing that $\mathscr{P}\{X_n + \cdots + X_{n+m-1} \ge mA\} > \delta^m$ and choosing $m$ so that $mA$ exceeds the length of $I$. The details are left to the reader.

Let $N^0 = \{0\} \cup N$ denote the set of positive integers. For a given sequence of r.v.'s $\{X_n, n \in N^0\}$ let us denote by $\mathscr{F}_I$ the Borel field generated by $\{X_n, n \in I\}$, where $I$ is a subset of $N^0$, such as $[0, n]$, $(n, \infty)$, or $\{n\}$. Thus $\mathscr{F}_{\{n\}}$, $\mathscr{F}_{[0,n]}$, and $\mathscr{F}_{(n,\infty)}$ have been denoted earlier by $\mathscr{F}\{X_n\}$, $\mathscr{F}_n$, and $\mathscr{F}'_n$, respectively.

DEFINITION OF MARKOV PROCESS. The sequence of r.v.'s $\{X_n, n \in N^0\}$ is said to be a Markov process or to possess the *"Markov property"* iff for every $n \in N^0$ and every $B \in \mathscr{B}^1$, we have

$$\mathscr{P}\{X_{n+1} \in B \mid X_0, \dots, X_n\} = \mathscr{P}\{X_{n+1} \in B \mid X_n\}. \tag{6}$$

This property may be verbally announced as: the conditional distribution (in the wide sense!) of each r.v. relative to all the preceding ones is the same as that relative to the last preceding one. Thus if $\{X_n\}$ is an independent process as defined in Chapter 8, then both the process itself and the process of the successive partial sums $\{S_n\}$ are Markov processes by Theorems 9.2.1 and 9.2.2. The latter category includes random walk as a particular case; note that in this case our notation $X_n$ rather than $S_n$ differs from that employed in Chapter 8.

Equation (6) is equivalent to the apparently stronger proposition: for every integrable $Y \in \mathscr{F}_{\{n+1\}}$, we have

$$\mathscr{E}\{Y \mid X_1, \dots, X_n\} = \mathscr{E}\{Y \mid X_n\}. \tag{6'}$$

It is clear that (6′) implies (6). To see the converse, let $Y_m$ be a sequence of simple r.v.'s belonging to $\mathscr{F}_{\{n+1\}}$ and increasing to $Y$. By (6) and property (ii) of conditional expectation in Sec. 9.1, (6′) is true when $Y$ is replaced by $Y_m$; hence by property (v) there, it is also true for $Y$.

The following remark will be repeatedly used. If $Y$ and $Z$ are integrable, $Z \in \mathscr{F}_1$, and

$$\int_\Lambda Y \, d\mathscr{P} = \int_\Lambda Z \, d\mathscr{P}$$

for each $\Lambda$ of the form

$$\bigcap_{j \in I_0} X_j^{-1}(B_j),$$

where $I_0$ is an arbitrary finite subset of $I$ and each $B_j$ is an arbitrary Borel set, then $Z = \mathscr{E}(Y \mid \mathscr{F}_1)$. This follows from the uniqueness of conditional expectation and a previous remark given before Theorem 9.1.3, since finite disjoint unions of sets of this form generate a field that generates $\mathscr{F}_I$.

If the index $n$ is regarded as a discrete time parameter, as is usual in the theory of stochastic processes, then $\mathscr{F}_{[0,n]}$ is the field of "the past and the present", while $\mathscr{F}_{(n,\infty)}$ is that of "the future"; whether the present is adjoined to the past or the future is often a matter of convenience. The Markov property just defined may be further characterized as follows.

**Theorem 9.2.4.** The Markov property is equivalent to either one of the two propositions below:

$$(7) \qquad \forall n \in N, \mathrm{M} \in \mathscr{F}_{(n,\infty)}: \mathscr{P}\{\mathrm{M} \mid \mathscr{F}_{[0,n]}\} = \mathscr{P}\{\mathrm{M} \mid X_n\}.$$

$$(8) \qquad \forall n \in N, \mathrm{M}_1 \in \mathscr{F}_{[0,n]}, \mathrm{M}_2 \in \mathscr{F}_{(n,\infty)}: \mathscr{P}\{\mathrm{M}_1\mathrm{M}_2 \mid X_n\} = \mathscr{P}\{\mathrm{M}_1 \mid X_n\}\mathscr{P}\{\mathrm{M}_2 \mid X_n\}.$$

These conclusions remain true if $\mathscr{F}_{(n,\infty)}$ is replaced by $\mathscr{F}_{[n,\infty)}$.

PROOF. To prove (7) implies (8), let $Y_i = 1_{\mathrm{M}_i}$, $i = 1, 2$. We then have

$$(9) \qquad \begin{aligned} \mathscr{P}\{\mathrm{M}_1 \mid X_n\}\mathscr{P}\{\mathrm{M}_2 \mid X_n\} &= \mathscr{E}\{Y_1 \mid X_n\}\mathscr{E}\{Y_2 \mid X_n\} \\ &= \mathscr{E}\{Y_1\mathscr{E}(Y_2 \mid X_n) \mid X_n\} \\ &= \mathscr{E}\{Y_1\mathscr{E}(Y_2 \mid \mathscr{F}_{[0,n]}) \mid X_n\} \\ &= \mathscr{E}\{\mathscr{E}(Y_1Y_2 \mid \mathscr{F}_{[0,n]}) \mid X_n\} \\ &= \mathscr{E}\{Y_1Y_2 \mid X_n\} = \mathscr{P}\{\mathrm{M}_1\mathrm{M}_2 \mid X_n\}, \end{aligned}$$

where the second and fourth equations follow from Theorems 9.1.3, the third from assumption (7), and the fifth from Theorem 9.1.5.

Conversely, to prove that (8) implies (7), let $\Lambda \in \mathscr{F}_{\{n\}}$, $\mathrm{M}_1 \in \mathscr{F}_{[0,n)}$, $\mathrm{M}_2 \in \mathscr{F}_{(n,\infty)}$. By the second equation in (9) applied to the fourth equation below, we have

$$\int_{\Lambda M_1} \mathscr{P}(M_2 \mid X_n)\, d\mathscr{P} = \int_{\Lambda M_1} \mathscr{E}(Y_2 \mid X_n)\, d\mathscr{P} = \int_{\Lambda} Y_1 \mathscr{E}(Y_2 \mid X_n)\, d\mathscr{P}$$

$$= \int_{\Lambda} \mathscr{E}\{Y_1 \mathscr{E}(Y_2 \mid X_n) \mid X_n\}\, d\mathscr{P}$$

$$= \int_{\Lambda} \mathscr{E}(Y_1 \mid X_n)\mathscr{E}(Y_2 \mid X_n)\, d\mathscr{P}$$

$$= \int_{\Lambda} \mathscr{P}(M_1 \mid X_n)\mathscr{P}(M_2 \mid X_n)\, d\mathscr{P}$$

$$= \int_{\Lambda} \mathscr{P}(M_1 M_2 \mid X_n)\, d\mathscr{P} = \mathscr{P}(\Lambda M_1 M_2).$$

Since disjoint unions of sets of the form $\Lambda M_1$ as specified above generate the Borel field $\mathscr{F}_{[0,n]}$, the uniqueness of $\mathscr{P}(M_2 \mid \mathscr{F}_{[0,n]})$ shows that it is equal to $\mathscr{P}(M_2 \mid X_n)$, proving (7).

Finally, we prove the equivalence of the Markov property and the proposition (7). Clearly the former is implied by the latter; to prove the converse we shall operate with conditional expectations instead of probabilities and use induction. Suppose that it has been shown that for every $n \in N$, and every bounded $f$ belonging to $\mathscr{F}_{[n+1,n+k]}$, we have

$$\mathscr{E}(f \mid \mathscr{F}_{[0,n]}) = \mathscr{E}(f \mid \mathscr{F}_{\{n\}}). \tag{10}$$

This is true for $k = 1$ by (6′). Let $g$ be bounded, $g \in \mathscr{F}_{[n+1,n+k+1]}$; we are going to show that (10) remains true when $f$ is replaced by $g$. For this purpose it is sufficient to consider a $g$ of the form $g_1 g_2$, where $g_1 \in \mathscr{F}_{[n+1,n+k]}$, $g_2 \in \mathscr{F}_{\{n+k+1\}}$, both bounded. The successive steps, in slow motion fashion, are as follows:

$$\mathscr{E}\{g \mid \mathscr{F}_{[0,n]}\} = \mathscr{E}\{\mathscr{E}(g \mid \mathscr{F}_{[0,n+k]}) \mid \mathscr{F}_{[0,n]}\} = \mathscr{E}\{g_1 \mathscr{E}(g_2 \mid \mathscr{F}_{[0,n+k]}) \mid \mathscr{F}_{[0,n]}\}$$

$$= \mathscr{E}\{g_1 \mathscr{E}(g_2 \mid \mathscr{F}_{\{n+k\}}) \mid \mathscr{F}_{[0,n]}\} = \mathscr{E}\{g_1 \mathscr{E}(g_2 \mid \mathscr{F}_{\{n+k\}}) \mid \mathscr{F}_{\{n\}}\}$$

$$= \mathscr{E}\{g_1 \mathscr{E}(g_2 \mid \mathscr{F}_{[n,n+k]}) \mid \mathscr{F}_{\{n\}}\} = \mathscr{E}\{\mathscr{E}(g_1 g_2 \mid \mathscr{F}_{[n,n+k]}) \mid \mathscr{F}_{\{n\}}\}$$

$$= \mathscr{E}\{g_1 g_2 \mid \mathscr{F}_{\{n\}}\} = \mathscr{E}\{g \mid \mathscr{F}_{\{n\}}\}.$$

It is left to the reader to scrutinize each equation carefully for explanation, except that the fifth one is an immediate consequence of the Markov property $\mathscr{E}\{g_2 \mid \mathscr{F}_{\{n+k\}}\} = \mathscr{E}\{g_2 \mid \mathscr{F}_{[0,n+k]}\}$ and (13) of Sec. 9.1. This establishes (7) for $M \in \bigcup_{k=1}^{\infty} \mathscr{F}_{(n,n+k)}$, which is a field generating $\mathscr{F}_{(n,\infty)}$. Hence (7) is true (why?). The last assertion of the theorem is left as an exercise.

The property embodied in (8) may be announced as follows: "The past and the future are conditionally independent given the present". In this form there is a symmetry that is not apparent in the other equivalent forms.

The Markov property has an extension known as the "strong Markov property". In the case discussed here, where the time index is $N^0$, it is an automatic consequence of the ordinary Markov property, but it is of great conceptual importance in applications. We recall the notions of an optional r.v. $\alpha$, the r.v. $X_\alpha$, and the fields $\mathscr{F}_\alpha$ and $\mathscr{F}'_\alpha$, which are defined for a general sequence of r.v.'s in Sec. 8.2. Note that here $\alpha$ may take the value 0, which is included in $N^0$. We shall give the extension in the form (7).

**Theorem 9.2.5.** Let $\{X_n, n \in N^0\}$ be a Markov process and $\alpha$ a finite optional r.v. relative to it. Then for each $\mathrm{M} \in \mathscr{F}'_\alpha$ we have

$$\mathscr{P}\{\mathrm{M} \mid \mathscr{F}_\alpha\} = \mathscr{P}\{\mathrm{M} \mid \alpha, X_\alpha\}. \tag{11}$$

PROOF. Since $\alpha \in \mathscr{F}_\alpha$ and $X_\alpha \in \mathscr{F}_\alpha$ (Exercise 2 of Sec. 8.2), the right member above belongs to $\mathscr{F}_\alpha$. To prove (11) it is then sufficient to verify that the right member satisfies the defining relation for the left, when M is of the form

$$\bigcap_{j=1}^{\ell}\{X_{\alpha+j} \in B_j\}, \qquad B_j \in \mathscr{B}^1, 1 \le j \le \ell; 1 \le \ell < \infty.$$

Put for each $n$,

$$\mathrm{M}_n = \bigcap_{j=1}^{\ell}\{X_{n+j} \in B_j\} \in \mathscr{F}_{(n,\infty)}.$$

Now the crucial step is to show that

$$\sum_{n=0}^{\infty}\mathscr{P}\{\mathrm{M}_n \mid X_n\}1_{\{\alpha=n\}} = \mathscr{P}\{\mathrm{M} \mid \alpha, X_\alpha\}. \tag{12}$$

By the lemma in the proof of Theorem 9.1.2, there exists a Borel measurable function $\varphi_n$ such that $\mathscr{P}\{\mathrm{M}_n \mid X_n\} = \varphi_n(X_n)$, from which it follows that the left member of (12) belongs to the Borel field generated by the two r.v.'s $\alpha$ and $X_\alpha$. Hence we shall prove (12) by verifying that its left member satisfies the defining relation for its right member, as follows. For each $m \in N$ and $B \in \mathscr{B}^1$, we have

$$\int_{\{\alpha=m;X_\alpha\in B\}}\sum_{n=0}^{\infty}\mathscr{P}\{\mathrm{M}_n \mid X_n\}1_{\{\alpha=n\}}\,d\mathscr{P} = \int_{\{\alpha=m;X_m\in B\}}\mathscr{P}\{\mathrm{M}_m \mid X_m\}\,d\mathscr{P}$$

$$= \int_{\{\alpha=m;X_m\in B\}}\mathscr{P}\{\mathrm{M}_m \mid \mathscr{F}_{[0,m]}\}\,d\mathscr{P} = \mathscr{P}\{\alpha = m; X_m \in B; \mathrm{M}_m\}$$

$$= \mathscr{P}\{\alpha = m; X_\alpha \in B; \mathrm{M}\},$$

where the second equation follows from an application of (7), and the third from the optionality of $\alpha$, namely

$$\{\alpha = m\} \in \mathscr{F}_{[0,m]}$$

This establishes (12).

Now let $\Lambda \in \mathscr{F}_\alpha$, then [cf. (3) of Sec. 8.2] we have

$$\Lambda = \bigcup_{n=0}^{\infty} \{(\alpha = n) \cap \Lambda_n\},$$

where $\Lambda_n \in \mathscr{F}_{[0,n]}$. It follows that

$$\mathscr{P}\{\Lambda \mathrm{M}\} = \sum_{n=0}^{\infty} \mathscr{P}\{\alpha = n; \Lambda_n; \mathrm{M}_n\} = \sum_{n=0}^{\infty} \int_{(\alpha=n)\cap\Lambda_n} \mathscr{P}\{\mathrm{M}_n \mid \mathscr{F}_{[0,n]}\}\, d\mathscr{P}$$

$$= \sum_{n=0}^{\infty} \int_{\Lambda} \mathscr{P}\{\mathrm{M}_n \mid X_n\} 1_{\{\alpha=n\}}\, d\mathscr{P} = \int_{\Lambda} \mathscr{P}\{\mathrm{M} \mid \alpha, X_\alpha\}\, d\mathscr{P},$$

where the third equation is by an application of (7) while the fourth is by (12). This being true for each $\Lambda$, we obtain (11). The theorem is proved.

When $\alpha$ is a constant $n$, it may be omitted from the right member of (11), and so (11) includes (7) as a particular case. It may be omitted also in the homogeneous case discussed below (because then the $\varphi_n$ above may be chosen to be independent of $n$).

There is a very general method of constructing a Markov process with given "transition probability functions", as follows. Let $P_0(\cdot)$ be an arbitrary p.m. on $(\mathscr{R}^1, \mathscr{B}^1)$. For each $n \geq 1$ let $P_n(\cdot, \cdot)$ be a function of the pair $(x, B)$ where $x \in \mathscr{R}^1$ and $B \in \mathscr{B}^1$, having the measurability properties below:

(a) for each $x$, $P_n(x, \cdot)$ is a p.m. on $\mathscr{B}^1$;
(b) for each $B$, $P_n(\cdot, B) \in \mathscr{B}^1$.

It is a consequence of Kolmogorov's extension theorem (see Sec. 3.3) that there exists a sequence of r.v.'s $\{X_n, n \in N^0\}$ on some probability space with the following "finite-dimensional joint distributions": for each $0 \leq \mathscr{L} < \infty$, $B_j \in \mathscr{B}^1$, $0 \leq j \leq n$:

$$\mathscr{P}\left\{\bigcap_{j=0}^{n} [X_j \in B_j]\right\} = \int_{B_0} P_0(dx_0) \int_{B_1} P_1(x_0, dx_1) \times \cdots \times \int_{B_n} P_n(x_{n-1}, dx_n). \tag{13}$$

There is no difficulty in verifying that (13) yields an $(n+1)$-dimensional p.m. on $(\mathscr{R}^{n+1}, \mathscr{B}^{n+1})$ and so also on each subspace by setting some of the $B_j$'s above to be $\mathscr{R}^1$, and that the resulting collection of p.m.'s are mutually consistent in the obvious sense [see (16) of Sec. 3.3]. But the actual construction of the process $\{X_n, n \in N^0\}$ with these given "marginal distributions" is omitted here, the procedure being similar to but somewhat more sophisticated than that given in Theorem 3.3.4, which is a particular case. Assuming the existence, we will now show that it has the Markov property by verifying (6) briefly. By Theorem 9.1.5, it will be sufficient to show that one version of the left member of (6) is given by $P_{n+1}(X_n, B)$, which belongs to $\mathscr{F}_{\{n\}}$ by condition (b) above. Let then

$$\Lambda = \bigcap_{j=0}^{n} [X_j \in B_j]$$

and $\mu^{(n+1)}$ be the $(n+1)$-dimensional p.m. of the random vector $(X_0, \ldots, X_n)$. It follows from Theorem 3.2.3 and (13) used twice that

$$\begin{aligned}\int_{\Lambda} P_{n+1}(X_n, B)\, d\mathscr{P} &= \int \cdots \int_{B_0 \times \cdots \times B_n} P_{n+1}(x_n, B) d\mu^{(n+1)} \\ &= \int \cdots \int_{B_0 \times \cdots \times B_n \times B} P_0(dx_0) \prod_{j=1}^{n+1} P_j(x_{j-1}, dx_j) \\ &= \mathscr{P}(\Lambda; X_{n+1} \in B).\end{aligned}$$

This is what was to be shown.

We call $P_0(\cdot)$ the initial distribution of the Markov process and $P_n(\cdot, \cdot)$ its "$n$th-stage transition probability function". The case where the latter is the same for all $n \geq 1$ is particularly important, and the corresponding Markov process is said to be "(temporally) homogeneous" or "with stationary transition probabilities". In this case we write, with $x = x_0$:

$$P^{(n)}(x, B) = \int \cdots \int_{\mathscr{R}^1 \times \cdots \times \mathscr{R}^1 \times B} \prod_{j=0}^{n-1} P(x_j, dx_{j+1}), \tag{14}$$

and call it the "$n$-step transition probability function"; when $n = 1$, the qualifier "1-step" is usually dropped. We also put $P^{(0)}(x, B) = 1_B(x)$. It is easy to see that

$$P^{(n+1)}(x, B) = \int_{\mathscr{R}^1} P^{(n)}(y, B) P^{(1)}(x, dy), \tag{15}$$

so that all $P^{(n)}$ are just the *iterates* of $P^{(1)}$.

It follows from Theorem 9.2.2 that for the Markov process $\{S_n, n \in N\}$ there, we have

$$P_n(x, B) = \mu_n(B - x).$$

In particular, a random walk is a homogeneous Markov process with the 1-step transition probability function

$$P^{(1)}(x, B) = \mu(B - x).$$

In the homogeneous case Theorem 9.2.4 may be sharpened to read like Theorem 8.2.2, which becomes then a particular case. The proof is left as an exercise.

**Theorem 9.2.6.** For a homogeneous Markov process and a finite r.v. $\alpha$ which is optional relative to the process, the pre-$\alpha$ and post-$\alpha$ fields are conditionally independent relative to $X_\alpha$, namely:

$$\forall \Lambda \in \mathscr{F}_\alpha, \mathrm{M} \in \mathscr{F}'_\alpha : \mathscr{P}\{\Lambda \mathrm{M} \mid X_\alpha\} = \mathscr{P}\{\Lambda \mid X_\alpha\}\mathscr{P}\{\mathrm{M} \mid X_\alpha\}.$$

Furthermore, the post-$\alpha$ process $\{X_{\alpha+n}, n \in N\}$ is a homogeneous Markov process with the same transition probability function as the original one.

Given a Markov process $\{X_n, n \in N^0\}$, the distribution of $X_0$ is a p.m. and for each $B \in \mathscr{B}^1$, there is according to Theorem 9.1.2 a Borel measurable function $\varphi_n(\cdot, B)$ such that

$$\mathscr{P}\{X_{n+1} \in B \mid X_n = x\} = \varphi_n(x, B).$$

It seems plausible that the function $\varphi_n(\cdot, \cdot)$ would correspond to the $n$-stage transition probability function just discussed. The trouble is that while condition (b) above may be satisfied for each $B$ by a particular choice of $\varphi_n(\cdot, B)$, it is by no means clear why the resulting collection for varying $B$ would satisfy condition (a). Although it is possible to ensure this by means of conditional distributions in the wide sense alluded to in Exercise 8 of Sec. 9.1, we shall not discuss it here (see Doob [16, chap. 2]).

The theory of Markov processes is the most highly developed branch of stochastic processes. Special cases such as Markov chains, diffusion, and processes with independent increments have been treated in many monographs, a few of which are listed in the Bibliography at the end of the book.

EXERCISES

***1.** Prove that the Markov property is also equivalent to the following proposition : if $t_1 < \cdots < t_n < t_{n+1}$ are indices in $N^0$ and $B_j$, $1 \le j \le n+1$, are Borel sets, then

$$\mathscr{P}\{X_{t_{n+1}} \in B_{n+1} \mid X_{t_1}, \ldots, X_{t_n}\} = \mathscr{P}\{X_{t_{n+1}} \in B_{n+1} \mid X_{t_n}\}.$$

In this form we can define a Markov process $\{X_t\}$ with a continuous parameter $t$ ranging in $[0, \infty)$.

**2.** Does the Markov property imply the following (notation as in Exercise 1):

$$\mathscr{P}\{X_{n+1} \in B_{n+1} \mid X_1 \in B_1, \ldots, X_n \in B_n\} = \mathscr{P}\{X_{n+1} \in B_{n+1} \mid X_n \in B_n\}?$$

**3.** Unlike independence, the Markov property is not necessarily preserved by a "functional": $\{f(X_n), n \in N^0\}$. Give an example of this, but show that it is preserved by each one-to-one Borel measurable mapping $f$.

$^\star$**4.** Prove the strong Markov property in the form of (8).

**5.** Prove Theorem 9.2.6.

$^\star$**6.** For the $P^{(n)}$ defined in (14), prove the "Chapman–Kolmogorov equations":

$$\forall m \in N, n \in N: P^{(m+n)}(x, B) = \int_{\mathscr{R}^1} P^{(m)}(x, dy) P^{(n)}(y, B).$$

**7.** Generalize the Chapman–Kolmogorov equation in the nonhomogeneous case.

$^\star$**8.** For the homogeneous Markov process constructed in the text, show that for each $f \geq 0$ we have

$$\mathscr{E}\{f(X_{m+n}) \mid X_m\} = \int_{\mathscr{R}^1} f(y) P^{(n)}(X_m, dy).$$

$^\star$**9.** Let $B$ be a Borel set, $f_1(x, B) = P(x, B)$, and define $f_n$ for $n \geq 2$ inductively by

$$f_n(x, B) = \int_{B^c} P(x, dy) f_{n-1}(y, B);$$

put $f(x, B) = \sum_{n=1}^{\infty} f_n(x, B)$. Prove that $f(X_n, B)$ is a version of the conditional probability $\mathscr{P}\{\bigcup_{j=n+1}^{\infty} [X_j \in B] \mid X_n\}$ for the homogeneous Markov process with transition probability function $P(\cdot, \cdot)$.

$^\star$**10.** Using the $f$ defined in Exercise 9, put

$$g(x, B) = f(x, B) - \sum_{n=1}^{\infty} \int_B P^{(n)}(x, dy)[1 - f(y, B)].$$

Prove that $g(X_n, B)$ is a version of the conditional probability

$$\mathscr{P}\{\limsup_j [X_j \in B] \mid X_n\}.$$

$^\star$**11.** Suppose that for a homogeneous Markov process the initial distribution has support in $N^0$ as a subset of $\mathscr{R}^1$, and that for each $i \in N^0$, the transition probability function $P(i, \cdot)$, also has support in $N^0$. Thus

$$\{P(i, j); (i, j) \in N^0 \times N^0\}$$

is an infinite matrix called the "transition matrix". Show that $P^{(n)}$ as a matrix is just the $n$th power of $P^{(1)}$. Express the probability $\mathscr{P}\{X_{t_k} = i_k, 1 \le k \le n\}$ in terms of the elements of these matrices. [This is the case of *homogeneous Markov chains*.]

**12.** A process $\{X_n, n \in N^0\}$ is said to possess the "$r$th-order Markov property", where $r \ge 1$, iff (6) is replaced by

$$\mathscr{P}\{X_{n+1} \in B \mid X_0, \dots, X_n\} = \mathscr{P}\{X_{n+1} \in B \mid X_n, \dots, X_{n-r+1}\}$$

for $n \ge r - 1$. Show that if $r < s$, then the $r$th-order Markov property implies the $s$th. The ordinary Markov property is the case $r = 1$.

**13.** Let $Y_n$ be the random vector $(X_n, X_{n+1}, \dots, X_{n+r-1})$. Then the vector process $\{Y_n, n \in N^0\}$ has the ordinary Markov property (trivially generalized to vectors) if and only if $\{X_n, n \in N^0\}$ has the $r$th-order Markov property.

**14.** Let $\{X_n, n \in N^0\}$ be an independent process. Let

$$S_n^{(1)} = \sum_{j=0}^{n} X_j, \quad S_n^{(r+1)} = \sum_{j=0}^{n} S_j^{(r)}$$

for $r \ge 1$. Then $\{S_n^{(r)}, n \in N^0\}$ has the $r$th-order Markov property. For $r = 2$, give an example to show that it need not be a Markov process.

**15.** If $\{S_n, n \in N\}$ is a random walk such that $\mathscr{P}\{S_1 \ne 0\}>0$, then for any finite interval $[a, b]$ there exists an $\epsilon < 1$ such that

$$\mathscr{P}\{S_j \in [a, b], 1 \le j \le n\} \le \epsilon^n.$$

This is just Exercise 6 of Sec. 5.5 again.]

**16.** The same conclusion is true if the random walk above is replaced by a homogeneous Markov process for which, e.g., there exist $\delta>0$ and $\eta>0$ such that $P(x, \mathscr{R}^1 - (x - \delta, x + \delta)) \ge \eta$ for every $x$.

## 9.3 Basic properties of smartingales

The sequence of sums of independent r.v.'s has motivated the generalization to a Markov process in the preceding section; in another direction it will now motivate a *martingale*. Changing our previous notation to conform with later usage, let $\{x_n, n \in N\}$ denote independent r.v.'s with mean zero and write $X_n = \sum_{j=1}^{n} x_j$ for the partial sum. Then we have

$$\begin{aligned}\mathscr{E}(X_{n+1} \mid x_1, \dots, x_n) &= \mathscr{E}(X_n + x_{n+1} \mid x_1, \dots, x_n) \\ &= X_n + \mathscr{E}(x_{n+1} \mid x_1, \dots, x_n) = X_n + \mathscr{E}(x_{n+1}) = X_n.\end{aligned}$$

Note that the conditioning with respect to $x_1, \dots, x_n$ may be replaced by conditioning with respect to $X_1, \dots, X_n$ (why?). Historically, the equation above led to the consideration of dependent r.v.'s $\{x_n\}$ satisfying the condition

$$(1) \qquad \mathscr{E}(x_{n+1} \mid x_1, \dots, x_n) = 0.$$

It is astonishing that this simple property should delineate such a useful class of stochastic processes which will now be introduced. In what follows, where the index set for $n$ is not specified, it is understood to be either $N$ or some initial segment $N_m$ of $N$.

DEFINITION OF MARTINGALE. The sequence of r.v.'s and B.F.'s $\{X_n, \mathscr{F}_n\}$ is called a *martingale* iff we have for each $n$:

(a) $\mathscr{F}_n \subset \mathscr{F}_{n+1}$ and $X_n \in \mathscr{F}_n$;
(b) $\mathscr{E}(|X_n|) < \infty$;
(c) $X_n = \mathscr{E}(X_{n+1} \mid \mathscr{F}_n)$, a.e.

It is called a *supermartingale* iff the "=" in (c) above is replaced by "$\ge$", and a *submartingale* iff it is replaced by "$\le$". For abbreviation we shall use the term *smartingale* to cover all three varieties. In case $\mathscr{F}_n = \mathscr{F}_{[1,n]}$ as defined in Sec. 9.2, we shall omit $\mathscr{F}_n$ and write simply $\{X_n\}$; more frequently however we shall consider $\{\mathscr{F}_n\}$ as given in advance and omitted from the notation.

Condition (a) is nowadays referred to as: $\{X_n\}$ is *adapted* to $\{\mathscr{F}_n\}$. Condition (b) says that all the r.v.'s are integrable; we shall have to impose stronger conditions to obtain most of our results. A particularly important one is the uniform integrability of the sequence $\{X_n\}$, which is discussed in Sec. 4.5. A weaker condition is given by

$$(2) \qquad \sup_n \mathscr{E}(|X_n|) < \infty;$$

when this is satisfied we shall say that $\{X_n\}$ is $L^1$-bounded. Condition (c) leads at once to the more general relation:

$$(3) \qquad n < m \Rightarrow X_n = \mathscr{E}(X_m \mid \mathscr{F}_n).$$

This follows from Theorem 9.1.5 by induction since

$$\mathscr{E}(X_m \mid \mathscr{F}_n) = \mathscr{E}(\mathscr{E}(X_m \mid \mathscr{F}_{m-1}) \mid \mathscr{F}_n) = \mathscr{E}(X_{m-1} \mid \mathscr{F}_n).$$

An equivalent form of (3) is as follows: for each $\Lambda \in \mathscr{F}_n$ and $n \leq m$, we have

$$\int_\Lambda X_n \, d\mathscr{P} = \int_\Lambda X_m \, d\mathscr{P}. \tag{4}$$

It is often safer to use the explicit formula (4) rather than (3), because conditional expectations can be slippery things to handle. We shall refer to (3) or (4) as the *defining relation* of a martingale; similarly for the "super" and "sub" varieties.

Let us observe that in the form (3) or (4), the definition of a smartingale is meaningful if the index set $N$ is replaced by any linearly ordered set, with "$<$" as the strict order. For instance, it may be an interval or the set of rational numbers in the interval. But even if we confine ourselves to a *discrete parameter* (as we shall do) there are other index sets to be considered below.

It is scarcely worth mentioning that $\{X_n\}$ is a supermartingale if and only if $\{-X_n\}$ is a submartingale, and that a martingale is both. However the extension of results from a martingale to a smartingale is not always trivial, nor is it done for the sheer pleasure of generalization. For it is clear that martingales are harder to come by than the other varieties. As between the super and sub cases, though we can pass from one to the other by simply changing signs, our force of habit may influence the choice. The next proposition is a case in point.

**Theorem 9.3.1.** Let $\{X_n, \mathscr{F}_n\}$ be a submartingale and let $\varphi$ be an increasing convex function defined on $\mathscr{R}^1$. If $\varphi(X_n)$ is integrable for every $n$, then $\{\varphi(X_n), \mathscr{F}_n\}$ is also a submartingale.

PROOF. Since $\varphi$ is increasing, and

$$X_n \leq \mathscr{E}\{X_{n+1} \mid \mathscr{F}_n\}$$

we have

$$\varphi(X_n) \leq \varphi(\mathscr{E}\{X_{n+1} \mid \mathscr{F}_n\}). \tag{5}$$

By Jensen's inequality (Sec. 9.1), the right member above does not exceed $\mathscr{E}\{\varphi(X_{n+1}) \mid \mathscr{F}_n\}$; this proves the theorem. As forewarned in 9.1, we have left out some "a.e." above and shall continue to do so.

**Corollary 1.** If $\{X_n, \mathscr{F}_n\}$ is a submartingale, then so is $\{X_n^+, \mathscr{F}_n\}$. Thus $\mathscr{E}(X_n^+)$ as well as $\mathscr{E}(X_n)$ is increasing with $n$.

**Corollary 2.** If $\{X_n, \mathscr{F}_n\}$ is a martingale, then $\{|X_n|, \mathscr{F}_n\}$ is a submartingale; and $\{|X_n|^p, \mathscr{F}_n\}$, $1 < p < \infty$, is a submartingale provided that every $X_n \in L^p$; similarly for $\{|X_n| \log^+ |X_n|, \mathscr{F}_n\}$ where $\log^+ x = (\log x) \vee 0$ for $x \geq 0$.

PROOF. For a martingale we have equality in (5) for *any* convex $\varphi$, hence we may take $\varphi(x) = |x|$, $|x|^p$ or $|x| \log^+ |x|$ in the proof above.

Thus for a martingale $\{X_n\}$, all three transmutations: $\{X_n^+\}$, $\{X_n^-\}$ and $\{|X_n|\}$ are submartingales. For a submartingale $\{X_n\}$, nothing is said about the last two.

**Corollary 3.** If $\{X_n, \mathscr{F}_n\}$ is a supermartingale, then so is $\{X_n \wedge A, \mathscr{F}_n\}$ where $A$ is any constant.

PROOF. We leave it to the reader to deduce this from the theorem, but here is a quick direct proof:

$$X_n \wedge A \geq \mathscr{E}(X_{n+1} \mid \mathscr{F}_n) \wedge \mathscr{E}(A \mid \mathscr{F}_n) \geq \mathscr{E}(X_{n+1} \wedge A \mid \mathscr{F}_n).$$

It is possible to represent any smartingale as a martingale plus or minus something special. Let us call a sequence of r.v.'s $\{Z_n, n \in N\}$ an *increasing process* iff it satisfies the conditions:

( **i**) $Z_1 = 0$; $Z_n \leq Z_{n+1}$ for $n \geq 1$;
( **ii**) $\mathscr{E}(Z_n) < \infty$ for each $n$.

It follows that $Z_\infty = \lim_{n\to\infty} \uparrow Z_n$ exists but may take the value $+\infty$; $Z_\infty$ is integrable if and only if $\{Z_n\}$ is $L^1$-bounded as defined above, which means here $\lim_{n\to\infty} \uparrow \mathscr{E}(Z_n) < \infty$. This is also equivalent to the uniform integrability of $\{Z_n\}$ because of (i). We can now state the result as follows.

**Theorem 9.3.2.** Any submartingale $\{X_n, \mathscr{F}_n\}$ can be written as

$$X_n = Y_n + Z_n, \tag{6}$$

where $\{Y_n, \mathscr{F}_n\}$ is a martingale, and $\{Z_n\}$ is an increasing process.

PROOF. From $\{X_n\}$ we define its *difference sequence* as follows:

$$x_1 = X_1, \quad x_n = X_n - X_{n-1}, \quad n \geq 2, \tag{7}$$

so that $X_n = \sum_{j=1}^n x_j$, $n \geq 1$ (cf. the notation in the first paragraph of this section). The defining relation for a submartingale then becomes

$$\mathscr{E}\{X_n \mid \mathscr{F}_{n-1}\} \geq 0,$$

with equality for a martingale. Furthermore, we put

$$y_1 = x_1, \qquad y_n = x_n - \mathscr{E}\{x_n \mid \mathscr{F}_{n-1}\}, \qquad Y_n = \sum_{j=1}^n y_j;$$

$$z_1 = 0, \qquad z_n = \mathscr{E}\{x_n \mid \mathscr{F}_{n-1}\}, \qquad Z_n = \sum_{j=1}^n z_j.$$

Then clearly $x_n = y_n + z_n$ and (6) follows by addition. To show that $\{Y_n, \mathscr{F}_n\}$ is a martingale, we may verify that $\mathscr{E}\{y_n \mid \mathscr{F}_{n-1}\} = 0$ as indicated a moment ago, and this is trivial by Theorem 9.1.5. Since each $z_n \geq 0$, it is equally obvious that $\{Z_n\}$ is an increasing process. The theorem is proved.

Observe that $Z_n \in \mathscr{F}_{n-1}$ for each $n$, by definition. This has important consequences; see Exercise 9 below. The decomposition (6) will be called *Doob's decomposition*. For a supermartingale we need only change the "+" there into "−", since $\{-Y_n, \mathscr{F}_n\}$ is a martingale. The following complement is useful.

**Corollary.** If $\{X_n\}$ is $L^1$-bounded [or uniformly integrable], then both $\{Y_n\}$ and $\{Z_n\}$ are $L^1$-bounded [or uniformly integrable].

PROOF. We have from (6):

$$\mathscr{E}(Z_n) \leq \mathscr{E}(|X_n|) - \mathscr{E}(Y_1)$$

since $\mathscr{E}(Y_n) = \mathscr{E}(Y_1)$. Since $Z_n \geq 0$ this shows that if $\{X_n\}$ is $L^1$-bounded, then so is $\{Z_n\}$; and $\{Y_n\}$ is too because

$$\mathscr{E}(|Y_n|) \leq \mathscr{E}(|X_n|) + \mathscr{E}(Z_n).$$

Next if $\{X_n\}$ is uniformly integrable, then it is $L^1$-bounded by Theorem 4.5.3, hence $\{Z_n\}$ is $L^1$-bounded and therefore uniformly integrable as remarked before. The uniform integrability of $\{Y_n\}$ then follows from the last-written inequality.

We come now to the fundamental notion of *optional sampling* of a smartingale. This consists in substituting certain random variables for the original index $n$ regarded as the time parameter of the process. Although this kind of thing has been done in Chapter 8, we will reintroduce it here in a slightly different way for the convenience of the reader. To begin with we adjoin a last index $\infty$ to the set $N$ and call it $N_\infty = \{1, 2, \dots, \infty\}$. This is an example of a linearly ordered set mentioned above. Next, adjoin $\mathscr{F}_\infty = \bigvee_{n=1}^{\infty} \mathscr{F}_n$ to $\{\mathscr{F}_n\}$.

A r.v. $\alpha$ taking values in $N_\infty$ is called *optional* (relative to $\{\mathscr{F}_n, n \in N_\infty\}$) iff for every $n \in N_\infty$ we have

$$\{\alpha \leq n\} \in \mathscr{F}_n. \tag{8}$$

Since $\mathscr{F}_n$ increases with $n$, the condition in (8) is unchanged if $\{\alpha \leq n\}$ is replaced by $\{\alpha = n\}$. Next, for an optional $\alpha$, the pre-$\alpha$ field $\mathscr{F}_\alpha$ is defined to be the class of all subsets $\Lambda$ of $\mathscr{F}_\infty$ satisfying the following condition: for each $n \in N_\infty$ we have

$$\Lambda \cap \{\alpha \leq n\} \in \mathscr{F}_n, \tag{9}$$

where again $\{\alpha \leq n\}$ may be replaced by $\{\alpha = n\}$. Writing then

(10) $$\Lambda_n = \Lambda \cap \{\alpha = n\},$$

we have $\Lambda_n \in \mathscr{F}_n$ and

$$\Lambda = \bigcup_n \Lambda_n = \bigcup_n [\{\alpha = n\} \cap \Lambda_n]$$

where the index $n$ ranges over $N_\infty$. This is (3) of Sec. 8.2. The reader should now do Exercises 1–4 in Sec. 8.2 to get acquainted with the simplest properties of optionality. Here are some of them which will be needed soon: $\mathscr{F}_\alpha$ is a B.F. and $\alpha \in \mathscr{F}_\alpha$; if $\alpha$ is optional then so is $\alpha \Lambda n$ for each $n \in N$; if $\alpha \leq \beta$ where $\beta$ is also optional then $\mathscr{F}_\alpha \subset \mathscr{F}_\beta$; in particular $\mathscr{F}_{\alpha \Lambda n} \subset \mathscr{F}_\alpha \cap \mathscr{F}_n$ and in fact this inclusion is an equation.

Next we assume $X_\infty$ has been defined and $X_\infty \in \mathscr{F}_\infty$. We then define $X_\alpha$ as follows:

(11) $$X_\alpha(\omega) = X_{\alpha(\omega)}(\omega);$$

in other words,

$$X_\alpha(\omega) = X_n(\omega) \quad \text{on} \quad \{\alpha = n\}, \quad n \in N_\infty.$$

This definition makes sense for any $\alpha$ taking values in $N_\infty$, but for an optional $\alpha$ we can assert moreover that

(12) $$X_\alpha \in \mathscr{F}_\alpha.$$

This is an exercise the reader should not miss; observe that it is a natural but nontrivial extension of the assumption $X_n \in \mathscr{F}_n$ for every $n$. Indeed, all the general propositions concerning optional sampling aim at the same thing, namely to make optional times behave like constant times, or again to enable us to substitute optional r.v.'s for constants. For this purpose conditions must sometimes be imposed either on $\alpha$ or on the smartingale $\{X_n\}$. Let us however begin with a perfect case which turns out to be also very important.

We introduce a class of martingales as follows. For any integrable r.v. $Y$ we put

(13) $$X_n = \mathscr{E}(Y \mid \mathscr{F}_n), \quad n \in N_\infty.$$

By Theorem 9.1.5, if $n \leq m$:

(14) $$X_n = \mathscr{E}\{\mathscr{E}(Y \mid \mathscr{F}_m) \mid \mathscr{F}_n\} = \mathscr{E}\{X_m \mid \mathscr{F}_n\}$$

which shows $\{X_n, \mathscr{F}_n\}$ is a martingale, not only on $N$ but also on $N_\infty$. The following properties extend both (13) and (14) to optional times.

**Theorem 9.3.3.** For any optional $\alpha$, we have

(15) $$X_\alpha = \mathscr{E}(Y \mid \mathscr{F}_\alpha).$$

If $\alpha \le \beta$ where $\beta$ is also optional, then $\{X_\alpha, \mathscr{F}_\alpha; X_\beta, \mathscr{F}_\beta\}$ forms a two-term martingale.

PROOF. Let us first show that $X_\alpha$ is integrable. It follows from (13) and Jensen's inequality that

$$|X_n| \le \mathscr{E}(|Y| \mid \mathscr{F}_n).$$

Since $\{\alpha = n\} \in \mathscr{F}_n$, we may apply this to get

$$\int_\Omega |X_\alpha|\, d\mathscr{P} = \sum_n \int_{\{\alpha=n\}} |X_n|\, d\mathscr{P} \le \sum_n \int_{\{\alpha=n\}} |Y|\, d\mathscr{P} = \int_\Omega |Y|\, d\mathscr{P} < \infty.$$

Next if $\Lambda \in \mathscr{F}_\alpha$, we have, using the notation in (10):

$$\int_\Lambda X_\alpha\, d\mathscr{P} = \sum_n \int_{\Lambda_n} X_n\, d\mathscr{P} = \sum_n \int_{\Lambda_n} Y\, d\mathscr{P} = \int_\Lambda Y\, d\mathscr{P},$$

where the second equation holds by (13) because $\Lambda_n \in \mathscr{F}_n$. This establishes (15). Now if $\alpha \le \beta$, then $\mathscr{F}_\alpha \subset \mathscr{F}_\beta$ and consequently by Theorem 9.1.5.

$$X_\alpha = \mathscr{E}\{\mathscr{E}(Y \mid \mathscr{F}_\beta) \mid \mathscr{F}_\alpha\} = \mathscr{E}\{X_\beta \mid \mathscr{F}_\alpha\},$$

which proves the second assertion of the theorem.

As an immediate corollary, if $\{\alpha_n\}$ is a sequence of optional r.v.'s such that

(16) $$\alpha_1 \le \alpha_2 \le \cdots \le \alpha_n \le \cdots,$$

then $\{X_{\alpha_n}, \mathscr{F}_{\alpha_n}\}$ is a martingale. This new martingale is obtained by sampling the original one at the optional times $\{\alpha_j\}$. We now proceed to extend the second part of Theorem 9.3.3 to a supermartingale. There are two important cases which will be discussed separately.

**Theorem 9.3.4.** Let $\alpha$ and $\beta$ be two bounded optional r.v.'s such that $\alpha \le \beta$. Then for any [super]martingale $\{X_n\}$, $\{X_\alpha, \mathscr{F}_\alpha; X_\beta, \mathscr{F}_\beta\}$ forms a [super]martingale.

PROOF. Let $\Lambda \in \mathscr{F}_\alpha$; using (10) again we have for each $k \ge j$:

$$\Lambda_j \cap \{\beta > k\} \in \mathscr{F}_k$$

because $\Lambda_j \in \mathscr{F}_j \subset \mathscr{F}_k$, whereas $\{\beta > k\} = \{\beta \leq k\}^c \in \mathscr{F}_k$. It follows from the defining relation of a supermartingale that

$$\int_{\Lambda_j \cap \{\beta > k\}} X_k \, d\mathscr{P} \geq \int_{\Lambda_j \cap \{\beta > k\}} X_{k+1} \, d\mathscr{P}$$

and consequently

$$\int_{\Lambda_j \cap \{\beta \geq k\}} X_k \, d\mathscr{P} \geq \int_{\Lambda_j \cap \{\beta = k\}} X_k \, d\mathscr{P} + \int_{\Lambda_j \cap \{\beta > k\}} X_{k+1} \, d\mathscr{P}$$

Rewriting this as

$$\int_{\Lambda_j \cap \{\beta \geq k\}} X_k \, d\mathscr{P} - \int_{\Lambda_j \cap \{\beta \geq k+1\}} X_{k+1} \, d\mathscr{P} \geq \int_{\Lambda_j \cap \{\beta = k\}} X_\beta \, d\mathscr{P};$$

summing over $k$ from $j$ to $m$, where $m$ is an upper bound for $\beta$; and then replacing $X_j$ by $X_\alpha$ on $\Lambda_j$, we obtain

$$(17) \quad \int_{\Lambda_j \cap \{\beta \geq j\}} X_\alpha \, d\mathscr{P} - \int_{\Lambda_j \cap \{\beta \geq m+1\}} X_{m+1} \, d\mathscr{P} \geq \int_{\Lambda_j \cap \{j \leq \beta \leq m\}} X_\beta \, d\mathscr{P}.$$

$$\int_{\Lambda_j} X_\alpha \, d\mathscr{P} \geq \int_{\Lambda_j} X_\beta \, d\mathscr{P}.$$

Another summation over $j$ from 1 to $m$ yields the desired result. In the case of a martingale the inequalities above become equations.

A particular case of a bounded optional r.v. is $\alpha_n = \alpha \Lambda n$ where $\alpha$ is an arbitrary optional r.v. and $n$ is a positive integer. Applying the preceding theorem to the sequence $\{\alpha_n\}$ as under Theorem 9.3.3, we have the following corollary.

**Corollary.** If $\{X_n, \mathscr{F}_n\}$ is a [super]martingale and $\alpha$ is an arbitrary optional r.v., then $\{X_{\alpha \Lambda n} \mathscr{F}_{\alpha \Lambda n}\}$ is a [super]martingale.

In the next basic theorem we shall assume that the [super]martingale is given on the index set $N_\infty$. This is necessary when the optional r.v. can take the value $+\infty$, as required in many applications; see the typical example in (5) of Sec. 8.2. It turns out that if $\{X_n\}$ is originally given only for $n \in N$, we may take $X_\infty = \lim_{n \to \infty} X_n$ to extend it to $N_\infty$ under certain conditions, see Theorems 9.4.5 and 9.4.6 and Exercise 6 of Sec. 9.4. A trivial case occurs when $\{X_n, \mathscr{F}_n; n \in N\}$ is a positive supermartingale; we may then take $X_\infty = 0$.

**Theorem 9.3.5.** Let $\alpha$ and $\beta$ be two arbitrary optional r.v.'s such that $\alpha \leq \beta$. Then the conclusion of Theorem 9.3.4 holds true for any supermartingale $\{X_n, \mathscr{F}_n; n \in N_\infty\}$.

*Remark.* For a martingale $\{X_n, \mathscr{F}_n; n \in N_\infty\}$ this theorem is contained in Theorem 9.3.3 since we may take the $Y$ in (13) to be $X_\infty$ here.

PROOF. (a) Suppose first that the supermartingale is positive with $X_\infty = 0$ a.e. The inequality (17) is true for every $m \in N$, but now the second integral there is positive so that we have

$$\int_{\Lambda_j} X_\alpha \, d\mathscr{P} \geq \int_{\Lambda_j \cap \{\beta \leq m\}} X_\beta \, d\mathscr{P}.$$

Since the integrands are positive, the integrals exist and we may let $m \to \infty$ and then sum over $j \in N$. The result is

$$\int_{\Lambda \cap \{\alpha < \infty\}} X_\alpha \, d\mathscr{P} \geq \int_{\Lambda \cap \{\beta < \infty\}} X_\beta \, d\mathscr{P}$$

which falls short of the goal. But we can add the inequality

$$\int_{\Lambda \cap \{\alpha = \infty\}} X_\alpha \, d\mathscr{P} = \int_{\Lambda \cap \{\alpha = \infty\}} X_\infty \, d\mathscr{P} = \int_{\Lambda \cap \{\beta = \infty\}} X_\infty \, d\mathscr{P} = \int_{\Lambda \cap \{\beta = \infty\}} X_\beta \, d\mathscr{P}$$

which is trivial because $X_\infty = 0$ a.e. This yields the desired

$$\int_{\Lambda} X_\alpha \, d\mathscr{P} \geq \int_{\Lambda} X_\beta \, d\mathscr{P}. \tag{18}$$

Let us show that $X_\alpha$ and $X_\beta$ are in fact integrable. Since $X_n \geq X_\infty$ we have $X_\alpha \leq \underline{\lim}_{n\to\infty} X_{\alpha \wedge n}$ so that by Fatou's lemma,

$$\mathscr{E}(X_\alpha) \leq \underline{\lim}_{n\to\infty} \mathscr{E}(X_{\alpha \wedge n}). \tag{19}$$

Since 1 and $\alpha \wedge n$ are two bounded optional r.v.'s satisfying $1 \leq \alpha \wedge n$; the right-hand side of (19) does not exceed $\mathscr{E}(X_1)$ by Theorem 9.3.4. This shows $X_\alpha$ is integrable since it is positive.

(b) In the general case we put

$$X'_n = \mathscr{E}\{X_\infty \mid \mathscr{F}_n\}, \quad X''_n = X_n - X'_n.$$

Then $\{X'_n, \mathscr{F}_n; n \in N_\infty\}$ is a martingale of the kind introduced in (13), and $X_n \geq X'_n$ by the defining property of supermartingale applied to $X_n$ and $X_\infty$. Hence the difference $\{X''_n, \mathscr{F}_n; n \in N\}$ is a positive supermartingale with $X''_\infty = 0$ a.e. By Theorem 9.3.3, $\{X'_\alpha, \mathscr{F}_\alpha; X'_\beta, \mathscr{F}_\beta\}$ forms a martingale; by case (a), $\{X''_\alpha, \mathscr{F}_\alpha; X''_\beta, \mathscr{F}_\beta\}$ forms a supermartingale. Hence the conclusion of the theorem follows simply by addition.

The two preceding theorems are the basic cases of *Doob's optional sampling theorem*. They do not cover all cases of optional sampling (see e.g.

Exercise 11 of Sec. 8.2 and Exercise 11 below), but are adequate for many applications, some of which will be given later.

Martingale theory has its intuitive background in gambling. If $X_n$ is interpreted as the gambler's capital at time $n$, then the defining property postulates that his expected capital after one more game, played with the knowledge of the entire past and present, is exactly equal to his current capital. In other words, his expected gain is zero, and in this sense the game is said to be "fair". Similarly a smartingale is a game consistently biased in one direction. Now the gambler may opt to play the game only at certain preferred times, chosen with the benefit of past experience and present observation, but without clairvoyance into the future. [The inclusion of the present status in his knowledge seems to violate raw intuition, but examine the example below and Exercise 13.] He hopes of course to gain advantage by devising such a "system" but Doob's theorem forestalls him, at least mathematically. We have already mentioned such an interpretation in Sec. 8.2 (see in particular Exercise 11 of Sec. 8.2; note that $\alpha + 1$ rather than $\alpha$ is the optional time there.) The present generalization consists in replacing a stationary independent process by a smartingale. The classical problem of "gambler's ruin" illustrates very well the ideas involved, as follows.

Let $\{S_n, n \in N^0\}$ be a random walk in the notation of Chapter 8, and let $S_1$ have the Bernoullian distribution $\frac{1}{2}\delta_1 + \frac{1}{2}\delta_{-1}$. It follows from Theorem 8.3.4, or the more elementary Exercise 15 of Sec. 9.2, that the walk will almost certainly leave the interval $[-a, b]$, where $a$ and $b$ are strictly positive integers; and since it can move only one unit a time, it must reach either $-a$ or $b$. This means that if we set

(20) $$\alpha = \min\{n \geq 1: S_n = -a\}, \quad \beta = \min\{n \geq 1: S_n = b\},$$

then $\gamma = \alpha \Lambda \beta$ is a finite optional r.v. It follows from the Corollary to Theorem 9.3.4 that $\{S_{\gamma \Lambda n}\}$ is a martingale. Now

(21) $$\lim_{n\to\infty} S_{\gamma \Lambda n} = S_\gamma \quad \text{a.e.}$$

and clearly $S_\gamma$ takes only the values $-a$ and $b$. The question is: with what probabilities? In the gambling interpretation: if two gamblers play a fair coin-tossing game and possess, respectively, $a$ and $b$ units of the constant stake as initial capitals, what is the probability of ruin for each?

The answer is immediate ("without any computation"!) if we show first that the two r.v.'s $\{S_1, S_\gamma\}$ form a martingale, for then

(22) $$\mathscr{E}(S_\gamma) = \mathscr{E}(S_1) = 0,$$

which is to say that

$$-a\mathscr{P}\{S_\gamma = -a\} + b\mathscr{P}\{S_\gamma = b\} = 0,$$

so that the probability of ruin is inversely proportional to the initial capital of the gambler, a most sensible solution.

To show that the pair $\{S_1, S_\gamma\}$ forms a martingale we use Theorem 9.3.5 since $\{S_{\gamma\wedge n}, n \in N_\infty\}$ is a bounded martingale. The more elementary Theorem 9.3.4 is inapplicable, since $\gamma$ is not bounded. However, there is a simpler way out in this case: (21) and the boundedness just mentioned imply that

$$\mathscr{E}(S_\gamma) = \lim_{n\to\infty} \mathscr{E}(S_{\gamma\wedge n}),$$

and since $\mathscr{E}(S_{\gamma\wedge 1}) = \mathscr{E}(S_1)$, (22) follows directly.

The ruin problem belonged to the ancient history of probability theory, and can be solved by elementary methods based on difference equations (see, e.g., Uspensky, *Introduction to mathematical probability*, McGraw-Hill, New York, 1937). The approach sketched above, however, has all the main ingredients of an elaborate modern theory. The little equation (22) is the prototype of a "harmonic equation", and the problem itself is a "boundary-value problem". The steps used in the solution — to wit: the introduction of a martingale, its optional stopping, its convergence to a limit, and the extension of the martingale property to include the limit with the consequent convergence of expectations — are all part of a standard procedure now ensconced in the general theory of Markov processes and the allied potential theory.

## EXERCISES

**1.** The defining relation for a martingale may be generalized as follows. For each optional r.v. $\alpha \le n$, we have $\mathscr{E}\{X_n \mid \mathscr{F}_\alpha\} = X_\alpha$. Similarly for a smartingale.

***2.** If $X$ is an integrable r.v., then the collection of (equivalence classes of) r.v.'s $\mathscr{E}(X \mid \mathscr{G})$ with $\mathscr{G}$ ranging over all Borel subfields of $\mathscr{F}$, is uniformly integrable.

**3.** Suppose $\{X_n^{(k)}, \mathscr{F}_n\}$, $k = 1, 2$, are two [super]martingales, $\alpha$ is a finite optional r.v., and $X_\alpha^{(1)} = [\ge] X_\alpha^{(2)}$. Define $X_n = X_n^{(1)} 1_{\{n\le\alpha\}} + X_n^{(2)} 1_{\{n>\alpha\}}$; show that $\{X_n, \mathscr{F}_n\}$ is a [super]martingale. [HINT: Verify the defining relation in (4) for $m = n + 1$.]

**4.** Suppose each $X_n$ is integrable and

$$\mathscr{E}\{X_{n+1} \mid X_1, \dots, X_n\} = n^{-1}(X_1 + \cdots + X_n)$$

then $\{(n^{-1})(X_1 + \cdots + X_n), n \in N\}$ is a martingale.

**5.** Every sequence of integrable r.v.'s is the sum of a supermartingale and a submartingale.

**6.** If $\{X_n, \mathscr{F}_n\}$ and $\{X'_n, \mathscr{F}_n\}$ are martingales, then so is $\{X_n + X'_n, \mathscr{F}_n\}$. But it may happen that $\{X_n\}$ and $\{X'_n\}$ are martingales while $\{X_n + X'_n\}$ is not.

[HINT: Let $x_1$ and $x'_1$ be independent Bernoullian r.v.'s; and $x_2 = x'_2 = +1$ or $-1$ according as $x_1 + x'_1 = 0$ or $\neq 0$; notation as in (7).]

**7.** Find an example of a positive martingale which is not uniformly integrable. [HINT: You win $2^n$ if it's heads $n$ times in a row, and you lose everything as soon as it's tails.]

**8.** Find an example of a martingale $\{X_n\}$ such that $X_n \to -\infty$ a.e. This implies that even in a "fair" game one player may be bound to lose an arbitrarily large amount if he plays long enough (and no limit is set to the liability of the other player). [HINT: Try sums of independent but not identically distributed r.v.'s with mean 0.]

***9.** Prove that if $\{Y_n, \mathscr{F}_n\}$ is a martingale such that $Y_n \in \mathscr{F}_{n-1}$, then for every $n$, $Y_n = Y_1$ a.e. Deduce from this result that Doob's decomposition (6) is unique (up to equivalent r.v.'s) under the condition that $Z_n \in \mathscr{F}_{n-1}$ for every $n \geq 2$. If this condition is not imposed, find two different decompositions.

**10.** If $\{X_n\}$ is a uniformly integrable submartingale, then for any optional r.v. $\alpha$ we have

(i) $\{X_{\alpha \wedge n}\}$ is a uniformly integrable submartingale;

(ii) $\mathscr{E}(X_1) \leq \mathscr{E}(X_\alpha) \leq \sup_n \mathscr{E}(X_n)$.

[HINT: $|X_{\alpha \wedge n}| \leq |X_\alpha| + |X_n|$.]

***11.** Let $\{X_n, \mathscr{F}_n; n \in N\}$ be a [super]martingale satisfying the following condition: there exists a constant $M$ such that for every $n \geq 1$:

$$\mathscr{E}\{|X_n - X_{n-1}| \,|\, \mathscr{F}_{n-1}\} \leq M \text{a.e.}$$

where $X_0 = 0$ and $\mathscr{F}_0$ is trivial. Then for any two optional r.v.'s $\alpha$ and $\beta$ such that $\alpha \leq \beta$ and $\mathscr{E}(\beta) < \infty$, $\{X_\alpha, \mathscr{F}_\alpha; X_\beta, \mathscr{F}_\beta\}$ is a [super]martingale. This is another case of optional sampling given by Doob, which includes Wald's equation (Theorem 5.5.3) as a special case. [HINT: Dominate the integrand in the second integral in (17) by $Y_\beta$ where $X_0 = 0$ and $Y_m = \sum_{n=1}^m |X_n - X_{n-1}|$. We have

$$\mathscr{E}(Y_\beta) = \sum_{n=1}^{\infty} \int_{\{\beta \geq n\}} |X_n - X_{n-1}| \, d\mathscr{P} \leq M\mathscr{E}(\beta).]$$

**12.** Apply Exercise 11 to the gambler's ruin problem discussed in the text and conclude that for the $\alpha$ in (20) we must have $\mathscr{E}(\alpha) = +\infty$. Verify this by elementary computation.

***13.** In the gambler's ruin problem take $b = 1$ in (20). Compute $\mathscr{E}(S_{\beta \wedge n})$ for a fixed $n$ and show that $\{S_0, S_{\beta \wedge n}\}$ forms a martingale. Observe that $\{S_0, S_\beta\}$ does not form a martingale and explain in gambling terms the effect of stopping

$\beta$ at $n$. This example shows why in optional sampling the option may be taken even with the knowledge of the present moment under certain conditions. In the case here the present (namely $\beta \wedge n$) may leave one no choice!

**14.** In the gambler's ruin problem, suppose that $S_1$ has the distribution

$$p\delta_1 + (1-p)\delta_{-1}, \quad p \neq \tfrac{1}{2};$$

and let $d = 2p - 1$. Show that $\mathscr{E}(S_\gamma) = d\mathscr{E}(\gamma)$. Compute the probabilities of ruin by using difference equations to deduce $\mathscr{E}(\gamma)$, and vice versa.

**15.** Prove that for any $L^1$-bounded smartingale $\{X_n, \mathscr{F}_n, n \in N_\infty\}$, and any optional $\alpha$, we have $\mathscr{E}(|X_\alpha|) < \infty$. [HINT: Prove the result first for a martingale, then use Doob's decomposition.]

***16.** Let $\{X_n, \mathscr{F}_n\}$ be a martingale: $x_1 = X_1$, $x_n = X_n - X_{n-1}$ for $n \geq 2$; let $v_n \in \mathscr{F}_{n-1}$ for $n \geq 1$ where $\mathscr{F}_0 = \mathscr{F}_1$; now put

$$T_n = \sum_{j=1}^{n} v_j x_j.$$

Show that $\{T_n, \mathscr{F}_n\}$ is a martingale provided that $T_n$ is integrable for every $n$. The martingale may be replaced by a smartingale if $v_n \geq 0$ for every $n$. As a particular case take $v_n = 1_{\{n \leq \alpha\}}$ where $\alpha$ is an optional r.v. relative to $\{\mathscr{F}_n\}$. What then is $T_n$? Hence deduce the Corollary to Theorem 9.3.4.

**17.** As in the preceding exercise, deduce a new proof of Theorem 9.3.4 by taking $v_n = 1_{\{\alpha < n \leq \beta\}}$.

## 9.4 Inequalities and convergence

We begin with two inequalities, the first of which is a generalization of Kolmogorov's inequality (Theorem 5.3.1).

**Theorem 9.4.1.** If $\{X_j, \mathscr{F}_j, j \in N_n\}$ is a submartingale, then for each real $\lambda$ we have

$$(1) \qquad \lambda \mathscr{P}\{\max_{1 \leq j \leq n} X_j \geq \lambda\} \leq \int_{\{\max_{1 \leq j \leq n} X_j \geq \lambda\}} X_n \, d\mathscr{P} \leq \mathscr{E}(X_n^+);$$

$$(2) \qquad \lambda \mathscr{P}\{\min_{1 \leq j \leq n} X_j \leq -\lambda\} \leq \mathscr{E}(X_n - X_1) - \int_{\{\min_{1 \leq j \leq n} X_j \leq -\lambda\}} X_n \, d\mathscr{P}$$
$$\leq \mathscr{E}(X_n^+) - \mathscr{E}(X_1).$$

PROOF. Let $\alpha$ be the first $j$ such that $X_j \geq \lambda$ if there is such a $j$ in $N_n$, otherwise let $\alpha = n$ (optional stopping at $n$). It is clear that $\alpha$ is optional;

since it takes only a finite number of values, Theorem 9.3.4 shows that the pair $\{X_\alpha, X_n\}$ forms a submartingale. If we write

$$\mathrm{M} = \{\max_{1 \le j \le n} X_j \ge \lambda\},$$

then $\mathrm{M} \in \mathscr{F}_\alpha$ (why?) and $X_\alpha \ge \lambda$ on M, hence the first inequality follows from

$$\lambda \mathscr{P}(\mathrm{M}) \le \int_{\mathrm{M}} X_\alpha \, d\mathscr{P} \le \int_{\mathrm{M}} X_n \, d\mathscr{P};$$

the second is just a cruder consequence.

Similarly let $\beta$ be the first $j$ such that $X_j \le -\lambda$ if there is such a $j$ in $N_n$, otherwise let $\beta = n$. Put also

$$\mathrm{M}_k = \{\min_{1 \le j \le k} X_j \le -\lambda\}.$$

Then $\{X_1, X_\beta\}$ is a submartingale by Theorem 9.3.4, and so

$$\begin{aligned} \mathscr{E}(X_1) \le \mathscr{E}(X_\beta) &= \int_{\{\beta \le n-1\}} X_\beta \, d\mathscr{P} + \int_{\mathrm{M}_{n-1}^c \mathrm{M}_n} X_n \, d\mathscr{P} + \int_{\mathrm{M}_n^c} X_n \, d\mathscr{P} \\ &\le -\lambda \mathscr{P}(\mathrm{M}_n) + \mathscr{E}(X_n) - \int_{\mathrm{M}_n} X_n \, d\mathscr{P}, \end{aligned}$$

which reduces to (2).

**Corollary 1.** If $\{X_n\}$ is a martingale, then for each $\lambda > 0$:

$$(3) \qquad \mathscr{P}\{\max_{1 \le j \le n} |X_j| \ge \lambda\} \le \frac{1}{\lambda} \int_{\{\max_{1 \le j \le n} |X_j| \ge \lambda\}} |X_n| \, d\mathscr{P} \le \frac{1}{\lambda} \mathscr{E}(|X_n|).$$

If in addition $\mathscr{E}(X_n^2) < \infty$ for each $n$, then we have also

$$(4) \qquad \mathscr{P}\{\max_{1 \le j \le n} |X_j| \ge \lambda\} \le \frac{1}{\lambda^2} \mathscr{E}(X_n^2).$$

These are obtained by applying the theorem to the submartingales $\{|X_n|\}$ and $\{X_n^2\}$. In case $X_n$ is the $S_n$ in Theorem 5.3.1, (4) is precisely the Kolmogorov inequality there.

**Corollary 2.** Let $1 \le m \le n$, $\Lambda_m \in \mathscr{F}_m$ and $\mathrm{M} = \{\max_{m \le j \le n} X_j \ge \lambda\}$, then

$$\lambda \mathscr{P}\{\Lambda_m \cap \mathrm{M}\} \le \int_{\Lambda_m \cap \mathrm{M}} X_n \, d\mathscr{P}.$$

This is proved just as (1) and will be needed later.

We now come to a new kind of inequality, which will be the tool for proving the main convergence theorem below. Given any sequence of r.v.'s $\{X_j\}$, for each sample point $\omega$, the convergence properties of the numerical sequence $\{X_j(\omega)\}$ hinge on the oscillation of the finite segments $\{X_j(\omega), j \in N_n\}$ as $n \to \infty$. In particular the sequence will have a limit, finite or infinite, if and only if the number of its oscillations between any two [rational] numbers $a$ and $b$ is finite (depending on $a$, $b$ and $\omega$). This is a standard type of argument used in measure and integration theory (cf. Exercise 10 of Sec. 4.2). The interesting thing is that for a smartingale, a sharp estimate of the expected number of oscillations is obtainable.

Let $a < b$. The number $\nu$ of "upcrossings" of the interval $[a, b]$ by a numerical sequence $\{x_1, \dots, x_n\}$ is defined as follows. Set

$$\alpha_1 = \min\{j: 1 \le j \le n, x_j \le a\},$$

$$\alpha_2 = \min\{j: \alpha_1 < j \le n, x_j \ge b\};$$

if either $\alpha_1$ or $\alpha_2$ is not defined because no such $j$ exists, we define $\nu = 0$. In general, for $k \ge 2$ we set

$$\alpha_{2k-1} = \min\{j: \alpha_{2k-2} < j \le n, x_j \le a\},$$

$$\alpha_{2k} = \min\{j: \alpha_{2k-1} < j \le n, x_j \ge b\};$$

if any one of these is undefined, then all the subsequent ones will be undefined. Let $\alpha_\ell$ be the last defined one, with $\ell = 0$ if $\alpha_1$ is undefined, then $\nu$ is defined to be $[\ell/2]$. Thus $\nu$ is the actual number of successive times that the sequence crosses from $\le a$ to $\ge b$. Although the exact number is not essential, since a couple of crossings more or less would make no difference, we must adhere to a rigid way of counting in order to be accurate below.

**Theorem 9.4.2.** Let $\{X_j, \mathscr{F}_j, j \in N_n\}$ be a submartingale and $-\infty < a < b < \infty$. Let $\nu_{[a,b]}^{(n)}(\omega)$ denote the number of upcrossings of $[a, b]$ by the sample sequence $\{X_j(\omega); j \in N_n\}$. We have then

$$\mathscr{E}\{\nu_{[a,b]}^{(n)}\} \le \frac{\mathscr{E}\{(X_n - a)^+\} - \mathscr{E}\{(X_1 - a)^+\}}{b - a} \le \frac{\mathscr{E}\{X_n^+\} + |a|}{b - a}. \tag{5}$$

PROOF. Consider first the case where $X_j \ge 0$ for every $j$ and $0 = a < b$, so that $\nu_{[a,b]}^{(n)}(\omega)$ becomes $\nu_{[0,b]}^{(n)}(\omega)$, and $X_{\alpha_j}(\omega) = 0$ if $j$ is odd, where $\alpha_j = \alpha_j(\omega)$ is defined as above with $x_j = X_j(\omega)$. For each $\omega$, the sequence $\alpha_j(\omega)$ is defined only up to $\ell(\omega)$, where $0 \le \ell(\omega) \le n$. But now we modify the definition so that $\alpha_j(\omega)$ is defined for $1 \le j \le n$ by setting it to be equal to $n$ wherever it was previously undefined. Since for some $\omega$, a previously defined $\alpha_j(\omega)$ may also be equal to $n$, this apparent confusion will actually simplify the formulas below. In the same vein we set $\alpha_0 \equiv 1$. Observe that $\alpha_n = n$ in any case, so that

$$X_n - X_1 = X_{\alpha_n} - X_{\alpha_0} = \sum_{j=0}^{n-1}(X_{\alpha_{j+1}} - X_{\alpha_j}) = \sum_{j \text{ even}} + \sum_{j \text{ odd}}.$$

If $j$ is odd and $j + 1 \le \ell(\omega)$, then

$$X_{\alpha_{j+1}}(\omega) \ge b > 0 = X_{\alpha_j}(\omega);$$

If $j$ is odd and $j = \ell(\omega)$, then

$$X_{\alpha_{j+1}}(\omega) = X_n(\omega) \ge 0 = X_{\alpha_j}(\omega);$$

if $j$ is odd and $\ell(\omega) < j$, then

$$X_{\alpha_{j+1}}(\omega) = X_n(\omega) = X_{\alpha_j}(\omega).$$

Hence in all cases we have

$$\begin{aligned} (6) \qquad \sum_{j \text{ odd}} (X_{\alpha_{j+1}}(\omega) - X_{\alpha_j}(\omega)) &\ge \sum_{\substack{j \text{ odd} \\ j+1 \le \ell(\omega)}} (X_{\alpha_{j+1}}(\omega) - X_{\alpha_j}(\omega)) \\ &\ge \left[\frac{\ell(\omega)}{2}\right] b = \nu_{[0,b]}^{(n)}(\omega) b. \end{aligned}$$

Next, observe that $\{\alpha_j, 0 \le j \le n\}$ as modified above is in general of the form $1 = \alpha_0 \le \alpha_1 < \alpha_2 < \cdots < \alpha_\ell \le \alpha_{\ell+1} = \cdots = \alpha_n = n$, and since constants are optional, this is an increasing sequence of optional r.v.'s. Hence by Theorem 9.3.4, $\{X_{\alpha_j}, 0 \le j \le n\}$ is a submartingale so that for each $j, 0 \le j \le n - 1$, we have $\mathscr{E}\{X_{\alpha_j+1} - X_{\alpha_j}\} \ge 0$ and consequently

$$\mathscr{E}\left\{\sum_{j \text{ even}} (X_{\alpha_{j+1}} - X_{\alpha_j})\right\} \ge 0.$$

Adding to this the expectations of the extreme terms in (6), we obtain

$$(7) \qquad \mathscr{E}(X_n - X_1) \ge \mathscr{E}(\nu_{[0,b]}^{(n)}) b,$$

which is the particular case of (5) under consideration.

In the general case we apply the case just proved to $\{(X_j - a)^+, j \in N_n\}$, which is a submartingale by Corollary 1 to Theorem 9.3.1. It is clear that the number of upcrossings of $[a, b]$ by the given submartingale is exactly that of $[0, b - a]$ by the modified one. The inequality (7) becomes the first inequality in (5) after the substitutions, and the second one follows since $(X_n - a)^+ \le X_n^+ + |a|$. Theorem 9.4.2 is proved.

The corresponding result for a supermartingale will be given below; but after such a painstaking definition of *upcrossing*, we may leave the dual definition of *downcrossing* to the reader.

**Theorem 9.4.3.** Let $\{X_j, \mathscr{F}_j; j \in N_n\}$ be a supermartingale and let $-\infty < ab < \infty$. Let $\tilde{\nu}^{(n)}_{[a,b]}$ be the number of downcrossings of $[a, b]$ by the sample sequence $\{X_j(\omega), j \in N_n\}$. We have then

$$\mathscr{E}\{\tilde{\nu}^{(n)}_{[a,b]}\} \le \frac{\mathscr{E}\{X_1 \wedge b\} - \mathscr{E}\{X_n \wedge b\}}{b - a} \tag{8}$$

PROOF. $\{-X_j, j \in N_n\}$ is a submartingale and $\tilde{\nu}^{(n)}_{[a,b]}$ is $\nu^{(n)}_{[-b,-a]}$ for this submartingale. Hence the first part of (5) becomes

$$\mathscr{E}\{\tilde{\nu}^{(n)}_{[a,b]}\} \le \frac{\mathscr{E}\{(-X_n + b)^+ - (-X_1 + b)^+\}}{-a - (-b)} = \frac{\mathscr{E}\{(b - X_n)^+ - (b - X_1)^+\}}{b - a}.$$

Since $(b - x)^+ = b - (b \wedge x)$ this is the same as in (8).

**Corollary.** For a positive supermartingale we have for $0 \le a < b < \infty$

$$\mathscr{E}\{\tilde{\nu}^{(n)}_{[a,b]}\} \le \frac{b}{b - a}.$$

G. Letta proved the sharper "dual":

$$\mathscr{E}\{\nu^{(n)}_{[a,b]}\} \le \frac{a}{b - a}.$$

(*Martingales et intégration stochastique*, Quaderni, Pisa, 1984, 48–49.)

The basic convergence theorem is an immediate consequence of the upcrossing inequality.

**Theorem 9.4.4.** If $\{X_n, \mathscr{F}_n; n \in N\}$ is an $L^1$-bounded submartingale, then $\{X_n\}$ converges a.e. to a finite limit.

*Remark.* Since

$$\mathscr{E}(|X_n|) = 2\mathscr{E}(X_n^+) - \mathscr{E}(X_n) \le 2\mathscr{E}(X_n^+) - \mathscr{E}(X_1),$$

the condition of $L^1$-boundedness is equivalent to the apparently weaker one below:

$$\sup_n \mathscr{E}(X_n^+) < \infty. \tag{9}$$

PROOF. Let $\nu_{[a,b]} = \lim_n \nu^{(n)}_{[a,b]}$. Our hypothesis implies that the last term in (5) is bounded in $n$; letting $n \to \infty$, we obtain $\mathscr{E}\{\nu_{[a,b]}\} < \infty$ for every $a$ and $b$, and consequently $\nu_{[a,b]}$ is finite with probability one. Hence, for each pair of rational numbers $a < b$, the set

$$\Lambda_{[a,b]} = \{\underline{\lim}_n X_n < a < b < \overline{\lim}_n X_n\}$$

is a null set; and so is the union over all such pairs. Since this union contains the set where $\underline{\lim}_n X_n < \overline{\lim}_n X_n$, the limit exists a.e. It must be finite a.e. by Fatou's lemma applied to the sequence $|X_n|$.

**Corollary.** Every uniformly bounded smartingale converges a.e. Every positive supermartingale and every negative submartingale converge a.e.

It may be instructive to sketch a direct proof of Theorem 9.4.4 which is done "by hand", so to speak. This is the original proof given by Doob (1940) for a martingale.

Suppose that the set $\Lambda_{[a,b]}$ above has probability $> \eta > 0$. For each $\omega$ in $\Lambda_{[a,b]}$, the sequence $\{X_n(\omega), n \in N\}$ takes an infinite number of values $< a$ and an infinite number of values $> b$. Let $1 = n_0 < n_1 < \dots$ and put

$$\Lambda_{2j-1} = \{\min_{n_{2j-2} \le i \le n_{2j-1}} X_i < a\}, \quad \Lambda_{2j} = \{\max_{n_{2j-1} < i \le n_{2j}} X_i > b\}.$$

Then for each $k$ it is possible to choose the $n_i$'s successively so that the differences $n_i - n_{i-1}$ for $1 \le i \le 2k$ are so large that "most" of $\Lambda_{[a,b]}$ is contained in $\bigcap_{i=1}^{2k} \Lambda_i$, so that

$$\mathscr{P}\left\{\bigcap_{i=1}^{2k} \Lambda_i\right\} > \eta.$$

Fixing an $n > n_{2k}$ and applying Corollary 2 to Theorem 9.4.1 to $\{-X_i\}$ as well as $\{X_i\}$, we have

$$a\mathscr{P}\left(\bigcap_{i=1}^{2j-1} \Lambda_i\right) \ge \int_{\bigcap_{i=1}^{2j-1}} X_{n_{2j-1}}\, d\mathscr{P} = \int_{\bigcap_{i=1}^{2j-1} \Lambda_i} X_n\, d\mathscr{P},$$

$$b\mathscr{P}\left(\bigcap_{i=1}^{2j} \Lambda_i\right) \le \int_{\bigcap_{i=1}^{2j} \Lambda_i} X_{n_{2j}}\, d\mathscr{P} = \int_{\bigcap_{i=1}^{2j} \Lambda_i} X_n\, d\mathscr{P},$$

where the equalities follow from the martingale property. Upon subtraction we obtain

$$(b-a)\mathscr{P}\left(\bigcap_{i=1}^{2j} \Lambda_i\right) - a\mathscr{P}\left(\bigcap_{i=1}^{2j-1} \Lambda_i \Lambda_{2j}^c\right) \le -\int_{\bigcap_{i=1}^{2j-1} \Lambda_i \Lambda_{2j}^c} X_n\, d\mathscr{P},$$

and consequently, upon summing over $1 \le j \le k$:

$$k(b-a)\eta - |a| \le \mathscr{E}(|X_n|).$$

This is impossible if $k$ is large enough, since $\{X_n\}$ is $L^1$-bounded.

Once Theorem 9.4.4 has been proved for a martingale, we can extend it easily to a positive or uniformly integrable supermartingale by using Doob's decomposition. Suppose $\{X_n\}$ is a positive supermartingale and $X_n = Y_n - Z_n$ as in Theorem 9.3.2. Then $0 \le Z_n \le Y_n$ and consequently

$$\mathscr{E}(Z_\infty) = \lim_{n\to\infty} \mathscr{E}(Z_n) \le \mathscr{E}(Y_1);$$

next we have

$$\mathscr{E}(Y_n) = \mathscr{E}(X_n) + \mathscr{E}(Z_n) \le \mathscr{E}(X_1) + \mathscr{E}(Z_\infty).$$

Hence $\{Y_n\}$ is an $L^1$-bounded martingale and so converges to a finite limit as $n \to \infty$. Since $Z_n \uparrow Z_\infty < \infty$ a.e., the convergence of $\{X_n\}$ follows. The case of a uniformly integrable supermartingale is just as easy by the corollary to Theorem 9.3.2.

It is trivial that a positive submartingale need not converge, since the sequence $\{n\}$ is such a one. The classical random walk $\{S_n\}$ (coin-tossing game) is an example of a martingale that does not converge (why?). An interesting and not so trivial consequence is that both $\mathscr{E}(S_n^+)$ and $\mathscr{E}(|S_n|)$ must diverge to $+\infty$! (Cf. Exercise 2 of Sec. 6.4.) Further examples are furnished by "stopped random walk". For the sake of concreteness, let us stay with the classical case and define $\gamma$ to be the first time the walk reaches $+1$. As in our previous discussion of the gambler's-ruin problem, the modified random walk $\{\tilde{S}_n\}$, where $\tilde{S}_n = S_{\gamma\wedge n}$, is still a martingale, hence in particular we have for each $n$:

$$\mathscr{E}(\tilde{S}_n) = \mathscr{E}(\tilde{S}_1) = \int_{\{\gamma=1\}} S_1 \, d\mathscr{P} + \int_{\{\gamma>1\}} S_1 \, d\mathscr{P} = \mathscr{E}(S_1) = 0.$$

As in (21) of Sec. 9.3 we have, writing $\tilde{S}_\infty = S_\gamma = 1$,

$$\lim_n \tilde{S}_n = \tilde{S}_\infty \quad \text{a.e.},$$

since $\gamma < \infty$ a.e., but this convergence now also follows from Theorem 9.4.4, since $\tilde{S}_n^+ \le 1$. Observe, however, that

$$\mathscr{E}(\tilde{S}_n) = 0 < 1 = \mathscr{E}(\tilde{S}_\infty).$$

Next, we change the definition of $\gamma$ to be the first time ($\ge 1$) the walk "returns" to 0, as usual supposing $S_0 \equiv 0$. Then $\tilde{S}_\infty = 0$ and we have indeed $\mathscr{E}(\tilde{S}_n) = \mathscr{E}(\tilde{S}_\infty)$. But for each $n$,

$$\int_{\{\tilde{S}_n>0\}} \tilde{S}_n \, d\mathscr{P} > 0 = \int_{\{\tilde{S}_n>0\}} \tilde{S}_\infty \, d\mathscr{P},$$

so that the "extended sequence" $\{\tilde{S}_1, \dots, \tilde{S}_n, \dots, \tilde{S}_\infty\}$ is no longer a martingale. These diverse circumstances will be dealt with below.

**Theorem 9.4.5.** The three propositions below are equivalent for a submartingale $\{X_n, \mathscr{F}_n; n \in N\}$:

(a) it is a uniformly integrable sequence;
(b) it converges in $L^1$;
(c) it converges a.e. to an integrable $X_\infty$ such that $\{X_n, \mathscr{F}_n; n \in N_\infty\}$ is a submartingale and $\mathscr{E}(X_n)$ converges to $\mathscr{E}(X_\infty)$.

PROOF. (a) $\Rightarrow$ (b): under (a) the condition in Theorem 9.4.4 is satisfied so that $X_n \to X_\infty$ a.e. This together with uniform integrability implies $X_n \to X_\infty$ in $L^1$ by Theorem 4.5.4 with $r = 1$.

(b) $\Rightarrow$ (c): under (b) let $X_n \to X_\infty$ in $L^1$, then $\mathscr{E}(|X_n|) \to \mathscr{E}(|X_\infty|) < \infty$ and so $X_n \to X_\infty$ a.e. by Theorem 9.4.4. For each $\Lambda \in \mathscr{F}_n$ and $n < n'$, we have

$$\int_\Lambda X_n \, d\mathscr{P} \leq \int_\Lambda X_{n'} \, d\mathscr{P}$$

by the defining relation. The right member converges to $\int_\Lambda X_\infty \, d\mathscr{P}$ by $L^1$-convergence and the resulting inequality shows that $\{X_n, \mathscr{F}_n; n \in N_\infty\}$ is a submartingale. Since $L^1$-convergence also implies convergence of expectations, all three conditions in (c) are proved.

(c) $\Rightarrow$ (a); under (c), $\{X_n^+, \mathscr{F}_n; n \in N_\infty\}$ is a submartingale; hence we have for every $\lambda > 0$:

$$(10) \qquad \int_{\{X_n^+ > \lambda\}} X_n^+ \, d\mathscr{P} \leq \int_{\{X_n^+ > \lambda\}} X_\infty^+ d\mathscr{P},$$

which shows that $\{X_n^+, n \in N\}$ is uniformly integrable. Since $X_n^+ \to X_\infty^+$ a.e., this implies $\mathscr{E}(X_n^+) \to \mathscr{E}(X_\infty^+)$. Since by hypothesis $\mathscr{E}(X_n) \to \mathscr{E}(X_\infty)$, it follows that $\mathscr{E}(X_n^-) \to \mathscr{E}(X_\infty^-)$. This and $X_n^- \to X_\infty^-$ a.e. imply that $\{X_n^-\}$ is uniformly integrable by Theorem 4.5.4 for $r = 1$. Hence so is $\{X_n\}$.

**Theorem 9.4.6.** In the case of a martingale, propositions (a) and (b) above are equivalent to (c$'$) or (d) below:

(c$'$) it converges a.e. to an integrable $X_\infty$ such that $\{X_n, \mathscr{F}_n; n \in N_\infty\}$ is a martingale;

(d) there exists an integrable r.v. $Y$ such that $X_n = \mathscr{E}(Y \mid \mathscr{F}_n)$ for each $n \in N$.

PROOF. (b) $\Rightarrow$ (c$'$) as before; (c$'$) $\Rightarrow$ (a) as before if we observe that $\mathscr{E}(X_n) = \mathscr{E}(X_\infty)$ for every $n$ in the present case, or more rapidly by considering $|X_n|$ instead of $X_n^+$ as below. (c$'$) $\Rightarrow$ (d) is trivial, since we may take the $Y$ in (d) to be the $X_\infty$ in (c$'$). To prove (d) $\Rightarrow$ (a), let $n < n'$, then by

Theorem 9.1.5:

$$\mathscr{E}(X_{n'} \mid \mathscr{F}_n) = \mathscr{E}(\mathscr{E}(Y \mid \mathscr{F}_{n'}) \mid \mathscr{F}_n) = \mathscr{E}(Y \mid \mathscr{F}_n) = X_n,$$

hence $\{X_n, \mathscr{F}_n, n \in N; Y, \mathscr{F}\}$ is a martingale by definition. Consequently $\{|X_n|, \mathscr{F}_n, n \in N; |Y|, \mathscr{F}\}$ is a submartingale, and we have for each $\lambda>0$:

$$\int_{\{|X_n|>\lambda\}} |X_n| \, d\mathscr{P} \leq \int_{\{|X_n|>\lambda\}} |Y| \, d\mathscr{P},$$

$$\mathscr{P}\{|X_n| > \lambda\} \leq \frac{1}{\lambda}\mathscr{E}(|X_n|) \leq \frac{1}{\lambda}\mathscr{E}(|Y|),$$

which together imply (a).

**Corollary.** Under (d), $\{X_n, \mathscr{F}_n, n \in N; X_\infty, \mathscr{F}_\infty; Y, \mathscr{F}\}$ is a martingale, where $X_\infty$ is given in (c′).

Recall that we have introduced martingales of the form in (d) earlier in (13) in Sec. 9.3. Now we know this class coincides with the class of uniformly integrable martingales.

We have already observed that the defining relation for a smartingale is meaningful on any linearly ordered (or even partially ordered) index set. The idea of extending the latter to a limit index is useful in applications to continuous-time stochastic processes, where, for example, a martingale may be defined on a dense set of real numbers in $(t_1, t_2)$ and extended to $t_2$. This corresponds to the case of extension from $N$ to $N_\infty$. The dual extension corresponding to that to $t_1$ will now be considered. Let $-N$ denote the set of strictly negative integers in their natural order, let $-\infty$ precede every element in $-N$, and denote by $-N_\infty$ the set $\{-\infty\} \cup (-N)$ in the prescribed order. If $\{\mathscr{F}_n, n \in -N\}$ is a decreasing (with decreasing $n$) sequence of Borel fields, their intersection $\bigcap_{n\in -N} \mathscr{F}_n$ will be denoted by $\mathscr{F}_{-\infty}$.

The convergence results for a submartingale on $-N$ are simpler because the right side of the upcrossing inequality (5) involves the expectation of the r.v. with the largest index, which in this case is the fixed $-1$ rather than the previous varying $n$. Hence for mere convergence there is no need for an extra condition such as (9).

**Theorem 9.4.7.** Let $\{X_n, n \in -N\}$ be a submartingale. Then

$$\lim_{n\to -\infty} X_n = X_{-\infty}, \quad \text{where} \quad -\infty \leq X_{-\infty} < \infty \quad \text{a.e.} \tag{11}$$

The following conditions are equivalent, and they are automatically satisfied in case of a martingale with "submartingale" replaced by "martingale" in (c):

(a) $\{X_n\}$ is uniformly integrable;
(b) $X_n \to X_{-\infty}$ in $L^1$;
(c) $\{X_n, n \in -N_\infty\}$ is a submartingale;
(d) $\lim_{n\to-\infty} \downarrow \mathscr{E}(X_n) > -\infty$.

PROOF. Let $\nu_{[a,b]}^{(n)}$ be the number of upcrossings of $[a, b]$ by the sequence $\{X_{-n,\ldots,X_{-1}}\}$. We have from Theorem 9.4.2:

$$\mathscr{E}\{\nu_{[a,b]}^{(n)}\} \le \frac{\mathscr{E}(X_{-1}^+) + |a|}{b-a}.$$

Letting $n \to \infty$ and arguing as the proof of Theorem 9.4.4, we conclude (11) by observing that

$$\mathscr{E}(X_{-\infty}^+) \le \varliminf_n \mathscr{E}(X_{-n}^+) \le \mathscr{E}(X_{-1}^+) < \infty.$$

The proofs of (a) $\Rightarrow$ (b) $\Rightarrow$ (c) are entirely similar to those in Theorem 9.4.5. (c) $\Rightarrow$ (d) is trivial, since $-\infty < \mathscr{E}(X_{-\infty}) \le \mathscr{E}(X_{-n})$ for each $n$. It remains to prove (d) $\Rightarrow$ (a). Letting $C$ denote the limit in (d), we have for each $\lambda>0$:

$$(12) \quad \lambda\mathscr{P}\{|X_n| > \lambda\} \le \mathscr{E}(|X_n|) = 2\mathscr{E}(X_n^+) - \mathscr{E}(X_n) \le 2\mathscr{E}(X_{-1}^+) - C < \infty.$$

It follows that $\mathscr{P}\{|X_n| > \lambda\}$ converges to zero uniformly in $n$ as $\lambda \to \infty$. Since

$$\int_{\{X_n^+>\lambda\}} X_n^+ \, d\mathscr{P} \le \int_{\{X_n^+>\lambda\}} X_{-1}^+ \, d\mathscr{P},$$

this implies that $\{X_n^+\}$ is uniformly integrable. Next if $n < m$, then

$$\begin{aligned} 0 \ge \int_{\{X_n<-\lambda\}} X_n \, d\mathscr{P} &= \mathscr{E}(X_n) - \int_{\{X_n\ge-\lambda\}} X_n \, d\mathscr{P} \\ &\ge \mathscr{E}(X_n) - \int_{\{X_n\ge-\lambda\}} X_m \, d\mathscr{P} \\ &= \mathscr{E}(X_n - X_m) + \mathscr{E}(X_m) - \int_{\{X_n\ge-\lambda\}} X_m \, d\mathscr{P} \\ &= \mathscr{E}(X_n - X_m) + \int_{\{X_n<-\lambda\}} X_m \, d\mathscr{P}. \end{aligned}$$

By (d), we may choose $-m$ so large that $\mathscr{E}(X_n - X_m) > -\epsilon$ for any given $\epsilon>0$ and for every $n < m$. Having fixed such an $m$, we may choose $\lambda$ so large that

$$\sup_n \int_{\{X_n<-\lambda\}} |X_m| \, d\mathscr{P} < \epsilon$$

by the remark after (12). It follows that $\{X_n^-\}$ is also uniformly integrable, and therefore (a) is proved.

The next result will be stated for the index set $\overline{N}$ of all integers in their natural order:

$$\overline{N} = \{\dots, -n, \dots, -2, -1, 0, 1, 2, \dots, n, \dots\}.$$

Let $\{\mathscr{F}_n\}$ be increasing B.F.'s on $\overline{N}$, namely: $\mathscr{F}_n \subset \mathscr{F}_m$ if $n \le m$. We may "close" them at both ends by adjoining the B.F.'s below:

$$\mathscr{F}_{-\infty} = \bigwedge_n \mathscr{F}_n, \quad \mathscr{F}_{\infty} = \bigvee_n \mathscr{F}_n.$$

Let $\{Y_n\}$ be r.v.'s indexed by $\overline{N}$. If the B.F.'s and r.v.'s are only given on $N$ or $-N$, they can be trivially extended to $\overline{N}$ by putting $\mathscr{F}_n = \mathscr{F}_1$, $Y_n = Y_1$ for all $n \le 0$, or $\mathscr{F}_n = \mathscr{F}_{-1}$, $Y_n = Y_{-1}$ for all $n \ge 0$. The following convergence theorem is very useful.

**Theorem 9.4.8.** Suppose that the $Y_n$'s are dominated by an integrable r.v.$Z$:

$$\sup_n |Y_n| \le Z; \tag{13}$$

and $\lim_n Y_n = Y_\infty$ or $Y_{-\infty}$ as $n \to \infty$ or $-\infty$. Then we have

$$\lim_{n\to\infty} \mathscr{E}\{Y_n \mid \mathscr{F}_n\} = \mathscr{E}\{Y_\infty \mid \mathscr{F}_\infty\}; \tag{14a}$$

$$\lim_{n\to-\infty} \mathscr{E}\{Y_n \mid \mathscr{F}_n\} = \mathscr{E}\{Y_{-\infty} \mid \mathscr{F}_{-\infty}\}. \tag{14b}$$

In particular for a fixed integrable r.v. $Y$, we have

$$\lim_{n\to\infty} \mathscr{E}\{Y \mid \mathscr{F}_n\} = \mathscr{E}\{Y \mid \mathscr{F}_\infty\}; \tag{15a}$$

$$\lim_{n\to-\infty} \mathscr{E}\{Y \mid \mathscr{F}_n\} = \mathscr{E}\{Y \mid \mathscr{F}_{-\infty}\}. \tag{15b}$$

where the convergence holds also in $L^1$ in both cases.

PROOF. We prove (15) first. Let $X_n = \mathscr{E}\{Y \mid \mathscr{F}_n\}$. For $n \in N$, $\{X_n, \mathscr{F}_n\}$ is a martingale already introduced in (13) of Sec. 9.3; the same is true for $n \in -N$. To prove (15a), we apply Theorem 9.4.6 to deduce (c′) there. It remains to identify the limit $X_\infty$ with the right member of (15a). For each $\Lambda \in \mathscr{F}_n$, we have

$$\int_\Lambda Y \, d\mathscr{P} = \int_\Lambda X_n \, d\mathscr{P} = \int_\Lambda X_\infty \, d\mathscr{P}.$$

Hence the equations hold also for $\Lambda \in \mathscr{F}_\infty$ (why?), and this shows that $X_\infty$ has the defining property of $\mathscr{E}(Y \mid \mathscr{F}_\infty$, since $X_\infty \in \mathscr{F}_\infty$. Similarly, the limit $X_{-\infty}$ in (15b) exists by Theorem 9.4.7; to identify it, we have by (c) there, for each $\Lambda \in \mathscr{F}_{-\infty}$:

$$\int_\Lambda X_{-\infty}\, d\mathscr{P} = \int_\Lambda X_n\, d\mathscr{P} = \int_\Lambda Y\, d\mathscr{P}.$$

This shows that $X_{-\infty}$ is equal to the right member of (15b).

We can now prove (14a). Put for $m \in N$:

$$W_m = \sup_{n\geq m} |Y_n - Y_\infty|;$$

then $|W_m| \leq 2Z$ and $\lim_{m\to\infty} W_m = 0$ a.e. Applying (15a) to $W_m$ we obtain

$$\overline{\lim_{n\to\infty}}\ \mathscr{E}\{|Y_n - Y_\infty| \big| \mathscr{F}_n\} \leq \lim_{n\to\infty} \mathscr{E}\{W_m \mid \mathscr{F}_n\} = \mathscr{E}\{W_m \mid \mathscr{F}_\infty\}.$$

As $m \to \infty$, the last term above converges to zero by dominated convergence (see (vii) of Sec. 9.1). Hence the first term must be zero and this clearly implies (14a). The proof of (14b) is completely similar.

Although the corollary below is a very special case we give it for historical interest. It is called Paul Lévy's zero-or-one law (1935) and includes Theorem 8.1.1 as a particular case.

**Corollary.** If $\Lambda \in \mathscr{F}_\infty$, then

(16) $$\lim_{n\to\infty} \mathscr{P}(\Lambda \mid \mathscr{F}_n) = 1_\Lambda \quad \text{a.e.}$$

The reader is urged to ponder over the intuitive meaning of this result and judge for himself whether it is "obvious" or "incredible".

## EXERCISES

***1.** Prove that for any smartingale, we have for each $\lambda>0$:

$$\lambda\mathscr{P}\{\sup_n |X_n| \geq \lambda\} \leq 3 \sup_n \mathscr{E}(|X_n|).$$

For a martingale or a positive or negative smartingale the constant 3 may be replaced by 1.

**2.** Let $\{X_n\}$ be a positive supermartingale. Then for almost every $\omega$, $X_k(\omega) = 0$ implies $X_n(\omega) = 0$ for all $n \geq k$. [This is the analogue of a minimum principle in potential theory.]

**3.** Generalize the upcrossing inequality for a submartingale $\{X_n, \mathscr{F}_n\}$ as follows:

$$\mathscr{E}\{\nu_{[a,b]}^{(n)} \mid \mathscr{F}_1\} \leq \frac{\mathscr{E}\{(X_n - a)^+ \mid \mathscr{F}_1\} - (X_1 - a)^+}{b - a}.$$

Similarly, generalize the downcrossing inequality for a positive supermartingale $\{X_n, \mathscr{F}_n\}$ as follows:

$$\mathscr{E}\{\tilde{\nu}_{[a,b]}^{(n)} \mid \mathscr{F}_1\} \leq \frac{X_1 \wedge b}{b-a}.$$

**★4.** As a sharpening of Theorems 9.4.2 and 9.4.3 we have, for a positive supermartingale $\{X_n, \mathscr{F}_n, n \in N\}$:

$$\mathscr{P}\{\nu_{[a,b]}^{(n)} \geq k\} \leq \frac{\mathscr{E}(X_1 \wedge a)}{b} \left(\frac{a}{b}\right)^{k-1},$$

$$\mathscr{P}\{\tilde{\nu}_{[a,b]}^{(n)} \geq k\} \leq \frac{\mathscr{E}(X_1 \wedge b)}{b} \left(\frac{a}{b}\right)^{k-1}.$$

These inequalities are due to Dubins. Derive Theorems 9.3.6 and 9.3.7 from them. [HINT:

$$\begin{aligned} b\mathscr{P}\{\alpha_{2j} < n\} &\leq \int_{\{\alpha_{2j}<n\}} X_{\alpha_{2j}} d\mathscr{P} \leq \int_{\{\alpha_{2j-1}<n\}} X_{\alpha_{2j}}\, d\mathscr{P} \\ &\leq \int_{\{\alpha_{2j-1}<n\}} X_{\alpha_{2j-1}}\, d\mathscr{P} \leq a\mathscr{P}\{\alpha_{2j-1} < n\} \end{aligned}$$

since $\{\alpha_{2j-1} < n\} \in \mathscr{F}_{\alpha_{2j-1}}$.]

**★5.** Every $L^1$-bounded martingale is the difference of two positive $L^1$-bounded martingales. This is due to Krickeberg. [HINT: Take one of them to be $\lim_{k\to\infty} \mathscr{E}\{X_k^+ \mid \mathscr{F}_n\}$.]

**★6.** A smartingale $\{X_n, \mathscr{F}_n; n \in N\}$ is said to be *closable* [on the right] iff there exists a r.v. $X_\infty$ such that $\{X_n, \mathscr{F}_n; n \in N_\infty\}$ is a smartingale of the same kind. Prove that if so then we can always take $X_\infty = \lim_{n\to\infty} X_n$. This supplies a missing link in the literature. [HINT: For a supermartingale consider $X_n = \mathscr{E}(X_\infty \mid \mathscr{F}_n) + Y_n$, then $\{Y_n, \mathscr{F}_n\}$ is a positive supermartingale so we may apply the convergence theorems to both terms of the decomposition.]

**7.** Prove a result for closability [on the left] which is similar to Exercise 6 but for the index set $-N$. Give an example to show that in case of $N$ we may have $\lim_{n\to\infty} \mathscr{E}(X_n) \neq \mathscr{E}(X_\infty)$, whereas in case of $-N$ closability implies $\lim_{n\to-\infty} \mathscr{E}(X_n) = \mathscr{E}(X_{-\infty})$.

**8.** Let $\{X_n, \mathscr{F}_n, n \in N\}$ be a submartingale and let $\alpha$ be a finite optional r.v. satisfying the conditions: (a) $\mathscr{E}(|X_\alpha|) < \infty$, and (b)

$$\varliminf_{n\to\infty} \int_{\{\alpha>n\}} |X_n|\, d\mathscr{P} = 0.$$

Then $\{X_{\alpha\wedge n}, \mathscr{F}_{\alpha\wedge n}; n \in N_\infty\}$ is a submartingale. [HINT: for $\Lambda \in \mathscr{F}_{\alpha\wedge n}$ bound $\int_\Lambda (X_\alpha - X_{\alpha\wedge n})\, d\mathscr{P}$ below by interposing $X_{\alpha\wedge m}$ where $n < m$.]

**9.** Let $\{X_n, \mathscr{F}_n; n \in N\}$ be a supermartingale satisfying the condition $\lim_{n\to\infty} \mathscr{E}(X_n) > -\infty$. Then we have the representation $X_n = X'_n + X''_n$ where $\{X'_n, \mathscr{F}_n\}$ is a martingale and $\{X''_n, \mathscr{F}_n\}$ is a positive supermartingale such that $\lim_{n\to\infty} X''_n = 0$ in $L^1$ as well as a.e. This is the analogue of F. Riesz's decomposition of a superharmonic function, $X'_n$ being the *harmonic* part and $X''_n$ the *potential* part. [HINT: Use Doob's decomposition $X_n = Y_n - Z_n$ and put $X''_n = Y_n - \mathscr{E}(Z_\infty \mid \mathscr{F}_n)$.]

**10.** Let $\{X_n, \mathscr{F}_n\}$ be a *potential*; namely a positive supermartingale such that $\lim_{n\to\infty} \mathscr{E}(X_n) = 0$; and let $X_n = Y_n - Z_n$ be the Doob decomposition [cf. (6) of Sec. 9.3]. Show that

$$X_n = \mathscr{E}(Z_\infty \mid \mathscr{F}_n) - Z_n.$$

***11.** If $\{X_n\}$ is a martingale or positive submartingale such that $\sup_n \mathscr{E}(X_n^2) < \infty$, then $\{X_n\}$ converges in $L^2$ as well as a.e.

**12.** Let $\{\xi_n, n \in N\}$ be a sequence of independent and identically distributed r.v.'s with zero mean and unit variance; and $S_n = \sum_{j=1}^n \xi_j$. Then for any optional r.v. $\alpha$ relative to $\{\xi_n\}$ such that $\mathscr{E}(\sqrt{\alpha}) < \infty$, we have $\mathscr{E}(|S_\alpha|) \le \sqrt{2}\mathscr{E}(\sqrt{\alpha})$ and $\mathscr{E}(S_\alpha) = 0$. This is an extension of Wald's equation due to Louis Gordon. [HINT: Truncate $\alpha$ and put $\eta_k = (S_k^2/\sqrt{k}) - (S_{k-1}^2/\sqrt{k-1})$; then

$$\mathscr{E}\{S_\alpha^2/\sqrt{\alpha}\} = \sum_{k=1}^{\infty} \int_{\{\alpha \ge k\}} \eta_k \, d\mathscr{P} \le \sum_{k=1}^{\infty} \mathscr{P}\{\alpha \ge k\}/\sqrt{k} \le 2\mathscr{E}(\sqrt{\alpha});$$

now use Schwarz's inequality followed by Fatou's lemma.]

The next two problems are meant to give an idea of the passage from discrete parameter martingale theory to the continuous parameter theory.

**13.** Let $\{X_t, \mathscr{F}_t; t \in [0, 1]\}$ be a continuous parameter supermartingale. For each $t \in [0, 1]$ and sequence $\{t_n\}$ decreasing to $t$, $\{X_{t_n}\}$ converges a.e. and in $L^1$. For each $t \in [0, 1]$ and sequence $\{t_n\}$ increasing to $t$, $\{X_{t_n}\}$ converges a.e. but not necessarily in $L^1$. [HINT: In the second case consider $X_{t_n} - \mathscr{E}\{X_t \mid \mathscr{F}_{t_n}\}$.]

***14.** In Exercise 13 let $Q$ be the set of rational numbers in [0, 1]. For each $t \in (0, 1)$ both limits below exist a.e.:

$$\lim_{\substack{s \uparrow t \\ s \in Q}} X_s, \qquad \lim_{\substack{s \downarrow t \\ s \in Q}} X_s.$$

[HINT: Let $\{Q_n, n \ge 1\}$ be finite subsets of $Q$ such that $Q_n \uparrow Q$; and apply the upcrossing inequality to $\{X_s, s \in Q_n\}$, then let $n \to \infty$.]

## 9.5 Applications

Although some of the major successes of martingale theory lie in the field of continuous-parameter stochastic processes, which cannot be discussed here, it has also made various important contributions within the scope of this work. We shall illustrate these below with a few that are related to our previous topics, and indicate some others among the exercises.

#### (I) The notions of "at least once" and "infinitely often"

These have been a recurring theme in Chapters 4, 5, 8, and 9 and play important roles in the theory of random walk and its generalization to Markov processes. Let $\{X_n, n \in N^0\}$ be an arbitrary stochastic process; the notation for fields in Sec. 9.2 will be used. For each $n$ consider the events:

$$\Lambda_n = \bigcup_{j=n}^{\infty} \{X_j \in B_j\},$$

$$\mathrm{M} = \bigcap_{n=1}^{\infty} \Lambda_n = \{X_j \in B_j \text{ i.o.}\},$$

where $B_n$ are arbitrary Borel sets.

**Theorem 9.5.1.** We have

$$\lim_{n\to\infty} \mathscr{P}\{\Lambda_{n+1} \mid \mathscr{F}_{[0,n]}\} = 1_{\mathrm{M}} \quad \text{a.e.,} \tag{1}$$

where $\mathscr{F}_{[0,n]}$ may be replaced by $\mathscr{F}_{\{n\}}$ or $X_n$ if the process is Markovian.

PROOF. By Theorem 9.4.8, (14a), the limit is

$$\mathscr{P}\{M \mid \mathscr{F}_{[0,\infty)}\} = 1_M.$$

The next result is a "principle of recurrence" which is useful in Markov processes; it is an extension of the idea in Theorem 9.2.3 (see also Exercises 15 and 16 of Sec. 9.2).

**Theorem 9.5.2.** Let $\{X_n, n \in N^0\}$ be a Markov process and $A_n$, $B_n$ Borel sets. Suppose that there exists $\delta>0$ such that for every $n$,

$$\mathscr{P}\{\bigcup_{j=n+1}^{\infty} [X_j \in B_j] \mid X_n\} \geq \delta \quad \text{a.e. on the set } \{X_n \in A_n\}; \tag{2}$$

then we have

$$\mathscr{P}\{[X_j \in A_j \text{ i.o.}]\backslash[X_j \in B_j \text{ i.o.}]\} = 0. \tag{3}$$

PROOF. Let $\Delta = \{X_j \in A_j \text{ i.o.}\}$ and use the notation $\Lambda_n$ and M above. We may ignore the null sets in (1) and (2). Then if $\omega \in \Delta$, our hypothesis implies that

$$\mathscr{P}\{\Lambda_{n+1} \mid X_n\}(\omega) \geq \delta \quad \text{i.o.}$$

In view of (1) this is possible only if $\omega \in \text{M}$. Thus $\Delta \subset \text{M}$, which implies (3).

The intuitive meaning of the preceding theorem has been given by Doeblin as follows: if the chance of a pedestrian's getting run over is greater than $\delta > 0$ each time he crosses a certain street, then he will not be crossing it indefinitely (since he will be killed first)! Here $\{X_n \in A_n\}$ is the event of the $n$th crossing, $\{X_n \in B_n\}$ that of being run over at the $n$th crossing.

### (II) Harmonic and superharmonic functions for a Markov process

Let $\{X_n, n \in N^0\}$ be a homogeneous Markov process as discussed in Sec. 9.2 with the transition probability function $P(\cdot, \cdot)$. An extended-valued function $f$ on $\mathscr{R}^1$ is said to be *harmonic* (*with respect to* $P$) iff it is integrable with respect to the measure $P(x, \cdot)$ for each $x$ and satisfies the following "harmonic equation";

$$\forall x \in \mathscr{R}^1: f(x) = \int_{\mathscr{R}^1} P(x, dy) f(y). \tag{4}$$

It is *superharmonic* (with respect to $P$) iff the "=" in (4) is replaced by "$\geq$"; in this case $f$ may take the value $+\infty$.

**Lemma.** If $f$ is [super]harmonic, then $\{f(X_n), n \in N^0\}$, where $X_0 \equiv x_0$ for some given $x_0$ in $\mathscr{R}^1$, is a [super]martingale.

PROOF. We have, recalling (14) of Sec. 9.2,

$$\mathscr{E}\{f(X_n)\} = \int_{\mathscr{R}^1} P^{(n)}(x_0, dy) f(y) < \infty,$$

as is easily seen by iterating (4) and applying an extended form of Fubini's theorem (see, e.g., Neveu [6]). Next we have, upon substituting $X_n$ for $x$ in (4):

$$f(X_n) = \int_{\mathscr{R}^1} P(X_n, dy) f(y) = \mathscr{E}\{f(X_{n+1}) \mid X_n\} = \mathscr{E}\{f(X_{n+1}) \mid \mathscr{F}_{[0,n]}\},$$

where the second equation follows by Exercise 8 of Sec. 9.2 and the third by Markov property. This proves the lemma in the harmonic case; the other case is similar. (Why not also the "sub" case?)

The most important example of a harmonic function is the $g(\cdot, B)$ of Exercise 10 of Sec. 9.2 for a given $B$; that of a superharmonic function is the

$f(\cdot, B)$ of Exercise 9 there. These assertions follow easily from their probabilistic meanings given in the cited exercises, but purely analytic verifications are also simple and instructive. Finally, if for some $B$ we have

$$\pi(x) = \sum_{n=0}^{\infty} P^{(n)}(x, B) < \infty$$

for every $x$, then $\pi(\cdot)$ is superharmonic and is called the "potential" of the set $B$.

**Theorem 9.5.3.** Suppose that the remote field of $\{X_n, n \in N^0\}$ is trivial. Then each bounded harmonic function is a constant a.e. with respect to each $\mu_n$, where $\mu_n$ is the p.m. of $X_n$.

PROOF. By Theorem 9.4.5, $f(X_n)$ converges a.e. to $Z$ such that

$$\{f(X_n), \mathscr{F}_{[0,n]}; Z, \mathscr{F}_{[0,\infty)}\}$$

is a martingale. Clearly $Z$ belongs to the remote field and so is a constant $c$ a.e. Since

$$f(X_n) = \mathscr{E}\{Z \mid \mathscr{F}_n\},$$

each $f(X_n)$ is the same constant $c$ a.e. Mapped into $\mathscr{R}^1$, the last assertion becomes the conclusion of the theorem.

#### (III) The supremum of a submartingale

The first inequality in (1) of Sec. 9.4 is of a type familiar in ergodic theory and leads to the result below, which has been called the "dominated ergodic theorem" by Wiener. In the case where $X_n$ is the sum of independent r.v.'s with mean 0, it is due to Marcinkiewicz and Zygmund. We write $||X||_p$ for the $L^p$-norm of $X$: $||X||_p^p = \mathscr{E}(|X|^p)$.

**Theorem 9.5.4.** Let $1 < p < \infty$ and $1/p + 1/q = 1$. Suppose that $\{X_n, n \in N\}$ is a positive submartingale satisfying the condition

$$\sup_n \mathscr{E}\{X_n^p\} < \infty. \tag{5}$$

Then $\sup_{n \in N} X_n \in L^p$ and

$$||\sup_n X_n||_p \leq q \sup_n ||X_n||_p. \tag{6}$$

PROOF. The condition (5) implies that $\{X_n\}$ is uniformly integrable (Exercise 8 of Sec. 4.5), hence by Theorem 9.4.5, $X_n \to X_\infty$ a.e. and $\{X_n, n \in N_\infty\}$ is a submartingale. Writing $Y$ for $\sup X_n$, we have by an obvious extension of the first equation in (1) of Sec. 9.4:

$$\forall\lambda{>}0\colon \lambda\mathscr{P}\{Y \geq \lambda\} \leq \int_{\{Y\geq\lambda\}} X_\infty \, d\mathscr{P}. \tag{7}$$

Now it turns out that such an inequality for any two r.v.'s $Y$ and $X_\infty$ implies the inequality $\|Y\|_p \leq q\|X_\infty\|_p$, from which (6) follows by Fatou's lemma. This is shown by the calculation below, where $G(\lambda) = \mathscr{P}\{Y \geq \lambda\}$.

$$\begin{aligned}
\mathscr{E}(Y^p) &= -\int_0^\infty \lambda^p \, dG(\lambda) \leq \int_0^\infty p\lambda^{p-1} G(\lambda)\, d\lambda \\
&\leq \int_0^\infty p\lambda^{p-1}\left[\frac{1}{\lambda}\int_{\{Y\geq\lambda\}} X_\infty \, d\mathscr{P}\right] d\lambda \\
&= \int_\Omega X_\infty \left[\int_0^Y p\lambda^{p-2}\, d\lambda\right] d\mathscr{P} \\
&= q\int_\Omega X_\infty Y^{p-1}\, d\mathscr{P} \leq q\|X_\infty\|_p \|Y^{p-1}\|_q \\
&= q\|X_\infty\|_p \{\mathscr{E}(Y^p)\}^{1/q}.
\end{aligned}$$

Since we do not yet know $\mathscr{E}(Y^p) < \infty$, it is necessary to replace $Y$ first with $Y \wedge m$, where $m$ is a constant, and verify the truth of (7) after this replacement, before dividing through in the obvious way. We then let $m \uparrow \infty$ and conclude (6).

The result above is false for $p = 1$ and is replaced by a more complicated one (Exercise 7 below).

#### (IV) Convergence of sums of independent r.v.'s

We return to Theorem 5.3.4 and complete the discussion there by showing that the convergence in distribution of the series $\sum_n X_n$ already implies its convergence a.e. This can also be proved by an analytic method based on estimation of ch.f.'s, but the martingale approach is more illuminating.

**Theorem 9.5.5.** If $\{X_n, n \in N\}$ is a sequence of independent r.v.'s such that $S_n = \sum_{j=1}^n X_j$ converges in distribution as $n \to \infty$, then $S_n$ converges a.e.

PROOF. Let $f_i$ be the ch.f. of $X_j$, so that

$$\varphi_n = \prod_{j=1}^n f_j$$

is the ch.f. of $S_n$. By the convergence theorem of Sec. 6.3, $\varphi_n$ converges everywhere to $\varphi$, the ch.f. of the limit distribution of $S_n$. We shall need only this fact for $|t| \leq t_0$, where $t_0$ is so small that $\varphi(t) \neq 0$ for $|t| \leq t_0$; then this is also true of $\varphi_n$ for all sufficiently large $n$. For such values of $n$ and a fixed $t$ with $|t| \leq t_0$, we define the complex-valued r.v. $Z_n$ as follows:

$$Z_n = \frac{e^{itS_n}}{\varphi_n(t)}. \tag{8}$$

Then each $Z_n$ is integrable; indeed the sequence $\{Z_n\}$ is uniformly bounded. We have for each $n$, if $\mathscr{F}_n$ denotes the Borel field generated by $S_1, \ldots, S_n$:

$$\begin{aligned}\mathscr{E}\{Z_{n+1} \mid \mathscr{F}_n\} &= \mathscr{E}\left\{\frac{e^{itS_n}}{\varphi_n(t)} \cdot \frac{e^{itX_{n+1}}}{f_{n+1}(t)} \middle| \mathscr{F}_n\right\} \\ &= \frac{e^{itS_n}}{\varphi_n(t)} \mathscr{E}\left\{\frac{e^{itX_{n+1}}}{f_{n+1}(t)} \middle| \mathscr{F}_n\right\} = \frac{e^{itS_n}}{\varphi_n(t)} \frac{f_{n+1}(t)}{f_{n+1}(t)} = Z_n,\end{aligned}$$

where the second equation follows from Theorem 9.1.3 and the third from independence. Thus $\{Z_n, \mathscr{F}_n\}$ is a martingale, in the sense that its real and imaginary parts are both martingales. Since it is uniformly bounded, it follows from Theorem 9.4.4 that $Z_n$ converges a.e. This means, for each $t$ with $|t| \leq t_0$, there is a set $\Omega_t$ with $\mathscr{P}(\Omega_t) = 1$ such that if $\omega \in \Omega_t$, then the sequence of complex numbers $e^{itS_n(\omega)}/\varphi_n(t)$ converges and so also does $e^{itS_n(\omega)}$. But how does one deduce from this the convergence of $S_n(\omega)$? The argument below may seem unnecessarily tedious, but it is of a familiar and indispensable kind in certain parts of stochastic processes.

Consider $e^{itS_n(\omega)}$ as a function of $(t, \omega)$ in the product space $T \times \Omega$, where $T = [-t_0, t_0]$, with the product measure $m \times \mathscr{P}$, where $m$ is the Lebesgue measure on $T$. Since this function is measurable in $(t, \omega)$ for each $n$, the set $C$ of $(t, \omega)$ for which $\lim_{n\to\infty} e^{itS_n(\omega)}$ exists is measurable with respect to $m \times \mathscr{P}$. Each section of $C$ by a fixed $t$ has full measure $\mathscr{P}(\Omega_t) = 1$ as just shown, hence Fubini's theorem asserts that almost every section of $C$ by a fixed $\omega$ must also have full measure $m(T) = 2t_0$. This means that there exists an $\tilde{\Omega}$ with $\mathscr{P}(\tilde{\Omega}) = 1$, and for each $\omega \in \tilde{\Omega}$ there is a subset $T_\omega$ of $T$ with $m(T_\omega) = m(T)$, such that if $t \in T_\omega$, then $\lim_{n\to\infty} e^{itS_n(\omega)}$ exists. Now we are in a position to apply Exercise 17 of Sec. 6.4 to conclude the convergence of $S_n(\omega)$ for $\omega \in \tilde{\Omega}$, thus finishing the proof of the theorem.

According to the preceding proof, due to Doob, the hypothesis of Theorem 9.5.5 may be further weakened to that the sequence of ch.f.'s of $S_n$ converges on a set of $t$ of strictly positive Lebesgue measure. In particular, if an infinite product $\Pi_n f_n$ of ch.f.'s converges on such a set, then it converges everywhere.

**(V) The strong law of large numbers**

Our next example is a new proof of the classical strong law of large numbers in the form of Theorem 5.4.2, (8). This basically different approach, which has more a measure-theoretic than an analytic flavor, is one of the striking successes of martingale theory. It was given by Doob (1949).

**Theorem 9.5.6.** Let $\{S_n, n \in N\}$ be a random walk (in the sense of Chapter 8) with $\mathscr{E}\{|S_1|\} < \infty$. Then we have

$$\lim_{n\to\infty} \frac{S_n}{n} = \mathscr{E}\{S_1\} \quad \text{a.e.}$$

PROOF. Recall that $S_n = \sum_{j=1}^{n} X_j$ and consider for $1 \leq k \leq n$:

$$\mathscr{E}\{X_k \mid S_n, S_{n+1}, \dots\} = \mathscr{E}\{X_k \mid \mathscr{G}_n\}, \tag{9}$$

where $\mathscr{G}_n$ is the Borel field generated by $\{S_j, j \geq n\}$. Thus

$$\mathscr{G}_n \downarrow \mathscr{G} = \bigwedge_{n\in N} \mathscr{G}_n$$

as $n$ increases. By Theorem 9.4.8, (15b), the right side of (9) converges to $\mathscr{E}\{X_k \mid \mathscr{G}\}$. Now $\mathscr{G}_n$ is also generated by $S_n$ and $\{X_j, j \geq n+1\}$, hence it follows from the independence of the latter from the pair $(X_k, S_n)$ and (3) of Sec. 9.2 that we have

$$\mathscr{E}\{X_k \mid S_n, X_{n+1}, X_{n+2}, \dots\} = \mathscr{E}\{X_k \mid S_n\} = \mathscr{E}\{X_1 \mid S_n\}, \tag{10}$$

the second equation by reason of symmetry (proof?). Summing over $k$ from 1 to $n$ and taking the average, we infer that

$$\frac{S_n}{n} = \mathscr{E}\{X_1 \mid \mathscr{G}_n\}$$

so that if $Y_{-n} = S_n/n$ for $n \in N$, $\{Y_n, n \in -N\}$ is a martingale. In particular,

$$\lim_{n\to\infty} \frac{S_n}{n} = \lim_{n\to\infty} \mathscr{E}\{X_1 \mid \mathscr{G}_n\} = \mathscr{E}\{X_1 \mid \mathscr{G}\},$$

where the second equation follows from the argument above. On the other hand, the first limit is a remote (even invariant) r.v. in the sense of Sec. 8.1, since for every $m \geq 1$ we have

$$\lim_{n\to\infty} \frac{S_n(\omega)}{n} = \lim_{n\to\infty} \frac{\sum_{j=m}^{n} X_j(\omega)}{n};$$

hence it must be a constant a.e. by Theorem 8.1.2. [Alternatively we may use Theorem 8.1.4, the limit above being even more obviously a permutable r.v.] This constant must be $\mathscr{E}\{\mathscr{E}\{X_1 \mid \mathscr{G}\}\} = \mathscr{G}\{X_1\}$, proving the theorem.

### (VI) Exchangeable events

The method used in the preceding example can also be applied to the theory of exchangeable events. The events $\{E_n, n \in N\}$ are said to be *exchangeable* iff for every $k \geq 1$, the probability of the joint occurrence of any $k$ of them is the same for every choice of $k$ members from the sequence, namely we have

$$\mathscr{P}\{E_{n_1} \cap \cdots \cap E_{n_k}\} = w_k, \quad k \in N; \tag{11}$$

for any subset $\{n_1, \ldots, n_k\}$ of $N$. Let us denote the indicator of $E_n$ by $e_n$, and put

$$N_n = \sum_{j=1}^{n} e_j;$$

then $N_n$ is the number of occurrences among the first $n$ events of the sequence. Denote by $\mathscr{G}_n$ the B.F. generated by $\{N_j, j \geq n\}$, and

$$\mathscr{G} = \bigwedge_{n \in N} \mathscr{G}_n.$$

Then the definition of exchangeability implies that if $n_j \leq n$ for $1 \leq j \leq k$, then

$$\mathscr{E}\left(\prod_{j=1}^{k} e_{n_j} \mid \mathscr{G}_n\right) = \mathscr{E}\left(\prod_{j=1}^{k} e_{n_j} \mid N_n\right) \tag{12}$$

and that this conditional expectation is the same for any subset $(n_1, \ldots, n_k)$ of $(1, \ldots, n)$. Put then $f_{n0} = 1$ and

$$f_{nk} = \sum_{(n_1, \ldots, n_k)} \prod_{j=1}^{k} e_{n_j}, \quad 1 \leq k \leq n,$$

where the sum is extended over all $\binom{n}{k}$ choices; this is the "elementary symmetric function" of degree $k$ formed by $e_1, \ldots, e_n$. Introducing an indeterminate $z$ we have the formal identity in $z$:

$$\sum_{j=0}^{n} f_{nj} z^j = \prod_{j=1}^{n} (1 + e_j z).$$

But it is trivial that $1 + e_j z = (1+z)^{e_j}$ since $e_j$ takes only the values 0 and 1, hence†

$$\sum_{j=0}^{n} f_{nj} z^j = \prod_{j=1}^{n} (1+z)^{e_j} = (1+z)^{N_n}.$$

From this we obtain by comparing the coefficients:

(13)
$$f_{nk} = \binom{N_n}{k}, \quad 0 \le k \le n.$$

It follows that the right member of (12) is equal to

$$\binom{N_n}{k} \Big/ \binom{n}{k}.$$

Letting $n \to \infty$ and using Theorem 9.4.8 (15b) in the left member of (12), we conclude that

(14)
$$\mathscr{E}\left(\prod_{j=1}^{k} e_{n_j} \mid \mathscr{G}\right) = \lim_{n\to\infty} \left(\frac{N_n}{n}\right)^k.$$

This is the key to the theory. It shows first that the limit below exists almost surely:

$$\lim_{n\to\infty} \frac{N_n}{n} = \eta,$$

and clearly $\eta$ is a r.v. satisfying $0 \le \eta \le 1$. Going back to (14) we have established the formula

(14′)
$$\mathscr{P}(E_{n_1} \cap \cdots \cap E_{n_k} \mid \mathscr{G}) = \eta^k, \quad k \in N;$$

and taking expectations we have identified the $w_k$ in (11):

$$w_k = \mathscr{E}(\eta^k).$$

Thus $\{w_k, k \in N\}$ is the sequence of moments of the distribution of $\eta$. This is de Finetti's theorem, as proved by D. G. Kendall. We leave some easy consequences as exercises below. An interesting classical example of exchangeable events is Pólya's urn scheme, see Rényi [24], and Chung [25].

#### (VII) Squared variation

Here is a small sample of the latest developments in martingale theory. Let $X = \{X_n, \mathscr{F}_n\}$ be a martingale; using the notation in (7) of Sec. 9.3, we put

$$Q_n^2 = Q_n^2(X) = \sum_{j=1}^{n} x_j^2.$$

† I owe this derivation to David Klarner.

The sequence $\{Q_n^2(X), n \in N\}$ associated with $X$ is called its *squared variation process* and is useful in many applications. We begin with the algebraic identity:

$$X_n^2 = \left(\sum_{j=1}^{n} x_j\right)^2 = \sum_{j=1}^{n} x_j^2 + 2\sum_{j=2}^{n} X_{j-1}x_j. \tag{15}$$

If $X_n \in L^2$ for every $n$, then all terms above are integrable, and we have for each $j$:

$$\mathscr{E}(X_{j-1}x_j) = \mathscr{E}(X_{j-1}\mathscr{E}(x_j \mid \mathscr{F}_{j-1})) = 0. \tag{16}$$

It follows that

$$\mathscr{E}(X_n^2) = \sum_{j=1}^{n} \mathscr{E}(x_j^2) = \mathscr{E}(Q_n^2).$$

When $X_n$ is the $n$th partial sum of a sequence of independent r.v.'s with zero mean and finite variance, the preceding formula reduces to the additivity of variances; see (6) of Sec. 5.1.

Now suppose that $\{X_n\}$ is a positive bounded supermartingale such that $0 \le X_n \le \lambda$ for all $n$, where $\lambda$ is a constant. Then the quantity of (16) is negative and bounded below by

$$\mathscr{E}(\lambda\mathscr{E}(x_j \mid \mathscr{F}_{j-1})) = \lambda\mathscr{E}(x_j) \le 0.$$

In this case we obtain from (15):

$$\lambda\mathscr{E}(X_n) \ge \mathscr{X}(X_n^2) \ge \mathscr{E}(Q_n^2) + 2\lambda\sum_{j=2}^{n} \mathscr{E}(x_j) = \mathscr{E}(Q_n^2) + 2\lambda[\mathscr{E}(X_n) = \mathscr{E}(X_1)];$$

so that

$$\mathscr{E}(Q_n^2) \le 2\lambda\mathscr{E}(X_1) \le 2\lambda^2. \tag{17}$$

If $X$ is a positive martingale, then $X\Lambda\lambda$ is a supermartingale of the kind just considered so that (17) is applicable to it. Letting $X^* = \sup_{1 \le n < \infty} X_n$,

$$\mathscr{P}\{Q_n(X) \ge \lambda\} \le \mathscr{P}\{X^* > \lambda\} + \mathscr{P}\{X^* \le \lambda; Q_n(X\Lambda\lambda) \ge \lambda\}.$$

By Theorem 9.4.1, the first term on the right is bounded by $\lambda^{-1}\mathscr{E}(X_1)$. The second term may be estimated by Chebyshev's inequality followed by (17) applied to $X\Lambda\lambda$:

$$\mathscr{P}\{Q_n(X\Lambda\lambda) \ge \lambda\} \le \frac{1}{\lambda^2}\mathscr{E}\{Q_n^2(X\Lambda\lambda)\} \le \frac{2}{\lambda}\mathscr{E}(X_1).$$

We have therefore established the inequality:

$$\mathscr{P}\{Q_n(X) \geq \lambda\} \leq \frac{3}{\lambda}\mathscr{E}(X_1) \tag{18}$$

for any positive martingale. Letting $n \to \infty$ and then $\lambda \to \infty$, we obtain

$$\sum_{j=1}^{\infty} x_j^2 = \lim_{n\to\infty} Q_n^2(X) < \infty \text{ a.e.}$$

Using Krickeberg's decomposition (Exercise 5 of Sec. 9.4) the last result extends at once to any $L^1$-bounded martingale. This was first proved by D. G. Austin. Similarly, the inequality (18) extends to any $L^1$-bounded martingale as follows:

$$\mathscr{P}\{Q_n(X) \geq \lambda\} \leq \frac{6}{\lambda}\sup_n \mathscr{E}(|X_n|). \tag{19}$$

The details are left as an exercise. This result is due to D. Burkholder. The simplified proofs given above are due to A. Garsia.

### (VIII) Derivation

Our final example is a feedback to the beginning of this chapter, namely to use martingale theory to obtain a Radon–Nikodym derivative. Let $Y$ be an integrable r.v. and consider, as in the proof of Theorem 9.1.1, the countably additive set function $\nu$ below:

$$\nu(\Lambda) = \int_\Lambda Y \, d\mathscr{P}. \tag{20}$$

For any countable measurable partition $\{\Delta_j^{(n)}, j \in N\}$ of $\Omega$, let $\mathscr{F}_n$ be the Borel field generated by it. Define the approximating function $X_n$ as follows:

$$X_n = \sum_j \frac{\nu(\Delta_j^{(n)})}{\mathscr{P}(\Delta_j^{(n)})} 1_{\Delta_j^{(n)}}, \tag{21}$$

where the fraction is taken to be zero if the denominator vanishes. According to the discussion of Sec. 9.1, we have

$$X_n = \mathscr{E}\{Y \mid \mathscr{F}_n\}.$$

Now suppose that the partitions become finer as $n$ increases so that $\{\mathscr{F}_n, n \in N\}$ is an increasing sequence of Borel fields. Then we obtain by Theorem 9.4.8:

$$\lim_{n\to\infty} X_n = \mathscr{E}\{Y \mid \mathscr{F}_\infty\}.$$

In particular if $Y \in \mathscr{F}_\infty$ we have obtained the Radon–Nikodym derivative $Y = d\nu/d\mathscr{P}$ as the almost everywhere limit of "increment ratios" over a "net":

$$Y(\omega) = \lim_{n\to\infty} \frac{\nu(\Delta_{j(\omega)}^{(n)})}{\mathscr{P}(\Delta_{j(\omega)}^{(n)})},$$

where $j(\omega)$ is the unique $j$ such that $\omega \in \Delta_j^{(n)}$.

If $(\Omega, \mathscr{F}, \mathscr{P})$ is $(\mathscr{U}, \mathscr{B}, m)$ as in Example 2 of Sec. 3.1, and $\nu$ is an arbitrary measure which is absolutely continuous with respect to the Lebesgue measure $m$, we may take the $n$th partition to be $0 = \xi_0^{(n)} < \cdots < \xi_n^{(n)} = 1$ such that

$$\max_{0\le k\le n-1} (\xi_{k+1}^{(n)} - \xi_k^{(n)}) \to 0.$$

For in this case $\mathscr{F}_\infty$ will contain each open interval and so also $\mathscr{B}$. If $\nu$ is not absolutely continuous, the procedure above will lead to the derivative of its absolutely continuous part (see Exercise 14 below). In particular, if $F$ is the d.f. associated with the p.m. $\nu$, and we put

$$f_n(x) = 2^n \left[F\left(\frac{k+1}{2^n}\right) - F\left(\frac{k}{2^n}\right)\right] \quad \text{for } \frac{k}{2^n} < x \le \frac{k+1}{2^n},$$

where $k$ ranges over all integers, then we have

$$\lim_{n\to\infty} f_n(x) = F'(x)$$

for almost all $x$ with respect to $m$; and $F'$ is the density of the absolutely continuous part of $F$; see Theorem 1.3.1. So we have come around to the beginning of this course, and the book is hereby ended.

## EXERCISES

$^\star$**1.** Suppose that $\{X_n, n \in N\}$ is a sequence of integer-valued r.v.'s having the following property. For each $n$, there exists a function $p_n$ of $n$ integers such that for every $k \in N$, we have

$$\mathscr{P}\{X_{k+j} = x_j, 1 \le j \le n\} = p_n(x_1, \ldots, x_n).$$

Define for a fixed $x_0$:

$$Z_n = \frac{p_{n+1}(x_0, X_1, \ldots, X_n)}{p_n(X_1, \ldots, X_n)}$$

if the denominator $>0$; otherwise $Z_n = 0$. Then $\{Z_n, n \in N\}$ is a martingale that converges a.e. and in $L^1$. [This is from information theory.]

**2.** Suppose that for each $n$, $\{X_j, 1 \leq j \leq n\}$ and $\{X'_j, 1 \leq j \leq n\}$ have respectively the $n$-dimensional probability density functions $p_n$ and $q_n$. Define

$$Y_n = \frac{q_n(X_1, \ldots, X_n)}{p_n(X_1, \ldots, X_n)}$$

if the denominator $>0$ and $= 0$ otherwise. Then $\{Y_n, n \in N\}$ is a supermartingale that converges a.e. [This is from statistics.]

**3.** Let $\{Z_n, n \in N^0\}$ be positive integer-valued r.v.'s such that $Z_0 \equiv 1$ and for each $n \geq 1$, the conditional distribution of $Z_n$ given $Z_0, \ldots, Z_{n-1}$ is that of $Z_{n-1}$ independent r.v.'s with the common distribution $\{p_k, k \in N^0\}$, where $p_1 < 1$ and

$$0 < m = \sum_{k=0}^{\infty} k p_k < \infty.$$

Then $\{W_n, n \in N^0\}$, where $W_n = Z_n/m^n$, is a martingale that converges, the limit being zero if $m \leq 1$. [This is from branching process.]

**4.** Let $\{X_n, n \in N\}$ be an arbitrary stochastic process and let $\mathscr{F}'_n$ be as in Sec. 8.1. Prove that the remote field is almost trivial if and only if for each $\Lambda \in \mathscr{F}_\infty$ we have

$$\lim_{n\to\infty} \sup_{\mathrm{M}\in\mathscr{F}'_n} |\mathscr{P}(\Lambda \mathrm{M}) - \mathscr{P}(\Lambda)\mathscr{P}(\mathrm{M})| = 0.$$

[HINT: Consider $\mathscr{P}(\Lambda \mid \mathscr{F}'_n)$ and apply 9.4.8. This is due to Blackwell and Freedman.]

***5.** In the notation of Theorem 9.6.2, suppose that there exists $\delta>0$ such that

$$\mathscr{P}\{X_j \in B_j \text{i.o.} \mid X_n\} \leq 1 - \delta \quad \text{a.e. on } \{X_n \in A_n\};$$

then we have

$$\mathscr{P}\{X_j \in A_j \quad \text{i.o.} \quad \text{and} \quad X_j \in B_j \text{ i.o.}\} = 0.$$

**6.** Let $f$ be a real bounded continuous function on $\mathscr{R}^1$ and $\mu$ a p.m. on $\mathscr{R}^1$ such that

$$\forall x \in \mathscr{R}^1: f(x) = \int_{\mathscr{R}^1} f(x+y)\mu(dy).$$

Then $f(x+s) = f(x)$ for each $s$ in the support of $\mu$. In particular, if $\mu$ is not of the lattice type, then $f$ is constant everywhere. [This is due to G. A. Hunt, who used it to prove renewal limit theorems. The approach was later rediscovered by other authors, and the above result in somewhat more general context is now referred to as Choquet–Deny's theorem.]

★**7.** The analogue of Theorem 9.5.4 for $p = 1$ is as follows: if $X_n \geq 0$ for all $n$, then

$$\mathscr{E}\{\sup_n X_n\} \leq \frac{e}{e-1}[1 + \sup_n \mathscr{E}\{X_n \log^+ X_n\}],$$

where $\log^+ x = \log x$ if $x \geq 1$ and 0 if $x \leq 1$.

★**8.** As an example of a more sophisticated application of the martingale convergence theorem, consider the following result due to Paul Lévy. Let $\{X_n, n \in N^0\}$ be a sequence of uniformly bounded r.v.'s, then the two series

$$\sum_n X_n \quad \text{and} \quad \sum_n \mathscr{E}\{X_n \mid X_1, \ldots, X_{n-1}\}$$

converge or diverge together. [HINT: Let

$$Y_n = X_n - \mathscr{E}\{X_n \mid X_1, \ldots, X_{n-1}\} \text{ and } Z_n = \sum_{j=1}^n Y_j.$$

Define $\alpha$ to be the first time that $Z_n > A$ and show that $\mathscr{E}(Z^+_{\alpha\wedge n})$ is bounded in $n$. Apply Theorem 9.4.4 to $\{Z_{\alpha\wedge n}\}$ for each $A$ to show that $Z_n$ converges on the set where $\overline{\lim}_n Z_n < \infty$; similarly also on the set where $\underline{\lim}_n Z_n > -\infty$. The situation is reminiscent of Theorem 8.2.5.]

**9.** Let $\{Y_k, 1 \leq k \leq n\}$ be independent r.v.'s with mean zero and finite variances $\sigma_k^2$;

$$S_k = \sum_{j=1}^k Y_j, \quad S_k^2 = \sum_{j=1}^k \sigma_j^2 > 0, \quad Z_k = S_k^2 - s_k^2$$

Prove that $\{Z_k, 1 \leq k \leq n\}$ is a martingale. Suppose now all $Y_k$ are bounded by a constant $A$, and define $\alpha$ and M as in the proof of Theorem 9.4.1, with the $X_k$ there replaced by the $S_k$ here. Prove that

$$S_n^2 \mathscr{P}(\mathrm{M}^c) \leq \mathscr{E}(S_\alpha^2) = \mathscr{E}(S_\alpha^2) \leq (\lambda + A)^2.$$

Thus we obtain

$$\mathscr{P}\{\max_{1\leq k\leq n} |S_k| \leq \lambda\} \leq \frac{(\lambda + A)^2}{S_n^2}.$$

an improvement on Exercise 3 of Sec. 5.3. [This is communicated by Doob.]

**10.** Let $\{X_n, n \in N\}$ be a sequence of independent, identically distributed r.v.'s with $\mathscr{E}(|X_1|) < \infty$; and $S_n = \sum_{j=1}^n X_j$. Define $\alpha = \inf\{n \geq 1: |X_n| > n\}$. Prove that if $\mathscr{E}((|S_\alpha|/\alpha)1_{\{\alpha<\infty\}}) < \infty$, then $\mathscr{E}(|X_1| \log^+ |X_1|) < \infty$. This is due to McCabe and Shepp. [HINT:

$$c_n = \prod_{j=1}^{n} \mathscr{P}\{|X_j| \leq j\} \to c>0;$$

$$\sum_{n=1}^{\infty} \frac{1}{n} \int_{\{\alpha=n\}} |X_n| \, d\mathscr{P} = \sum_{n=1}^{\infty} \frac{c_{n-1}}{n} \int_{\{|X_1|>n\}} |X_1| \, d\mathscr{P};$$

$$\sum_{n=1}^{\infty} \frac{1}{n} \int_{\{\alpha=n\}} |S_{n-1}| \, d\mathscr{P} < \infty.]$$

**11.** Deduce from Exercise 10 that $\mathscr{E}(\sup_n |S_n|/n) < \infty$ if and only if $\mathscr{E}(|X_1| \log^+ |X_1|) < \infty$. [HINT: Apply Exercise 7 to the martingale $\{\dots, S_n/n, \dots, S_2/2, S_1\}$ in Example (V).]

**12.** In Example (VI) show that (i) $\mathscr{G}$ is generated by $\eta$; (ii) the events $\{E_n, n \in N\}$ are conditionally independent given $\eta$; (iii) for any $l$ events $E_{n_j}$, $1 \leq j \leq l$ and any $k \leq l$ we have

$$\mathscr{P}\{E_{n_1} \cap \dots \cap E_{n_k} \cap E^c_{n_{k+1}} \cap \dots \cap E^c_{n_l}\} = \int_0^1 \binom{l}{k} x^k (1-x)^{l-k} G(dx)$$

where $G$ is the distributions of $\eta$.

**13.** Prove the inequality (19).

***14.** Prove that if $\nu$ is a measure on $\mathscr{F}_\infty$ that is singular with respect to $\mathscr{P}$, then the $X_n$'s in (21) converge a.e. to zero. [HINT: Show that

$$\nu(\Lambda) = \int_\Lambda X_n \, d\mathscr{P} \quad \text{for} \quad \Lambda \in \mathscr{F}_m, m \leq n,$$

and apply Fatou's lemma. $\{X_n\}$ is a supermartingale, not necessarily a martingale!]

**15.** In the case of $(\mathscr{U}, \mathscr{B}, m)$, suppose that $\nu = \delta_1$ and the $n$th partition is obtained by dividing $\mathscr{U}$ into $2^n$ equal parts: what are the $X_n$'s in (21)? Use this to "explain away" the St. Peterburg paradox (see Exercise 5 of Sec. 5.2).

### Bibliographical Note

Most of the results can be found in Chapter 7 of Doob [17]. Another useful account is given by Meyer [20]. For an early but stimulating account of the connections between random walks and partial differential equations, see

A. Khintchine, *Asymptotische Gesetze der Wahrscheinlichkeitsrechnung*. Springer-Verlag, Berlin, 1933.

Theorems 9.5.1 and 9.5.2 are contained in

Kai Lai Chung, *The general theory of Markov processes according to Doeblin*, Z. Wahrscheinlichkeitstheorie **2** (1964), 230–254.

For Theorem 9.5.4 in the case of an independent process, see

J. Marcinkiewicz and A. Zygmund, *Sur les fonctions independants*, Fund. Math. **29** (1937), 60–90,

which contains other interesting and useful relics.

The following article serves as a guide to the recent literature on martingale inequalities and their uses:

D. L. Burkholder, *Distribution function inequalities for martingales*, Ann. Probability **1** (1973), 19–42.

# Supplement: Measure and Integral

For basic mathematical vocabulary and notation the reader is referred to §1.1 and §2.1 of the main text.

## 1 Construction of measure

Let $\Omega$ be an abstract space and $\mathscr{S}$ its total Borel field, then $A \in \mathscr{S}$ means $A \subset \Omega$.

DEFINITION 1. A function $\mu^*$ with domain $\mathscr{S}$ and range in $[0, \infty]$ is an *outer measure* iff the following properties hold:

(a) $\mu^*(\phi) = 0$;
(b) (monotonicity) if $A_1 \subset A_2$, then $\mu^*(A_1) \leq \mu^*(A_2)$;
(c) (subadditivity) if $\{A_j\}$ is a countable sequence of sets in $\mathscr{S}$, then

$$\mu^*\left(\bigcup_j A_j\right) \leq \sum_j \mu^*(A_j).$$

DEFINITION 2. Let $\mathscr{F}_0$ be a field in $\Omega$. A function $\mu$ with domain $\mathscr{F}_0$ and range in $[0, \infty]$ is a *measure on* $\mathscr{F}_0$ iff (a) and the following property hold:

(d) (additivity) if $\{B_j\}$ is a countable sequence of disjoint sets in $\mathscr{F}_0$ and $\bigcup_j B_j \in \mathscr{F}_0$, then

$$\mu\left(\bigcup_j B_j\right) = \sum_j \mu(B_j). \tag{1}$$

Let us show that the properties (b) and (c) for outer measure hold for a measure $\mu$, provided all the sets involved belong to $\mathscr{F}_0$.

If $A_1 \in \mathscr{F}_0$, $A_2 \in \mathscr{F}_0$, and $A_1 \subset A_2$, then $A_1^c A_2 \in \mathscr{F}_0$ because $\mathscr{F}_0$ is a field; $A_2 = A_1 \cup A_1^c A_2$ and so by (d):

$$\mu(A_2) = \mu(A_1) + \mu(A_1^c A_2) \geq \mu(A_1).$$

Next, if each $A_j \in \mathscr{F}_0$, and furthermore if $\bigcup_j A_j \in \mathscr{F}_0$ (this must be assumed for a countably infinite union because it is not implied by the definition of a field!), then

$$\bigcup_j A_j = A_1 \cup A_1^c A_2 \cup A_1^c A_2^c A_3 \cup \ldots$$

and so by (d), since each member of the disjoint union above belongs to $\mathscr{F}_0$:

$$\begin{aligned}\mu\left(\bigcup_j A_j\right) &= \mu(A_1) + \mu(A_1^c A_2) + \mu(A_1^c A_2^c A_3) + \cdots \\ &\leq \mu(A_1) + \mu(A_2) + \mu(A_3) + \cdots\end{aligned}$$

by property (b) just proved.

The symbol $N$ denotes the sequence of natural numbers (strictly positive integers); when used as index set, it will frequently be omitted as understood. For instance, the index $j$ used above ranges over $N$ or a finite segment of $N$.

Now let us suppose that the field $\mathscr{F}_0$ is a Borel field to be denoted by $\mathscr{F}$ and that $\mu$ is a measure on it. Then if $A_n \in \mathscr{F}$ for each $n \in N$, the countable union $\bigcup_n A_n$ and countable intersection $\bigcap_n A_n$ both belong to $\mathscr{F}$. In this case we have the following fundamental properties.

(e) (increasing limit) if $A_n \subset A_{n+1}$ for all $n$ and $A_n \uparrow A = \bigcup_n A_n$, then

$$\lim_n \uparrow \mu(A_n) = \mu(A).$$

(f) (decreasing limit) if $A_n \supset A_{n+1}$ for all $n$, $A_n \downarrow A = \bigcap_n A_n$, and for some $n$ we have $\mu(A_n) < \infty$, then

$$\lim_n \downarrow \mu(A_n) = \mu(A).$$

The additional assumption in (f) is essential. For a counterexample let $A_n = (n, \infty)$ in $R$, then $A_n \downarrow \phi$ the empty set, but the measure (length!) of $A_n$ is $+\infty$ for all $n$, while $\phi$ surely must have measure 0. See §3 for formalities of this trivial example. It can even be made discrete if we use the *counting measure* # of natural numbers: let $A_n = \{n, n+1, n+2, \ldots\}$ so that $\#(A_n) = +\infty$, $\#(\bigcap_n A_n) = 0$.

Beginning with a measure $\mu$ on a field $\mathscr{F}_0$, not a Borel field, we proceed to construct a measure on the Borel field $\mathscr{F}$ generated by $\mathscr{F}_0$, namely the minimal Borel field containing $\mathscr{F}_0$ (see §2.1). This is called an extension of $\mu$ from $\mathscr{F}_0$ to $\mathscr{F}$, when the notation $\mu$ is maintained. Curiously, we do this by first constructing an outer measure $\mu^*$ on the total Borel field $\mathscr{S}$ and then showing that $\mu^*$ is in truth a measure on a certain Borel field to be denoted by $\mathscr{F}^*$ that contains $\mathscr{F}_0$. Then of course $\mathscr{F}^*$ must contain the minimal $\mathscr{F}$, and so $\mu^*$ restricted to $\mathscr{F}$ is an extension of the original $\mu$ from $\mathscr{F}_0$ to $\mathscr{F}$. But we have obtained a further extension to $\mathscr{F}^*$ that is in general "larger" than $\mathscr{F}$ and possesses a further desirable property to be discussed.

DEFINITION 3. Given a measure $\mu$ on a field $\mathscr{F}_0$ in $\Omega$, we define $\mu^*$ on $\mathscr{S}$ as follows, for any $A \in \mathscr{S}$:

$$\text{(2)} \qquad \mu^*(A) = \inf\left\{\sum_j \mu(B_j) \middle| B_j \in \mathscr{F}_0 \text{ for all } j \text{ and } \bigcup_j B_j \supset A\right\}.$$

A countable (possibly finite) collection of sets $\{B_j\}$ satisfying the conditions indicated in (2) will be referred to below as a "covering" of $A$. The infimum taken over all such coverings exists because the single set $\Omega$ constitutes a covering of $A$, so that

$$0 \le \mu^*(A) \le \mu^*(\Omega) \le +\infty.$$

It is not trivial that $\mu^*(A) = \mu(A)$ if $A \in \mathscr{F}_0$, which is part of the next theorem.

**Theorem 1.** We have $\mu^* = \mu$ on $\mathscr{F}_0$; $\mu^*$ on $\mathscr{S}$ is an outer measure.

PROOF. Let $A \in \mathscr{F}_0$, then the single set $A$ serves as a covering of $A$; hence $\mu^*(A) \le \mu(A)$. For any covering $\{B_j\}$ of $A$, we have $AB_j \in \mathscr{F}_0$ and

$$\bigcup_j AB_j = A \in \mathscr{F}_0;$$

hence by property (c) of $\mu$ on $\mathscr{F}_0$ followed by property (b):

$$\mu(A) = \mu\left(\bigcup_j AB_j\right) \le \sum_j \mu(AB_j) \le \sum_j \mu(B_j).$$

It follows from (2) that $\mu(A) \le \mu^*(A)$. Thus $\mu^* = \mu$ on $\mathscr{F}_0$.

To prove $\mu^*$ is an outer measure, the properties (a) and (b) are trivial. To prove (c), let $\epsilon > 0$. For each $j$, by the definition of $\mu^*(A_j)$, there exists a covering $\{B_{jk}\}$ of $A_j$ such that

$$\sum_k \mu(B_{jk}) \le \mu^*(A_j) + \frac{\epsilon}{2^j}.$$

The double sequence $\{B_{jk}\}$ is a covering of $\bigcup_j A_j$ such that

$$\sum_j \sum_k \mu(B_{jk}) \le \sum_j \mu^*(A_j) + \epsilon.$$

Hence for any $\epsilon > 0$:

$$\mu^*\left(\bigcup_j A_j\right) \le \sum_j \mu^*(A_j) + \epsilon$$

that establishes (c) for $\mu^*$, since $\epsilon$ is arbitrarily small.

With the outer measure $\mu^*$, a class of sets $\mathscr{F}^*$ is associated as follows.

DEFINITION 4. A set $A \subset \Omega$ belongs to $\mathscr{F}^*$ iff for every $Z \subset \Omega$ we have

$$\mu^*(Z) = \mu^*(AZ) + \mu^*(A^c Z). \tag{3}$$

If in (3) we change "=" into "≤", the resulting inequality holds by (c); hence (3) is equivalent to the reverse inequality when "=" is changed into "≥".

**Theorem 2.** $\mathscr{F}^*$ is a Borel field and contains $\mathscr{F}_0$. On $\mathscr{F}^*$, $\mu^*$ is a measure.

PROOF. Let $A \in \mathscr{F}_0$. For any $Z \subset \Omega$ and any $\epsilon > 0$, there exists a covering $\{B_j\}$ of $Z$ such that

$$\sum_j \mu(B_j) \le \mu^*(Z) + \epsilon. \tag{4}$$

Since $AB_j \in \mathscr{F}_0$, $\{AB_j\}$ is a covering of $AZ$; $\{A^c B_j\}$ is a covering of $A^c Z$; hence

$$\mu^*(AZ) \le \sum_j \mu(AB_j), \quad \mu^*(A^c Z) \le \sum_j \mu(A^c B_j). \tag{5}$$

Since $\mu$ is a measure on $\mathscr{F}_0$, we have for each $j$:

$$\mu(AB_j) + \mu(A^c B_j) = \mu(B_j). \tag{6}$$

It follows from (4), (5), and (6) that

$$\mu^*(AZ) + \mu^*(A^cZ) \leq \mu^*(Z) + \epsilon.$$

Letting $\epsilon \downarrow 0$ establishes the criterion (3) in its "$\geq$" form. Thus $A \in \mathscr{F}^*$, and we have proved that $\mathscr{F}_0 \subset \mathscr{F}^*$.

To prove that $\mathscr{F}^*$ is a Borel field, it is trivial that it is closed under complementation because the criterion (3) is unaltered when $A$ is changed into $A^c$. Next, to show that $\mathscr{F}^*$ is closed under union, let $A \in \mathscr{F}^*$ and $B \in \mathscr{F}^*$. Then for any $Z \subset \Omega$, we have by (3) with $A$ replaced by $B$ and $Z$ replaced by $ZA$ or $ZA^c$:

$$\mu^*(ZA) = \mu^*(ZAB) + \mu^*(ZAB^c);$$
$$\mu^*(ZA^c) = \mu^*(ZA^cB) + \mu^*(ZA^cB^c).$$

Hence by (3) again as written:

$$\mu^*(Z) = \mu^*(ZAB) + \mu^*(ZAB^c) + \mu^*(ZA^cB) + \mu^*(ZA^cB^c).$$

Applying (3) with $Z$ replaced by $Z(A \cup B)$, we have

$$\begin{aligned}\mu^*(Z(A \cup B)) &= \mu^*(Z(A \cup B)A) + \mu^*(Z(A \cup B)A^c)\\ &= \mu^*(ZA) + \mu^*(ZA^cB)\\ &= \mu^*(ZAB) + \mu^*(ZAB^c) + \mu^*(ZA^cB).\end{aligned}$$

Comparing the two preceding equations, we see that

$$\mu^*(Z) = \mu^*(Z(A \cup B)) + \mu^*(Z(A \cup B)^c).$$

Hence $A \cup B \in \mathscr{F}^*$, and we have proved that $\mathscr{F}^*$ is a field.

Now let $\{A_j\}$ be an infinite sequence of sets in $\mathscr{F}^*$; put

$$B_1 = A_1, \quad B_j = A_j \backslash \left( \bigcup_{i=1}^{j-1} A \right) \text{ for } j \geq 2.$$

Then $\{B_j\}$ is a sequence of disjoint sets in $\mathscr{F}^*$ (because $\mathscr{F}^*$ is a field) and has the same union as $\{A_j\}$. For any $Z \subset \Omega$, we have for each $n \geq 1$:

$$\begin{aligned}\mu^*\left(Z\bigcup_{j=1}^{n} B_j\right) &= \mu^*\left(Z\left(\bigcup_{j=1}^{n} B_j\right)B_n\right) + \mu^*\left(Z\left(\bigcup_{j=1}^{n} B_j\right)B_n^c\right)\\ &= \mu^*(ZB_n) + \mu^*\left(Z\bigcup_{j=1}^{n-1} B_j\right)\end{aligned}$$

because $B_n \in \mathscr{F}^*$. It follows by induction on $n$ that

$$\mu^*\left(Z\bigcup_{j=1}^{n}B_j\right) = \sum_{j=1}^{n}\mu^*(ZB_j). \tag{7}$$

Since $\bigcup_{j=1}^{n} B_j \in \mathscr{F}^*$, we have by (7) and the monotonicity of $\mu^*$:

$$\begin{aligned}\mu^*(Z) &= \mu^*\left(Z\bigcup_{j=1}^{n}B_j\right) + \mu^*\left(Z\left(\bigcup_{j=1}^{n}B_j\right)^c\right)\\ &\geq \sum_{j=1}^{n}\mu^*(ZB_j) + \mu^*\left(Z\left(\bigcup_{j=1}^{\infty}B_j\right)^c\right).\end{aligned}$$

Letting $n \uparrow \infty$ and using property (c) of $\mu^*$, we obtain

$$\mu^*(Z) \geq \mu^*\left(Z\bigcup_{j=1}^{\infty}B_j\right) + \mu^*\left(Z\left(\bigcup_{j=1}^{\infty}B_j\right)^c\right)$$

that establishes $\bigcup_{j=1}^{\infty} B_j \in \mathscr{F}^*$. Thus $\mathscr{F}^*$ is a Borel field.

Finally, let $\{B_j\}$ be a sequence of disjoint sets in $\mathscr{F}^*$. By the property (b) of $\mu^*$ and (7) with $Z = \Omega$, we have

$$\mu^*\left(\bigcup_{j=1}^{\infty}B_j\right) \geq \limsup_n \mu^*\left(\bigcup_{j=1}^{n}B_j\right) = \lim_n \sum_{j=1}^{n}\mu^*(B_j) = \sum_{j=1}^{\infty}\mu^*(B_j).$$

Combined with the property (c) of $\mu^*$, we obtain the countable additivity of $\mu^*$ on $\mathscr{F}^*$, namely the property (d) for a measure:

$$\mu^*\left(\bigcup_{j=1}^{\infty}B_j\right) = \sum_{j=1}^{\infty}\mu^*(B_j).$$

The proof of Theorem 2 is complete.

## 2 Characterization of extensions

We have proved that

$$\mathscr{S} \supset \mathscr{F}^* \supset \mathscr{F} \supset \mathscr{F}_0,$$

where some of the "$\supset$" may turn out to be "$=$". Since we have extended the measure $\mu$ from $\mathscr{F}_0$ to $\mathscr{F}^*$ in Theorem 2, what for $\mathscr{F}$? The answer will appear in the sequel.

The triple $(\Omega, \mathscr{F}, \mu)$ where $\mathscr{F}$ is a Borel field of subsets of $\Omega$, and $\mu$ is a measure on $\mathscr{F}$, will be called a *measure space*. It is qualified by the adjective "finite" when $\mu(\Omega) < \infty$, and by the noun "probability" when $\mu(\Omega) = 1$.

A more general case is defined below.

DEFINITION 5. A measure $\mu$ on a field $\mathscr{F}_0$ (not necessarily Borel field) is said to be $\sigma$-finite iff there exists a sequence of sets $\{\Omega_n, n \in N\}$ in $\mathscr{F}_0$ such that $\mu(\Omega_n) < \infty$ for each $n$, and $\bigcup_n \Omega_n = \Omega$. In this case the measure space $(\Omega, \mathscr{F}, \mu)$, where $\mathscr{F}$ is the minimal Borel field containing $\mathscr{F}_0$, is said to be "$\sigma$-finite on $\mathscr{F}_0$".

**Theorem 3.** Let $\mathscr{F}_0$ be a field and $\mathscr{F}$ the Borel field generated by $\mathscr{F}_0$. Let $\mu_1$ and $\mu_2$ be two measures on $\mathscr{F}$ that agree on $\mathscr{F}_0$. If one of them, hence both are $\sigma$-finite on $\mathscr{F}_0$, then they agree on $\mathscr{F}$.

PROOF. Let $\{\Omega_n\}$ be as in Definition 5. Define a class $\mathscr{C}$ of subsets of $\Omega$ as follows:

$$\mathscr{C} = \{A \subset \Omega : \mu_1(\Omega_n A) = \mu_2(\Omega_n A) \text{ for all } n \in N\}.$$

Since $\Omega_n \in \mathscr{F}_0$, for any $A \in \mathscr{F}_0$ we have $\Omega_n A \in \mathscr{F}_0$ for all $n$; hence $\mathscr{C} \supset \mathscr{F}_0$. Suppose $A_k \in \mathscr{C}$, $A_k \subset A_{k+1}$ for all $k \in N$ and $A_k \uparrow A$. Then by property (e) of $\mu_1$ and $\mu_2$ as measures on $\mathscr{F}$, we have for each $n$:

$$\mu_1(\Omega_n A) = \lim_k \uparrow \mu_1(\Omega_n A_k) = \lim_k \uparrow \mu_2(\Omega_n A_k) = \mu_2(\Omega_n A).$$

Thus $A \in \mathscr{C}$. Similarly by property (f), and the hypothesis $\mu_1(\Omega_n) = \mu_2(\Omega_n) < \infty$, if $A_k \in \mathscr{C}$ and $A_k \downarrow A$, then $A \in \mathscr{C}$. Therefore $\mathscr{C}$ is closed under both increasing and decreasing limits; hence $\mathscr{C} \supset \mathscr{F}$ by Theorem 2.1.2 of the main text. This implies for any $A \in \mathscr{F}$:

$$\mu_1(A) = \lim_n \uparrow \mu_1(\Omega_n A) = \lim_n \uparrow \mu_2(\Omega_n A) = \mu_2(A)$$

by property (e) once again. Thus $\mu_1$ and $\mu_2$ agree on $\mathscr{F}$.

It follows from Theorem 3 that under the $\sigma$-finite assumption there, the outer measure $\mu^*$ in Theorem 2 restricted to the minimal Borel field $\mathscr{F}$ containing $\mathscr{F}_0$ is the unique extension of $\mu$ from $\mathscr{F}_0$ to $\mathscr{F}$. What about the more extensive extension to $\mathscr{F}^*$? We are going to prove that it is also unique when a further property is imposed on the extension. We begin by defining two classes of special sets in $\mathscr{F}$.

DEFINITION 6. Given the field $\mathscr{F}_0$ of sets in $\Omega$, let $\mathscr{F}_{0\sigma\delta}$ be the collection of all sets of the form $\bigcap_{m=1}^{\infty} \bigcup_{n=1}^{\infty} B_{mn}$ where each $B_{mn} \in \mathscr{F}_0$, and $\mathscr{F}_{0\delta\sigma}$ be the collection of all sets of the form $\bigcup_{m=1}^{\infty} \bigcap_{n=1}^{\infty} B_{mn}$ where each $B_{mn} \in \mathscr{F}_0$.

Both these collections belong to $\mathscr{F}$ because the Borel field is closed under countable union and intersection, and these operations may be iterated, here twice only, for each collection. If $B \in \mathscr{F}_0$, then $B$ belongs to both $\mathscr{F}_{0\sigma\delta}$ and $\mathscr{F}_{0\delta\sigma}$ because we can take $B_{mn} = B$. Finally, $A \in \mathscr{F}_{0\sigma\delta}$ if and only if $A^c \in \mathscr{F}_{0\delta\sigma}$ because

$$\left(\bigcap_m \bigcup_n B_{mn}\right)^c = \bigcup_m \bigcap_n B_{mn}^c.$$

**Theorem 4.** Let $A \in \mathscr{F}^*$. There exists $B \in \mathscr{F}_{0\sigma\delta}$ such that

$$A \subset B; \quad \mu^*(A) = \mu^*(B).$$

If $\mu$ is $\sigma$-finite on $\mathscr{F}_0$, then there exists $C \in \mathscr{F}_{0\delta\sigma}$ such that

$$C \subset A; \quad \mu^*(C) = \mu^*(A).$$

PROOF. For each $m$, there exists $\{B_{mn}\}$ in $\mathscr{F}$ such that

$$A \subset \bigcup_n B_{mn}; \quad \sum_n \mu^*(B_{mn}) \leq \mu^*(A) + \frac{1}{m}.$$

Put

$$B_m = \bigcup_n B_{mn}; \quad B = \bigcap_m B_m;$$

then $A \subset B$ and $B \in \mathscr{F}_{0\sigma\delta}$. We have

$$\mu^*(B) \leq \mu^*(B_m) \leq \sum_n \mu^*(B_{mn}) \leq \mu^*(A) + \frac{1}{m}.$$

Letting $m \uparrow \infty$ we see that $\mu^*(B) \leq \mu^*(A)$; hence $\mu^*(B) = \mu^*(A)$. The first assertion of the theorem is proved.

To prove the second assertion, let $\Omega_n$ be as in Definition 5. Applying the first assertion to $\Omega_n A^c$, we have $B_n \in \mathscr{F}_{0\sigma\delta}$ such that

$$\Omega_n A^c \subset B_n; \quad \mu^*(\Omega_n A^c) = \mu^*(B_n).$$

Hence we have

$$\Omega_n A^c \subset \Omega_n B_n; \quad \mu^*(\Omega_n A^c) = \mu^*(\Omega_n B_n).$$

Taking complements with respect to $\Omega_n$, we have since $\mu^*(\Omega_n) < \infty$:

$$\Omega_n A \supset \Omega_n B_n^c;$$

$$\mu^*(\Omega_n A) = \mu^*(\Omega_n) - \mu^*(\Omega_n A^c) = \mu^*(\Omega_n) - \mu^*(\Omega_n B_n) = \mu^*(\Omega_n B_n^c).$$

Since $\Omega_n \in \mathscr{F}_0$ and $B_n^c \in \mathscr{F}_{0\delta\sigma}$, it is easy to verify that $\Omega_n B_n^c \in \mathscr{F}_{0\delta\sigma}$ by the distributive law for the intersection with a union. Put

$$C = \bigcup_n \Omega_n B_n^c.$$

It is trivial that $C \in \mathscr{F}_{0\delta\sigma}$ and

$$A = \bigcup_n \Omega_n A \supset C.$$

Consequently, we have

$$\begin{aligned} \mu^*(A) \geq \mu^*(C) &\geq \liminf_n \mu^*(\Omega_n B_n^c) \\ &= \liminf_n \mu^*(\Omega_n A) = \mu^*(A), \end{aligned}$$

the last equation owing to property (e) of the measure $\mu^*$. Thus $\mu^*(A) = \mu^*(C)$, and the assertion is proved.

The measure $\mu^*$ on $\mathscr{F}^*$ is constructed from the measure $\mu$ on the field $\mathscr{F}_0$. The restriction of $\mu^*$ to the minimal Borel field $\mathscr{F}$ containing $\mathscr{F}_0$ will henceforth be denoted by $\mu$ instead of $\mu^*$.

In a general measure space $(\Omega, \mathscr{G}, \nu)$, let us denote by $\mathscr{N}(\mathscr{G}, \nu)$ the class of all sets $A$ in $\mathscr{G}$ with $\nu(A) = 0$. They are called the *null sets* when $\mathscr{G}$ and $\nu$ are understood, or $\nu$-null sets when $\mathscr{G}$ is understood. Beware that if $A \subset B$ and $B$ is a null set, it does not follow that $A$ is a null set because $A$ may not be in $\mathscr{G}$! This remark introduces the following definition.

DEFINITION 7. The *measure space* $(\Omega, \mathscr{G}, \nu)$ is called *complete* iff any subset of a null set is a null set.

**Theorem 5.** The following three collections of subsets of $\Omega$ are idential:

**(i)** $A \subset \Omega$ and the outer measure $\mu^*(A) = 0$;

**(ii)** $A \in \mathscr{F}^*$ and $\mu^*(A) = 0$;

**(iii)** $A \subset B$ where $B \in \mathscr{F}$ and $\mu(B) = 0$.

It is the collection $\mathscr{N}(\mathscr{F}^*, \mu^*)$.

PROOF. If $\mu^*(A) = 0$, we will prove $A \in \mathscr{F}^*$ by verifying the criterion (3). For any $Z \subset \Omega$, we have by properties (a) and (b) of $\mu^*$:

$$0 \leq \mu^*(ZA) \leq \mu^*(A) = 0; \quad \mu^*(ZA^c) \leq \mu^*(Z);$$

and consequently by property (c):

$$\mu^*(Z) = \mu^*(ZA \cup ZA^c) \le \mu^*(ZA) + \mu^*(ZA^c) \le \mu^*(Z).$$

Thus (3) is satisfied and we have proved that (i) and (ii) are equivalent.

Next, let $A \in \mathscr{F}^*$ and $\mu^*(A) = 0$. Then we have by the first assertion in Theorem 4 that there exists $B \in \mathscr{F}$ such that $A \subset B$ and $\mu^*(A) = \mu(B)$. Thus $A$ satisfies (iii). Conversely, if $A$ satisfies (iii), then by property (b) of outer measure: $\mu^*(A) \le \mu^*(B) = \mu(B) = 0$, and so (i) is true.

As consequence, any subset of a $(\mathscr{F}^*, \mu^*)$-null set is a $(\mathscr{F}^*, \mu^*)$-null set. This is the first assertion in the next theorem.

**Theorem 6.** The measure space $(\Omega, \mathscr{F}^*, \mu^*)$ is complete. Let $(\Omega, \mathscr{G}, \nu)$ be a complete measure space; $\mathscr{G} \supset \mathscr{F}_0$ and $\nu = \mu$ on $\mathscr{F}_0$. If $\mu$ is $\sigma$-finite on $\mathscr{F}_0$ then

$$\mathscr{G} \supset \mathscr{F}^* \text{ and } \nu = \mu^* \text{ on } \mathscr{F}^*.$$

PROOF. Let $A \in \mathscr{F}^*$, then by Theorem 4 there exists $B \in \mathscr{F}$ and $C \in \mathscr{F}$ such that

$$C \subset A \subset B; \quad \mu(C) = \mu^*(A) = \mu(B). \tag{8}$$

Since $\nu = \mu$ on $\mathscr{F}_0$, we have by Theorem 3, $\nu = \mu$ on $\mathscr{F}$. Hence by (8) we have $\nu(B - C) = 0$. Since $A - C \subset B - C$ and $B - C \in \mathscr{G}$, and $(\Omega, \mathscr{G}, \nu)$ is complete, we have $A - C \in \mathscr{G}$ and so $A = C \cup (A - C) \in \mathscr{G}$.

Moreover, since $C$, $A$, and $B$ belong to $\mathscr{G}$, it follows from (8) that

$$\mu(C) = \nu(C) \le \nu(A) \le \nu(B) = \mu(B)$$

and consequently by (8) again $\nu(A) = \mu(A)$. The theorem is proved.

To summarize the gist of Theorems 4 and 6, if the measure $\mu$ on the field $\mathscr{F}_0$ is $\sigma$-finite on $\mathscr{F}_0$, then $(\mathscr{F}, \mu)$ is its unique extension to $\mathscr{F}$, and $(\mathscr{F}^*, \mu^*)$ is its minimal complete extension. Here one is tempted to change the notation $\mu$ to $\mu_0$ on $\mathscr{F}_0$!

We will complete the picture by showing how to obtain $(\mathscr{F}^*, \mu^*)$ from $(\mathscr{F}, \mu)$, reversing the order of previous construction. Given the measure space $(\Omega, \mathscr{F}, \mu)$, let us denote by $\mathscr{C}$ the collection of subsets of $\Omega$ as follows: $A \in \mathscr{C}$ iff there exists $B \in \mathscr{N}(\mathscr{F}, \mu)$ such that $A \subset B$. Clearly $\mathscr{C}$ has the "hereditary" property: if $A$ belongs to $\mathscr{C}$, then all subsets of $A$ belong to $\mathscr{C}$; $\mathscr{C}$ is also closed under countable union. Next, we define the collection

$$\overline{\mathscr{F}} = \{A \subset \Omega \mid A = B - C \text{ where } B \in \mathscr{F}, C \in \mathscr{C}\}. \tag{9}$$

where the symbol "$-$" denotes *strict difference* of sets, namely $B - C = BC^c$ where $C \subset B$. Finally we define a function $\overline{\mu}$ on $\overline{\mathscr{F}}$ as follows, for the $A$

shown in (9):

$$\overline{\mu}(A) = \mu(B). \tag{10}$$

We will legitimize this definition and with the same stroke prove the monotonicity of $\overline{\mu}$. Suppose then

$$B_1 - C_1 \subset B_2 - C_2, \quad B_i \in \mathscr{F}, C_i \in \mathscr{C}, i = 1, 2. \tag{11}$$

Let $C_1 \subset D \in \mathscr{N}(\mathscr{F}, \mu)$. Then $B_1 \subset B_2 \cup D$ and so $\mu(B_1) \leq \mu(B_2 \cup D) = \mu(B_2)$. When the $\subset$ in (11) is "=", we can interchange $B_1$ and $B_2$ to conclude that $\mu(B_1) = \mu(B_2)$, so that the definition (10) is legitimate.

**Theorem 7.** $\overline{\mathscr{F}}$ is a Borel field and $\overline{\mu}$ is a measure on $\overline{\mathscr{F}}$.

PROOF. Let $A_n \in \overline{\mathscr{F}}$, $n \in N$; so that $A_n = B_n C_n^c$ as in (9). We have then

$$\bigcap_{n=1}^{\infty} A_n = \left(\bigcap_{n=1}^{\infty} B_n\right) \cap \left(\bigcup_{n=1}^{\infty} C_n\right)^c.$$

Since the class $\mathscr{C}$ is closed under countable union, this shows that $\overline{\mathscr{F}}$ is closed under countable intersection. Next let $C \subset D$, $D \in \mathscr{N}(\mathscr{F}, \mu)$; then

$$\begin{aligned} A^c &= B^c \cup C = B^c \cup BC = B^c \cup (B(D - (D - C))) \\ &= (B^c \cup BD) - B(D - C). \end{aligned}$$

Since $B(D - C) \subset D$, we have $B(D - C) \in \mathscr{C}$; hence the above shows that $\overline{\mathscr{F}}$ is also closed under complementation and therefore is a Borel field. Clearly $\overline{\mathscr{F}} \supset \mathscr{F}$ because we may take $C = \phi$ in (9).

To prove $\overline{\mu}$ is countably additive on $\overline{\mathscr{F}}$, let $\{A_n\}$ be disjoint sets in $\overline{\mathscr{F}}$. Then

$$A_n = B_n - C_n, \quad B_n \in \mathscr{F}, C_n \in \mathscr{C}.$$

There exists $D$ in $\mathscr{N}(\mathscr{F}, \mu)$ containing $\bigcup_{n=1}^{\infty} C_n$. Then $\{B_n - D\}$ are disjoint and

$$\bigcup_{n=1}^{\infty} (B_n - D) \subset \bigcup_{n=1}^{\infty} A_n \subset \bigcup_{n=1}^{\infty} B_n.$$

All these sets belong to $\overline{\mathscr{F}}$ and so by the monotonicity of $\overline{\mu}$:

$$\overline{\mu}\left(\bigcup_n (B_n - D)\right) \leq \overline{\mu}\left(\bigcup_n A_n\right) \leq \overline{\mu}\left(\bigcup_n B_n\right).$$

Since $\overline{\mu} = \mu$ on $\mathscr{F}$, the first and third members above are equal to, respectively:

$$\mu\left(\bigcup_n (B_n - D)\right) = \sum_n \mu(B_n - D) = \sum_n \mu(B_n) = \sum_n \overline{\mu}(A_n);$$

$$\mu\left(\bigcup_n B_n\right) \leq \sum_n \mu(B_n) = \sum_n \overline{\mu}(A_n).$$

Therefore we have

$$\overline{\mu}\left(\bigcup_n A_n\right) = \sum_n \overline{\mu}(A_n).$$

Since $\overline{\mu}(\phi) = \overline{\mu}(\phi - \phi) = \mu(\phi) = 0$, $\overline{\mu}$ is a measure on $\overline{\mathscr{F}}$.

**Corollary.** In truth: $\overline{\mathscr{F}} = \mathscr{F}^*$ and $\overline{\mu} = \mu^*$.

PROOF. For any $A \in \mathscr{F}^*$, by the first part of Theorem 4, there exists $B \in \mathscr{F}$ such that

$$A = B - (B - A), \quad \mu^*(B - A) = 0.$$

Hence by Theorem 5, $B - A \in \mathscr{C}$ and so $A \in \overline{\mathscr{F}}$ by (9). Thus $\mathscr{F}^* \subset \overline{\mathscr{F}}$. Since $\mathscr{F} \subset \mathscr{F}^*$ and $\mathscr{C} \in \mathscr{F}^*$ by Theorem 6, we have $\overline{\mathscr{F}} \subset \mathscr{F}^*$ by (9). Hence $\overline{\mathscr{F}} = \mathscr{F}^*$. It follows from the above that $\mu^*(A) = \mu(B) = \overline{\mu}(A)$. Hence $\mu^* = \overline{\mu}$ on $\mathscr{F}^* = \overline{\mathscr{F}}$.

The question arises naturally whether we can extend $\mu$ from $\mathscr{F}_0$ to $\mathscr{F}$ directly without the intervention of $\mathscr{F}^*$. This is indeed possible by a method of transfinite induction originally conceived by Borel; see an article by LeBlanc and G.E. Fox: "On the extension of measure by the method of Borel", Canadian Journal of Mathematics, 1956, pp. 516–523. It is technically lengthier than the method of outer measure expounded here.

Although the case of a countable space $\Omega$ can be treated in an obvious way, it is instructive to apply the general theory to see what happens.

Let $\Omega = N \cup \omega$; $\mathscr{F}_0$ is the minimal field (not Borel field) containing each singleton $n$ in $N$, but not $\omega$. Let $N_f$ denote the collection of all finite subsets of $N$; then $\mathscr{F}_0$ consists of $N_f$ and the complements of members of $N_f$ (with respect to $\Omega$), the latter all containing $\omega$. Let $0 \leq \mu(n) < \infty$ for all $n \in N$; a measure $\mu$ is defined on $\mathscr{F}_0$ as follows:

$$\mu(A) = \sum_{n \in A} \mu(n) \text{ if } A \in N_f; \quad \mu(A^c) = \mu(\Omega) - \mu(A).$$

We must still define $\mu(\Omega)$. Observe that by the properties of a measure, we have $\mu(\Omega) \geq \sum_{n \in N} \mu(n) = s$, say.

Now we use Definition 3 to determine the outer measure $\mu^*$. It is easy to see that for any $A \subset N$, we have

$$\mu^*(A) = \sum_{n \in A} \mu(n).$$

In particular $\mu^*(N) = s$. Next we have

$$\mu^*(\omega) = \inf_{A \in N_f} \mu(A^c) = \mu(\Omega) - \sup_{A \in N_f} \mu(A) = \mu(\Omega) - s$$

provided $s < \infty$; otherwise the inf above is $\infty$. Thus we have

$$\mu^*(\omega) = \mu(\Omega) - s \text{ if } \mu(\Omega) < \infty; \mu^*(\omega) = \infty \text{ if } \mu(\Omega) = \infty.$$

It follows that for any $A \subset \Omega$:

$$\mu^*(A) = \sum_{n \in A} \mu^*(n)$$

where $\mu^*(n) = \mu(n)$ for $n \in N$. Thus $\mu^*$ is a measure on $\mathscr{S}$, namely, $\mathscr{F}^* = \mathscr{S}$.

But it is obvious that $\mathscr{F} = \mathscr{S}$ since $\mathscr{F}$ contains $N$ as countable union and so contains $\omega$ as complement. Hence $\mathscr{F} = \mathscr{F}^* = \mathscr{S}$.

If $\mu(\Omega) = \infty$ and $s = \infty$, the extension $\mu^*$ of $\mu$ to $\mathscr{S}$ is not unique, because we can define $\mu(\omega)$ to be any positive number and get an extension. Thus $\mu$ is not $\sigma$-finite on $\mathscr{F}_0$, by Theorem 3. But we can verify this directly when $\mu(\Omega) = \infty$, whether $s = \infty$ or $s < \infty$. Thus in the latter case, $\mu(\omega) = \infty$ is also the unique extension of $\mu$ from $\mathscr{F}_0$ to $\mathscr{S}$. This means that the condition of $\sigma$-finiteness on $\mathscr{F}_0$ is only a sufficient and not a necessary condition for the unique extension.

As a ramification of the example above, let $\Omega = N \cup \omega_1 \cup \omega_2$, with two extra points adjoined to $N$, but keep $\mathscr{F}_0$ as before. Then $\mathscr{F}(=\mathscr{F}^*)$ is strictly smaller than $\mathscr{S}$ because neither $\omega_1$ nor $\omega_2$ belongs to it. From Definition 3 we obtain

$$\mu^*(\omega_1 \cup \omega_2) = \mu^*(\omega_1) = \mu^*(\omega_2).$$

Thus $\mu^*$ is not even two-by-two additive on $\mathscr{S}$ unless the three quantities above are zero. The two points $\omega_1$ and $\omega_2$ form an inseparable couple. We leave it to the curious reader to wonder about other possibilities.

## 3 Measures in *R*

Let $R = (-\infty, +\infty)$ be the set of real members, alias the real line, with its Euclidean topology. For $-\infty \le a < b \le +\infty$,

(12) $$(a, b] = \{x \in R: a < x \le b\}$$

is an interval of a particular shape, namely open at left end and closed at right end. For $b = +\infty$, $(a, +\infty] = (a, +\infty)$ because $+\infty$ is not in $R$. By choice of the particular shape, the complement of such an interval is the union of two intervals of the same shape:

$$(a, b]^c = (-\infty, a] \cup (b, \infty].$$

When $a = b$, of course $(a, a] = \phi$ is the empty set. A finite or countably infinite number of such intervals may merge end to end into a single one as illustrated below:

$$(13) \qquad (0, 2] = (0, 1] \cup (1, 2]; \quad (0, 1] = \bigcup_{n=1}^{\infty} \left( \frac{1}{n+1}, \frac{1}{n} \right].$$

Apart from this possibility, the representation of $(a, b]$ is unique.

The minimal Borel field containing all $(a, b]$ will be denoted by $\mathscr{B}$ and called *the Borel field* of $R$. Since a bounded open interval is the countable union of intervals like $(a, b]$, and any open set in $R$ is the countable union of (disjount) bounded open intervals, the Borel field $\mathscr{B}$ contains all open sets; hence by complementation it contains all closed sets, in particular all compact sets. Starting from one of these collections, forming countable union and countable intersection successively, a countable number of times, one can build up $\mathscr{B}$ through a transfinite induction.

Now suppose a measure $m$ has been defined on $\mathscr{B}$, subject to the sole assumption that its value for a finite (alias bounded) interval be finite, namely if $-\infty < a < b < +\infty$, then

$$(14) \qquad 0 \le m((a, b]) < \infty.$$

We associate a point function $F$ on $R$ with the set function $m$ on $\mathscr{B}$, as follows:

$$(15) \quad F(0) = 0; \quad F(x) = m((0, x]) \text{ for } x > 0; \quad F(x) = -m((x, 0]) \text{ for } x < 0.$$

This function may be called the "generalized distribution" for $m$. We see that $F$ is finite everywhere owing to (14), and

$$(16) \qquad m((a, b]) = F(b) - F(a).$$

$F$ is increasing (viz. nondecreasing) in $R$ and so the limits

$$F(+\infty) = \lim_{x \to +\infty} F(x) \le +\infty, \qquad F(-\infty) = \lim_{x \to -\infty} F(x) \ge -\infty$$

both exist. We shall write $\infty$ for $+\infty$ sometimes. Next, $F$ has unilateral limits everywhere, and is right-continuous:

$$F(x-) \le F(x) = F(x+).$$

The right-continuity follows from the monotone limit properties ($e$) and ($f$) of $m$ and the primary assumption (14). The measure of a single point $x$ is given by

$$m(x) = F(x) - F(x-).$$

We shall denote a point and the set consisting of it (singleton) by the same symbol.

The simplest example of $F$ is given by $F(x) \equiv x$. In this case $F$ is continuous everywhere and (14) becomes

$$m((a, b]) = b - a.$$

We can replace $(a, b]$ above by $(a, b)$, $[a, b)$ or $[a, b]$ because $m(x) = 0$ for each $x$. This measure is the *length* of the line-segment from $a$ to $b$. It was in this classic case that the following extension was first conceived by Émile Borel (1871–1956).

We shall follow the methods in §§1–2, due to H. Lebesgue and C. Carathéodory. Given $F$ as specified above, we are going to construct a measure $m$ on $\mathscr{B}$ and a larger Borel field $\mathscr{B}^*$ that fulfills the prescription (16).

The first step is to determine the minimal field $\mathscr{B}_0$ containing all $(a, b]$. Since a field is closed under finite union, it must contain all sets of the form

$$B = \bigcup_{j=1}^{n} I_j, \quad I_j = (a_j,\ b_j],\ 1 \leq j \leq n; \quad n \in N. \tag{17}$$

Without loss of generality, we may suppose the intervals $I_j$ to be disjoint, by merging intersections as illustiated by

$$(1, 3] \cup (2, 4] = (1, 4].$$

Then it is clear that the complement $B^c$ is of the same form. The union of two sets like $B$ is also of the same form. Thus the collection of all sets like $B$ already forms a field and so it must be $\mathscr{B}_0$. Of course it contains (includes) the empty set $\phi = (a, a]$ and $R$. However, it does not contain any $(a, b)$ except $R$, $[a, b)$, $[a, b]$, or any single point!

Next we define a measure $m$ on $\mathscr{B}_0$ satisfying (16). Since the condition ($d$) in Definition 2 requires it to be finitely additive, there is only one way: for the generic $B$ in (17) with disjoint $I_j$ we must put

$$m(B) = \sum_{j=1}^{n} m(I_j) = \sum_{j=1}^{n} (F(b_j) - F(a_j)). \tag{18}$$

Having so defined $m$ on $\mathscr{B}_0$, we must now prove that it satisfies the condition ($d$) in toto, in order to proclaim it to be a measure on $\mathscr{B}_0$. Namely, if

$\{B_k, 1 \le k \le l \le \infty\}$ is a finite or countable sequence of disjoint sets in $\mathscr{B}_0$, we must prove

$$m\left(\bigcup_{k=1}^{l} B_k\right) = \sum_{k=1}^{l} m(B_k). \tag{19}$$

whenever $l$ is finite, and moreover when $l = \infty$ and the union $\cup_{k=1}^{\infty} B_k$ happens to be in $\mathscr{B}_0$.

The case for a finite $l$ is really clear. If each $B_k$ is represented as in (17), then the disjoint union of a finite number of them is represented in a similar manner by pooling together all the disjoint $I_j$'s from the $B_k$'s. Then the equation (19) just means that a finite double array of numbers can be summed in two orders.

If that is so easy, what is the difficulty when $l = \infty$? It turns out, as Borel saw clearly, that the crux of the matter lies in the following fabulous "banality."

**Borel's lemma.** If $-\infty \le a < b \le +\infty$ and

$$(a, b] = \bigcup_{j=1}^{\infty} (a_j, b_j], \tag{20}$$

where $a_j < b_j$ for each $j$, and the intervals $(a_j, b_j]$ are disjoint, then we have

$$F(b) - F(a) = \sum_{j=1}^{\infty} \left(F(b_j) - F(a_j)\right). \tag{21}$$

PROOF. We will first give a transfinite argument that requires knowledge of ordinal numbers. But it is so intuitively clear that it can be appreciated without that prerequisite. Looking at (20) we see there is a unique index $j$ such that $b_j = b$; name that index $k$ and rename $a_k$ as $c_1$. By removing $(a_k, b_k] = (c_1, b]$ from both sides of (20) we obtain

$$(a, c_1] = \bigcup_{\substack{j=1 \\ j \ne k}}^{\infty} (a_j, b_j]. \tag{22}$$

This small but giant step shortens the original $(a, b]$ to $(a, c_1]$. Obviously we can repeat the process and shorten it to $(a, c_2]$ where $a \le c_2 < c_1 = b$, and so by mathematical induction we obtain a sequence $a \le c_n < \cdots < c_2 < c_1 = b$.

Needless to say, if for some $n$ we have $c_n = a$, then we have accomplished our purpose, but this cannot happen under our specific assumptions because we have not used up all the infinite number of intervals in the union.

Therefore the process must go on *ad infinitum*. Suppose then $c_n > c_{n+1}$ for all $n \in N$, so that $c_\omega = \lim_n \downarrow c_n$ exists, then $c_\omega \geq a$. If $c_\omega = a$ (which can easily happen, see (13)), then we are done and (21) follows, although the terms in the series have been gathered step by step in a (possibly) different order. What if $c_\omega > a$? In this case there is a unique $j$ such that $b_j = c_\omega$; rename the corresponding $a_j$ as $c_{\omega 1}$. We have now

$$(a, c_\omega] = \bigcup_{j=1}^{\infty} (a'_j, b'_j], \tag{23}$$

where the $(a'_j, b'_j]$'s are the leftovers from the original collection in (20) after an infinite number of them have been removed in the process. The interval $(c_{\omega 1}, c_\omega]$ is contained in the reduced new collection and we can begin a new process by first removing it from both sides of (23), then the next, to be denoted by $[c_{\omega 2}, c_{\omega 1}]$, and so on. If for some $n$ we have $c_{\omega n} = a$, then (21) is proved because at each step a term in the sum is gathered. Otherwise there exists the limit $\lim_n \downarrow c_{\omega n} = c_{\omega\omega} \geq a$. If $c_{\omega\omega} = a$, then (21) follows in the limit. Otherwise $c_{\omega\omega}$ must be equal to some $b_j$ (why?), and the induction goes on. Let us spare ourselves of the cumbersome notation for the successive well-ordered ordinal numbers. But will this process stop after a countable number of steps, namely, does there exist an ordinal number $\alpha$ of countable cardinality such that $c_\alpha = a$? The answer is "yes" because there are only countably many intervals in the union (20).

The preceding proof (which may be made logically formal) reveals the possibly complex structure hidden in the "order-blind" union in (20). Borel in his *Thèse* (1894) adopted a similar argument to prove a more general result that became known as his Covering Theorem (see below). A proof of the latter can be found in any text on real analysis, without the use of ordinal numbers. We will use the covering theorem to give another proof of Borel's lemma, for the sake of comparison (and learning).

This second proof establishes the equation (21) by two inequalities in opposite direction. The first inequality is easy by considering the first $n$ terms in the disjoint union (20):

$$F(b) - F(a) \geq \sum_{j=1}^{n} (F(b_j) - F(a_j)).$$

As $n$ goes to infinity we obtain (21) with the "=" replaced by "≥".

The other half is more subtle: the reader should pause and think why? The previous argument with ordinal numbers tells the story.

**Borel's covering theorem.** Let $[a, b]$ be a compact interval, and $(a_j, b_j)$, $j \in N$, be bounded open intervals, which may intersect arbitrarily, such that

$$[a, b] \subset \bigcup_{j=1}^{\infty}(a_j, b_j). \tag{24}$$

Then there exists a finite integer $l$ such that when $l$ is substituted for $\infty$ in the above, the inclusion remains valid.

In other words, a finite subset of the original infinite set of open intervals suffices to do the covering. This theorem is also called the Heine–Borel Theorem; see Hardy [1] (in the general Bibliography) for two proofs by Besicovitch.

To apply (24) to (20), we must alter the shape of the intervals $(a_j, b_j]$ to fit the picture in (24).

Let $-\infty < a < b < \infty$; and $\epsilon > 0$. Choose $a'$ in $(a, b)$, and for each $j$ choose $b'_j > b_j$ such that

$$F(a') - F(a) < \frac{\epsilon}{2}; \quad F(b'_j) - F(b_j) < \frac{\epsilon}{2^{j+1}}. \tag{25}$$

These choices are possible because $F$ is right continuous; and now we have

$$[a', b] \subset \bigcup_{j=1}^{\infty}(a_j, b'_j)$$

as required in (24). Hence by Borel's theorem, there exists a finite $l$ such that

$$[a', b] \subset \bigcup_{j=1}^{l}(a_j, b'_j). \tag{26}$$

From this it follows "easily" that

$$F(b) - F(a') \leq \sum_{j=1}^{l}(F(b'_j) - F(a_j)). \tag{27}$$

We will spell out the proof by induction on $l$. When $l = 1$ it is obvious. Suppose the assertion has been proved for $l - 1$, $l \geq 2$. From (26) as written, there is $k$, $1 \leq k \leq l$, such that $a_k < a' < b'_k$ and so

$$F(b'_k) - F(a') \leq F(b'_k) - F(a_k). \tag{28}$$

If we intersect both sides of (26) with the complement of $(a_k, b'_k)$, we obtain

$$[b'_k, b] \subset \bigcup_{\substack{j=1\\ j\neq k}}^{l} (a_j, b'_j).$$

Here the number of intervals on the right side is $l - 1$; hence by the induction hypothesis we have

$$F(b) - F(b'_k) \leq \sum_{\substack{j=1\\ j\neq k}}^{l} (F(b'_j) - F(a_j)).$$

Adding this to (28) we obtain (27), and the induction is complete. It follows from (27) and (25) that

$$F(b) - F(a) \leq \sum_{j=1}^{l} (F(b_j) - F(a_j)) + \epsilon.$$

Beware that the $l$ above depends on $\epsilon$. However, if we change $l$ to $\infty$ (back to infinity!) then the infinite series of course does not depend on $\epsilon$. Therefore we can let $\epsilon \to 0$ to obtain (21) when the "=" there is changed to "$\leq$", namely the other half of Borel's lemma, for finite $a$ and $b$.

It remains to treat the case $a = -\infty$ and/or $b = +\infty$. Let

$$(-\infty, b] \subset \bigcup_{j=1}^{\infty} (a_j, b_j].$$

Then for any $a$ in $(-\infty, b)$, (21) holds with "=" replaced by "$\leq$". Letting $a \to -\infty$ we obtain the desired result. The case $b = +\infty$ is similar. Q.E.D.

In the following, all $I$ with subscripts denote intervals of the shape $(a, b]$; $\sum$ denotes union of disjoint sets. Let $B \in \mathscr{B}_0$; $B_j \in \mathscr{B}_0$, $j \in N$. Thus

$$B = \sum_{i=1}^{n} I_i\ ; \quad B_j = \sum_{k=1}^{n_j} I_{jk}.$$

Suppose

$$B = \sum_{j=1}^{\infty} B_j$$

so that

$$\sum_{i=1}^{n} I_i = \sum_{j=1}^{\infty} \sum_{k=1}^{n_j} I_{jk}. \tag{29}$$

We will prove

$$(30) \qquad \sum_{i=1}^{n} m(I_i) = \sum_{j=1}^{\infty} \sum_{k=1}^{n_j} m(I_{jk}).$$

For $n = 1$, (29) is of the form (20) since a countable set of sets can be ordered as a sequence. Hence (30) follows by Borel's lemma. In general, simple geometry shows that each $I_i$ in (29) is the union of a subcollection of the $I_{jk}$'s. This is easier to see if we order the $I_i$'s in algebraic order and, after merging where possible, separate them at nonzero distances. Therefore (30) follows by adding $n$ equations, each of which results from Borel's lemma.

This completes the proof of the countable additivity of $m$ on $\mathscr{B}_0$, namely (19) is true as stipulated there for $l = \infty$ as well as $l < \infty$.

The general method developed in §1 can now be applied to $(R, \mathscr{B}_0, m)$. Substituting $\mathscr{B}_0$ for $\mathscr{F}_0$, $m$ for $\mu$ in Definition 3, we obtain the outer measure $m^*$. It is remarkable that the countable additivity of $m$ on $\mathscr{B}_0$, for which two painstaking proofs were given above, is used exactly in one place, at the beginning of Theorem 1, to prove that $m^* = m$ on $\mathscr{B}_0$. Next, we define the Borel field $\mathscr{B}^*$ as in Definition 4. By Theorem 6, $(R, \mathscr{B}^*, m^*)$ is a complete measure space. By Definition 5, $m$ is $\sigma$-finite on $\mathscr{B}_0$ because $(-n, n] \uparrow (-\infty, \infty)$ as $n \uparrow \infty$ and $m((-n, n])$ is finite by our primary assumption (14). Hence by Theorem 3, the restriction of $m^*$ to $\mathscr{B}$ is the unique extension of $m$ from $\mathscr{B}_0$ to $\mathscr{B}$.

In the most important case where $F(x) \equiv x$, the measure $m$ on $\mathscr{B}_0$ is the *length*: $m((a, b]) = b - a$. It was Borel who, around the turn of the twentieth century, first conceived of the notion of a countably additive "length" on an extensive class of sets, now named after him: *the Borel field* $\mathscr{B}$. A member of this class is called a *Borel set*. The larger Borel field $\mathscr{B}^*$ was first constructed by Lebesgue from an outer and an inner measure (see pp. 28–29 of main text). The latter was later bypassed by Carathéodory, whose method is adopted here. A member of $\mathscr{B}^*$ is usually called *Lebesgue-measurable*. The intimate relationship between $\mathscr{B}$ and $\mathscr{B}^*$ is best seen from Theorem 7.

The generalization to a generalized distribution function $F$ is sometimes referred to as Borel–Lebesgue–Stieltjes. See §2.2 of the main text for the special case of a probability distribution.

The generalization to a Euclidean space of higher dimension presents no new difficulty and is encumbered with tedious geometrical "baggage".

It can be proved that the cardinal number of all Borel sets is that of the real numbers (viz. all points in $R$), commonly denoted by $\mathfrak{c}$ (the *continuum*). On the other hand, if $Z$ is a Borel set of cardinal $\mathfrak{c}$ with $m(Z) = 0$, such as the Cantor ternary set (p. 13 of main text), then by the remark preceding Theorem 6, all subsets of $Z$ are Lebesgue-measurable and hence their totality has cardinal $2^C$ which is strictly greater than c (see e.g. [3]). It follows that there are incomparably more Lebesgue-measurable sets than Borel sets.

It is however not easy to exhibit a set in $\mathscr{B}^*$ but not in $\mathscr{B}$; see Exercise No. 15 on p. 15 of the main text for a clue, but that example uses a non-Lebesgue-measurable set to begin with.

Are there non-Lebesgue-measurable sets? Using the Axiom of Choice, we can "define" such a set rather easily; see example [3] or [5]. However, Paul Cohen has proved that the axiom is independent of the other logical axioms known as Zermelo–Fraenkel system commonly adopted in mathematics; and Robert Solovay has proved that in a certain model without the axiom of choice, all sets of real numbers are Lebesgue-measurable. In the notation of Definition 1 in §1 in this case, $\mathscr{B}^* = \mathscr{S}$ and the outer measure $m^*$ is a measure on $\mathscr{S}$.

N.B. Although no explicit invocation is made of the axiom of choice in the main text of this book, a weaker version of it under the prefix "countable" must have been casually employed on the q.t. Without the latter, allegedly it is impossible to show that the union of a countable collection of countable sets is countable. This kind of logical finesse is beyond the scope of this book.

## 4 Integral

The measure space $(\Omega, \mathscr{F}, \mu)$ is fixed. A function $f$ with domain $\Omega$ and range in $R^* = [-\infty, +\infty]$ is called *$\mathscr{F}$-measurable* iff for each real number $c$ we have

$$\{f \le c\} = \{\omega \in \Omega: f(\omega) \le c\} \in \mathscr{F}.$$

We write $f \in \mathscr{F}$ in this case. It follows that for each set $A \in \mathscr{B}$, namely a Borel set, we have

$$\{f \in A\} \in \mathscr{F};$$

and both $\{f = +\infty\}$ and $\{f = -\infty\}$ also belong to $\mathscr{F}$. Properties of measurable functions are given in Chapter 3, although the measure there is a probability measure.

A function $f \in \mathscr{F}$ with range a countable set in $[0, \infty]$ will be called a *basic function*. Let $\{a_j\}$ be its range (which may include "$\infty$"), and $A_j = \{f = a_j\}$. Then the $A_j$'s are disjoint sets with union $\Omega$ and

$$f = \sum_j a_j 1_{A_j} \tag{31}$$

where the sum is over a countable set of $j$.

We proceed to define an integral for functions in $\mathscr{F}$, in three stages, beginning with basic functions.

DEFINITION 8(a). For the basic function $f$ in (31), its *integral* is defined to be

$$E(f) = \sum_j a_j \mu(A_j) \tag{32}$$

and is also denoted by

$$\int f \, d\mu = \int_\Omega f(\omega) \mu(d\omega).$$

If a term in (32) is $0.\infty$ or $\infty.0$, it is taken to be 0. In particular if $f \equiv 0$, then $E(0) = 0$ even if $\mu(\Omega) = \infty$. If $A \in \mathscr{F}$ and $\mu(A) = 0$, then the basic function

$$\infty.1_A + 0.1_{A^c}$$

has integral equal to

$$\infty.0 + 0.\mu(A^c) = 0.$$

We list some of the properties of the integral.

**(i)** Let $\{B_j\}$ be a countable set of disjoint sets in $\mathscr{F}$, with union $\Omega$ and $\{b_j\}$ arbitrary positive numbers or $\infty$, not necessarily distinct. Then the function

$$\sum_j b_j 1_{B_j} \tag{33}$$

is basic, and its integral is equal to

$$\sum_j b_j \mu(B_j).$$

PROOF. Collect all equal $b_j$'s into $a_j$ and the corresponding $B_j$'s into $A_j$ as in (31). The result follows from the theorem on double series of positive terms that it may be summed in any order to yield a unique sum, possibly $+\infty$.

**(ii)** If $f$ and $g$ are basic and $f \le g$, then

$$E(f) \le E(g).$$

In particular if $E(f) = +\infty$, then $E(g) = +\infty$.

PROOF. Let $f$ be as in (31) and $g$ as in (33). The doubly indexed set $\{A_j \cap B_k\}$ are disjoint and their union is $\Omega$. We have using (i):

$$E(f) = \sum_j \sum_k a_j \mu(A_j \cap B_k);$$

$$E(g) = \sum_k \sum_j b_k \mu(A_j \cap B_k).$$

The order of summation in the second double series may be reversed, and the result follows by the countable additivity of $\mu$.

**(iii)** If $f$ and $g$ are basic functions, $a$ and $b$ positive numbers, then $af + bg$ is basic and

$$E(af + bg) = aE(f) + bE(g).$$

PROOF. It is trivial that $af$ is basic and

$$E(af) = aE(f).$$

Hence it is sufficient to prove the result for $a = b = 1$. Using the double decomposition in (ii), we have

$$E(f + g) = \sum_j \sum_k (a_j + b_k)\mu(A_j \cap B_k).$$

Splitting the double series in two and then summing in two orders, we obtain the result.

It is good time to state a general result that contains the double series theorem used above and some other version of it that will be used below.

**Double Limit Lemma.** Let $\{C_{jk}; j \in N, k \in N\}$ be a doubly indexed array of real numbers with the following properties:

(a) for each fixed $j$, the sequence $\{C_{jk}; k \in N\}$ is increasing in $k$;
(b) for each fixed $k$, the sequence $\{C_{jk}; j \in N\}$ is increasing in $j$.

Then we have

$$\lim_j \uparrow \lim_k \uparrow C_{jk} = \lim_k \uparrow \lim_j \uparrow C_{jk} \leq +\infty.$$

The proof is surprisingly simple. Both repeated limits exist by fundamental analysis. Suppose first that one of these is the finite number $C$. Then for any $\epsilon > 0$, there exist $j_0$ and $k_0$ such that $C_{j_0 k_0} > C - \epsilon$. This implies that the other limit $> C - \epsilon$. Since $\epsilon$ is arbitrary and the two indices are interchangeable, the two limits must be equal. Next if the $C$ above is $+\infty$, then changing $C - \epsilon$ into $\epsilon^{-1}$ finishes the same argument.

As an easy exercise, the reader should derive the cited theorem on double series from the Lemma.

Let $A \in \mathscr{F}$ and $f$ be a basic function. Then the product $1_A f$ is a basic function and its integral will be denoted by

$$E(A; f) = \int_A f(\omega)\mu(d\omega) = \int_A f \, d\mu. \tag{34}$$

**(iv)** Let $A_n \in \mathscr{F}$, $A_n \subset A_{n+1}$ for all $n$ and $A = \cup_n A_n$. Then we have

$$\lim_n E(A_n; f) = E(A; f). \tag{35}$$

PROOF. Denote $f$ by (33), so that $1_A f = \sum_j b_j 1_{AB_j}$. By (i),

$$E(A; f) = \sum_j b_j \mu(AB_j)$$

with a similar equation where $A$ is replaced by $A_n$. Since $\mu(A_n B_j) \uparrow \mu(AB_j)$ as $n \uparrow \infty$, and $\sum_{j=1}^{m} \uparrow \sum_{j=1}^{\infty}$ as $m \uparrow \infty$, (35) follows by the double limit theorem.

Consider now an increasing sequence $\{f_n\}$ of basic functions, namely, $f_n \leq f_{n+1}$ for all $n$. Then $f = \lim_n \uparrow f_n$ exists and $f \in \mathscr{F}$, but of course $f$ need not be basic; and its integral has yet to be defined. By property (ii), the numerical sequence $E(f_n)$ is increasing and so $\lim_n \uparrow E(f_n)$ exists, possibly equal to $+\infty$. It is tempting to define $E(f)$ to be that limit, but we need the following result to legitimize the idea.

**Theorem 8.** Let $\{f_n\}$ and $\{g_n\}$ be two increasing sequences of basic functions such that

$$\lim_n \uparrow f_n = \lim_n \uparrow g_n \tag{36}$$

(everywhere in $\Omega$). Then we have

$$\lim_n \uparrow E(f_n) = \lim_n \uparrow E(g_n). \tag{37}$$

PROOF. Denote the common limit function in (36) by $f$ and put

$$A = \{\omega \in \Omega: f(\omega) > 0\},$$

then $A \in \mathscr{F}$. Since $0 \leq g_n \leq f$, we have $1_{A^c} g_n = 0$ identically; hence by property (iii):

$$E(g_n) = E(A; g_n) + E(A^c; g_n) = E(A; g_n). \tag{38}$$

Fix an $n$ and put for each $k \in N$:

$$A_k = \left\{\omega \in \Omega: f_k(\omega) > \frac{n-1}{n} g_n(\omega)\right\}.$$

Since $f_k \leq f_{k+1}$, we have $A_k \subset A_{k+1}$ for all $k$. We are going to prove that

$$\bigcup_{k=1}^{\infty} A_k = A. \tag{39}$$

If $\omega \in A_k$, then $f(\omega) \geq f_k(\omega) > [(n-1)/n]g_n(\omega) \geq 0$; hence $\omega \in A$. On the other hand, if $\omega \in A$ then

$$\lim_k \uparrow f_k(\omega) = f(\omega) \geq g_n(\omega)$$

and $f(\omega) > 0$; hence there exists an index $k$ such that

$$f_k(\omega) > \frac{n-1}{n} g_n(\omega)$$

and so $\omega \in A_k$. Thus (39) is proved. By property (ii), since

$$f_k \geq 1_{A_k} f_k \geq 1_{A_k} \left( \frac{n-1}{n} g_n \right)$$

we have

$$E(f_k) \geq E(A_k; f_k) \geq \frac{n-1}{n} E(A_k; g_n).$$

Letting $k \uparrow \infty$, we obtain by property (iv):

$$\lim_k \uparrow E(f_k) \geq \frac{n-1}{n} \lim_k E(A_k; g_n)$$
$$= \frac{n-1}{n} E(A; g_n) = \frac{n-1}{n} E(g_n)$$

where the last equation is due to (38). Now let $n \uparrow \infty$ to obtain

$$\lim_k \uparrow E(f_k) \geq \lim_n \uparrow E(g_n).$$

Since $\{f_n\}$ and $\{g_n\}$ are interchangeable, (37) is proved.

**Corollary.** Let $f_n$ and $f$ be basic functions such that $f_n \uparrow f$, then $E(f_n) \uparrow E(f)$.

PROOF. Take $g_n = f$ for all $n$ in the theorem.

The class of positive $\mathscr{F}$-measurable functions will be denoted by $\mathscr{F}_+$. Such a function can be approximated in various ways by basic functions. It is nice to do so by an increasing sequence, and of course we should approximate all functions in $\mathscr{F}_+$ in the same way. We choose a particular mode as follows.

Define a function on $[0, \infty]$ by the (uncommon) symbol ) ]:

$$)0] = 0; \quad )\infty] = \infty;$$
$$)x] = n - 1 \text{ for } x \in (n-1, n],\ n \in N.$$

Thus $)\pi] = 3$, $)4] = 3$. Next we define for any $f \in \mathscr{F}_+$ the approximating sequence $\{f^{(m)}\}$, $m \in N$, by

$$f^{(m)}(\omega) = \frac{)2^m f(\omega)]}{2^m}. \tag{40}$$

Each $f^{(m)}$ is a basic function with range in the set of *dyadic* (*binary*) numbers: $\{k/2^m\}$ where $k$ is a nonnegative integer or $\infty$. We have $f^{(m)} \leq f^{(m+1)}$ for all $m$, by the magic property of bisection. Finally $f^{(m)} \uparrow f$ owing to the *left*-continuity of the function $x \to )x]$.

DEFINITION 8(b). For $f \in \mathscr{F}_+$, its integral is defined to be

$$E(f) = \lim_m \uparrow E(f^{(m)}). \tag{41}$$

When $f$ is basic, Definition 8(b) is consistent with 8(a), by Corollary to Theorem 8. The extension of property (ii) of integrals to $\mathscr{F}_+$ is trivial, because $f \leq g$ implies $f^{(m)} \leq g^{(m)}$. On the contrary, $(f+g)^{(m)}$ is not $f^{(m)} + g^{(m)}$, but since $f^{(m)} + g^{(m)} \uparrow (f+g)$, it follows from Theorem 8 that

$$\lim_m \uparrow E(f^{(m)} + g^{(m)}) = \lim_m \uparrow E((f+g)^{(m)})$$

that yields property (iii) for $\mathscr{F}_+$, together with $E(af^{(m)}) \uparrow aE(f)$, for $a \geq 0$.

Property (iv) for $\mathscr{F}_+$ will be given in an equivalent form as follows.

(iv) For $f \in \mathscr{F}_+$, the function of sets defined on $\mathscr{F}$ by

$$A \to E(A; f)$$

is a measure.

PROOF. We need only prove that if $A = \cup_{n=1}^{\infty} A_n$ where the $A_n$'s are disjoint sets in $\mathscr{F}$, then

$$E(A; f) = \sum_{n=1}^{\infty} E(A_n; f).$$

For a basic $f$, this follows from properties (iii) and (iv). The extension to $\mathscr{F}_+$ can be done by the double limit theorem and is left as an exercise.

There are three fundamental theorems relating the convergence of functions with the convergence of their integrals. We begin with Beppo Levi's theorem on Monotone Convergence (1906), which is the extension of Corollary to Theorem 8 to $\mathscr{F}_+$.

**Theorem 9.** Let $\{f_n\}$ be an increasing sequence of functions in $\mathscr{F}_+$ with limit $f: f_n \uparrow f$. Then we have

$$\lim_n \uparrow E(f_n) = E(f) \leq +\infty.$$

PROOF. We have $f \in \mathscr{F}_+$; hence by Definition 8(b), (41) holds. For each $f_n$, we have, using analogous notation:

$$\lim_m \uparrow E(f_n^{(m)}) = E(f_n). \tag{42}$$

Since $f_n \uparrow f$, the numbers $)2^m f_n(\omega)] \uparrow )2^m f(\omega)]$ as $n \uparrow \infty$, owing to the left continuity of $x \to )x]$. Hence by Corollary to Theorem 8,

$$\lim_n \uparrow E(f_n^{(m)}) = E(f^{(m)}). \tag{43}$$

It follows that

$$\lim_m \uparrow \lim_n \uparrow E(f_n^{(m)}) = \lim_m \uparrow E(f^{(m)}) = E(f).$$

On the other hand, it follows from (42) that

$$\lim_n \uparrow \lim_m \uparrow E(f_n^{(m)}) = \lim_n \uparrow E(f_n).$$

Therefore the theorem is proved by the double limit lemma.

From Theorem 9 we derive Lebesgue's theorem in its pristine *positive* guise.

**Theorem 10.** Let $f_n \in \mathscr{F}_+$, $n \in N$. Suppose

(a) $\lim_n f_n = 0$;
(b) $E(\sup_n f_n) < \infty$.

Then we have

$$\lim_n E(f_n) = 0. \tag{44}$$

PROOF. Put for $n \in N$:

$$g_n = \sup_{k \geq n} f_k. \tag{45}$$

Then $g_n \in \mathscr{F}_+$, and as $n \uparrow \infty$, $g_n \downarrow \limsup_n f_n = 0$ by (a); and $g_1 = \sup_n f_n$ so that $E(g_1) < \infty$ by (b).

Now consider the sequence $\{g_1 - g_n\}$, $n \in N$. This is increasing with limit $g_1$. Hence by Theorem 9, we have

$$\lim_n \uparrow E(g_1 - g_n) = E(g_1).$$

By property (iii) for $F_+$,

$$E(g_1 - g_n) + E(g_n) = E(g_1).$$

Substituting into the preceding relation and cancelling the finite $E(g_1)$, we obtain $E(g_n) \downarrow 0$. Since $0 \le f_n \le g_n$, so that $0 \le E(f_n) \le E(g_n)$ by property (ii) for $F_+$, (44) follows.

The next result is known as Fatou's lemma, of the same vintage 1906 as Beppo Levi's. It has the virtue of "no assumptions" with the consequent one-sided conclusion, which is however often useful.

**Theorem 11.** Let $\{f_n\}$ be an arbitrary sequence of functions in $f_+$. Then we have

$$E(\liminf_n f_n) \le \liminf_n E(f_n). \tag{46}$$

PROOF. Put for $n \in N$:

$$g_n = \inf_{k \ge n} f_k,$$

then

$$\liminf_n f_n = \lim_n \uparrow g_n.$$

Hence by Theorem 9,

$$E(\liminf_n f_n) = \lim_n \uparrow E(g_n). \tag{47}$$

Since $g_n \le f_n$, we have $E(g_n) \le E(f_n)$ and

$$\liminf_n E(g_n) \le \liminf_n E(f_n).$$

The left member above is in truth the right member of (47); therefore (46) follows as a milder but neater conclusion.

We have derived Theorem 11 from Theorem 9. Conversely, it is easy to go the other way. For if $f_n \uparrow f$, then (46) yields $E(f) \le \lim_n \uparrow E(f_n)$. Since $f \ge f_n$, $E(f) \ge \lim_n \uparrow E(f_n)$; hence there is equality.

We can also derive Theorem 10 directly from Theorem 11. Using the notation in (45), we have $0 \le g_1 - f_n \le g_1$. Hence by condition (a) and (46),

$$\begin{aligned} E(g_1) = E(\liminf_n (g_1 - f_n)) &\le \liminf_n (E(g_1) - E(f_n)) \\ &= E(g_1) - \limsup_n E(f_n) \end{aligned}$$

that yields (44).

The three theorems 9, 10, and 11 are intimately woven.

We proceed to the final stage of the integral. For any $f \in \mathscr{F}$ with range in $[-\infty, +\infty]$, put

$$f^+ = \begin{cases} f & \text{on} \quad \{f \geq 0\}, \\ 0 & \text{on} \quad \{f < 0\}; \end{cases} \qquad f^- = \begin{cases} -f & \text{on} \quad \{f \leq 0\}, \\ 0 & \text{on} \quad \{f > 0\}. \end{cases}$$

Then $f^+ \in \mathscr{F}_+$, $f^- \in \mathscr{F}_+$, and

$$f = f^+ - f^-; \ |f| = f^+ + f^-$$

By Definition 8(b) and property (iii):

$$E(|f|) = E(f^+) + E(f^-). \tag{48}$$

DEFINITION 8(c). For $f \in \mathscr{F}$, its integral is defined to be

$$E(f) = E(f^+) - E(f^-), \tag{49}$$

provided the right side above is defined, namely not $\infty - \infty$. We say $f$ is integrable, or $f \in L^1$, iff both $E(f^+)$ and $E(f^-)$ are finite; in this case $E(f)$ is a finite number. When $E(f)$ exists but $f$ is not integrable, then it must be equal to $+\infty$ or $-\infty$, by (49).

A set $A$ in $\Omega$ is called a *null set* iff $A \in \mathscr{F}$ and $\mu(A) = 0$. A mathematical proposition is said to hold *almost everywhere*, or *a.e.* iff there is a null set $A$ such that it holds outside $A$, namely in $A^c$.

A number of important observations are collected below.

**Theorem 12.** **(i)** The function $f$ in $\mathscr{F}$ is integrable if and only if $|f|$ is integrable; we have

$$|E(f)| \leq E(|f|). \tag{50}$$

**(ii)** For any $f \in \mathscr{F}$ and any null set $A$, we have

$$E(A; f) = \int_A f \, d\mu = 0; \quad E(f) = E(A^c; f) = \int_{A^c} f \, d\mu. \tag{51}$$

**(iii)** If $f \in L^1$, then the set $\{\omega \in \Omega : |f(\omega)| = \infty\}$ is a null set.

**(iv)** If $f \in L^1$, $g \in \mathscr{F}$, and $|g| \leq |f|$ a.e., then $g \in L^1$.

**(v)** If $f \in \mathscr{F}$, $g \in \mathscr{F}$, and $g = f$ a.e., then $E(g)$ exists if and only if $E(f)$ exists, and then $E(g) = E(f)$.

**(vi)** If $\mu(\Omega) < \infty$, then any a.e. bounded $\mathscr{F}$-measurable function is integrable.

PROOF. (i) is trivial from (48) and (49); (ii) follows from

$$1_A |f| \leq 1_A . \infty$$

so that

$$0 \le E(1_A|f|) \le E(1_A.\infty) = \mu(A).\infty = 0.$$

This implies (51).

To prove (iii), let

$$A(n) = \{|f| \ge n\}.$$

Then $A(n) \in \mathscr{F}$ and

$$n\mu(A(n)) = E(A(n); n) \le E(A(n); |f|) \le E(|f|).$$

Hence

$$\mu(A(n)) \le \frac{1}{n}E(|f|). \tag{52}$$

Letting $n \uparrow \infty$, so that $A(n) \downarrow \{|f| = \infty\}$; since $\mu(A(1)) \le E(|f|) < \infty$, we have by property (f) of the measure $\mu$:

$$\mu(\{|f| = \infty\}) = \lim_n \downarrow \mu(A(n)) = 0.$$

To prove (iv), let $|g| \le |f|$ on $A^c$, where $\mu(A) = 0$. Then

$$|g| \le 1_A.\infty + 1_{A^c}.|f|$$

and consequently

$$E(|g|) \le \mu(A).\infty + E(A^c; |f|) \le 0.\infty + E(|f|) = E(|f|).$$

Hence $g \in L^1$ if $f \in L^1$.

The proof of (v) is similar to that of (iv) and is left as an exercise. The assertion (vi) is a special case of (iv) since a constant is integrable when $\mu(\Omega) < \infty$.

*Remark.* A case of (52) is known as Chebyshev's inequality; see p. 48 of the main text. Indeed, it can be strengthened as follows:

$$\lim_n n\mu(A(n)) \le \lim_n E(A(n); |f|) = 0. \tag{53}$$

This follows from property (f) of the measure

$$A \to E(A; |f|);$$

see property (iv) of the integral for $\mathscr{F}_+$.

There is also a strengthening of (ii), as follows.

If $B_k \in \mathscr{F}$ and $\mu(B_k) \to 0$ as $k \to \infty$, then

$$\lim_k E(B_k; f) = 0.$$

To prove this we may suppose, without loss of generality, that $f \in \mathscr{F}_+$. We have then

$$E(B_k; f) = E(B_k \cap A(n); f) + E(B_k \cap A(n)^c; f)$$
$$\leq E(A(n); f) + E(B_k)n.$$

Hence

$$\limsup_k E(B_k; f) \leq E(A(n); f)$$

and the result follows by letting $n \to \infty$ and using (53).

It is convenient to define, for any $f$ in $\mathscr{F}$, a class of functions denoted by $C(f)$, as follows: $g \in C(f)$ iff $g = f$ a.e. When $(\Omega, \mathscr{F}, \mu)$ is a complete measure space, such a $g$ is automatically in $\mathscr{F}$. To see this, let $B = \{g \neq f\}$. Our definition of "a.e." means only that $B$ is a subset of a null set $A$; in plain English this does not say whether $g$ is equal to $f$ or not anywhere in $A - B$. However if the measure is complete, then any subset of a null set is also a null set, so that not only the set $B$ but all its subsets are null, hence in $\mathscr{F}$. Hence for any real number $c$,

$$\{g \leq c\} = \{g = f; g \leq c\} \cup \{g \neq f; g \leq c\}$$

belongs to $\mathscr{F}$, and so $g \in \mathscr{F}$.

A member of $C(f)$ may be called a *version* of $f$, and may be substituted for $f$ wherever a null set "does not count". This is the point of (iv) and (v) in Theorem 12. Note that when the measure space is complete, the assumption "$g \in \mathscr{F}$" there may be omitted. A particularly simple version of $f$ is the following *finite version*:

$$\overline{f} = \begin{cases} f & \text{on} \quad \{|f| < \infty\}, \\ 0 & \text{on} \quad \{|f| = \infty\}; \end{cases}$$

where 0 may be replaced by some other number, e.g., by 1 in $E(\log f)$.

In functional analysis, it is the class $C(f)$ rather than an individual $f$ that is a member of $L^1$.

As examples of the preceding remarks, let us prove properties (ii) and (iii) for integrable functions.

(ii) if $f \in L^1$, $g \in L^1$, and $f \leq g$ a.e., then

$$E(f) \leq E(g).$$

PROOF. We have, except on a null set:

$$f^+ - f^- \leq g^+ - g^-$$

but we cannot transpose terms that may be $+\infty$! Now substitute finite versions of $f$ and $g$ in the above (without changing their notation) and then transpose as follows:

$$f^+ + g^- \le g^+ + f^-.$$

Applying properties (ii) and (iii) for $\mathscr{F}_+$, we obtain

$$E(f^+) + E(g^-) \le E(g^+) + E(f^-).$$

By the assumptions of $L^1$, all the four quantities above are finite numbers. Transposing back we obtain the desired conclusion.

(iii) if $f \in L^1$, $g \in L^1$, then $f + g \in L^1$, and

$$E(f + g) = E(f) + E(g).$$

Let us leave this as an exercise. If we assume only that both $E(f)$ and $E(g)$ exist and that the right member in the equation above is defined, namely not $(+\infty) + (-\infty)$ or $(-\infty) + (+\infty)$, does $E(f + g)$ then exist and equal to the sum? We leave this as a good exercise for the curious, and return to Theorem 10 in its practical form.

Theorem $10^{\Delta}$. Let $f_n \in \mathscr{F}$; suppose

(a) $\lim_n f_n = f$ a.e.;

(b) there exists $\varphi \in L^1$ such that for all $n$:

$$|f_n| \le \varphi \quad \text{a.e.}$$

Then we have

(c) $\lim_n E(|f_n - f|) = 0$.

PROOF. observe first that

$$|\lim_n f_n| \le \sup_n |f_n|;$$

$$|f_n - f| \le |f_n| + |f| \le 2 \sup_n |f_n|;$$

provided the left members are defined. Since the union of a countable collection of null sets is a null set, under the hypotheses (a) and (b) there is a null set $A$ such that on $\Omega - A$, we have $\sup_n |f_n| \le \varphi$ hence by Theorem 12 (iv), all $|f_n|$, $|f|$, $|f_n - f|$ are integrable, and therefore we can substitute their finite versions without affecting their integrals, and moreover $\lim_n |f_n - f| = 0$ on $\Omega - A$. (Remember that $f_n - f$ need not be defined before the substitutions!). By using Theorem 12 (ii) once more if need be, we obtain the conclusion (c) from the positive version of Theorem 10.

This theorem is known as Lebesgue's dominated convergence theorem, vintage 1908. When $\mu(\Omega) < \infty$, any constant $C$ is integrable and may be used for $\varphi$; hence in this case the result is called bounded convergence theorem.

Curiously, the best known part of the theorem is the corollary below with a fixed $B$.

**Corollary.** We have

$$\lim_n \int_B f_n \, d\mu = \int_B f \, d\mu$$

uniformly in $B \in \mathscr{F}$.

This is trivial from (c), because, in alternative notation:

$$|E(B; f_n) - E(B; f)| \leq E(B; |f_n - f|) \leq E(|f_n - f|).$$

In the particular case where $B = \Omega$, the Corollary contains a number of useful results such as the integration term by term of power series or Fourier series. A glimpse of this is given below.

## 5 Applications

The general theory of integration applied to a probability space is summarized in §§3.1–3.2 of the main text. The specialization to $R$ expounded in §3 above will now be described and illustrated.

A function $f$ defined on $R$ with range in $[-\infty, +\infty]$ is called a *Borel function* iff $f \in \mathscr{B}$; it is called a *Lebesgue-measurable* function iff $f \in \mathscr{B}^*$. The domain of definition $f$ may be an arbitrary Borel set or Lebesgue-measurable set $D$. This case is reduced to that for $D = R$ by extending the definition of $f$ to be zero outside $D$. The integral of $f \in \mathscr{B}^*$ corresponding to the measure $m^*$ constructed from $F$ is denoted by

$$E(f) = \int_{-\infty}^{\infty} f(x) \, dF(x).$$

In case $F(x) \equiv x$, this is called the *Lebesgue integral* of $f$; in this case the usual notation is, for $A \in \mathscr{B}^*$:

$$\int_A f(x) \, dx = E(A; f).$$

Below are some examples of the application of preceding theorems to classic analysis.

**Example 1.** Let $I$ be a bounded interval in $R$; $\{u_k\}$ a sequence of functions on $I$; and for $x \in I$:

$$s_n(x) = \sum_{k=1}^{n} u_k(x), \quad n \in N.$$

Suppose the infinite series $\sum_k u_k(x)$ converges $I$; then in the usual notation:

$$\lim_{n\to\infty}\sum_{k=1}^{n} u_k(x) = \sum_{k=1}^{\infty} u_k(x) = s(x)$$

exists and is finite. Now suppose each $u_k$ is Lebesgue-integrable, then so is each $s_n$, by property (iii) of the integral; and

$$\int_I s_n(x)\,dx = \sum_{k=1}^{n}\int_I u_k(x)\,dx.$$

Question: does the numerical series above converge? and if so is the sum of integrals equal to the integral of the sum:

$$\sum_{k=1}^{\infty}\int_I u_k(x)\,dx = \int_I \sum_{k=1}^{\infty} u_k(x)\,dx = \int_I s(x)\,dx?$$

This is the problem of integration term by term.

A very special but important case is when the interval $I = [a, b]$ is compact and the functions $u_k$ are all continuous in $I$. If we assume that the series $\sum_{k=1}^{\infty} u_k(x)$ converges *uniformly* in $I$, then it follows from elementary analysis that the sequence of partial sums $\{s_n(x)\}$ is totally bounded, that is,

$$\sup_n \sup_x |s_n(x)| = \sup_x \sup_n |s_n(x)| < \infty.$$

Since $m(I) < \infty$, the bounded convergence theorem applies to yield

$$\lim_n \int_I s_n(x)\,dx = \int_I \lim_n s_n(x)\,dx.$$

The Taylor series of an analytic function always converges uniformly and absolutely in any compact subinterval of its interval of convergence. Thus the result above is fruitful.

Another example of term-by-term integration goes back to Theorem 8.

**Example 2.** Let $u_k \geq 0$, $u_k \in L^1$, then

$$\int_a^b \left(\sum_{k=1}^{\infty} u_k\right) d\mu = \sum_{k=1}^{\infty}\int_a^b u_k\,d\mu. \tag{54}$$

Let $f_n = \sum_{k=1}^{n} u_k$, then $f_n \in L^1$, $f_n \uparrow f = \sum_{k=1}^{\infty} u_k$. Hence by monotone convergence

$$E(f) = \lim_n E(f_n)$$

that is (54).

When $u_k$ is general the preceding result may be applied to $|u_k|$ to obtain

$$\int \left( \sum_{k=1}^{\infty} |u_k| \right) d\mu = \sum_{k=1}^{\infty} \int |u_k| \, d\mu.$$

If this is finite, then the same is true when $|u_k|$ is replaced by $u_k^+$ and $u_k^-$. It then follows by subtraction that (54) is also true. This result of term-by-term integration may be regarded as a special case of the Fubini–Tonelli theorem (pp. 63–64), where one of the measures is the counting measure on $N$.

For another perspective, we will apply the Borel–Lebesgue theory of integral to the older Riemann context.

**Example 3.** Let $(I, \mathscr{B}^*, m)$ be as in the preceding example, but let $I = [a, b]$ be compact. Let $f$ be a continuous function on $I$. Denote by $P$ a partition of $I$ as follows:

$$a = x_0 < x_1 < x_2 < \cdots < x_n = b;$$

and put

$$\delta(P) = \max_{1 \le k \le n} (x_k - x_{k-1}).$$

For each $k$, choose a point $\xi_k$ in $[x_{k-1}, x_k]$, and define a function $f_P$ as follows:

$$f_P(x) = \begin{cases} \xi_1, & \text{for } x \in [x_0, x_1], \\ \xi_k, & \text{for } x \in (x_{k-1}, x_k],\ 2 \le k \le n. \end{cases}$$

Particular choices of $\xi_k$ are: $\xi_k = f(x_{k-1}); \xi_k = f(x_k)$;

$$\text{(55)} \qquad \xi_k = \min_{x_{k-1} \le x \le x_k} f(x); \quad \xi_k = \max_{x_{k-1} \le x \le x_k} f(x).$$

The $f_P$ is called a *step function*; it is an approximant of $f$. It is not basic by Definition 8(a) but $f_P^+$ and $f_P^-$ are. Hence by Definitions 8(a) and 8(c), we have

$$E(f_P) = \sum_{k=1}^{n} f(\xi_k)(x_k - x_{k-1}).$$

The sum above is called a *Riemann sum*; when the $\xi_k$ are chosen as in (55), they are called *lower* and *upper* sums, respectively.

Now let $\{P(n), n \in N\}$ be a sequence of partitions such that $\delta(P(n)) \to 0$ as $n \to \infty$. Since $f$ is continuous on a compact set, it is bounded. It follows that there is a constant $C$ such that

$$\sup_{n \in N} \sup_{x \in I} |f_{P(n)}(x)| < C.$$

Since I is bounded, we can apply the bounded convergence theorem to conclude that

$$\lim_n E(f_{P(n)}) = E(f).$$

The finite existence of the limit above signifies the Riemann-integrability of $f$, and the limit is then its *Riemann-integral* $\int_a^b f(x)\,dx$. Thus we have proved that a continuous function on a compact interval is Riemann-integrable, and its Riemann-integral is equal to the Lebesgue integral. Let us recall that in the new theory, any bounded measurable function is integrable over any bounded measurable set. For example, the function

$$\sin\frac{1}{x}, \quad x \in (0, 1]$$

being bounded by 1 is integrable. But from the strict Riemannian point of view it has only an "improper" integral because (0, 1] is not closed and the function is not continuous on [0, 1], indeed it is not definable there. Yet the limit

$$\lim_{\epsilon\downarrow 0}\int_\epsilon^1 \sin\frac{1}{x}\,dx$$

exists and can be *defined* to be $\int_0^1 \sin(1/x)\,dx$, As a matter of fact, the Riemann sums do converge despite the unceasing oscillation of $f$ between 0 and 1 as $x \downarrow 0$.

**Example 4.** The Riemann integral of a function on $(0, \infty)$ is called an "infinite integral" and is definable as follows:

$$\int_0^\infty f(x)\,dx = \lim_{n\to\infty}\int_0^n f(x)\,dx$$

when the limit exists and is finite. A famous example is

$$f(x) = \frac{\sin x}{x}, \quad x \in (0, \infty). \tag{56}$$

This function is bounded by 1 and is continuous. It can be extended to $[0, \infty)$ by defining $f(0) = 1$ by continuity. A cute calculation (see §6.2 of main text) yields the result (useful in Optics):

$$\lim_n \int_0^n \frac{\sin x}{x}\,dx = \frac{\pi}{2}.$$

By contrast, the function $|f|$ is not Lebesgue-integrable. To show this, we use trigonometry:

$$\left(\frac{\sin x}{x}\right)^+ \geq \frac{1}{\sqrt{2}}\frac{1}{(2n+1)\pi} = C_n \text{ for } x \in \left(2n\pi + \frac{\pi}{4}, 2n\pi + \frac{3\pi}{4}\right) = I_n.$$

Thus for $x > 0$:

$$\left(\frac{\sin x}{x}\right)^+ \geq \sum_{n=1}^\infty C_n 1_{I_n}(x).$$

The right member above is a basic function, with its integral:

$$\sum_n C_n m(I_n) = \sum_n \frac{\pi}{\sqrt{2}(2n+1)2} = +\infty.$$

It follows that $E(f^+) = +\infty$. Similarly $E(f^-) = +\infty$; therefore by Definition 8(c) $E(f)$ does not exist! This example is a splendid illustration of the following Non-Theorem.

Let $f \in \mathscr{B}$ and $f_n = f 1_{(0,n)}$, $n \in N$. Then $f_n \in \mathscr{B}$ and $f_n \to f$ as $n \to \infty$. Even when the $f_n'$s are "totally bounded", it does not follow that

$$\lim_n E(f_n) = E(f); \tag{57}$$

indeed $E(f)$ may not exist.

On the other hand, if we assume, in addition, either (a) $f \geq 0$; or (b) $E(f)$ exists, in particular $f \in L^1$; then the limit relation will hold, by Theorems 9 and 10, respectively. The next example falls in both categories.

**Example 5.** The square of the function $f$ in (56):

$$f(x)^2 = \left(\frac{\sin x}{x}\right)^2, \quad x \in \mathbf{R}$$

is integrable in the Lebesgue sense, and is also improperly integrable in the Riemann sense.

We have

$$f(x)^2 \leq 1_{(-1,+1)} + 1_{(-\infty,-1)\cup(+1,+\infty)} \frac{1}{x^2}$$

and the function on the right side is integrable, hence so is $f^2$.

Incredibly, we have

$$\int_0^\infty \left(\frac{\sin x}{x}\right)^2 dx = \frac{\pi}{2} = (RI) \int_0^\infty \frac{\sin x}{x}\, dx,$$

where we have inserted an "*RI*" to warn against taking the second integral as a Lebesgue integral. See §6.2 for the calculation. So far as I know, nobody has explained the *equality* of these two integrals.

**Example 6.** The most notorious example of a simple function that is not Riemann-integrable and that baffled a generation of mathematicians is the function $1_Q$, where $Q$ is the set of rational numbers. Its Riemann sums can be made to equal any real number between 0 and 1, when we confine $Q$ to the unit interval (0, 1). The function is so totally discontinuous that the Riemannian way of approximating it, *horizontally* so to speak, fails utterly. But of course it is ludicrous even to consider this indicator function rather than the set $Q$ itself. There was a historical reason for this folly: integration was regarded as the inverse operation to differentiation, so that to integrate was meant to "find the primitive" whose derivative is to be the integrand, for example,

$$\int^\xi x\, dx = \frac{\xi^2}{2}, \quad \frac{d}{d\xi}\frac{\xi^2}{2} = \xi;$$

$$\int^\xi \frac{1}{x}\, dx = \log \xi, \quad \frac{d}{d\xi}\log \xi = \frac{1}{\xi}.$$

A primitive is called "indefinite integral", and $\int_1^2 (1/x)\,dx$ e.g. is called a "definite integral." Thus the unsolvable problem was to find $\int_0^\xi 1_Q(x)\,dx$, $0 < \xi < 1$.

The notion of measure as length, area, and volume is much more ancient than Newton's fluxion (derivative), not to mention the primitive measure of counting with fingers (and toes). The notion of "countable additivity" of a measure, although seemingly natural and facile, somehow did not take hold until Borel saw that

$$m(Q) = \sum_{q \in Q} m(q) = \sum_{q \in Q} 0 = 0.$$

There can be no question that the "length" of a single point $q$ is zero. Euclid gave it "zero dimension".

This is the beginning of MEASURE. An INTEGRAL is a weighted measure, as is obvious from Definition 8(a). The rest is approximation, *vertically* as in Definition 8(b), and convergence, as in all analysis.

As for the connexion with differentiation, Lebesgue made it, and a clue is given in §1.3 of the main text.

# General bibliography

[The five divisions below are merely a rough classification and are not meant to be mutually exclusive.]


### 1. Basic analysis

[1] G. H. Hardy, *A course of pure mathematics*, 10th ed. Cambridge University Press, New York, 1952.

[2] W. Rudin, *Principles of mathematical analysis*, 2nd ed. McGraw-Hill Book Company, New York, 1964.

[3] I. P. Natanson, *Theory of functions of a real variable* (translated from the Russian). Frederick Ungar Publishing Co., New York, 1955.

### 2. Measure and integration

[4] P. R. Halmos, *Measure theory*. D. Van Nostrand Co., Inc., Princeton, N.J., 1956.

[5] H. L. Royden, *Real analysis*. The Macmillan Company, New York, 1963.

[6] J. Neveu, *Mathematical foundations of the calculus of probability*. Holden-Day, Inc., San Francisco, 1965.

### 3. Probability theory

[7] Paul Lévy, *Calcul des probabilités*. Gauthier-Villars, Paris, 1925.

[8] A. Kolmogoroff, *Grundbegriffe der Wahrscheinlichkeitsrechnung*. Springer-Verlag, Berlin, 1933.

[9] M. Fréchet, *Généralités sur les probabilités. Variables aléatoires.* Gauthier-Villars, Paris, 1937.

[10] H. Cramér, *Random variables and probability distributions*, 3rd ed. Cambridge University Press, New York, 1970 [1st ed., 1937].

[11] Paul Lévy, *Théorie de l'addition des variables aléatoires*, 2nd ed. Gauthier-Villars, Paris, 1954 [1st ed., 1937].

[12] B. V. Gnedenko and A. N. Kolmogorov, *Limit distributions for sums of independent random variables* (translated from the Russian). Addison-Wesley Publishing Co., Inc., Reading, Mass., 1954.

[13] William Feller, *An introduction to probability theory and its applications*, vol. 1 (3rd ed.) and vol. 2 (2nd ed.). John Wiley & Sons, Inc., New York, 1968 and 1971 [1st ed. of vol. 1, 1950].

[14] Michel Loève, *Probability theory*, 3rd ed. D. Van Nostrand Co., Inc., Princeton, N.J., 1963.

[15] A. Rényi, *Probability theory.* North-Holland Publishing Co., Amsterdam, 1970.

[16] Leo Breiman, *Probability.* Addison-Wesley Publishing Co., Reading, Mass., 1968.

### 4. Stochastic processes

[17] J. L. Doob, *Stochastic processes.* John Wiley & Sons, Inc., New York, 1953.

[18] Kai Lai Chung, *Markov chains with stationary transition probabilities*, 2nd ed. Springer-Verlag, Berlin, 1967 [1st ed., 1960].

[19] Frank Spitzer, *Principles of random walk.* D. Van Nostrand Co., Princeton, N.J., 1964.

[20] Paul–André Meyer, *Probabilités et potential.* Herman (Editions Scientifiques), Paris, 1966.

[21] G. A. Hunt, *Martingales et processus de Markov.* Dunod, Paris, 1966.

[22] David Freedman, *Brownian motion and diffusion.* Holden-Day, Inc., San Francisco, 1971.

### 5. Supplementary reading

[23] Mark Kac, *Statistical independence in probability, analysis and number theory*, Carus Mathematical Monograph 12. John Wiley & Sons, Inc., New York, 1959.

[24] A. Rényi, *Foundations of probability.* Holden-Day, Inc., San Francisco, 1970.

[25] Kai Lai Chung, *Elementary probability theory with stochastic processes.* Springer-Verlag, Berlin, 1974.

# Index

**D**



**E**



**F**

## G



## H



## I



## J



## K



## L



## M